Additive Manufacturing (AM) for Advanced Materials and Structures: Green and Intelligent Development Trend

Additive Manufacturing (AM) for Advanced Materials and Structures: Green and Intelligent Development Trend

Editors

Hao Yi
Huajun Cao
Menglin Liu
Le Jia

MDPI • Basel • Beijing • Wuhan • Barcelona • Belgrade • Manchester • Tokyo • Cluj • Tianjin

Editors

Hao Yi
College of Mechanical and
Vehicle Engineering
Chongqing University
Chongqing
China

Huajun Cao
College of Mechanical and
Vehicle Engineering
Chongqing University
Chongqing
China

Menglin Liu
College of Mechanical and
Vehicle Engineering
Chongqing University
Chongqing
China

Le Jia
College of Mechanical and
Vehicle Engineering
Chongqing University
Chongqing
China

Editorial Office
MDPI
St. Alban-Anlage 66
4052 Basel, Switzerland

This is a reprint of articles from the Special Issue published online in the open access journal *Crystals* (ISSN 2073-4352) (available at: www.mdpi.com/journal/crystals/special_issues/additive_manufacturing_green).

For citation purposes, cite each article independently as indicated on the article page online and as indicated below:

LastName, A.A.; LastName, B.B.; LastName, C.C. Article Title. *Journal Name* **Year**, *Volume Number*, Page Range.

ISBN 978-3-0365-6335-0 (Hbk)
ISBN 978-3-0365-6334-3 (PDF)

Contents

About the Editors

Hao Yi

Dr. Hao Yi is an Associate Professor (Doctoral Supervisor) at the College of Mechanical and Vehicle Engineering at Chongqing University, China. He serves as an Editorial Board Member of *Virtual and Physical Prototyping, International Journal of Precision Engineering and Manufacturing-Green Technology, Proceedings of the Institution of Mechanical Engineers, Part C: Journal of Mechanical Engineering Science*, etc. His main research interests focus on 3D Printing and Additive Manufacturing, Green Manufacturing, Production Research, etc.

Huajun Cao

Dr. Huajun Cao is a Full Professor in the College of Mechanical and Vehicle Engineering, Chongqing University, China. His research interests mainly include green manufacturing and remanufacturing, green and intelligent factories, and high-speed dry machining technology and equipment.

Menglin Liu

Menglin Liu is currently a Ph.D. in the College of Mechanical and Vehicle Engineering, Chongqing University, Chongqing, China. His research activities are mainly focused on additive manufacturing and 3D printing.

Le Jia

Le Jia is currently a Ph.D. in the College of Mechanical and Vehicle Engineering, Chongqing University, Chongqing, China. His research activities are mainly focused on additive manufacturing and 3D printing.

Editorial

Additive Manufacturing (AM) for Advanced Materials and Structures: Green and Intelligent Development Trend

Menglin Liu [1,2], Hao Yi [1,2,3,*] and Huajun Cao [1,2]

1 State Key Laboratory of Mechanical Transmission, Chongqing University, Chongqing 400044, China
2 College of Mechanical and Vehicle Engineering, Chongqing University, Chongqing 400044, China
3 Chongqing Key Laboratory of Metal Additive Manufacturing (3D Printing), Chongqing University, Chongqing 400044, China
* Correspondence: haoyi@cqu.edu.cn

Citation: Liu, M.; Yi, H.; Cao, H. Additive Manufacturing (AM) for Advanced Materials and Structures: Green and Intelligent Development Trend. *Crystals* **2023**, *13*, 92. https://doi.org/10.3390/cryst13010092

Received: 30 December 2022
Accepted: 3 January 2023
Published: 4 January 2023

Additive manufacturing (AM) is an emerging and rapidly evolving technology that has revolutionized the way products are developed, fabricated and commercialized. This has enabled the disruption of long-running manufacturing processes, leading to economic and societal change. Many design and manufacturing technologies are receiving widespread attention to advance AM technology towards high efficiency, high precision, high performance, and low cost in an environmentally friendly manner. This Special Issue focuses on exploring topical issues in additive manufacturing processes, material design, structure design, process planning, and performance evaluation. The call for articles for this Special Issue resulted in an enthusiastic response from the research community, who contributed an excellent series of high quality and technically diverse manuscripts. This Special Issue "Additive Manufacturing (AM) for Advanced Materials and Structures: Green and Intelligent Development Trend" covers topics surrounding the structural design of complex components, the integrated, advanced design for the preparation and manufacture of high-performance materials, and performance optimization; containing a mix of 20 communications, original articles, and review articles.

In the ever-expanding field of additive manufacturing processes, Chao et al. [1] have proposed a novel high-resolution fused deposition 3D printing technique based on electric field-driven (EFD) jet deposition. An experimental approach based this process was devised to print polycaprolactone (PCL) porous scaffold structures. To explore the application prospects of this technique in the fabrication of microchannel structures, Chao et al. [2] have successfully printed waxy structures with a size of tens of microns.

Laser-based additive manufacturing processes are of long-standing interest among emerging additive manufacturing processes. For example, in laser powder bed fusion (LPBF) technology, scholars have conducted extensive research into structural design, material and part properties, and process strategies. With regard to structural design, Li et al. [3] have developed an optimization method for a body-centered cubic with Z support (BCCZ) lattice based on parametric modeling. The designed BCCZ structures were able to maintain their strength whilst also retaining light weights. Ma et al. [4] have prepared diamond lattice structures with different material distributions using selected laser melting techniques. The mechanical behavior of the structures was investigated under quasi-static and dynamic loading, and the gradient sheet diamond (GSHD) was found to possess the highest yield strength. With regard to material and part properties, Shen et al. [5] used LPBF to fabricate SiC-reinforced Al–Zn–Mg–Cu composites in situ. The results showed that the organization of the composites was regulated, the matrix grains were refined, and the grain orientation growth was suppressed. In a similar fashion, to improve the mechanical properties of the AlSi10Mg alloy prepared by LPBF, Lu et al. [6] have investigated the effect of nano-Si3N4 reinforcement on the densification behavior, microstructure, and tensile properties of AlSi10Mg. It was revealed that the tensile and yield strengths of the composites steadily increased with increasing nano-Si_3N_4 content, while the elongation decreased. Li et al. [7] have analyzed the fracture behavior of 316L stainless steel fabricated with defects by selective laser melting using a near-field kinetic approach. They demonstrated that crack sprouting is caused by the defects and crack branching contributes to complex multi-crack extensions. The effect of scanning strategy on the quality of the manufactured part is also important in process optimization. Cao et al. [8] have examined the effect of the transient temperature field of the molten layer in LPBF under linear and annular laser scanning strategies on the forming accuracy and quality of the manufactured part. Their analysis identified that annular scanning was more suitable than linear bi-directional scanning for the high-precision fabrication of thin-walled Fe–Cr–Al overlays. Other laser-based additive manufacturing processes such as laser cladding have also been the focus of attention. Wang et al. [9] used this technique to produce WC (hard tungsten carbide) Co–Cr alloy coatings with different mass fractions on 316L substrates. It was found that laser cladding of the Co–Cr–WC composite layer could significantly improve the wear and corrosion resistance of the 316L substrate. Lasers also have the important ability to fabricate micro/nanostructures. Du et al. [10] have manufactured Se-doped silicon thin films by irradiating Si–Se bilayer-coated silicon with femtosecond (fs) and picosecond (ps) lasers. Their work revealed that the changes brought about by ps laser processing are significant for ultrafast laser processing of brass-doped silicon in silicon-based integrated circuits. Ultrafast lasers can effectively process special materials and improve the mechanical properties of parts, giving them the advantage over short pulse lasers and continuous wave lasers. Finally, Wu et al. [11] have reviewed the interaction mechanisms between ultrafast lasers and metallic materials and discussed the current status and challenges of ultrafast laser application in the formation of special materials.

Aside from laser-based additive manufacturing processes, fused deposition modeling (FDM), projection stereolithography, and resistive additive manufacturing are also discussed. Tura et al. [12] have used adaptive neuro-fuzzy methods and artificial neural networks to predict the tensile strength of ABS parts manufactured using fused deposition models. The results showed that an enhanced mechanical strength can be achieved by optimizing the process parameters. Based on FDM technology, Yang et al. [13] have explored an additive manufacturing process based on continuous carbon fiber-reinforced polylactic acid (PLA) composite prepreg filaments, resulting in the direct additive manufacture of lightweight and high-strength composite honeycomb load-bearing structures. Regarding stereolithography, Wen et al. [14] introduced a structure optimization-based compensation method to improve the geometric accuracy of microstructures printed by projective stereolithography. As for resistive additive manufacturing, Li et al. [15] have optimized the relative process parameters and analyzed their effects on the morphology of coating formation.

The concept of additive manufacturing can also be extended to high-performance coating preparation, which has received increasing attention in recent years. Zhang et al. [16] have researched the tribological properties of different crystalline diamond coatings prepared by the microwave plasma chemical vapor deposition (MPCVD) method in dry and seawater environments, providing important insights into the wear behavior of diamond coatings in seawater. Zhou et al. [17] investigated the nanomechanical properties of Ni-Co-Al-Ce coatings fabricated by velocity oxygen-fuel (HVOF) spraying, providing vital predictions for the erosion resistance of MCrAlY coatings. Li et al. [18] have synthesized two types of cemented carbide tools based on WC–Co–Zr and WC–Ni–Zr, namely cemented carbide tools and functional gradient cemented carbide (FGC) tools with FCC-phase ZrN-rich surfaces, which have potential in future hybrid additive/subtractive manufacturing applications.

Green, intelligent, and high-performance manufacturing processes are the focus of this Special Issue; therefore, several other studies on manufacturing trends are herein presented. The influence of inclusions on the mechanical properties of spring steels is significant; thus, Li et al. [19] have investigated the effect of alkalinity and Al_2O_3 content on slag viscosity and structure to explore their effect on inclusion removal from steel. The generation of coal fly ash (CFA) in manufacturing is a serious barrier in the development of eco-friendly process manufacturing. Qi et al. [20] have constructed three different regression models to quickly and accurately predict the generation of CFA, thus saving time in planning of CFA disposal.

This Special Issue, "Additive Manufacturing (AM) for Advanced Materials and Structures: Green and Intelligent Development Trend," can be considered as a review of the progress of additive manufacturing over the past year in the areas of advanced materials, structural design, and manufacturing processes.

Conflicts of Interest: The author declares no conflict of interest.

References

1. Chao, Y.; Yi, H.; Cao, F.; Lu, S.; Ma, L. Experimental Analysis of Polycaprolactone High-Resolution Fused Deposition Manufacturing-Based Electric Field-Driven Jet Deposition. *Crystals* **2022**, *12*, 1660. [CrossRef]
2. Chao, Y.; Yi, H.; Cao, F.; Li, Y.; Cen, H.; Lu, S. Experimental Analysis of Wax Micro-Droplet 3D Printing Based on a High-Voltage Electric Field-Driven Jet Deposition Technology. *Crystals* **2022**, *12*, 277. [CrossRef]
3. Li, H.; Yang, W.; Ma, Q.; Qian, Z.; Yang, L. Specific Sensitivity Analysis and Imitative Full Stress Method for Optimal BCCZ Lattice Structure by Additive Manufacturing. *Crystals* **2022**, *12*, 1844. [CrossRef]
4. Ma, Q.; Li, Z.; Li, J. Compression Behavior of SLM-Prepared 316L Shwartz Diamond Structures under Dynamic Loading. *Crystals* **2022**, *12*, 447. [CrossRef]
5. Shen, Z.; Li, N.; Wang, T.; Wu, Z. Effects of SiC Content on Wear Resistance of Al-Zn-Mg-Cu Matrix Composites Fabricated via Laser Powder Bed Fusion. *Crystals* **2022**, *12*, 1801. [CrossRef]
6. Lu, Z.; Han, Y.; Gao, Y.; Cao, F.; Zhang, H.; Miao, K.; Deng, X.; Li, D. Effect of Nano-Si_3N_4 Reinforcement on the Microstructure and Mechanical Properties of Laser-Powder-Bed-Fusioned $AlSi_{10}Mg$ Composites. *Crystals* **2022**, *12*, 366. [CrossRef]
7. Li, H.; Zhang, J. The Fracture Behavior of 316L Stainless Steel with Defects Fabricated by SLM Additive Manufacturing. *Crystals* **2021**, *11*, 1542. [CrossRef]
8. Cao, F.; Zhang, H.; Zhou, H.; Han, Y.; Li, S.; Ran, Y.; Zhang, J.; Miao, K.; Lu, Z.; Li, D. The Effect of Scanning Strategies on FeCrAl Nuclear Thin-Wall Cladding Manufacturing Accuracy by Laser Powder Bed Fusion. *Crystals* **2022**, *12*, 1197. [CrossRef]
9. Wang, X.; Wang, M. Research on the Preparation Process and Performance of a Wear-Resistant and Corrosion-Resistant Coating. *Crystals* **2022**, *12*, 591. [CrossRef]
10. Du, L.; Liu, S.; Yin, J.; Pang, S.; Yi, H. Comparison of Dopant Incorporation and Near-Infrared Photoresponse for Se-Doped Silicon Fabricated by fs Laser and ps Laser Irradiation. *Crystals* **2022**, *12*, 1589. [CrossRef]
11. Wu, Y.; Chen, Y.; Kong, L.; Jing, Z.; Liang, X. A Review on Ultrafast-Laser Power Bed Fusion Technology. *Crystals* **2022**, *12*, 1480. [CrossRef]
12. Tura, A.D.; Lemu, H.G.; Mamo, H.B. Experimental Investigation and Prediction of Mechanical Properties in a Fused Deposition Modeling Process. *Crystals* **2022**, *12*, 844. [CrossRef]
13. Yang, L.; Zheng, D.; Jin, G.; Yang, G. Fabrication and Formability of Continuous Carbon Fiber Reinforced Resin Matrix Composites Using Additive Manufacturing. *Crystals* **2022**, *12*, 649. [CrossRef]
14. Wen, C.; Chen, Z.; Chen, Z.; Zhang, B.; Cheng, Z.; Yi, H.; Jiang, G.; Huang, J. Improvement of the Geometric Accuracy for Microstructures by Projection Stereolithography Additive Manufacturing. *Crystals* **2022**, *12*, 819. [CrossRef]
15. Li, S.; Ma, K.; Xu, C.; Yang, L.; Lu, B. Numerical Analysis and Experimental Verification of Resistance Additive Manufacturing. *Crystals* **2022**, *12*, 193. [CrossRef]
16. Zhang, H.; Song, H.; Pang, M.; Yang, G.; Ji, F.; Jiang, N.; Nishimura, K. Tribological Performance of Microcrystalline Diamond (MCD) and Nanocrystalline Diamond (NCD) Coating in Dry and Seawater Environment. *Crystals* **2022**, *12*, 1345. [CrossRef]
17. Zhou, F.; Guo, D.; Xu, B.; Wang, Y.; Wang, Y. Nanomechanical Characterization of High-Velocity Oxygen-Fuel NiCoCrAlYCe Coating. *Crystals* **2022**, *12*, 1246. [CrossRef]
18. Li, S.; Xue, X.; Chen, J.; Lu, T.; Zhao, Z.; Deng, X.; Lu, Z.; Wang, Z.; Li, Z.; Qu, Z. Study on the Properties of Coated Cutters on Functionally Graded WC-Co/Ni-Zr Substrates with FCC Phase Enriched Surfaces. *Crystals* **2021**, *11*, 1538. [CrossRef]
19. Li, Y.; Chen, C.; Hu, H.; Yang, H.; Sun, M.; Jiang, Z. Effect of Slag Adjustment on Inclusions and Mechanical Properties of Si-Killed 55SiCr Spring Steel. *Crystals* **2022**, *12*, 1721. [CrossRef]
20. Qi, C.; Wu, M.; Lu, X.; Zhang, Q.; Chen, Q. Comparison and Determination of Optimal Machine Learning Model for Predicting Generation of Coal Fly Ash. *Crystals* **2022**, *12*, 556. [CrossRef]

Article

Specific Sensitivity Analysis and Imitative Full Stress Method for Optimal BCCZ Lattice Structure by Additive Manufacturing

Haonan Li [1], Weidong Yang [1,*], Qianchao Ma [1], Zhihan Qian [1] and Li Yang [2,*]

[1] School of Mechanical Engineering, Hebei University of Technology, Tianjin 300401, China
[2] Department of Industrial Engineering, University of Louisville, Louisville, KY 40292, USA
* Correspondence: yangweidong@hebut.edu.cn (W.Y.); li.yang.1@louisville.edu (L.Y.)

Abstract: Additive manufacturing (AM) can quickly and easily obtain lattice structures with light weight and excellent mechanical properties. Body-centered cubic (BCC) lattice structure is a basic type of lattice structure. BCC with Z strut (BCCZ) lattice structure is a derivative structure of BCC lattice structure, and it has good adaptability to AM. Generally, the thickness of each pillar in the BCCZ lattice structure is uniform, which results in the uneven stress distribution of each pillar. This makes the potential of light weight and high strength of the BCCZ lattice structure not fully played, and the utilization rate of materials can be further improved. This paper designs an optimization method. Through the structural analysis of a BCCZ lattice structure, an optimization method of a BCCZ lattice structure based on parametric modeling parameters is presented. The section radius of all pillars in the BCCZ lattice is taken as a design variable, and the specific sensitivity analysis method and simulated full stress optimization idea are successively used to determine the optimal section radius of each pillar. Finally, the corresponding model is designed and samples are manufactured by LPBF technology for simulation and experimental verification. The results of simulation and experiment show that the strength limit of the optimized parts increased by 18.77% and 18.43%, respectively, compared with that before optimization.

Keywords: additive manufacturing; lattice structures; BCCZ; specific sensitivity analysis; imitative full stress method

Citation: Li, H.; Yang, W.; Ma, Q.; Qian, Z.; Yang, L. Specific Sensitivity Analysis and Imitative Full Stress Method for Optimal BCCZ Lattice Structure by Additive Manufacturing. *Crystals* **2022**, *12*, 1844. https://doi.org/10.3390/cryst12121844

Academic Editors: Hao Yi, Huajun Cao, Menglin Liu and Le Jia

Received: 24 November 2022
Accepted: 14 December 2022
Published: 16 December 2022

Publisher's Note: MDPI stays neutral with regard to jurisdictional claims in published maps and institutional affiliations.

1. Introduction

Based on the principle of discrete stacking, AM realizes model design through computer aided design (CAD) and direct manufacturing driven by 3D data [1]. As an advanced manufacturing technology, AM has subverted the traditional manufacturing concept and has incomparable unique advantages in manufacturing complexity and special structures, bringing conceptual innovation to the entire manufacturing industry [2]. Compared with traditional manufacturing methods, AM makes up for the vacancy of traditional manufacturing methods, enabling parts that in the past were difficult to manufacture or could not be manufactured or had high manufacturing costs to be processed and manufactured [3]. One of the most outstanding advantages of AM is that it increases the degree of freedom of design, which is very helpful for the manufacturing of lattice structures. In the past two decades, AM has been rapidly developed and applied. Powder bed fusion (PBF) is a group of AM techniques. When equipped with lasers as energy sources, the processes are also known as laser powder bed fusion (LPBF). LPBF is also commercially known as selective laser melting (SLM) or direct metal laser melting, which is shown in Figure 1 [4]. LPBF can fabricate complex components by melting metal powders layer by layer using a high-energy laser beam and generally can be used for producing complicated parts [5]. The forming size is getting smaller and smaller, reaching the millimeter level. The printing material has also achieved the transition from the past titanium alloy to the present aluminum alloy [6]. At the same time, relevant enterprises around the world have also begun to increase the research and development of AM equipment, and have made relatively successful progress, and are gradually realizing the commercialization of AM equipment [7]. With the upgrading and development of AM equipment, the gap between structural design and manufacturing has been further narrowed, and the manufacturing problems of millimeter scale complex structures have been solved. Structural design benefiting from AM has gradually entered the perspective of designers [8]. The design method for AM reduces the manufacturing constraints in the traditional design concept, making the forming of millimeter scale complex structures more simple and direct forming possible, which helps to achieve parametric control of complex structures [9]. The lattice structure benefits from the development of AM, and its shortcomings that were difficult to manufacture in the past have been solved to a large extent. It has many characteristics such as light weight, high strength, excellent energy absorption, good thermal performance, sound insulation and noise reduction, and biocompatibility [10–13]. Different from the traditional civil construction industry, the lattice structure oriented to AM is usually at the millimeter level. The appearance of AM makes the lattice structure design break through the size constraints and can realize the design and manufacturing at the millimeter level [14]. In recent years, successful cases of the combination of lattice structure design and AM have been emerging, and the feasibility of this scheme is constantly being verified. The emergence and development of AM provide a new solution to the problem that the lattice structure has been difficult to be manufactured by traditional manufacturing methods, and provide support and help for the design of millimeter level lattice structure [15–17].

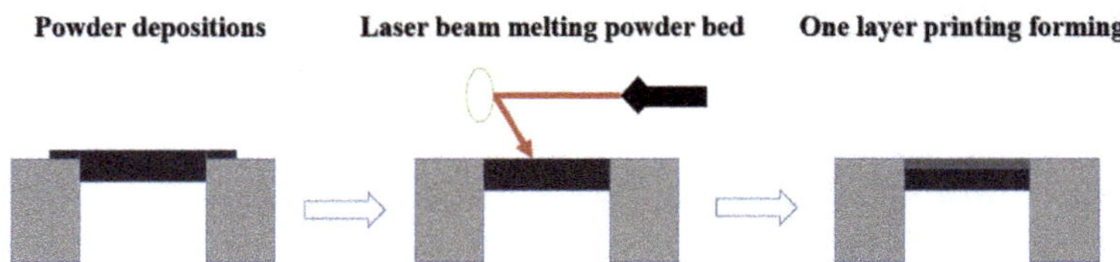

Figure 1. Laser powder bed fusion process.

The research on the combination scheme of AM and lattice structure design in aerospace [18], automotive industry [19], biomedicine [20] and other aspects shows that AM and lattice structure design show a very broad development prospect in the field of light weight. With the help of AM, the design model of lattice structure has been greatly guaranteed for manufacturing, and the lattice structure design for AM also provides researchers with greater imagination space. At the same time, the steps of the entire AM process are more concise, which can greatly save manufacturing time and reduce manufacturing costs from the design end of the computer equipment to the manufacturing equipment [21]. The lattice structure produced by AM can meet the size and shape requirements and achieve millimeter level precision manufacturing, greatly ensure the performance requirements, reduce material loss and improve material utilization. In economic development, large machines (such as aircraft, automobiles, etc.), after using the parts produced by the lattice structure design scheme for AM, reduce their own weight and energy consumption, so as to reduce all kinds of pollution caused by energy consumption and help to achieve green development with AM as the core [22].

BCC lattice structure is characterized by simple topology, strong designability, strong manufacturability, and excellent mechanical properties [23]. Moreover, it is worth noting that the success rate of BCC lattice structure in AM is very high in millimeter scale manufacturing, which is mainly due to the tilt angle of the pillar in the BCC lattice structure. Leary et al. [24,25] quantified manufacturability of aluminum lattice strut elements by experiment. They set four possible strut inclination angles: 0, 35.26, 45 and 90 degrees in a unit cell with equal side length and found that the manufacturability of struts at other angles was well beyond 0 degrees. BCCZ lattice structure is a derivative structure of BCC lattice structure. On the basis of retaining the advantages of BCC lattice structure with strong design, good manufacturing and a high success rate in AM, its related performance has been greatly improved compared with BCC lattice structure due to the existence of a strengthened Z pillar [26–28]. Chua et al. [29] used the finite element analysis method to study the mechanical properties of lattice structure in order to avoid cracks and pores that may occur in the process of AM. Chen et al. [30] used the MIST optimization algorithm to optimize the pillar size inside the cell. Jin et al. [31] completed the parametric design by obtaining the cell length diameter ratio distribution matrix and mapping the matrix to the pillar size of the unit cell. In the above research, the manufacturability and optimization methods of struts in lattice structures were studied. However, in the current research, the optimization design of the BCCZ lattice structure for AM needs to be strengthened. At present, the thickness of each pillar in the BCZZ lattice structure is uniform, which easily leads to uneven stress distribution of each pillar. The advantage of the BCCZ lattice structure in light weight and great strength has not been fully utilized, and the utilization rate of materials can be further improved in order to further take advantage of the combination of AM and lattice structure to better release constraints. This paper takes BCCZ lattice structure as the research object. On the basis of a structural analysis of the BCCZ lattice structure, the section radius of all pillars in the BCCZ lattice structure is taken as the design variable. A BCCZ lattice structure optimization method based on the combination of specific sensitivity analysis and the idea of imitative full stress is proposed to maximize the material utilization of each pillar in the BCCZ lattice structure. Last, the optimization method in this paper is explained and verified by finite element simulation and experiment. Through the designed method in this paper, the dangerous stress in the BCCZ structure can be controlled within the allowable stress range, and the strength of the whole structure can be improved. The designed BCCZ structure can maintain better strength on the basis of light weight.

2. Analysis of BCCZ Lattice Structure

The lattice structure can be seen as a porous structure composed of many cells according to the set distribution rules. Chen et al. [32] summarized that the unit cell had an important impact on the characteristics of lattice structures, such as mechanical response, specific surface area, stiffness, pore size. Therefore, the unit cell is very representative in the lattice structure analysis [33]. Therefore, this section analyzes the BCCZ lattice structure from the perspective of the BCCZ unit cell.

2.1. BCCZ Unit Cell Topology Analysis

The BCCZ lattice structure is generally composed of BCCZ cells through X, Y and Z arrays. Figure 2 shows the BCCZ lattice structure and its corresponding unit cell structure diagram. The BCCZ lattice structure in Figure 2a is generated by BCCZ cells in Figure 2b through repeated arrays in X, Y and Z directions, and contains $5 \times 3 \times 2$ cells, respectively.

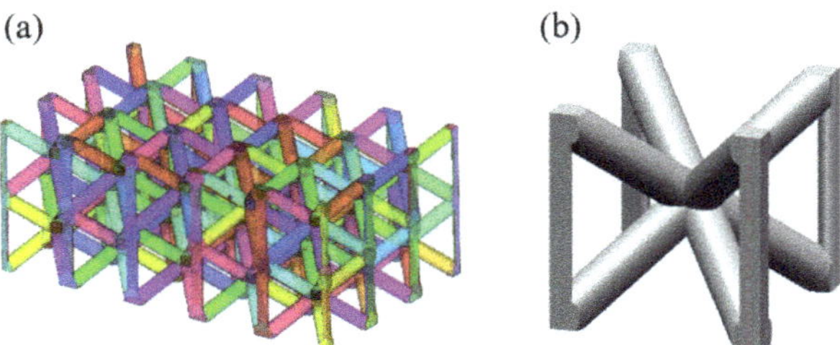

Figure 2. (a) BCCZ lattice structure diagram; (b) BCCZ unit cell diagram.

As shown in Figure 2a, the BCCZ lattice structure composed of BCCZ cells can be regarded as a micro truss structure, whose structural elements are composed of pillars and nodes. The topological structure of the BCCZ unit cell has little change compared with the BCC unit cell. Figure 3 shows the topology of the BCC unit cell and BCCZ unit cell, as well as the numbering sequence of nodes and pillars inside the unit cells. The BCC unit cell central node (N9) and 8 outer nodes (N1–N8) are connected in the form of pillars, and the location of the central node is determined by Formula (1). On the basis of inheriting the topological configuration of the BCCZ unit cell, the BCCZ unit cell can be constructed by adding a strengthened Z pillar in the direction parallel to the load. That is, on the basis of node N1–N9, add P9 to P12 between nodes N1 to N5, N2 to N6, N3 to N7, and N4 to N8.

$$
\begin{cases}
x_9 = \frac{1}{8} \sum_{i=1}^{8} x_i \\
y_9 = \frac{1}{8} \sum_{i=1}^{8} y_i \\
z_9 = \frac{1}{8} \sum_{i=1}^{8} z_i
\end{cases}
\tag{1}
$$

where (x_9, y_9, z_9) is the coordinate of central node N9, and (x_i, y_i, z_i) corresponds to the coordinate of outer nodes N1–N8.

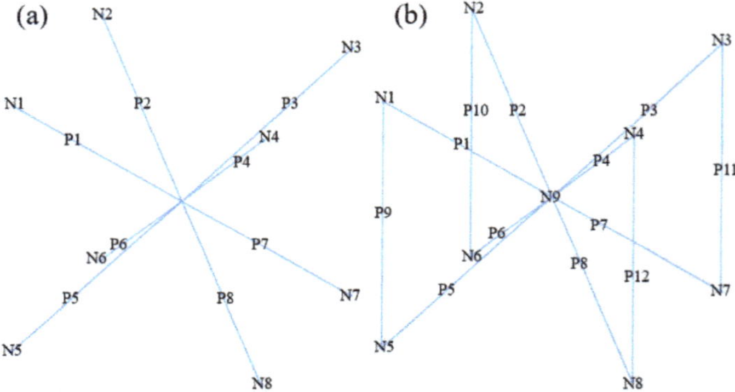

Figure 3. Topological structure comparison diagram of BCC unit cell and BCCZ unit cell and numbering sequence of each pillar: (**a**) BCC; (**b**) BCCZ.

2.2. Parametric Modeling

The topological analysis of Section 2.1 is helpful to realize the parametric modeling of the BCCZ unit cell, and then the parametric modeling of the BCCZ lattice structure can be realized. In the current research, the modeling precondition is to set the section shape of the BCCZ lattice structure pillar as a circle, and the external geometry of BCCZ unit cell as a cube [34].

Figure 4 shows the BCCZ lattice structure decomposition steps. The BCCZ lattice structure can be divided into many BCCZ cells. The BCCZ unit cell can be divided into eight inclined pillars of cylinders and four outer pillars of 1/4 cylinders. Among them, each outer pillar intersects two inclined pillars, and eight inclined pillars intersect at the central node. The side length of the cell is set as S, the length of the inclined pillar is $l_i (i = 1, 2, \ldots\ldots, 8)$, the section radius of the inclined pillar is $r_i (i = 1, 2, \ldots\ldots, 8)$, the length of the outer pillar is $l_i' (i = 9, 10, 11, 12)$, and the section radius of the outer pillar is $r_i' (i = 9, 10, 11, 12)$. Formulas (2)–(4) show the fixed geometric relationship in the BCCZ unit cell:

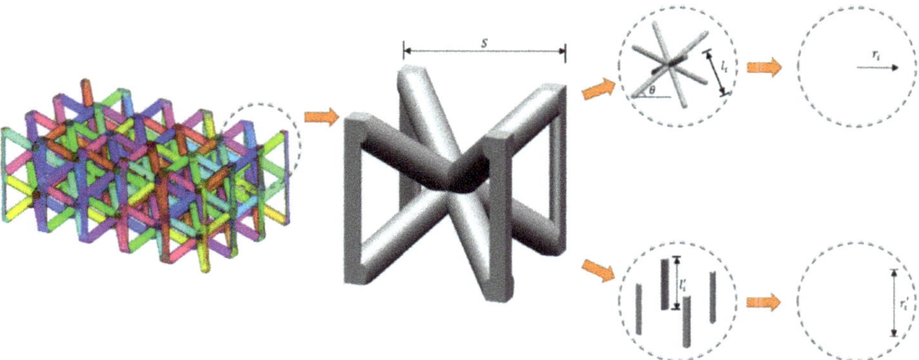

Figure 4. Analysis of design parameters of BCCZ lattice structure.

(1)   Inclined pillar length l_i and unit cell side length S:

$$l_i = \tfrac{\sqrt{3}}{2}S \qquad (2)$$

(2)   Outer pillar length l_i' and unit cell side length S:

$$l_i' = S \qquad (3)$$

(3)   Angle between inclined pillar and horizontal direction θ:

$$\theta = \sin^{-1}\tfrac{1}{\sqrt{3}} \qquad (4)$$

To sum up, the parametric modeling of the BCCZ unit cell can be completed through unit cell side length S, the section radius $r_i(i = 1,2,\ldots\ldots,8)$ of the inclined pillar and the section radius $r_i'(i = 9,10,11,12)$ of the outer pillar. By assigning the above design parameters, BCCZ cells with different structural parameters can be obtained. The BCCZ unit cell is an outstanding representative of the BCCZ lattice structure. According to the structural relationship between the two, the structural parameters of the BCCZ lattice structure can be summarized as unit cell side length S, the section radius $r_i(i = 1,2,\ldots\ldots,8)$ of the inclined pillar, the section radius $r_i'(i = 9,10,11,12)$ of the outer pillar, X, Y and the number of cells in Z direction x, y and z.

3. Optimization Method

The current optimization of lattice structure can be summarized in the following two aspects. On the one hand, a few researchers changed the cell strut into a curve by removing the constraint that the section radius of the cell strut is equal everywhere. On the macro level, it reflects that the shape of the lattice structure pillar has changed, and the optimization at the cell level has been realized by reducing the impact of stress concentration, which has realized the innovation of cell configuration to a certain extent [35]. On the other hand, because it is difficult to propose new unit cell configurations, most researchers still use various optimization methods (such as topology optimization) to optimize and design lattice structures with better performance on the basis of existing unit cells [36,37]. According to the above two kinds of optimization ideas, this paper synthesizes their respective characteristics, and on the basis of the existing cell elements and the analysis in Section 2, takes the section radius of each pillar of BCCZ lattice structure as an independent design variable and relieves the constraint that the section radius between the pillars is equal. A BCCZ lattice structure optimization method combining specific sensitivity analysis and imitative full stress design method is proposed. Figure 5 shows the design process of this method.

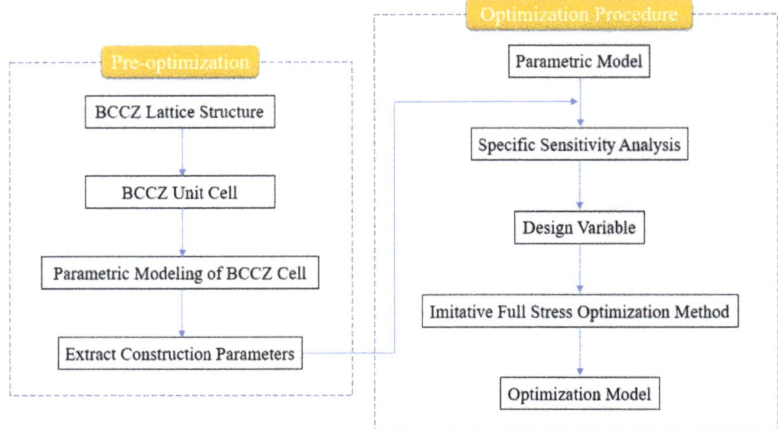

Figure 5. Procedures for the optimization design of specific sensitivity analysis and imitative full stress design.

3.1. Specific Sensitivity Analysis

According to the analysis in Section 2.2, the number of pillars of the BCCZ lattice structure is large, which leads to many variables to be designed. As a result, BCCZ lattice structure optimization is difficult to design, and complex design problems and a reasonable optimization design scheme are difficult to determine. To solve this problem, this paper uses the method of specific sensitivity analysis to distinguish the design variables and solve the optimization direction of each design variable, so as to carry out the follow-up optimization work.

The concept of sensitivity is often used in rigid frame structures in practical projects. It can be used to express the degree of influence on a certain performance of the overall structure when a variable changes [38]. In this paper, the design variable is the section radius of each pillar of the BCCZ lattice structure, that is, the section radius r_i of inclined pillar and the section radius r_i' of outer pillar, and the influence performance can be a mechanical property of the BCCZ lattice structure [39]. The mathematical model of sensitivity is shown in Formula (5).

$$P = p\left(r_1, \ldots, r_i, \ldots, r_m, r_1', \ldots, r_i', \ldots, \ldots r_n'\right) \tag{5}$$

where P is the performance function of BCCZ lattice structure, m is the number of inclined pillars and n is the number of outer pillars.

The change in the BCCZ lattice structure performance caused by the increment dr_i or dr_i' of any pillar section radius change is dP, namely:

$$dP = \frac{\partial p}{\partial r_1}dr_1 + \ldots + \frac{\partial p}{\partial r_i}dr_i + \ldots + \frac{\partial p}{\partial r_m}dr_m + \frac{\partial p}{\partial r_1'}dr_1' + \ldots + \frac{\partial p}{\partial r_i'}dr_i' + \ldots + \frac{\partial p}{\partial r_n'}dr_n' = \sum_{i=1}^{m}\frac{\partial p}{\partial r_i}dr_i + \sum_{i=5}^{n}\frac{\partial p}{\partial r_i'}dr_i' \tag{6}$$

where the partial derivative of the performance function P of the BCCZ lattice structure to the section radius r_i or r_i' of the pillar is the performance sensitivity of the BCCZ lattice structure.

The performance sensitivity analysis can be used to determine the design variables that have a significant impact on the overall performance of the BCCZ lattice structure among the BCCZ lattice structure optimization design variables, and formulate an optimization design scheme to improve the optimization efficiency. However, the advantage of the BCCZ lattice structure over strength cannot be fully exploited by analyzing performance sensitivity only. Xu et al. [40] designed some lattice structures with different relative densities (0.15–0.5) and carried out experimental monitoring. They systematically studied the effect of relative density on the compressive properties of lattice structures, including mechanical properties, failure mechanisms and energy absorption capacity. The results show that the compressibility of the lattice structure changes significantly with the change of the relative density, and the performance of the lattice structure can usually be improved by simply increasing the section radius. In other words, when the section radius is very sensitive to the performance and mass, if the design variable is optimized, the BCCZ lattice structure performance will be improved as well as the BCCZ lattice structure mass. That is, the reason for the BCCZ lattice structure performance improvement may be achieved by adding materials, which violates the original optimization intention of greatly improving the material utilization. In this paper, the concept of specific sensitivity is introduced to find out the performance sensitivity of design variables under unit mass and the mass sensitivity of design variables under unit performance.

In this paper, the specific sensitivity is defined as the ratio of the sensitivity of the BCCZ lattice structure performance to the pillar section radius and the sensitivity of the BCCZ lattice structure mass to the pillar section radius, namely:

$$\eta_i = \frac{\varepsilon_i}{\alpha_i} \tag{7}$$

where η_i is the specific sensitivity, ε_i is the performance sensitivity and α_i is the mass sensitivity.

Because it is difficult to obtain the sensitivity value directly through the theoretical formula, at the same time, there is a complex stress concentration phenomenon at the node of the BCCZ lattice structure, and the coupling between design variables is also complex, there are many problems and difficulties in applying the analytical method directly [41,42]. Therefore, this paper used finite element analysis tools to extract the values of performance sensitivity and mass sensitivity. The following example is to illustrate the method.

According to Formula (6), if the variable is x_i, the performance function is p_i; when the variable is increased by one unit, the performance function corresponding to x_{i+1} is p_{i+1}.

$$\begin{cases} dP = p_{i+1} - p_i \\ dx_i = x_{i+1} - x_i = 1 \end{cases} \tag{8}$$

That is, when the mass unit changes,

$$\varepsilon_i = \frac{\partial p}{\partial x_i} = dP \tag{9}$$

The values of p_{i+1} and p_i are extracted by finite element analysis tools, and the performance sensitivity value of any design variable x_i can be obtained by Formulas (8) and (9).

The mass sensitivity can also be obtained by the above method, and the specific sensitivity can be obtained by Formula (7). At the same time, in order to make the specific sensitivity value more accurate, this paper obtains the final specific sensitivity value by setting multiple groups of data and curve fitting multiple groups of results. According to the specific sensitivity value obtained, the optimal design scheme can be further formulated on the basis of determining the design variables. Its purpose is:

(1) Increasing the design variable with the maximum specific sensitivity can achieve the maximum performance improvement with the minimum mass increase.
(2) Reducing the design variable with the minimum specific sensitivity can achieve the minimum performance degradation with the maximum mass reduction.

3.2. Imitative Full Stress Optimization Method

Through the analysis of Section 3.1, the design variables and their optimization directions can be determined, and the optimization design scheme can be preliminarily obtained. For the design variables to be optimized, this paper refers to the idea of imitative full stress design and proposes an optimization method for the cross section dimensions of each pillar of the BCCZ lattice structure. The basic idea of imitative full stress is to make the most unfavorable stress in structural members approach or reach the allowable stress of materials, so as to make full use of materials. The mathematical model of this method is established as follows.

1. Optimization design variables: the BCCZ lattice structure section radius r_i of inclined pillar and section radius r_i' of outer pillar.

$$R = \left(r_1, \ldots, r_m, r_1', \ldots r_n'\right) \tag{10}$$

2. Objective function: the overall mass of the BCCZ lattice structure. The objective of optimization is to minimize the objective function.

$$\min\ W \approx \frac{\sqrt{3}}{2} \sum_{i=1}^{m} \pi r_i^2 S + \sum_{i=1}^{n} \pi r_i'^2 S \tag{11}$$

3. Constraint 1: the strength of each pillar approximates the allowable stress value.

$$C(x) = \sigma_{imax} - [\sigma_i] \cong 0 \tag{12}$$

In the formula, $\cong$ is a symbol to describe the imitative full stress, which means that the value on the left of the symbol is infinitely close to but not more than the value on the right of the symbol, σ_{imax} is the maximum von Mises stress of pillar i under a certain condition, and $[\sigma_i]$ is the allowable stress of pillar materials of the BCCZ lattice structure.

4. Constraint 2: The value range of each pillar radius of the BCCZ lattice structure shall be between the minimum manufacturing size and the maximum space size of AM.

$$\begin{cases} r_{i_{min}} \le r_i \le r_{i_{max}} \\ r_{i_{min}}' \le r_i' \le r_{i_{max}}' \end{cases} \tag{13}$$

where $r_{i_{min}}$ and $r_{i_{min}}'$ are the minimum manufacturing dimensions of AM, set as 0.5 mm according to [24,25,43], and $r_{i_{max}}$ and $r_{i_{max}}'$ are the maximum space dimensions of AM. Considering the application scope of the pillar, 30% of the pillar length is selected as the upper limit of the section radius.

Therefore, the steps of the imitative full stress optimization method in this paper are as follows.

1. In the initial stage, set a reasonable set of section radius of each pillar.

$$R^0 = \left\{r_1^0, \ldots, r_m^0, r_1'^0, \ldots r_n'^0\right\} \tag{14}$$

where R^0 represents the set of initial section radius of each pillar, and $r_i^0 \left(r_i'^0\right)$ represents the initial value of pillar i.

2. Structural analysis. Similar to the difficulties encountered in applying the analytical method to the specific sensitivity analysis, this paper chooses to use the finite element analysis tool to obtain the maximum von Mises stress value and the minimum von Mises stress value of each pillar of the BCCZ lattice structure, namely σ_{imax} and σ_{imin}, and sort them in descending and ascending order, respectively, and set

$$\begin{cases} \sigma_{max} = max\{\sigma_{imax}\}\,(i = 1, 2, \cdots, m+n) \\ \sigma_{min} = min\{\sigma_{imin}\}\,(i = 1, 2, \cdots, m+n) \end{cases} \tag{15}$$

where σ_{max} and σ_{min} are the maximum von Mises stress and minimum von Mises stress of the BCCZ lattice structure as a whole, respectively.

3. Strength verification. Compare the σ_{max} obtained in step 2 with the allowable stress $[\sigma_i]$, and enter the corresponding iteration cycles 4 and 5 according to the comparison results.

4. When $\sigma_{max} > [\sigma_i]$, it indicates that the overall strength of the BCCZ lattice structure under the current section radius of each pillar is too large, and the section radius of the pillar where the maximum von Mises stress is located must be increased, and then go to step 2 again to update the structural analysis results. When the condition $\sigma_{max} \leq [\sigma_i]$ is met, jump out of the cycle and output the value of the current section radius of each pillar. If the end of cycle condition cannot be satisfied all the time, the optimization is terminated because there is no feasible solution to the problem.

5. When $\sigma_{max} \leq [\sigma_i]$, it indicates that the materials of some pillars may not be fully utilized under the current section radius of each pillar, so it is necessary to find out the corresponding pillar and reduce the section radius of the corresponding pillar, and turn to step 2 again. According to the new structure analysis results, it can be divided into the following two cases:

    (1) $\sigma_{max} \leq [\sigma_i]$: It shows that the material is still not fully utilized under the current section radius of each pillar, so the section radius of the pillar where the current minimum von Mises stress is located must be reduced, and go to step 2. If $\sigma_{max} \leq [\sigma_i]$ is still satisfied until the minimum value of the optimization variable is obtained, the minimum value is output.

    (2) $\sigma_{max} > [\sigma_i]$: It shows that the pillar with the reduced section radius in the previous step has reached the material utilization limit. It is necessary to sort out the results of the structural analysis in the previous step according to the minimum von Mises stress value, find the pillar with the next von Mises stress and reduce its section radius, and then turn to step 2. If the section radius of all optimized pillars is reduced once by continuous traversal, the cycle will be skipped and the section radius of each pillar meeting $\sigma_{max} \leq [\sigma_i]$ will be output.

6. The values of each pillar obtained in Cycle 4 and Cycle 5 constitute the final optimization results.

$$R^m = \left\{ r_1^u, \ldots, r_m^u, r_5^v, \ldots r_n^v \right\} \tag{16}$$

where u and v are the number of iterations in the two cycles, respectively.

Figure 6 shows the step flow chart of the imitative full stress optimization method.

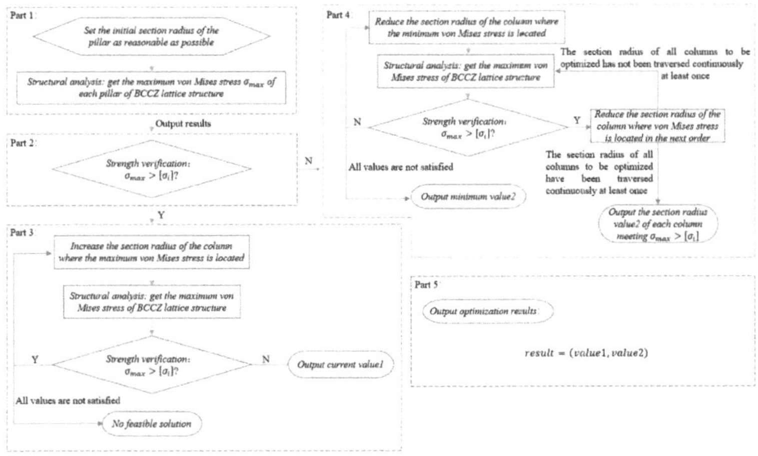

Figure 6. Flow chart of imitative full stress optimization method.

4. Example

4.1. Models for the Test

In this paper, numerical examples are given to verify the effectiveness of the proposed optimization method for the BCCZ lattice structure parameters. The optimization goal is to improve the compressive strength of the test model. As unit cell is very representative in the study of lattice structure, the BCCZ unit cell was selected as a test case to illustrate. The BCCZ unit cell size was set as 5 mm, the initial section radius of each pillar was 0.5 mm, and the allowable stress $[\sigma_i]$ was set as 120 MPa. The numbering sequence of the BCCZ unit cell is shown in Figure 3. The BCCZ unit cell numerical model was designed based on this, and the model diagram is shown in Figure 7.

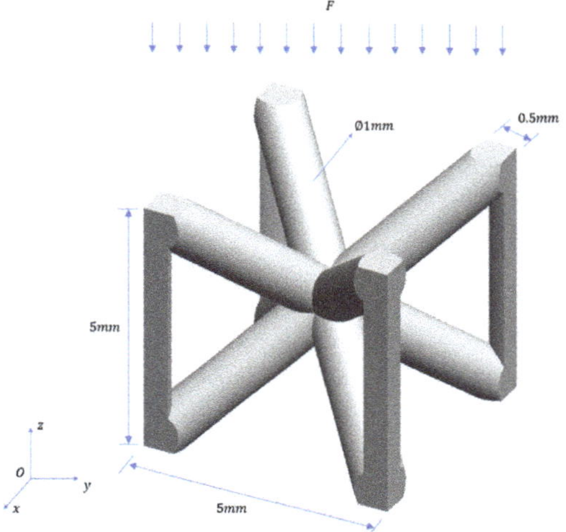

Figure 7. Test case—BCCZ unit cell initial model.

4.2. Optimization Results

According to the analysis conclusion in Section 2.2, the BCCZ unit cell test case was analyzed for specific sensitivity, and each pillar was taken as an independent variable to evaluate the impact of the section radius of each pillar on the overall performance. The results of specific sensitivity from pillar P1 to pillar P12 are shown in Table 1.

Table 1. Specific sensitivity η_i result of pillars P1 to P12.

Pillar	η_i
P1	−0.37
P2	−0.12
P3	−0.47
P4	−0.32
P5	−0.39
P6	−0.13
P7	−0.63
P8	−0.39
P9	−2.06
P10	−3.88
P11	−5.18
P12	−4.97

According to the specific sensitivity analysis results shown in Table 1, the following optimization design scheme can be obtained according to the specific sensitivity analysis results processing method in Section 3.1.

(1) The section radius of P1–P8 is reduced, so that the mass of BCCZ unit cell is reduced without great influence on the strength of the BCCZ unit cell.
(2) Increase the section radius of P9–P12 so as to improve the strength of BCCZ unit cell under the condition that the mass of BCCZ unit cell is basically unchanged.

According to the above design scheme, the optimization of the BCCZ unit cell can be divided into two directions. First, the initial BCCZ unit cell was analyzed by the finite element method. At this time, the maximum equivalent stress was greater than the allowable stress. The BCCZ unit cell strength needs to be improved according to the optimal design scheme (2). After the imitative full stress optimization iteration at this stage, the section radius of the pillars P9–P12 was increased, making the maximum equivalent stress less than the allowable stress, so that the test case met the strength requirements. Secondly, after the strength requirements were met, the pursuit of minimizing the overall mass was started. At this time, the mass of BCCZ unit cell should be reduced according to the optimal design scheme (1) under the framework where the maximum equivalent stress is less than the allowable stress. After the imitative full stress optimization iteration at this stage, the minimum section radius of the P1–P8 pillars under the strength requirements was calculated. Finally, the optimization results of the two stages were combined to output the optimization results of the imitative full stress of pillars P1–P12. The final optimization results are shown in Table 2 and Figure 8.

Table 2. Results of BCCZ unit cell optimized by imitative full stress.

Pillar	Section Radius (mm)
P1	1.1
P2	1.3
P3	1.1
P4	1.3
P5	1
P6	0.9
P7	1
P8	0.9
P9	1
P10	0.9
P11	1
P12	0.9

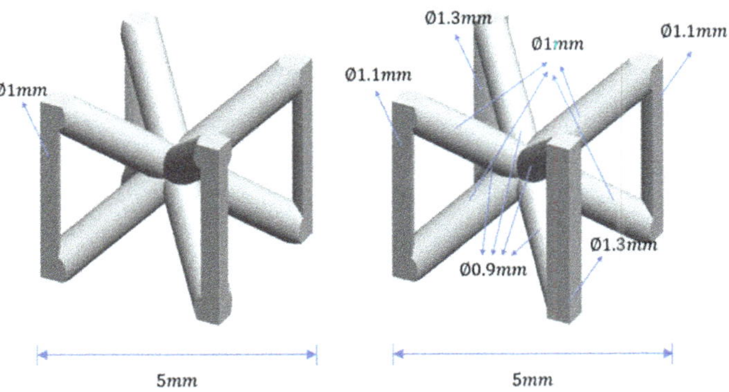

Figure 8. BCCZ unit cell comparison before and after imitative full stress optimization.

5. Numerical Simulation and Experimental Verification

In order to verify the effectiveness of the proposed optimization method, the optimized BCCZ unit cell obtained in Section 4.2 was simulated and verified by experiments. In terms of the manufacturing of experimental samples, Yu et al. [44], Tang et al. [45] and Dong et al. [46] used, respectively, the electron beam powder bed fusion (EBPBF), binder jetting (BJ) and fused deposition modeling (FDM) techniques to manufacture corresponding experimental samples for their own research experiments. In order to ensure that the samples have high accuracy, this paper selected LPBF technology to manufacture the corresponding experimental samples. In recent years, $AlSi_{10}Mg$ material has been used more and more for the fabrication of AM'ed parts, which has good molding effect and cheaper cost, and low-density and high strength. At the same time, it is suitable for manufacturing lattice structures [24,47]. The experimental sample in this paper was formed by LPBF with a light weight structure, considering the economic rationality. Therefore, light material $AlSi_{10}Mg$ alloy powder with good welding formability, thermal conductivity and corrosion resistance was selected [7]. The chemical composition and physicomechanical properties of the $AlSi_{10}Mg$ alloy powder used for simulation and experiment are shown in Tables 3 and 4.

Table 3. Chemical composition of $AlSi_{10}Mg$.

Material	Main Elements			Impurity							
	Al	Si	Mg	Fe	Cu	Mn	Ni	Zn	Pb	Sn	Ti
$AlSi_{10}Mg$	bal	11.7	0.39	0.15	0.05	0.45	0.05	0.10	0.05	0.05	0.15

Table 4. Physicomechanical properties of $AlSi_{10}Mg$.

Material	Density [kg/m^3]	Young's Modulus [MPa]	Yield Strength [MPa]
$AlSi_{10}Mg$	2670	75,000	350

5.1. Numerical Simulation

In order to compare with the experiment and simulate the quasi-static compression process in the actual experiment, this paper designed the BCCZ unit cell simulation model before and after optimization, as shown in Figure 9. The pressing plate was set at both ends of the BCCZ unit cell to simulate the pressure head in the experiment, and the ANSYS Workbench was used for simulation analysis.

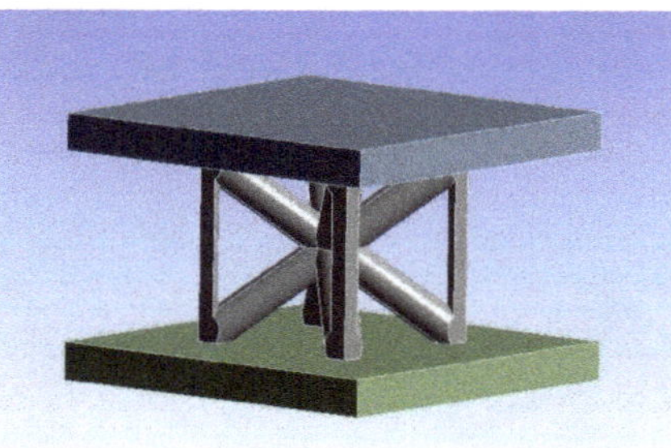

Figure 9. Simulation model of BCCZ unit cell.

According to the experimental conditions of quasi-static compression, the BCCZ unit cell and the compression platform were in static friction contact during compression, and the friction coefficient between metals was 0.15. The upper platen applied a one-way vertical displacement load of 1.5 mm, the lower platen was set as a fixed constraint, and large deformation was opened during simulation. After the use of the ANSYS intelligent grid division, through comparison, it was found that further increasing the number of grids had little impact on the simulation results. This showed that the grid was independent at this time, so this paper used the intelligent grid generation function of ANSYS for simulation analysis. The results of the simulation analysis are shown in Table 5.

Table 5. Simulation analysis results of BCCZ unit cell before and after optimization.

	Strength Limit (MPa)		
BCCZ	Initial 668.72	After optimization 794.22	Optimize efficiency 18.77%

5.2. Sample Preparation

The LiM-X260A metal 3D printer produced by Tianjin LiM Laser Technology Co., Ltd. (Tianjin, China) was selected as the experimental equipment for manufacturing experimental samples, and the equipment is shown in Figure 10. The equipment has the advantages of high precision, high forming efficiency, high automation, high reliability, high efficiency and stability. Table 6 shows the process manufacturing parameters of manufacturing samples. During manufacturing, the outer pillars of the samples were placed vertically with the printing plane, as shown in Figure 11.

Figure 10. LiM-X260A metal 3D printer produced by Tianjin LiM Laser Technology Co., Ltd.

Table 6. Process parameters for sample manufacturing.

Process Parameters	Numerical Value
Laser power	370 W
Delamination thickness	30 μm
Scanning speed	1300 mm/s
Laser wavelength	1060~1080 nm
Spot diameter	85 μm
Oxygen content	≤1000 ppm

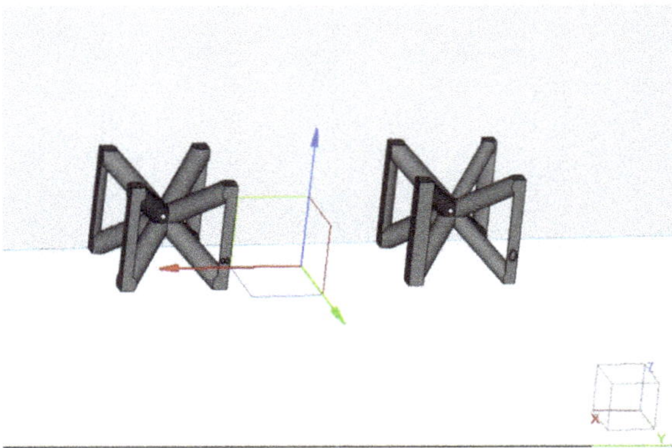

Figure 11. Schematic diagram of sample placement during AM. Left: before optimization, Right: after optimization.

We printed three experimental samples for each digital model, and the BCCZ unit cell structure physical diagram manufactured by the LPBF process is shown in Figure 12. The overall printing of the sample was complete, and there was no incomplete phenomenon such as bar fracture. According to the measurement, all dimensions were consistent with the theoretical dimensions.

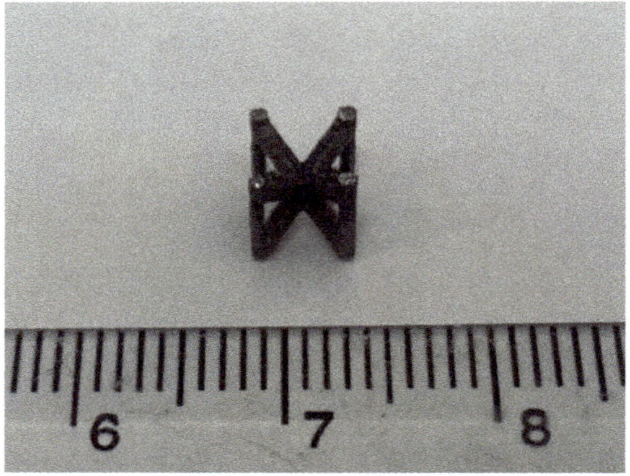

Figure 12. Printed sample.

5.3. Experimental Process and Results

For the experimental samples manufactured by LPBF technology, the WDW-20 electronic universal testing machine was selected for a quasi-static compression test to quantify the compressive strength of the BCCZ unit cells [48]. During the compression test, we set the loading speed of the indenter to 1 mm/min to simulate the quasi-static conditions. In the software, we set the force sensor to detect that the force reached 0.001 KN and recorded the deformation data. During the test, we observed the load displacement curve and recorded the test process with a camera. When the sample was about to be compacted, we stopped the machine and saved the experimental data. Figure 13 shows the experimental process of the compression experiment.

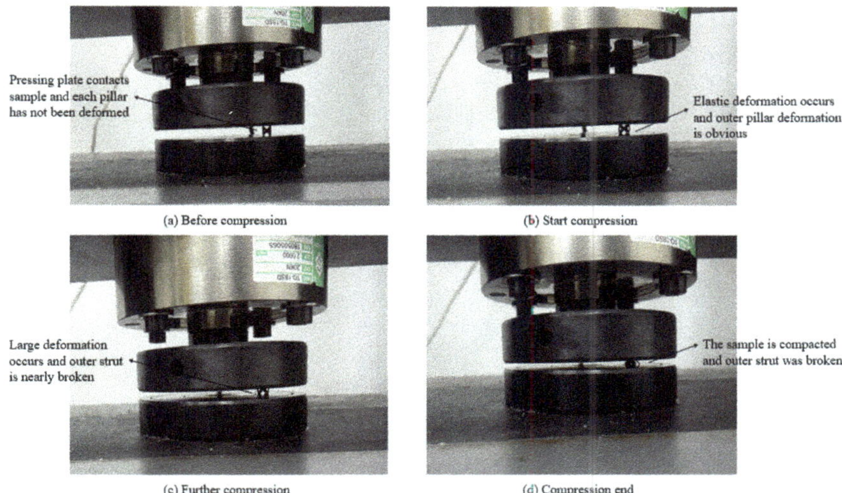

Figure 13. Experimental process of quasi-static compression experiment.

In the compression experiment, the BCCZ unit cell was deformed due to the falling of the pressing plate, and these results were obtained through the sensor. At the same time, the load at each moment could be captured through the force sensor, so that multiple groups of corresponding experimental data could be obtained. All the compression experiments were completed, and the load displacement curve was drawn after sorting out several groups of experimental data. In this paper, the contact area between the compression platform and the sample was read through the theoretical model. After processing, the stress–strain curve of the sample before and after optimization was obtained. For the same type of BCCZ unit cell model, three repeated tests were carried out, and it was found that the curve change trend of the three experiments was basically consistent. Figure 14 shows the stress—strain response of the BCCZ unit cell under compression before and after optimization.

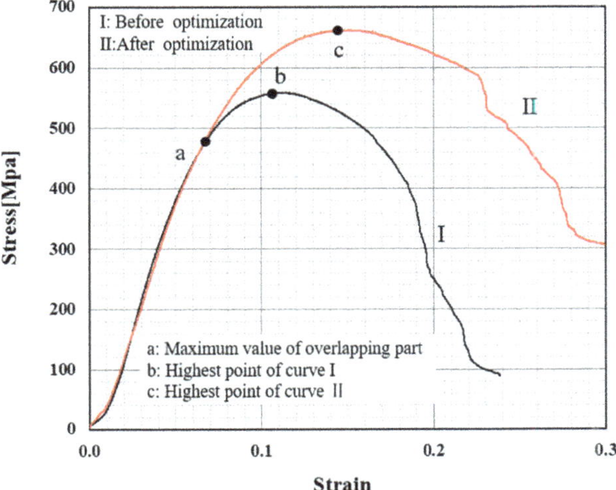

Figure 14. Stress–strain curves of BCCZ unit cell before and after optimization obtained through quasi-static compression experiments. Black: before optimization, it corresponds to Figure 8 (**left**); Red: after optimization, it corresponds to Figure 8 (**right**).

The following conclusions can be drawn by observing the stress–strain curve shown in Figure 14:

- The BCCZ unit cell compression mainly goes through four stages: linear elastic stage, yield stage, strengthening stage and failure stage. The origin in Figure 14 corresponds to the state shown in Figure 13a;
- The material is linear elastic. When the elastic modulus reaches the elastic limit, the BCCZ cell begins to yield plastic. At the Oa curve, the stress–strain curves of BCCZ cells before and after optimization are basically consistent. This stage corresponds to Figure 13b;
- The maximum compressive strength limit (point c) of the BCCZ unit cell after optimization is significantly higher than that of the BCCZ unit cell before optimization (point b), which indicates that the strength of the BCCZ unit cell obtained through the optimization method in this paper has been improved, and the optimization method in this paper has an obvious effect on strength optimization. The specific data are shown in Table 7;
- The BCCZ unit cell will have curve fluctuation in the later failure stage, which may be caused by the alternate fracture of the strengthened Z pillar. This stage corresponds to Figure 13c,d.

Table 7. Comparison of strength and mass of BCCZ cells before and after optimization after three quasi-static compression tests.

	Strength Limit (MPa)			Mass (10–5 kg)		
	Initial	After optimization	Optimize efficiency	Initial	After optimization	Optimize efficiency
BCCZ	558.40	661.31	18.43%	6.64	6.47	0.03%

6. Conclusions

In order to solve the problem of uniform pillar thickness and uneven stress distribution in the BCCZ lattice structure, so as to better play the performance advantages of the BCCZ lattice structure in light weight and high strength, and improve the utilization rate of materials. This paper proposes an optimization method of BCCZ lattice structure based on parametric modeling parameters. The main work of this paper is as follows.

(1) The structural parameters of the BCCZ lattice structure are analyzed. From the perspective of a unit cell, the construction elements of BCCZ lattice structure are explored, and it is concluded that the necessary elements for BCCZ lattice structure modeling are cell side length S, section radius $r_i(i = 1, 2, \ldots\ldots, 8)$ of inclined pillar, section radius $r_i'(i = 9, 10, 11, 12)$ of outer pillar, X, Y, and the number of cells in Z direction x, y and z. Through the above construction elements, the BCCZ lattice structure model can be built.

(2) Taking the construction parameters of the BCCZ lattice structure as independent variables, the finite element analysis method is used to reduce the influence of stress concentration at the nodes of the BCCZ lattice structure and simplify the calculation of the interaction between variables. Assuming that the cell side length S is unchanged, the section radius of each pillar of BCCZ lattice structure is optimized. The specific sensitivity analysis is used to simplify the number of design variables and give the optimization directions of different design variables. The subsequent optimization design scheme is given according to the specific sensitivity analysis results. Based on the optimization design scheme and the idea of imitative full stress, the BCCZ lattice structure is iterated for strength optimization to improve the material utilization rate of the BCCZ lattice structure and obtain the optimal section radius of each pillar in the structure.

(3) A test case is designed to verify the optimization method. BCCZ unit cell is selected for the test case, with the goal of improving its compressive strength. The simulation

and quasi-static compression experiments are used to compare the BCCZ unit cell before and after optimization. The comparison results show that the strength limit of the BCCZ unit cell after optimization is increased by 18.77% and 18.43%, respectively, indicating that the optimization results have strong consistency.

The advantage of the optimization method in this paper is that the stress can be selected according to the setting, and the results of the section radius of each pillar of BCCZ lattice structure with the minimum mass meeting the strength requirements can be output. In the test case in this paper, the objective of simulation is compressive strength. Under other load conditions, the optimization method in this paper also has reference value. Besides the BCCZ lattice structure, this optimization method can also be used for reference for other node structure lattice structures. In addition, the experimental results of this optimization method also prove the advantages of non-uniform design, and non-uniform lattice structures can have higher strength than uniform lattice structures. These two kinds of situations will be considered in our next work.

Author Contributions: Conceptualization, H.L. and W.Y.; methodology, H.L.; validation, H.L. and Z.Q.; formal analysis, Q.M. and H.L.; investigation, H.L. and Z.Q.; resources, W.Y.; writing—original draft preparation, H.L. and Q.M.; writing—review and editing, W.Y. and L.Y.; supervision, W.Y.; funding acquisition, W.Y. All authors have read and agreed to the published version of the manuscript.

Funding: The authors are thankful for the support from the National Natural Science Foundation of China, under the Grant No. 52175313, and Key Science and Technology Project of Hebei Province, China, under the Grant No. 21284901Z.

Data Availability Statement: Not applicable.

Acknowledgments: These experiments were completed in the Advanced Materials Testing and Analysis Center of HEBUT. At the same time, thanks to the scientific research platform provided by the National Engineering Research Center for Technical Innovation Method and Tool. The authors thank them for their help.

Conflicts of Interest: The authors declare no conflict of interest.

References

1. Guo, N.; Leu Ming, C. Additive manufacturing: Technology, applications and research needs. *Front. Mech. Eng.* **2013**, *8*, 215–243. [CrossRef]
2. Gu, D.D.; Shi, X.Y.; Poprawe, R.; Bourell, D.L.; Setchi, R.; Zhu, J.H. Material-structure-performance integrated laser-metal additive manufacturing. *Science* **2021**, *372*, eabg1487. [CrossRef] [PubMed]
3. Safavi, M.S.; Bordbar-Khiabani, A.; Khalil-Allafi, J.; Mozafari, M.; Visai, L. Additive Manufacturing: An Opportunity for the Fabrication of Near-Net-Shape NiTi Implants. *J. Manuf. Mater. Process.* **2022**, *6*, 65. [CrossRef]
4. Sing, S.L. Perspectives on Additive Manufacturing Enabled Beta-Titanium Alloys for Biomedical Applications. *Int. J. Bioprint.* **2022**, *8*, 478. [CrossRef]
5. Zhang, H.; Gu, D.; Dai, D. Laser printing path and its influence on molten pool configuration, microstructure and mechanical properties of laser powder bed fusion processed rare earth element modified Al-Mg alloy. *Virtual Phys. Prototyp.* **2022**, *17*, 308–328. [CrossRef]
6. Maconachie, T.; Leary, M.; Tran, P.; Harris, J.; Liu, Q.; Lu, G.X.; Ruan, D.; Faruque, O.; Brandt, M. The effect of topology on the quasi-static and dynamic behaviour of SLM AlSi10Mg lattice structures. *Int. J. Adv. Manuf. Technol.* **2022**, *118*, 4085–4104. [CrossRef]
7. Aboulkhair, N.T.; Simonelli, M.; Parry, L.; Ashcroft, I.; Tuck, C.; Hague, R. 3D printing of Aluminium alloys: Additive Manufacturing of Aluminium alloys using selective laser melting. *Prog. Mater. Sci.* **2019**, *106*, 45. [CrossRef]
8. Kumar, A.; Collini, L.; Daurel, A.; Jeng, J.Y. Design and additive manufacturing of closed cells from supportless lattice structure. *Addit. Manuf.* **2020**, *33*, 10. [CrossRef]
9. Vaissier, B.; Pernot, J.P.; Chougrani, L.; Veron, P. Parametric design of graded truss lattice structures for enhanced thermal dissipation. *Comput. Aided Des.* **2019**, *115*, 1–12. [CrossRef]
10. Al-Saedi, D.S.J.; Masood, S.H.; Faizan-Ur-Rab, M.; Alomarah, A.; Ponnusamy, P. Mechanical properties and energy absorption capability of functionally graded F2BCC lattice fabricated by SLM. *Mater. Des.* **2018**, *144*, 32–44. [CrossRef]
11. Catchpole-Smith, S.; Selo, R.R.J.; Davis, A.W.; Ashcroft, I.A.; Tuck, C.J.; Clare, A. Thermal conductivity of TPMS lattice structures manufactured via laser powder bed fusion. *Addit. Manuf.* **2019**, *30*, 9. [CrossRef]

12. Chen, Y.; Zhao, B.H.; Liu, X.N.; Hu, G.K. Highly anisotropic hexagonal lattice material for low frequency water sound insulation. *Extreme Mech. Lett.* **2020**, *40*, 7. [CrossRef]

13. Zhang, L.C.; Chen, L.Y. A Review on Biomedical Titanium Alloys: Recent Progress and Prospect. *Adv. Eng. Mater.* **2019**, *21*, 29. [CrossRef]

14. Tamburrino, F.; Graziosi, S.; Bordegoni, M. The Design Process of Additively Manufactured Mesoscale Lattice Structures: A Review. *J. Comput. Inf. Sci. Eng.* **2018**, *18*, 040801. [CrossRef]

15. Cutolo, A.; Engelen, B.; Desmet, W.; Van Hooreweder, B. Mechanical properties of diamond lattice Ti-6Al-4V structures produced by laser powder bed fusion: On the effect of the load direction. *J. Mech. Behav. Biomed. Mater.* **2020**, *104*, 11. [CrossRef] [PubMed]

16. Vrana, R.; Jaros, J.; Koutny, D.; Nosek, J.; Zikmund, T.; Kaiser, J.; Palousek, D. Contour laser strategy and its benefits for lattice structure manufacturing by selective laser melting technology. *J. Manuf. Process.* **2022**, *74*, 640–657. [CrossRef]

17. Vuksanovich, B.; Gygi, C.; Cortes, P.; Chavez, J.; MacDonald, E.; Ohara, R.; Du Plessis, A. Non-Destructive Inspection of Sacrificial 3D Sand-Printed Molds with Geometrically Complex Lattice Cavities. *Int. J. Met.* **2022**, *16*, 1091–1100. [CrossRef]

18. Jia, D.J.; Li, F.C.; Zhang, Y. 3D-printing process design of lattice compressor impeller based on residual stress and deformation. *Sci. Rep.* **2020**, *10*, 11. [CrossRef]

19. Yin, S.; Chen, H.Y.; Wu, Y.B.; Li, Y.B.; Xu, J. Introducing composite lattice core sandwich structure as an alternative proposal for engine hood. *Compos. Struct.* **2018**, *201*, 131–140. [CrossRef]

20. du Plessis, A.; Yadroitsava, I.; Yadroitsev, I.; le Rouxa, S.G.; Blaine, D.C. Numerical comparison of lattice unit cell designs for medical implants by additive manufacturing. *Virtual Phys. Prototyp.* **2018**, *13*, 266–281. [CrossRef]

21. Abdulhameed, O.; Al-Ahmari, A.; Ameen, W.; Mian, S.H. Additive manufacturing: Challenges, trends, and applications. *Adv. Mech. Eng.* **2019**, *11*, 27. [CrossRef]

22. Wang, J.; Li, Y.; Hu, G.; Yang, M.S. Lightweight Research in Engineering: A Review. *Appl. Sci.* **2019**, *9*, 24. [CrossRef]

23. Lei, H.S.; Li, C.L.; Meng, J.X.; Zhou, H.; Liu, Y.B.; Zhang, X.Y.; Wang, P.D.; Fang, D.N. Evaluation of compressive properties of SLM-fabricated multi-layer lattice structures by experimental test and mu-CT-based finite element analysis. *Mater. Des.* **2019**, *169*, 15. [CrossRef]

24. Leary, M.; Mazur, M.; Elambasseril, J.; McMillan, M.; Chirent, T.; Sun, Y.; Qian, M.; Easton, M.; Brandt, M. Selective laser melting (SLM) of AlSi12Mg lattice structures. *Mater. Des.* **2016**, *98*, 344–357. [CrossRef]

25. Leary, M.; Mazur, M.; Williams, H.; Yang, E.; Alghamdi, A.; Lozanovski, B.; Zhang, X.Z.; Shidid, D.; Farahbod-Sternahl, L.; Witt, G.; et al. Inconel 625 lattice structures manufactured by selective laser melting (SLM): Mechanical properties, deformation and failure modes. *Mater. Des.* **2018**, *157*, 179–199. [CrossRef]

26. Maskery, I.; Hussey, A.; Panesar, A.; Aremu, A.; Tuck, C.; Ashcroft, I.; Hague, R. An investigation into reinforced and functionally graded lattice structures. *J. Cell. Plast.* **2017**, *53*, 151–165. [CrossRef]

27. Wang, S.; Wang, J.; Xu, Y.J.; Zhang, W.H.; Zhu, J.H. Compressive behavior and energy absorption of polymeric lattice structures made by additive manufacturing. *Front. Mech. Eng.* **2020**, *15*, 319–327. [CrossRef]

28. Umer, R.; Barsoum, Z.; Jishi, H.Z.; Ushijima, K.; Cantwell, W.J. Analysis of the compression behaviour of different composite lattice designs. *J. Compos Mater.* **2018**, *52*, 715–729. [CrossRef]

29. Chua, C.; Sing, S.L.; Chua, C.K. Characterisation of in-situ alloyed titanium-tantalum lattice structures by laser powder bed fusion using finite element analysis. *Virtual Phys. Prototyp.* **2022**, *18*, e2138463. [CrossRef]

30. Chen, W.; Zheng, X.; Liu, S. Finite-Element-Mesh Based Method for Modeling and Optimization of Lattice Structures for Additive Manufacturing. *Materials* **2018**, *11*, 2073. [CrossRef]

31. Jin, X.; Li, G.X.; Zhang, M. Optimal design of three-dimensional non-uniform nylon lattice structures for selective laser sintering manufacturing. *Adv. Mech. Eng.* **2018**, *10*, 1687814018790833. [CrossRef]

32. Chen, L.-Y.; Liang, S.-X.; Liu, Y.; Zhang, L.-C. Additive manufacturing of metallic lattice structures: Unconstrained design, accurate fabrication, fascinated performances, and challenges. *Mater. Sci. Eng. R Rep.* **2021**, *146*, 100648. [CrossRef]

33. Zargarian, A.; Esfahanian, M.; Kadkhodapour, J.; Ziaei-Rad, S.; Zamani, D. On the fatigue behavior of additive manufactured lattice structures. *Theor. Appl. Fract. Mech.* **2019**, *100*, 225–232. [CrossRef]

34. Li, P.Y.; Ma, Y.E.; Sun, W.B.; Qian, X.D.; Zhang, W.H.; Wang, Z.H. Fracture and failure behavior of additive manufactured Ti6Al4V lattice structures under compressive load. *Eng. Fract. Mech.* **2021**, *244*, 12. [CrossRef]

35. Zhao, M.; Zhang, D.Z.; Li, Z.H.; Zhang, T.; Zhou, H.L.; Ren, Z.H. Design, mechanical properties, and optimization of BCC lattice structures with taper struts. *Compos. Struct.* **2022**, *295*, 15. [CrossRef]

36. Zhang, L.; Song, B.; Yang, L.; Shi, Y.S. Tailored mechanical response and mass transport characteristic of selective laser melted porous metallic biomaterials for bone scaffolds. *Acta Biomater.* **2020**, *112*, 298–315. [CrossRef] [PubMed]

37. Liu, Y.; Zhuo, S.R.; Xiao, Y.N.; Zheng, G.L.; Dong, G.Y.; Zhao, Y.F. Rapid Modeling and Design Optimization of Multi-Topology Lattice Structure Based on Unit-Cell Library. *J. Mech. Des.* **2020**, *142*, 15. [CrossRef]

38. Niu, X.P.; Wang, R.Z.; Liao, D.; Zhu, S.P.; Zhang, X.C.; Keshtegar, B. Probabilistic modeling of uncertainties in fatigue reliability analysis of turbine bladed disks. *Int. J. Fatigue* **2021**, *142*, 11. [CrossRef]

39. Tang, Y.; Dong, G.; Zhou, Q.; Zhao, Y.F. Lattice Structure Design and Optimization With Additive Manufacturing Constraints. *IEEE Trans. Autom. Sci. Eng.* **2018**, *15*, 1546–1562. [CrossRef]

40. Xu, Y.L.; Li, T.T.; Cao, X.Y.; Tan, Y.Q.; Luo, P.H. Compressive Properties of 316L Stainless Steel Topology-Optimized Lattice Structures Fabricated by Selective Laser Melting. *Adv. Eng. Mater.* **2021**, *23*, 15. [CrossRef]

41. Li, D.; Qin, R.; Xu, J.; Zhou, J.; Chen, B. Improving Mechanical Properties and Energy Absorption of Additive Manufacturing Lattice Structure by Struts' Node Strengthening. *Acta Mech. Solida Sin.* **2022**, 1–17. [CrossRef]
42. Mokryakov, V.V. Strength analysis of an elastic plane containing a square lattice of circular holes under mechanical loading. *Mech. Solids* **2014**, *49*, 568–577. [CrossRef]
43. Tanlak, N.; De Lange, D.F.; Van Paepegem, W. Numerical prediction of the printable density range of lattice structures for additive manufacturing. *Mater. Des.* **2017**, *133*, 549–558. [CrossRef]
44. Yu, G.J.; Li, X.; Dai, L.S.; Xiao, L.J.; Song, W.D. Compressive properties of imperfect Ti-6Al-4V lattice structure fabricated by electron beam powder bed fusion under static and dynamic loadings. *Addit. Manuf.* **2022**, *49*, 18. [CrossRef]
45. Tang, Y.; Zhou, Y.; Hoff, T.; Garon, M.; Zhao, Y.F. Elastic modulus of 316 stainless steel lattice structure fabricated via binder jetting process. *Mater. Sci. Technol.* **2016**, *32*, 648–656. [CrossRef]
46. Dong, G.Y.; Wijaya, G.; Tang, Y.L.; Zhao, Y.F. Optimizing process parameters of fused deposition modeling by Taguchi method for the fabrication of lattice structures. *Addit. Manuf.* **2018**, *19*, 62–72. [CrossRef]
47. Martin, J.H.; Yahata, B.D.; Hundley, J.M.; Mayer, J.A.; Schaedler, T.A.; Pollock, T.M. 3D printing of high-strength aluminium alloys. *Nature* **2017**, *549*, 365–369. [CrossRef]
48. Feng, Q.X.; Tang, Q.; Liu, Y.; Setchi, R.; Soe, S.; Ma, S.; Bai, L. Quasi-static analysis of mechanical properties of Ti6Al4V lattice structures manufactured using selective laser melting. *Int. J. Adv. Manuf. Technol.* **2018**, *94*, 2301–2313. [CrossRef]

Communication

Effects of SiC Content on Wear Resistance of Al-Zn-Mg-Cu Matrix Composites Fabricated via Laser Powder Bed Fusion

Zhigang Shen [1], Ning Li [2,3,*], Ting Wang [2,3] and Zhisheng Wu [1]

[1] School of Materials Science and Engineering, Taiyuan University of Science and Technology, Taiyuan 030024, China
[2] State Key Laboratory of Advanced Welding and Joining, Harbin Institute of Technology at Weihai, Weihai 264209, China
[3] Shandong Provincial Key Laboratory of Special Welding Technology, Harbin Institute of Technology at Weihai, Weihai 264209, China
* Correspondence: 19b909130@stu.hit.edu.cn; Tel./Fax: +86-0631-5687196

Abstract: In this paper, in situ SiC-reinforced Al-Zn-Mg-Cu composites were fabricated by laser powder bed fusion (LPBF). The effects of SiC content on the microstructure, phase composition, microhardness, and wear resistance of as-printed composites were preliminarily investigated. Results show that the microstructure was regulated, the matrix grains were refined, and the tendency to orientation grain growth was suppressed. SiC particles reacted in situ with the Al matrix to produce Si, Al_4C_3, and Al_4SiC_4 phases. The microhardness and wear resistance of as-printed composites increased with SiC content due to the fine grain strengthening of the matrix and the second phase strengthening of precipitates and reinforcements.

Keywords: laser powder bed fusion; aluminum matrix composites; microstructure evolution; microhardness; wear resistance

Citation: Shen, Z.; Li, N.; Wang, T.; Wu, Z. Effects of SiC Content on Wear Resistance of Al-Zn-Mg-Cu Matrix Composites Fabricated via Laser Powder Bed Fusion. *Crystals* **2022**, *12*, 1801. https://doi.org/10.3390/cryst12121801

Academic Editors: Hao Yi, Huajun Cao, Menglin Liu and Le Jia

Received: 23 November 2022
Accepted: 8 December 2022
Published: 10 December 2022

Publisher's Note: MDPI stays neutral with regard to jurisdictional claims in published maps and institutional affiliations.

1. Introduction

Aluminum matrix composites (AMCs) have attracted much attention because of their higher specific strength, specific stiffness, and wear resistance compared with aluminum alloys [1,2]. Silicon carbide (SiC) particles are used as reinforcement for AMCs due to their high hardness, wear resistance, and good metallurgical compatibility with aluminum alloy [3,4]. High-strength Al-Zn-Mg-Cu alloy is a pivotal raw material for structural parts, but its high cracking susceptibility during crystallization limits its application [5,6]. SiC-reinforced aluminum matrix composites are the most popular and representative of this system [7]. The mature methods for preparing AMCs include melt stirring, squeeze casting, pressurized infiltration, and vacuum infiltration [8]. However, the mechanical properties of AMCs are often affected by the segregation and settling of SiC particles and weak interfacial bonding between SiC particles and the matrix [9]. Many scholars have proposed improved methods to prepare AMCs with higher performance [10]. Laser powder bed fusion (LPBF) is an innovative strategy for fabricating AMCs due to the advantages of high precision, adjustable raw powder compositions, and direct formability of components [11,12]. Gu et al. [13,14] have prepared AlSi10Mg alloy, AlN/AlSi10Mg composites, and SiC/AlSi10Mg composites by LPBF. Results show that AlSi10Mg alloy has good printability, and the mechanical properties of as-printed composites can be optimized by ceramic reinforcements. However, the low-strength Al-Si alloys could not meet the actual performance requirements. The high-strength Al-Zn-Mg-Cu alloys are a better matrix for the aluminum matrix composites. In this study, SiC-reinforced Al-Zn-Mg-Cu composites were fabricated via LPBF. The effects of SiC content on the microstructure, microhardness, and wear resistance of as-printed composites were investigated. This study made a preliminary attempt to prepare wear-resistant Al-Zn-Mg-Cu composites by LPBF.

2. Materials and Methods

Commercial Al-Zn-Mg-Cu alloy powders (D_{50} = 38 μm) with a composition of Al-5.64Zn-2.31Mg-1.40Cu-0.32Fe-0.22Cr-0.08Mn-0.05Si (wt.%) and SiC ceramic powders (D_{50} = 10 μm) were used as raw materials. The morphologies and particle size histograms are shown in Figure 1a,b, respectively. The homogeneous composite powders with different SiC fractions (0 wt.%~4 wt.%) were prepared by a planetary shaker-mixer, and the mixing time was two hours. The morphology of 4 wt.% SiC/Al-Zn-Mg-Cu composite powder is shown in Figure 1c. The sphericity of Al-Zn-Mg-Cu alloy powders was not damaged, and SiC particles were uniformly dispersed in Al-Zn-Mg-Cu powders. Table 1 lists the sample labels of the as-printed composites and the corresponding composite powder compositions. Samples for metallographic and performance characterization were printed directly on EOS M290 (Germany, EOS). The schematic of sample orientation is illustrated in Figure 1d. The optimized LPBF parameters for obtaining the high-density as-printed composites are as follows: laser power 340 W, laser scanning speed 800 mm/s, layer thickness 30 μm, and hatch distance 100 μm. The strip-scanning method was adopted with a strip width of 8 mm and a rotation angle of 67°. The schematic of the laser scanning strategy is demonstrated in Figure 1e. Metallographic samples were sanded and mechanically polished layer by layer and then etched with Keller's reagent before being observed. Scanning electron microscopy (SEM, Zeiss) was conducted to scrutinize the microstructure, and the accompanying electron backscattering diffraction (EBSD, EDAX) was employed to analyze the crystallographic features. Samples for EBSD need to be electropolished with a 10% perchloric acid alcohol solution. X-ray diffraction (XRD, DX-2700) was used to analyze the phase composition. Under the loading time of 10 s and loading amount of 100 g, the microhardness of metallographic specimens was measured by Vickers (HV-1000). Wear resistance was investigated on the friction and wear tester (HF-1000), and the schematic is presented in Figure 1f. Before the test, the samples were processed into circular samples with a diameter of 10 mm, polished with 2000 # sandpaper, and cleaned with alcohol. The parameters used are 500 g (load), 560 r/min (rotational speed), and 5 mm (friction diameter). Silicon nitride balls with a diameter of 6 mm and a hardness of 20 GPa were used as the anti-abrasive material. Three groups of tests were conducted to ensure accuracy. The mass before and after testing was measured to obtain the loss of abrasive debris and used to characterize wear resistance.

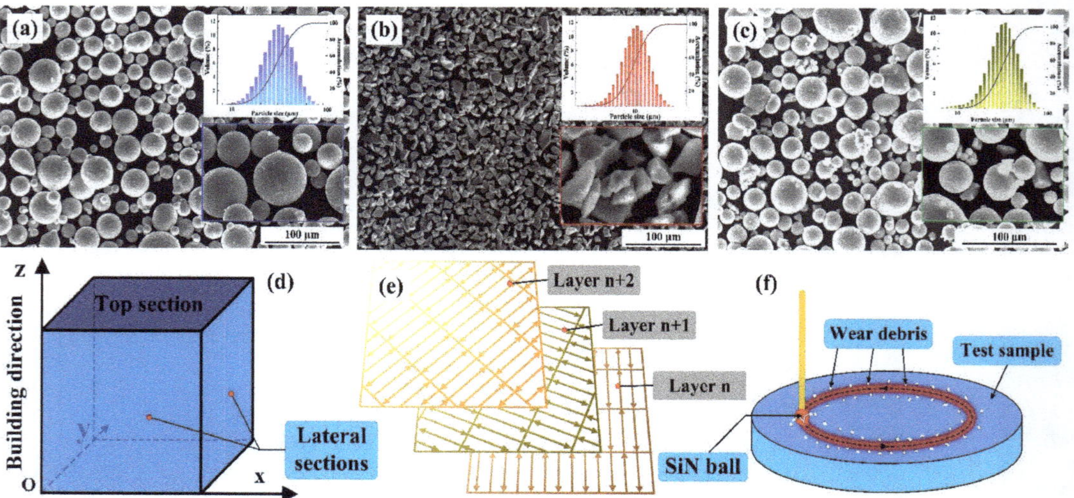

Figure 1. Surface morphologies of (**a**) Al-Zn-Mg-Cu alloy powders, (**b**) SiC powders, and (**c**) 4 wt.% SiC/Al-Zn-Mg-Cu composite powders with insert showing the particle size distribution histogram. Schematics of (**d**) sample orientation, (**e**) laser scanning strategy, and (**f**) friction and wear test.

Table 1. Sample labels of as-printed Al-Zn-Mg-Cu alloy and SiC/Al-Zn-Mg-Cu composites.

Sample Label	Powder Composition
S0	Al-Zn-Mg-Cu alloy
S1	Al-Zn-Mg-Cu alloy + 1 wt.% SiC reinforcement
S2	Al-Zn-Mg-Cu alloy + 2 wt.% SiC reinforcement
S3	Al-Zn-Mg-Cu alloy + 3 wt.% SiC reinforcement
S4	Al-Zn-Mg-Cu alloy + 4 wt.% SiC reinforcement

3. Results and Discussion

3.1. Morphologies and Microstructure

Figure 2 shows the SEM images of as-printed samples with different SiC contents. Fusion lines, cracks, pores, and grain boundaries were scrutinized in the as-printed Al-Zn-Mg-Cu alloy (Figure 2a). These cracks were typical solidification cracks that cracked along the grain boundaries [15]. From the insert (Figure 2b), no precipitates were found at the grain boundaries, and poor intergranular bonding was revealed. Figure 2c presents the microstructure of the S2 sample, where the number of cracks was significantly reduced. A small number of intergranular precipitates were found at the grain boundaries. When the SiC content was 4 wt.%, no cracks were observed within the as-printed composites, as shown in Figure 2d. The irregular SiC reinforcements were uniformly embedded in the matrix without evident agglomeration. Figure 2f shows the XRD diffraction patterns of as-printed samples with different SiC contents. Without SiC reinforcement modification, only the Al phase was detected in the as-printed S0 sample. The as-printed SiC-reinforced Al-Zn-Mg-Cu composites consisted of the Al, SiC, Mg_2Si, Al_4C_3, Al_4SiC_4, and Si phases. During the LPBF process, SiC particles reacted in situ with the Al matrix as follows [16,17].

$$4Al + 4SiC = 3Si + Al_4SiC_4 \tag{1}$$

$$4Al + 3SiC = 3Si + Al_4C_3 \tag{2}$$

Al_4C_3, Al_4SiC_4, and Si phases were generated in the molten pool, and some of the generated Si reacted with Mg to form the Mg_2Si phase. Short rod-like Al_4SiC_4 and granular Si-eutectic phases were observed to fill the grain boundaries, as shown in Figure 2e.

Figure 3 illustrates the grain maps, grain size distribution histograms, and pole figures (PFs) of the as-printed S0 and S4 samples, revealing the effects of SiC reinforcement on the matrix grains. The unidentified black regions in Figure 3a,d are cracks and SiC reinforcement, respectively. The coarse columnar crystals (Figure 3a) were refined into fine columnar and equiaxed crystals (Figure 3d) with SiC particles. The size distribution of matrix grains followed unimodal distribution, and the average size was refined from 37.15 μm (S0 sample) to 20.50 μm (S4 sample). Figure 3c,f compare the effect of SiC reinforcement on crystallization textures of the as-printed materials, where A1 is the building direction (BD). The fiber texture of the S0 sample along the [001] crystal orientation parallel to the BD was observed in Figure 3c, predicting the preferential growth of Al grains in the as-printed Al-Zn-Mg-Cu alloy. The maximum value of multiple uniform densities (MUD) was 5.427, which appeared in the $(\bar{1}00)$ crystal plane of the [001] pole figure. The matrix texture was weakened by incorporating SiC particles. For the as-printed S4 sample, the maximum value of MUD was 4.646. For the unmodified Al-Zn-Mg-Cu alloy, the grains were nucleated by attaching to the anterior molten pool and solidified in the building direction. The heterogeneous nucleation effect was noticeable when SiC ceramic particles were introduced.

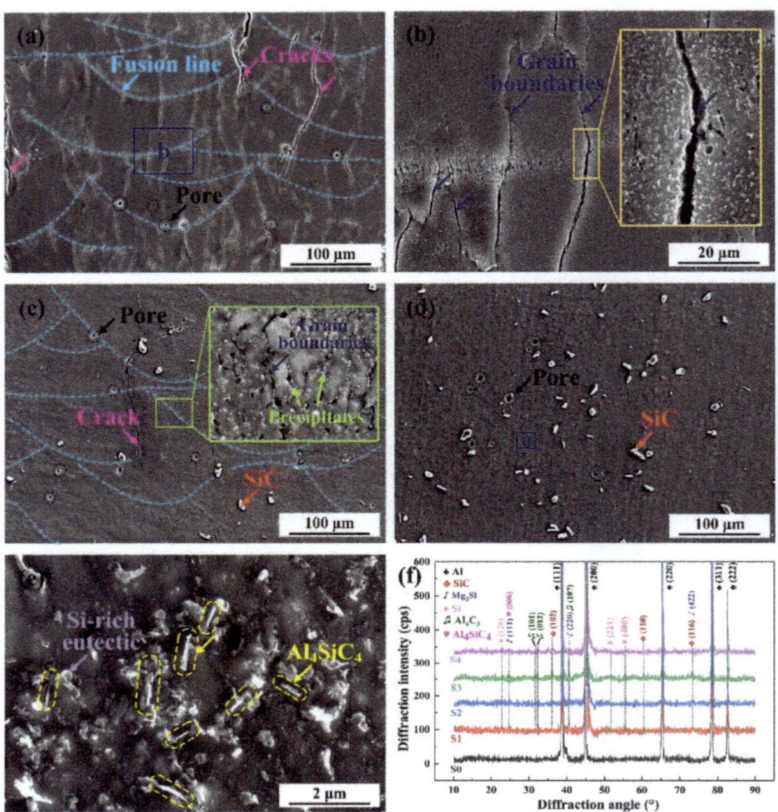

Figure 2. SEM images of the as-printed (**a**,**b**) S0 sample, (**c**) S2 sample, and (**d**,**e**) S4 sample. (**f**) XRD diffraction patterns of the as-printed samples.

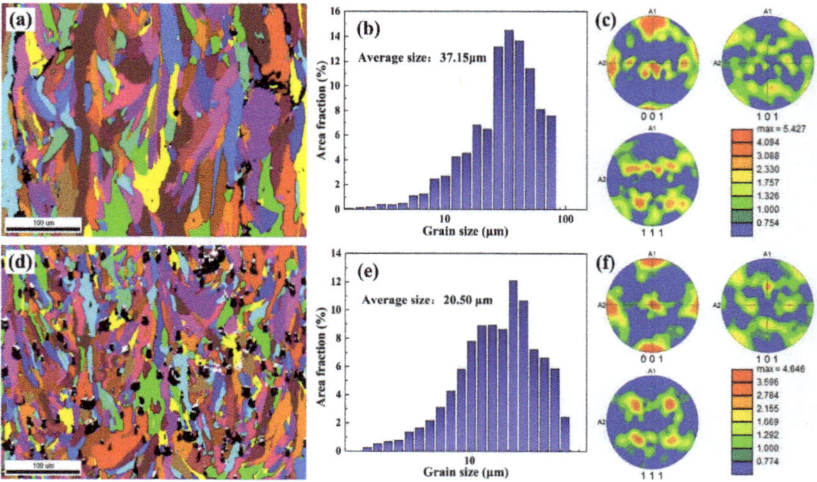

Figure 3. Grain maps, grain size distribution histograms, and PFs along the lateral section of the as-printed (**a**–**c**) S0 and (**d**–**f**) S4 samples.

3.2. Microhardness

Figure 4 indicates the microhardness of as-printed samples in the lateral and top sections, revealing the effect of SiC particles on the microhardness. The microhardness of the unmodified S0 sample in the side and top sections was 93 ± 5 $HV_{0.1}$ and 102 ± 10 $HV_{0.1}$, respectively. The microhardness gradually increased with the incorporation of SiC reinforcement. The microhardness of the S4 sample was 156.8 ± 6.4 $HV_{0.1}$ and 161.4 ± 10.5 $HV_{0.1}$, respectively. Fine grain strengthening of the matrix and particle strengthening of the reinforcement and precipitates were the main reasons for the increased microhardness. The heterogeneity of microhardness was observed along the lateral and top sections, and the top section was higher than the side section. This phenomenon was related to the directional growth of the matrix grains and gradually decreased with increasing SiC content. For the as-printed S0 sample, the lateral section was composed of coarse columnar crystals, and the top section was the fine equiaxed crystal. Fine grain strengthening could be responsible for the difference in microhardness. With the introduction of SiC reinforcement, the columnar grain on the side was gradually refined, and the difference in microhardness was gradually reduced.

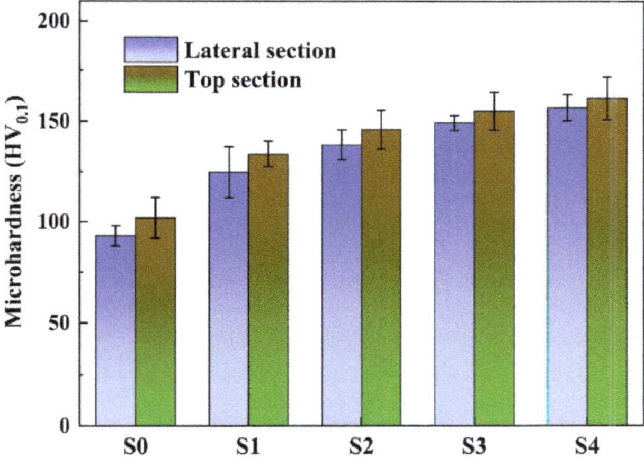

Figure 4. Microhardness histograms of as-printed S0-S4 samples in the lateral and top sections.

3.3. Wear Behavior

The effect of SiC reinforcement on the wear resistance of as-printed AMCs was investigated, and the results are shown in Figure 5. The coefficient of friction (COF) curves are shown in Figure 5a. COF curves showed a similar evolution, characterized by dramatic fluctuations and gradually decreasing with time. Fluctuations might be due to cracks and SiC reinforcement, which could lead to the stripping of the matrix and reinforcement from the sample during friction. The COF curves tended to be stable as the debris with poor binding to the matrix gradually fell off. The average COF values of S0 to S4 samples were 0.507, 0.473, 0.389, 0.348, and 0.288, respectively. The weight loss of the as-printed materials is shown in Figure 5b. The average weight loss for the as-printed S0 to S4 samples were 14.3 ± 1.7 mg, 13.8 ± 1.8 mg, 12.7 ± 2 mg, 9.9 ± 1.2 mg, and 7.6 ± 0.9 mg, respectively. Combining the results of COF and weight loss, the wear resistance of as-printed composites was reinforced with the incorporation of SiC reinforcement. The increase in wear resistance of as-printed composites was mainly due to the synergistic effect of matrix and reinforcement. The hardened SiC ceramic particles could resist the abrasive pressing and improve the deformation resistance of the as-printed composites. The strength of the matrix was increased due to crack inhibition and grain refinement.

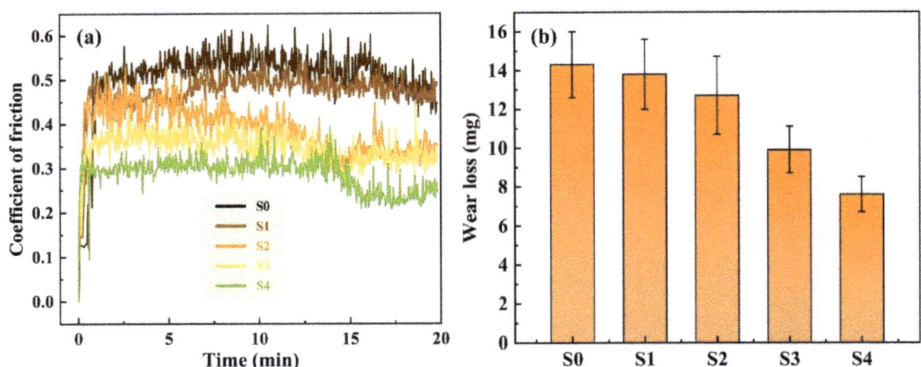

Figure 5. Results of friction and wear. (**a**) The variation curves of friction coefficient vs. sliding time and (**b**) the weight loss of the as-printed S0–S4 samples.

4. Conclusions

In this paper, the in situ SiC-reinforced Al-Zn-Mg-Cu composites were prepared by LPBF. The microhardness and wear resistance were reinforced while suppressing the hot cracks. SiC particles reacted in situ with the Al matrix to form Al_4SiC_4, Al_4C_3, and Si phases in the molten pool, which precipitated near the grain boundary during solidification. Crack suppression was mainly due to grain refinement, disordered grain growth, and grain boundary structure optimization. The fine grain strengthening of the Al matrix and the second phase strengthening of precipitates and reinforcement were the main reasons for the increase in microhardness and wear resistance.

Author Contributions: Investigation, data curation, Z.S.; investigation, data curation, and writing—original draft preparation, N.L.; supervision, T.W.; equipment management, Z.W. All authors have read and agreed to the published version of the manuscript.

Funding: Please add: This research was funded by the Shandong Provincial Key Research and Development Program, grant number 2019JZZY010439 and the National Natural Science Foundation of China, grant number 52175308.

Data Availability Statement: Not applicable.

Conflicts of Interest: The authors declare no conflict of interest.

References

1. Dong, B.-X.; Li, Q.-Y.; Shu, S.-L.; Duan, X.-Z.; Zou, Q.; Han, X.; Yang, H.-Y.; Qiu, F.; Jiang, Q.-C. Investigation on the elevated-temperature tribological behaviors and mechanism of Al-Cu-Mg composites reinforced by in-situ size-tunable TiB_2-TiC particles. *Tribol. Int.* **2023**, *177*, 107943. [CrossRef]
2. Samal, P.; Vundavilli, P.R.; Meher, A.; Mahapatra, M.M. Recent progress in aluminum metal matrix composites: A review on processing, mechanical and wear properties. *J. Manuf. Process.* **2020**, *59*, 131–152. [CrossRef]
3. Mao, Y.; Li, J.; Vivek, A.; Daehn, G.S. High strength impact welding of 7075 Al to a SiC-reinforced aluminum metal matrix composite. *Mater. Lett.* **2021**, *303*, 130549. [CrossRef]
4. Hu, C.Q.; Du, H.L. Fretting fatigue behaviours of SiC reinforced aluminium alloy matrix composite and its monolithic alloy. *Mater. Sci. Eng. A* **2022**, *847*, 143347. [CrossRef]
5. Klein, T.; Schnall, M.; Gomes, B.; Warczok, P.; Fleischhacker, D.; Morais, P.J. Wire-arc additive manufacturing of a novel high-performance Al-Zn-Mg-Cu alloy: Processing, characterization and feasibility demonstration. *Addit. Manuf.* **2021**, *37*, 101663. [CrossRef]
6. Zhu, Z.; Ng, F.L.; Seet, H.L.; Lu, W.; Liebscher, C.H.; Rao, Z.; Raabe, D.; Nai, S.M.L. Superior mechanical properties of a selective-laser-melted AlZnMgCuScZr alloy enabled by a tunable hierarchical microstructure and dual-nanoprecipitation. *Mater. Today* **2022**, *52*, 90–101. [CrossRef]
7. Ye, T.; Xu, Y.; Ren, J. Effects of SiC particle size on mechanical properties of SiC particle reinforced aluminum metal matrix composite. *Mater. Sci. Eng. A* **2019**, *753*, 146–155. [CrossRef]

8.   Imran, M.; Khan, A.R.A. Characterization of Al-7075 metal matrix composites: A review. *J. Mater. Res. Technol.* **2019**, *8*, 3347–3356. [CrossRef]

9.   Hasan, S.T.; Beynon, J.H.; Faulkner, R.G. Role of segregation and precipitates on interfacial strengthening mechanisms in SiC reinforced aluminium alloy when subjected to thermomechanical processing. *J. Mater. Process. Technol.* **2004**, *153–154*, 758–764. [CrossRef]

10.   Jiang, J.; Wang, Y. Microstructure and mechanical properties of the semisolid slurries and rheoformed component of nano-sized SiC/7075 aluminum matrix composite prepared by ultrasonic-assisted semisolid stirring. *Mater. Sci. Eng. A* **2015**, *639*, 350–358. [CrossRef]

11.   Chen, W.; Xu, L.; Zhang, Y.; Han, Y.; Zhao, L.; Jing, H. Additive manufacturing of high-performance 15-5PH stainless steel matrix composites. *Virtual Phys. Prototyp.* **2022**, *17*, 366–381. [CrossRef]

12.   Koh, H.K.; Moo, J.G.S.; Sing, S.L.; Yeong, W.Y. Use of Fumed Silica Nanostructured Additives in Selective Laser Melting and Fabrication of Steel Matrix Nanocomposites. *Materials* **2022**, *15*, 1869. [CrossRef] [PubMed]

13.   Dai, D.; Gu, D.; Xia, M.; Ma, C.; Chen, H.; Zhao, T.; Hong, C.; Gasser, A.; Poprawe, R. Melt spreading behavior, microstructure evolution and wear resistance of selective laser melting additive manufactured AlN/AlSi$_{10}$Mg nanocomposite. *Surf. Coatings Technol.* **2018**, *349*, 279–288. [CrossRef]

14.   Chang, F.; Gu, D.; Dai, D.; Yuan, P. Selective laser melting of in-situ Al$_4$SiC$_4$ + SiC hybrid reinforced Al matrix composites: Influence of starting SiC particle size. *Surf. Coatings Technol.* **2015**, *272*, 15–24. [CrossRef]

15.   Li, G.; Li, X.; Guo, C.; Zhou, Y.; Tan, Q.; Qu, W.; Li, X.; Hu, X.; Zhang, M.-X.; Zhu, Q. Investigation into the effect of energy density on densification, surface roughness and loss of alloying elements of 7075 aluminium alloy processed by laser powder bed fusion. *Opt. Laser Technol.* **2022**, *147*, 107621. [CrossRef]

16.   Lu, Q.; Ou, Y.; Zhang, P.; Yan, H. Fatigue performance and material characteristics of SiC/AlSi$_{10}$Mg composites by selective laser melting, Mater. *Sci. Eng. A.* **2022**, *858*, 144163. [CrossRef]

17.   Guo, B.; Chen, B.; Zhang, X.; Cen, X.; Wang, X.; Song, M.; Ni, S.; Yi, J.; Shen, T.; Du, Y. Exploring the size effects of Al$_4$C$_3$ on the mechanical properties and thermal behaviors of Al-based composites reinforced by SiC and carbon nanotubes. *Carbon* **2018**, *135*, 224–235. [CrossRef]

Article

Effect of Slag Adjustment on Inclusions and Mechanical Properties of Si-Killed 55SiCr Spring Steel

Yang Li, Changyong Chen *, Hao Hu, Hao Yang, Meng Sun and Zhouhua Jiang *

Department of Special Steel Metallurgy, School of Metallurgy, Northeastern University, Shenyang 110819, China
* Correspondence: 1610456@stu.neu.edu.cn (C.C.); jiangzh@smm.neu.edu.cn (Z.J.); Tel.: +024-8367-8691 (C.C.); +024-8368-6453 (Z.J.)

Abstract: The effects of the Al_2O_3 content and basicity of $CaO-SiO_2-Al_2O_3-10$ wt.% MgO refining slag on inclusions removal in 55SiCr spring steel were investigated. The viscosity of slag was studied using a viscometer, while the microstructure investigation involved using a water-quenching furnace and a Fourier-transform infrared spectrometer. The influence mechanism of the slag adjustment on inclusions was explored through thermodynamic calculations and kinetic analysis. The results indicated that the viscosity of the molten slag increased gradually with the content of Al_2O_3 increasing due to it increasing the degree of polymerization of the slag network structure, especially the $[AlO_4]^{5-}$ and [Si-O-Si] structures. In contrast, the viscosity of molten slag experienced the opposite pattern, with the basicity of molten slag increasing. This was due to the fact that Ca^{2+} can significantly reduce the degree of polymerization of a slag network structure, especially the percentages of the $[SiO_4]^{4-}$, $[AlO_4]^{5-}$ and [Si-O-Si] network structures. Finally, the changes in physical properties and structure of slag significantly affected the removal effect of the inclusions in molten steel. As a result, the number, size distribution, composition distribution and morphology of the inclusions displayed significant changes when the content of Al_2O_3 increased from 3 wt.% to 12 wt.% and the basicity of the slag gradually increased from 0.5 to 1.2.

Keywords: 55SiCr steel; spring steel; refining slag; non-metallic inclusions; high temperature viscosity

Citation: Li, Y.; Chen, C.; Hu, H.; Yang, H.; Sun, M.; Jiang, Z. Effect of Slag Adjustment on Inclusions and Mechanical Properties of Si-Killed 55SiCr Spring Steel. *Crystals* **2022**, *12*, 1721. https://doi.org/10.3390/cryst12121721

Academic Editors: Hao Yi, Huajun Cao, Menglin Liu and Le Jia

Received: 5 November 2022
Accepted: 23 November 2022
Published: 27 November 2022

1. Introduction

High strength, fatigue resistance and impact resistance are important properties of spring steels [1,2]. Non-metallic inclusion is one of the most important factors that cause fatigue fractures in spring steel [3,4]. Inclusions with high hardness and melting points, such as alumina and spinel, often act as crack sources of fatigue failure [5–7]. The total oxygen (T.O) of steel, as well as the type, number, size, morphology and distribution of inclusions in steel, play essential roles in spring steel cleanness, and an improvement in cleanliness can effectively prolong the fatigue life of the steel [8–10].

Ladle furnace (LF) slag refining is a widely used technology to control inclusions in spring steel production [11,12]. Appropriate optimization of the refining slag composition can reduce the T.O and the number and size of inclusions in spring steel, as well as control the composition of inclusions located in the low melting area [13,14]. Zhang et al. [15] studied the low-melting-point region (at 1673 K) in a $MnO-CaO-SiO_2-Al_2O_3$ system with the largest area when the Al_2O_3 content in this system was 25 wt.%. He et al. [16] showed that the inclusions are plastic at the end of the refining process, the basicity $(R = wt.\%CaO/wt.\% SiO_2)$ of refining slag is in the range of 1.0~1.2 and the Al_2O_3 content of slag is in the range of 3~9 wt.%. Yang et al. [17] showed that the inclusions located in the low-melting-point region when the basicity was in the range of 1.00 to 1.19 had a C/A value (wt.% CaO/wt.% Al_2O_3) above 9 at 1673 K. Similar results were mentioned in other studies [18–20].

Wu et al. [21] studied the effect of refining slag with low basicity on the inclusions in 55SiCr suspension spring steel. The results indicate that the composition of the $CaO-SiO_2-Al_2O_3$ ternary system inclusions is located at the center of the low-melting-point zone, and the plastic deformation ability of the inclusions is good. Nevertheless, the authors did not study the influence of viscosity and structure of refining slag on removing non-metallic inclusions in spring steel.

Du et al. [22] studied the influence of refining slag with high basicity on inclusions in 55SiCr suspension spring steel. The results showed that the number of inclusions decreased sharply as the basicity of the slag gradually increased, and the diameter of most of the inclusions was less than 10 μm. Similarly, the authors did not study the influence of the alkalinity of refining slag on its viscosity and structure.

Non-metallic inclusions have significant effects on many mechanical properties of suspension spring steel, including the strength, plasticity, toughness and fatigue properties. Li et al. [23] studied the effect of inclusions on the tensile fracture properties of 55CrSi spring steel. The results demonstrated that interior inclusions have a significant effect on the ductility and a minimal effect on the tensile strength of spring steel. However, the authors only selected the samples obtained under a single smelting condition as the research object, and the change rules of mechanical properties of spring steel treated with refining slag with different compositions have not been studied.

As for the influence of basicity and Al_2O_3 content on the inclusions in spring steel, most researchers only made a one-sided analysis from the perspective of thermodynamics. Few researchers elaborated on the influence of basicity and Al_2O_3 content on inclusions in steel from the perspective of dynamics according to their influence on the physical properties and structure of slag. Therefore, in this study, the effects of basicity and Al_2O_3 content on the viscosity and structure of slag were studied in detail, and their effects on the removal of inclusions in steel were explored. In addition, the effects of inclusions on the mechanical properties of spring steel are discussed.

2. Materials and Methods

2.1. Materials

The effect of the Al_2O_3 content and basicity in refining slag on the inclusions in spring steel was investigated in a $MoSi_2$ high-temperature resistance furnace. Table 1 shows the main chemical composition of the 55SiCr spring steel.

Table 1. Chemical compositions of the 55SiCr spring steel (wt.%) used in this experiment.

Elements	C	Si	Mn	Cr	Ni	P	S	V
Range	0.55–0.59	1.40–1.60	0.60–0.80	0.60–0.80	0.20–0.30	≤ 0.012	≤ 0.008	0.08–0.20
Target	0.57	1.50	0.70	0.70	0.25	≤ 0.012	≤ 0.008	0.15

2.2. Experimental Equipment and Procedure

Seven sample types were created by treating different synthetic LF refining slags with four different Al_2O_3 contents of 3.0 wt.%, 5.0 wt.%, 8.0 wt.% and 12.0 wt.% and four basicities of 0.5, 0.8, 1.0 and 1.2, as shown in Table 2.

Table 2. Chemical compositions of the low-basicity refining slag (wt.%).

Heat	CaO	SiO$_2$	MgO	Al$_2$O$_3$	R
1$^{\#}$	38.7	48.3	10.0	3.0	0.8
2$^{\#}$	37.8	47.2	10.0	5.0	0.8
3$^{\#}$	36.4	45.5	10.0	8.0	0.8
4$^{\#}$	34.7	43.3	10.0	12.0	0.8
5$^{\#}$	28.3	56.7	10.0	5.0	0.5
2$^{\#}$	37.8	47.2	10.0	5.0	0.8
6$^{\#}$	42.5	42.5	10.0	5.0	1.0
7$^{\#}$	46.3	38.7	10.0	5.0	1.2

Note: R = CaO/SiO$_2$.

Experiments were carried out in a MoSi$_2$ high-temperature resistance furnace. An argon atmosphere was kept in the experiments all the time, blowing from the bottom of the furnace tube to the top. The experimental procedures were carried out as follows. First, a 1.00 kg steel rod was placed into a MgO crucible with a 60×10^{-3} m inner diameter and an 80×10^{-3} m depth. Then, the crucible was placed in a graphite crucible to prevent liquid metal from leaking. After the whole crucible was placed in the chamber, the power was switched on and the furnace was heated to the experimental temperature of 1873 K. Alloys were added into the molten steel when the temperature reached 1873 K, and the molten steel was deoxidized using Si. After that, 0.05 kg of synthetic LF refining slag powder was put into the surface of the molten steel. The refining time was constant at 45 min for all of the experiments.

A direct reading spectrometer was utilized to detect the compositions of Si, Mn, Cr, V, Mo, Ni, Al, P and S. For C and S, an infrared C/S analyzer was applied. Furthermore, a LECO® TC 500 O$_2$/N$_2$ analyzer was selected to detect O and N. The ASPEX (FEI Company, Hillsboro, USA) was used to indicate the number, size and compositional distribution of the inclusions. Finally, SEM-EDS was selected to analyze the morphology of the inclusions.

The sample treatment method for inclusion observation and mechanical properties is shown in Figure 1. A cylindrical ingot was cut into two semi-cylinders along the diameter. One was forged and a heat treatment was conducted for a tensile test, and the other was used to acquire samples for inclusion observation. The forging process started at 1200 °C after heat preservation for 2 h, and was finally air-cooled to room temperature The cross-section of the forging cylinder had an 18 mm diameter. Forged steel was austenized at 880 °C for 30 min and oil-quenched to room temperature, followed by tempering at 450 °C for 120 min. The atmosphere of the heat treatment process was air. For the mechanical property characterization, tensile tests were conducted on a Shimadzu AGS-X100KN (Shimadzu, Kyodo, Japan) electronic tensile testing machine following standard GB/T 228.1-2010 (ISO 6892-1:2009, MOD) [24].

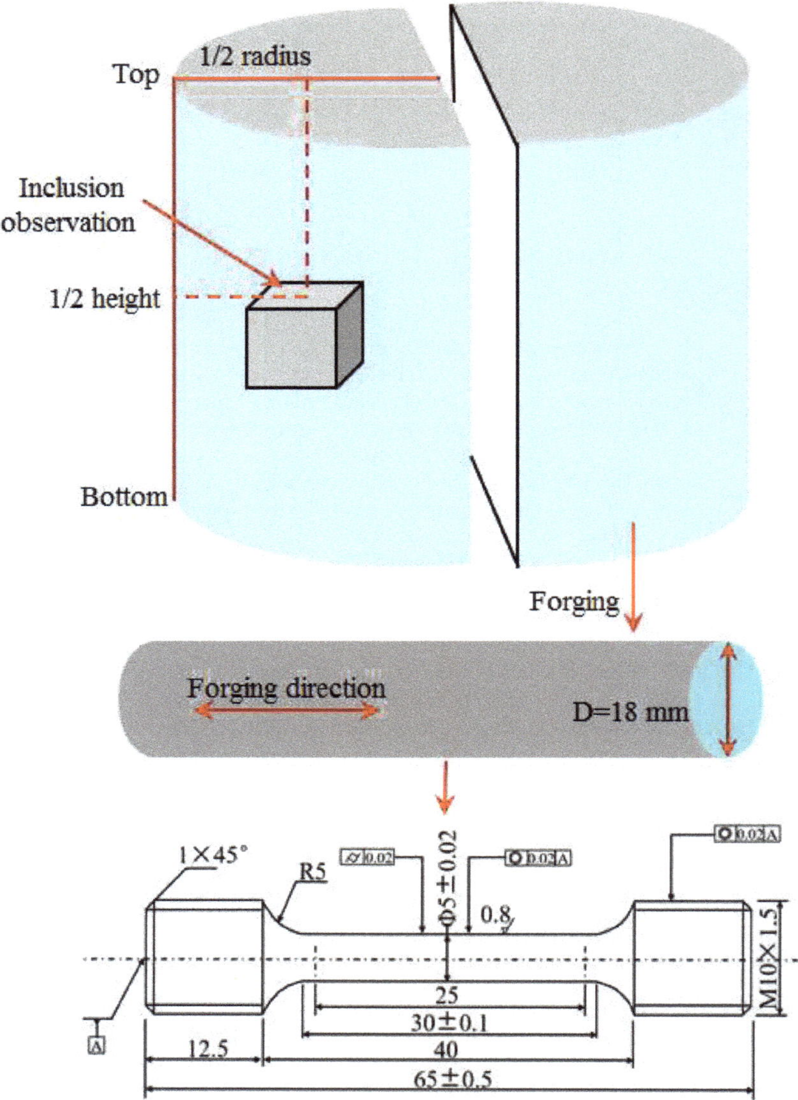

Figure 1. Sample treatment method for the inclusion observation and mechanical properties testing.

2.3. Equipment and Specific Experimental Steps for the Viscosity Measurements of Refining Slag

In this study, a Brookfield DVT rotary viscometer was selected to measure the viscosity of the slag. A schematic diagram of the equipment is shown in Figure 2. The error range of the viscosity measurement was ±1%, and the reproducibility of the experimental data was ±0.2%. The crucible material used in the experiment was molybdenum, and the size of the crucible was Φ 31 mm × 61 mm. The material of the adopted rotor was molybdenum, and its size was Φ 17 mm × 25 mm, with an angle of 120° at the top and tail of the rotor. During the experiment, the distance between the top of the rotor and the molybdenum crucible was 3 mm. In order to ensure that the slag liquid level did not exceed 3~5 mm from the tail of the rotor during the experiment, 65 g of slag was weighed for each experiment.

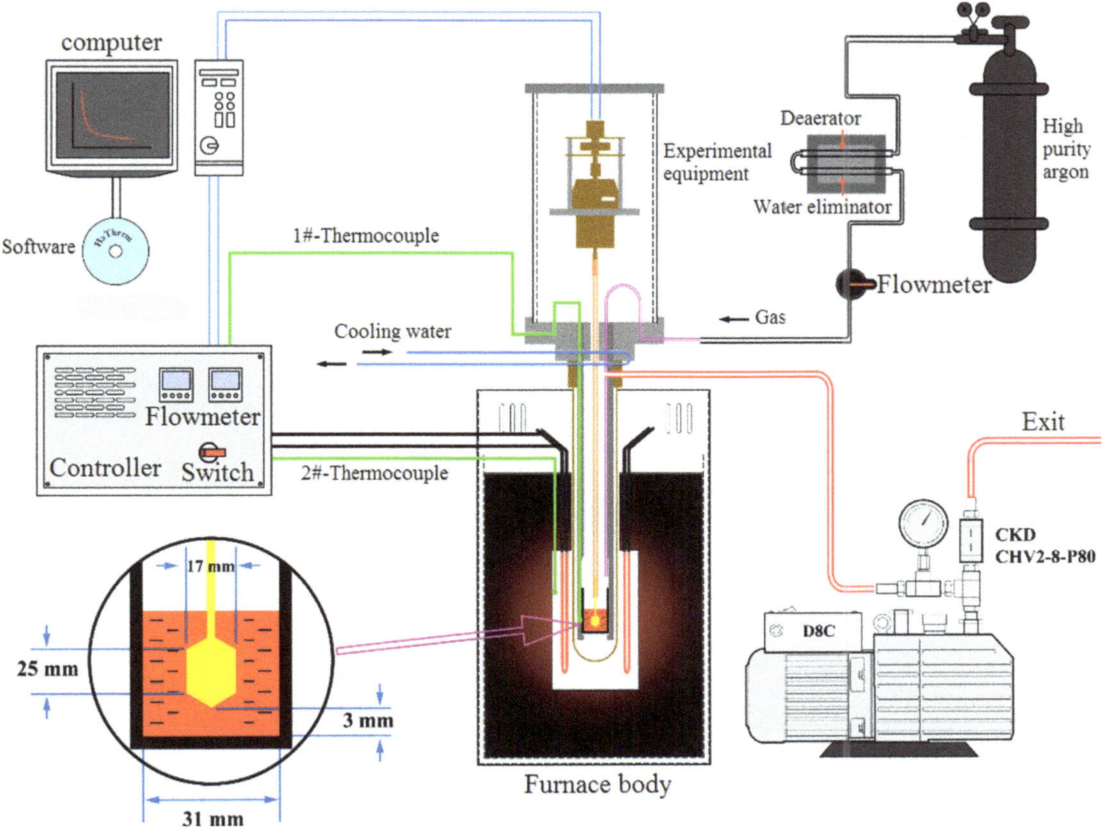

Figure 2. Diagram of the high-temperature viscometer and its auxiliary device.

The detailed steps of the viscosity test were as follows: (1) Preparation of experiment: 65 g of slag was prepared and put into the molybdenum crucible after the compression. (2) Zero adjustment of the viscometer: the viscometer was left to idle for the zero calibration without hanging the rotor. (3) Power on and temperature increase: after the zero calibration, the equipment was sealed, the power was turned on, the temperature was increased and the vacuum pumping began to work. After a vacuum was achieved, the shielding Ar gas was introduced at the flow rate of 100 mL·min^{-1} (to prevent the rotor fluctuation caused by excessive air flow and experimental error). Cooling water was fed into the circulation when the furnace temperature reached 673 K. (4) Measuring the viscosity: the temperature was kept constant for half an hour when the temperature reached 1873 K, and then the rotor was lowered to 3 mm from the bottom of the molybdenum crucible and the slag viscosity was measured at the speed of 100 r·min^{-1}, where viscosity data was recorded every 10 s and continued for 5 min. (5) Viscosity measurement during the cooling process: the temperature was decreased by 20 K each time after the viscosity at 1873 K was measured. The temperature was kept constant for 10 min, and then the viscosity of the slag was measured at this temperature at the speed of 100 r min^{-1}. (6) Finally, the temperature was increased again to 1773 K, and the rotor was lifted to stop measuring the viscosity when the torque was greater than 100%. The viscometer was powered off and cooled down. The switches of the Ar gas and cooling water were closed when the furnace temperature reached 873 K and 673 K, respectively.

2.4. Apparatus and Specific Experimental Steps for the Structure Measurement of Refining Slag

A schematic diagram of the water bath quenching equipment is shown in Figure 3. The graphite crucible (20 mm diameter, 40 mm height) containing the refined slag sample was hung in the furnace with platinum wire while ensuring it was located in the constant temperature zone. A basin of ice water was placed at the vertical bottom of the furnace. The power was turned on and the switches for the cooling water and argon gas were turned on at the same time. The temperature increased gradually following the program that was set in the equipment. The furnace was kept at a constant temperature for 1 hour when the temperature reached 1873 K, and then the switch for the platinum wire was released to make the graphite crucible quickly fall into the ice water at the bottom of the furnace. The slag could maintain its structure in the molten state because the liquid refining slag was rapidly cooled from 1873 K to 273 K.

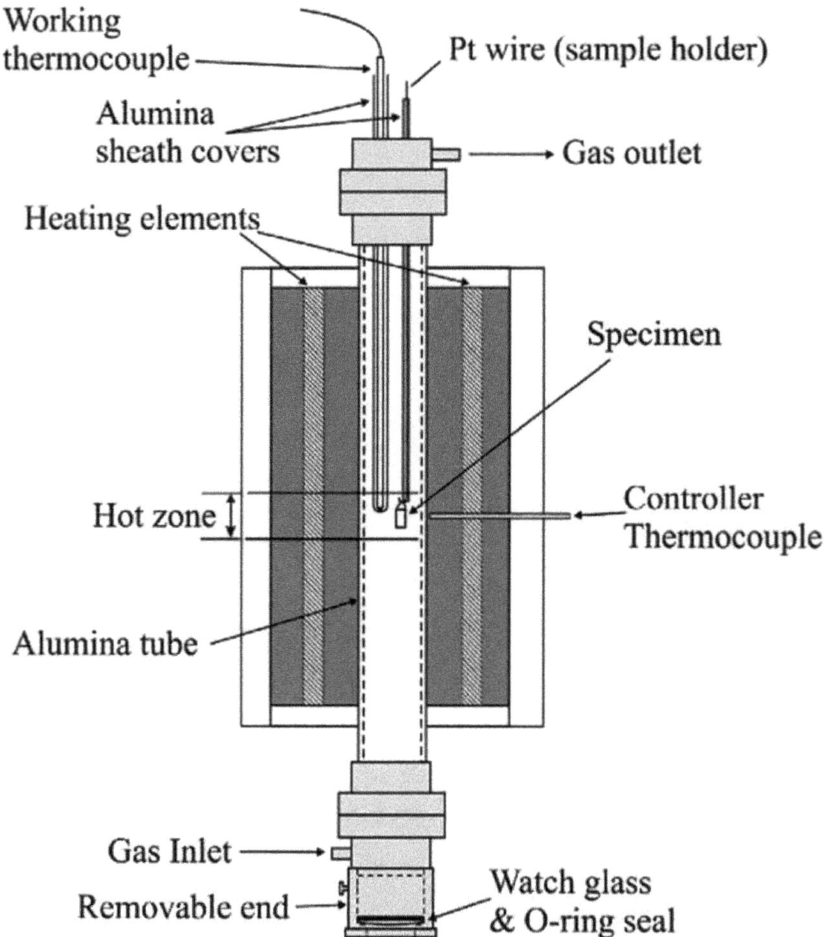

Figure 3. Schematic diagram of the vertical resistance furnace used in the water quenching experiment.

The refining slag was dried after it was removed from ice water, and then it was ground to a particle size below 200 mesh. Finally, the structure of the refining slag was detected using Fourier-transform infrared spectrometry (FTIR).

3. Results and Discussion

3.1. Number and Size of Inclusions

The chemical compositions of the 55SiCr steels are shown in Table 3.

Table 3. Chemical compositions of the 55SiCr steels (wt.%).

	C	Si	Mn	Cr	Ni	V	T.O	N	[Al]s	P	S
1[#]	0.577	1.446	0.729	0.732	0.267	0.172	0.0016	0.0039	0.0012	0.0084	0.0053
2[#]	0.582	1.480	0.741	0.746	0.252	0.163	0.0019	0.0028	0.0013	0.0077	0.0058
3[#]	0.570	1.478	0.713	0.715	0.276	0.163	0.0021	0.0032	0.0014	0.0076	0.0050
4[#]	0.556	1.498	0.683	0.697	0.251	0.162	0.0023	0.0027	0.0015	0.0092	0.0064
5[#]	0.560	1.452	0.705	0.716	0.257	0.158	0.0020	0.0034	0.0013	0.0075	0.0056
2[#]	0.582	1.480	0.741	0.746	0.252	0.163	0.0019	0.0028	0.0013	0.0077	0.0058
6[#]	0.575	1.427	0.750	0.721	0.266	0.163	0.0017	0.0034	0.0013	0.0072	0.0050
7[#]	0.564	1.512	0.676	0.716	0.248	0.152	0.0015	0.0036	0.0012	0.0096	0.0060

The T.O and [Al]s contents increased from 0.0016 wt.% to 0.0023 wt.% and from 0.0012 wt.% to 0.0015 wt.%, respectively, as the content of Al_2O_3 increased from 3 wt.% to 12 wt.%. In contrast, the T.O content decreased from 0.0020 wt.% to 0.0015 wt.% as the basicity increased from 0.5 to 1.2.

The number and size distribution of inclusions in the 1[#]~7[#] steel samples is shown in Table 4. Obviously, the quantity density gradually increased from 8.81 to 8.96, while the percentage of inclusions with sizes smaller than 5 μm increased from 61% to 77% with increasing Al_2O_3. In contrast, the quantity density gradually decreased from 8.92 to 8.54, while the percentage of inclusions with a size smaller than 5 μm decreased from 76% to 55% as the basicity increased.

Table 4. Results of the inclusions in the 1[#]~7[#] cast samples.

No.	Total Number	Area (mm^2)	Quantity Density
1[#]	705	80	8.81
2[#]	692	80	8.65
3[#]	701	80	8.76
4[#]	717	80	8.96
5[#]	714	80	8.92
2[#]	692	80	8.65
6[#]	687	80	8.56
7[#]	683	80	8.54

The size distribution of the inclusions in the 1[#]~7[#] steel samples is shown in Table 5. For the 1[#]~4[#] samples with different Al_2O_3 contents, the percentage of inclusions with a diameter larger than 10 μm decreased from 13% to 4%. In contrast, the percentage of inclusions with a diameter smaller than 2 μm increased from 20% to 33%. For the 2[#], 5[#], 6[#] and 7[#] samples with different basicities, the percentage of inclusions with a diameter larger than 10 μm increased from 6% to 19%. In contrast, the percentage of inclusions with a diameter smaller than 2 μm increased from 31% to 17%.

Table 5. The size distribution of the inclusions in the 1#~7# steel samples (percentage).

	<1 μm	1–2 μm	2–5 μm	5–10 μm	10–15 μm	>15 μm
1#	2	18	41	26	7	6
2#	3	22	41	23	6	5
3#	5	24	44	21	4	2
4#	7	26	44	19	3	1
5#	6	25	45	18	4	2
2#	3	22	41	23	6	5
6#	2	18	40	24	10	6
7#	1	16	38	26	11	8

3.2. Composition and Morphology of Typical Inclusions

SEM-EDS was selected to analyze the composition and morphology of the inclusions. Several randomly selected inclusions were tested and analyzed regarding the composition of the inclusions. The mapping method of typical composite inclusions was carried out to accurately analyze the elemental distribution and structure of the structural inclusions. According to the results, four kinds of typical inclusions were observed in the steel samples, namely, $CaO–SiO_2–Al_2O_3–MgO$, $CaO–SiO_2–Al_2O_3–MgO–MnS$, MnS, and SiO_2, as shown in Figure 4.

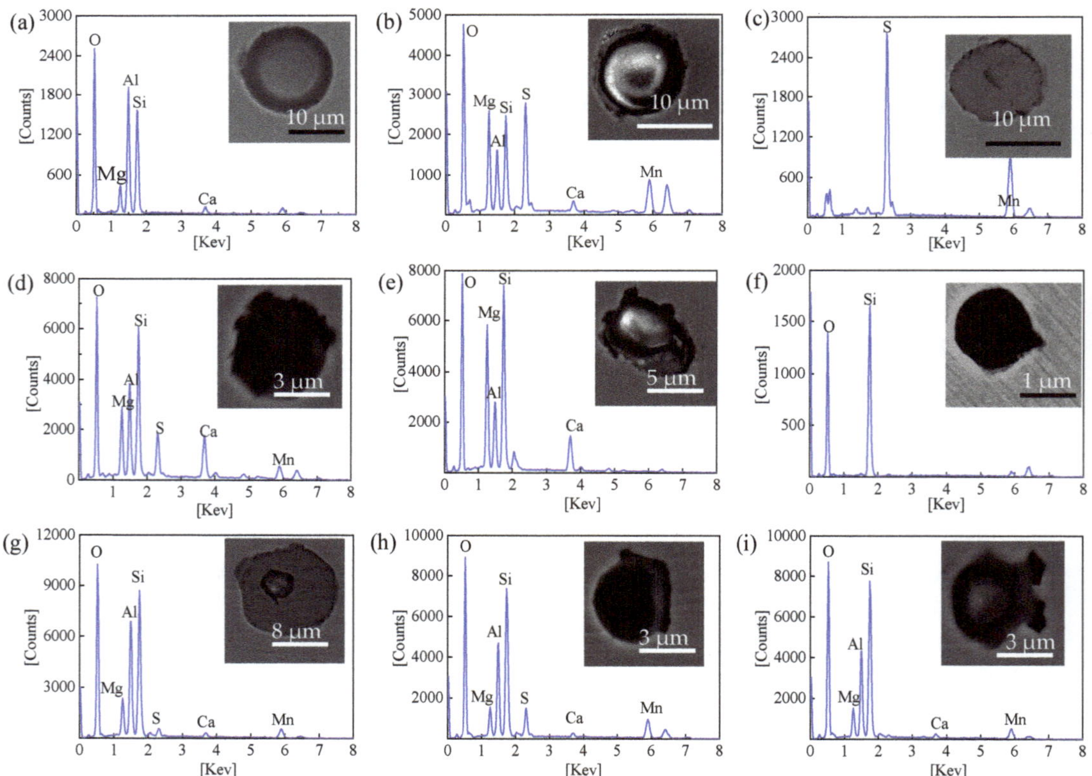

Figure 4. *Cont.*

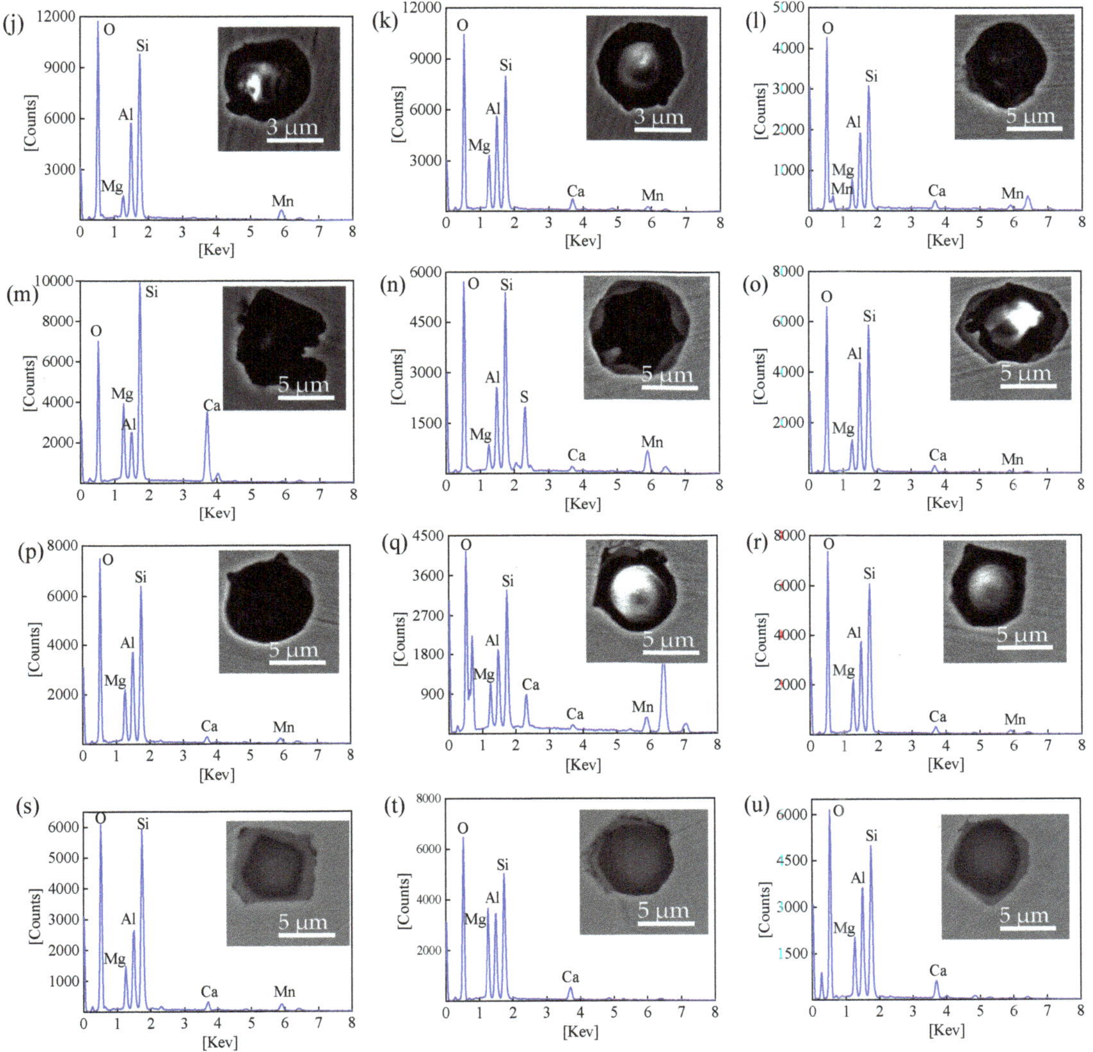

Figure 4. The typical inclusions in steel samples. $1^{\#}$: (**a–c**); $2^{\#}$: (**d–f**); $3^{\#}$: (**g–i**); $4^{\#}$: (**j–l**); $5^{\#}$: (**m–o**); $6^{\#}$: (**p–r**); $7^{\#}$: (**s–u**).

Most of the compounds contained oxide inclusions, especially the CaO–SiO_2–Al_2O_3–MgO inclusions with a diameter of 10 μm. The MnS inclusions generally formed during solidification with a diameter larger than 5 μm. In contrast, the SiO_2 inclusions had a smaller diameter of approximately 1 μm.

The multi-component composite inclusions were generally uniform; that is, the elements in the inclusions were evenly distributed in the whole inclusion without delamination, as shown in Figure 5. Generally, these inclusions would not cause obvious harm to spring steel. In addition, the structure of the layered composite inclusions was usually a homogeneous composite with a layer of MnS inclusions wrapped around the edges. This kind of inclusion will be separated during the hot rolling and cold drawing of spring steel due to the composition and plasticity of the inner and outer layers being different; they would appear as long strips along the rolling direction or drawing direction.

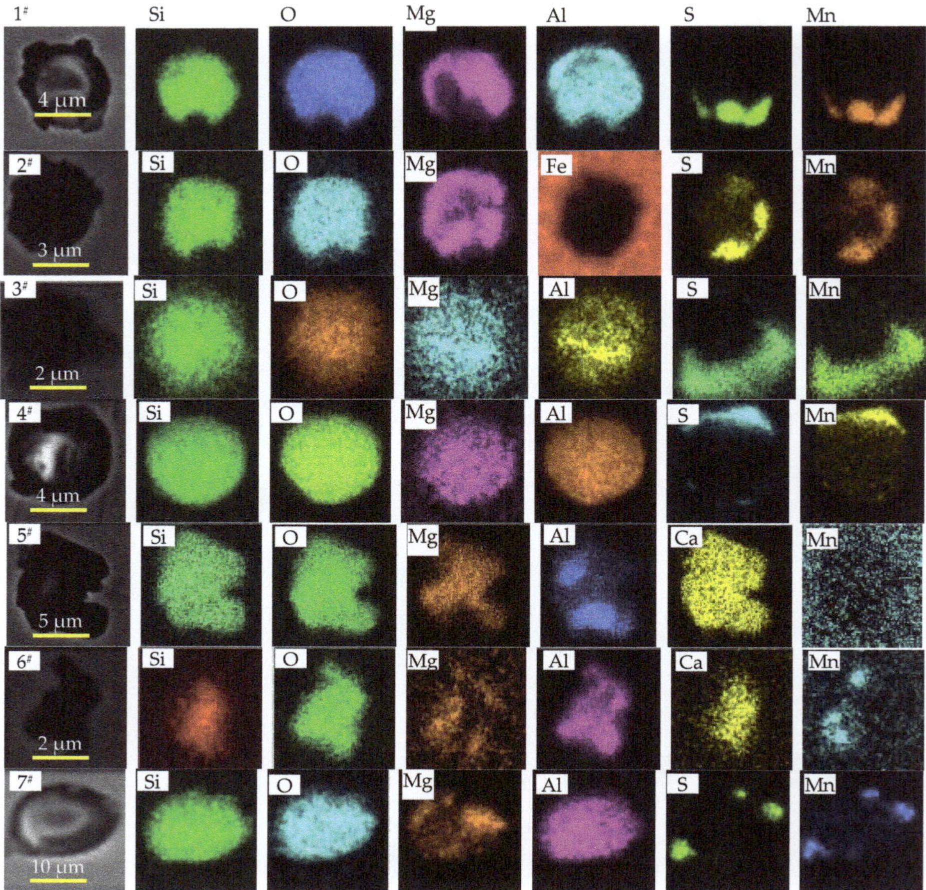

Figure 5. Elemental mapping of inclusions in steel samples.

3.3. Composition Distribution of Typical Inclusions

The composition distribution of inclusions overlayed on a phase diagram with different Al_2O_3 contents and different basicities in the refining slag are shown in Figures 6 and 7, respectively. The "small symbols" are the compositions of each inclusion in a ternary phase diagram, and the "colored square" is the average composition of all the inclusions in the $1^{\#} \sim 7^{\#}$ steel samples. The detail composition of typical inclusions in samples were shows in Table 6.

Figure 6 shows that the content of $Al_2O_{3(inc)}$ in the inclusions and aluminosilicate inclusions had a tendency of increasing with the content of $Al_2O_{3(slag)}$ increasing in the slag. The compound oxide inclusions were mainly concentrated in low-melting-point regions for all of the steel samples. In detail, the average contents of inclusions were $SiO_{2(inc)}$: 48.27%, $CaO_{(inc)}$+$MgO_{(inc)}$: 34.39%, $Al_2O_{3(inc)}$: 17.34%; $SiO_{2(inc)}$: 48.87%, $CaO_{(inc)}$+$MgO_{(inc)}$: 30.01%, $Al_2O_{3(inc)}$: 21.12%; $SiO_{2(inc)}$: 53.46%, $CaO_{(inc)}$+$MgO_{(inc)}$: 17.44%, $Al_2O_{3(inc)}$: 29.10%, $SiO_{2(inc)}$: 49.63%, $CaO_{(inc)}$+$MgO_{(inc)}$: 18.40% and $Al_2O_{3(inc)}$: 31.97%.

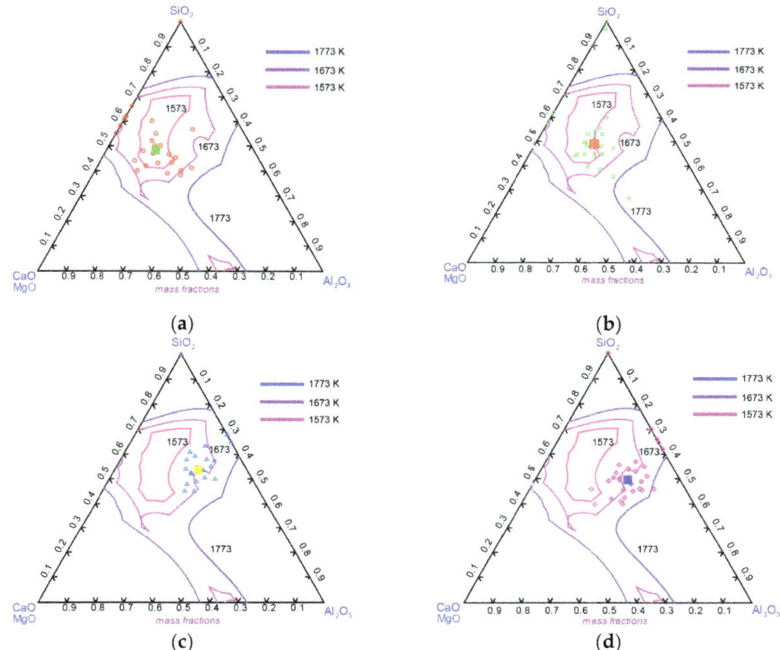

Figure 6. Inclusion distribution overlaid on a phase diagram with the refining slags at fixed Al_2O_3 contents of (**a**) 3 wt.%, (**b**) 5 wt.%, (**c**) 8 wt.% and (**d**) 12 wt.%.

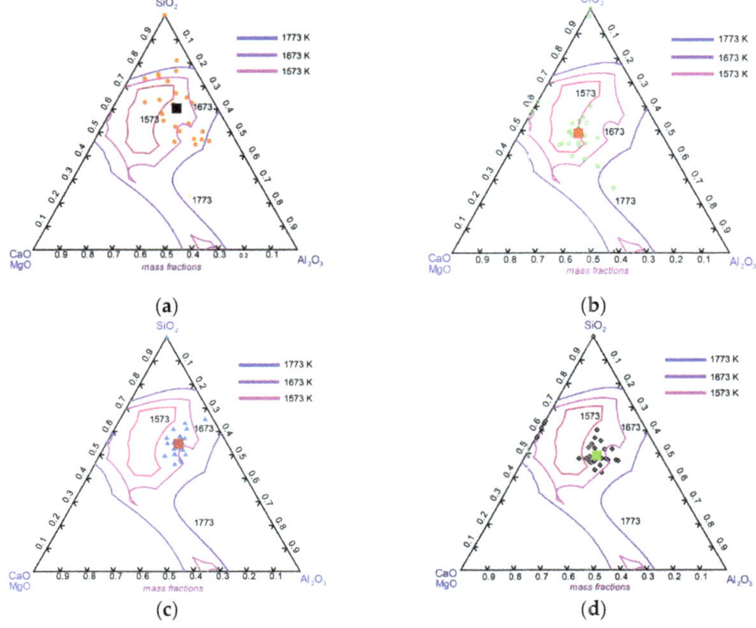

Figure 7. Inclusion distribution overlaid on a phase diagram with the refining slags with different basicities of (**a**) 0.5, (**b**) 0.8, (**c**) 1.0 and (**d**) 1.2.

Table 6. Chemical compositions of typical inclusions in the experimental steel samples (mass fraction, %).

Exp.	No.	Size (μm)	CaO	Al$_2$O$_3$	SiO$_2$	MgO	MnO
1$^\#$ (3wt.% Al$_2$O$_3$, R = 0.8)	1	5.7	3	28	35	19	15
	2	6.3	1	30	40	28	1
	3	4.8	4	21	43	32	0
	4	7.1	1	30	37	29	3
	5	6.5	8	26	45	21	0
	6	2.3	5	23	51	11	10
	7	4.2	8	25	43	24	0
2$^\#$ (5wt.% Al$_2$O$_3$, R = 0.8)	1	3.3	15	21	38	13	13
	2	2.9	4	45	27	24	0
	3	5.8	1	31	32	26	10
	4	7.6	4	15	25	21	35
	5	4.7	10	17	43	27	3
	6	4.1	8	15	40	27	10
	7	4.2	4	20	45	31	0
3$^\#$ (8wt.% Al$_2$O$_3$, R = 0.8)	1	5.2	2	18	44	5	31
	2	6.6	2	24	46	6	22
	3	4.3	2	24	54	7	13
	4	2.8	6	31	49	12	2
	5	4.7	5	26	40	13	16
	6	5.2	3	27	51	9	10
	7	6.7	2	23	57	6	12
4$^\#$ (12 wt.% Al$_2$O$_3$, R = 0.8)	1	3.7	0	26	56	5	13
	2	4.6	6	26	49	12	7
	3	4.1	5	29	50	16	0
	4	6.3	3	25	40	16	16
	5	5.5	2	23	33	13	29
	6	3.7	5	27	52	8	8
	7	4.3	3	31	49	11	6
5$^\#$ (5 wt.% Al$_2$O$_3$, R = 0.5)	1	5.6	2	19	49	5	25
	2	4.2	1	26	34	6	33
	3	2.9	1	19	41	5	34
	4	4.7	1	24	26	5	44
	5	4.4	3	27	40	7	23
	6	5.1	2	20	33	6	39
	7	6.0	2	35	44	9	10
6$^\#$ (5 wt.% Al$_2$O$_3$, R = 1.0)	1	5.2	2	24	50	0	24
	2	6.0	4	25	51	12	8
	3	4.9	4	24	53	10	9
	4	4.1	1	27	53	16	3
	5	3.8	3	25	55	10	7
	6	2.6	1	19	38	6	36
	7	5.5	4	22	53	9	12
7$^\#$ (5 wt.% Al$_2$O$_3$, R = 1.2)	1	3.3	1	21	52	24	2
	2	4.7	9	22	44	16	9
	3	5.8	6	27	50	13	4
	4	2.8	1	20	29	9	41
	5	6.0	12	22	48	10	8
	6	4.6	3	18	52	9	18
	7	2.7	3	25	47	25	0

Figure 7 shows that most inclusions were located in the low-melting-point region, the inclusions distribution was dispersed in the 5$^\#$ steel sample with a basicity of 0.5 in the slag, and the inclusions distribution was concentrated in the 2$^\#$, 6$^\#$ and 7$^\#$ steel samples with basicities of 0.8, 1.0 and 1.2 in the slag, respectively. In detail, the average contents of inclusions in these steel samples were SiO$_{2(inc)}$: 60.26%, CaO$_{(inc)}$+MgO$_{(inc)}$: 15.49%,

$Al_2O_{3(inc)}$: 24.25%; $SiO_{2(inc)}$: 48.87%, $CaO_{(inc)}+MgO_{(inc)}$: 30.01%, $Al_2O_{3(inc)}$: 21.12%; $SiO_{2(inc)}$: 54.56%, $CaO_{(inc)}+MgO_{(inc)}$: 18.27%, $Al_2O_{3(inc)}$: 27.17%, $SiO_{2(inc)}$: 49.06%, $CaO_{(inc)}+MgO_{(inc)}$: 24.34% and $Al_2O_{3(inc)}$: 26.60%.

3.4. Mechanical Properties of the Experiment Steels

The mechanical properties of the experimental steels are shown in Table 7. It is obvious that the tensile strength gradually increased from 1357.83 MPa to 1437.04 MPa as the content of Al_2O_3 in the slag increased from 3 wt.% (1#) to 12 wt.% (4#). In contrast, the reduction in area and elongation slightly decreased from 27.58% and 10.24% to 24.31% and 9.36%, respectively.

Table 7. Mechanical properties of the experiment steels.

No.	1#	2#	3#	4#	5#	2#	6#	7#
R_m (MPa)	1357.83	1390.64	1397.40	1437.04	1442.12	1390.64	1386.41	1367.84
Ψ (%)	27.58	25.52	25.24	24.31	24.23	25.52	25.78	27.32
δ (%)	10.24	9.60	9.56	9.36	9.32	9.60	9.63	9.96

The fracture morphologies are shown in Figure 8. From the macroscopic appearance of the fractures, the fractures of the four samples were typically cup-shaped and were divided into a fiber area, radiation area and shear lip area from the center to the edge. The radiation areas of the fractures of samples 1# and 2# were relatively flat, and there were few secondary microcracks, showing certain brittle fracture characteristics. In particular, many long and deep but directionless cracks appeared in the radiation area of sample 1#, as shown in the white triangle area in the figure. In addition, the area of the radiation area tended to gradually increase; in contrast, the area of the shear lip gradually decreased. Moreover, the shape of the radiation area also gradually changed from an irregular shape and ellipse to a more regular circle. Finally, the number of secondary microcracks in the radiation area of the fracture surface of samples 3# and 4# significantly increased, radiating from the center to the edge along the radial direction. Furthermore, the length and depth of the secondary crack gradually became more uniform.

Comparing the microstructures of the four steel samples' fractures, it was found that they all contained three kinds of microstructures, namely, secondary microcracks, tear dimples and fine equiaxed dimples. There was no significant difference in the fracture morphology of the four steel samples.

The tensile strength decreased gradually from 1442.12 MPa to 1367.84 MPa as the basicity of slag increased from 0.5 (5#) to 1.2 (7#). In contrast, the reduction in area and elongation slightly increased from 24.23% to 27.32% and from 9.36% to 9.96%, respectively.

The fracture morphologies of samples 5#, 2#, 6# and 7# are shown in Figure 9. By comparing the macro morphology of these four samples, it can be seen that all the fracture morphologies were typical cup-shaped vertebrae, which was their common point. The 5# sample had the largest radiation area, and the shape tended to be a regular circle. Secondary microcracks with high density were evenly distributed in the radiation area. This indicated that the sample's structure was relatively uniform. The radiation area of sample 2# was an irregular oval with a small area. In contrast, the shear lip area was large. Moreover, the fracturing of the 2# sample was relatively flat, with only a small number of secondary cracks, which generally presented the characteristics of brittle fracture. The fiber area at the fracture of the 6# sample was near the center of the circle, and the shape of the radiation area was a regular circle with a large area. The regularity of the fracture morphology of sample 7# was the worst, where the fiber area was far away from the center of the circle, the morphology of the radiation area was irregular, and the radiation area was very uneven with a large number and distribution of long and deep cracks. In addition, the area of the shear lip was large.

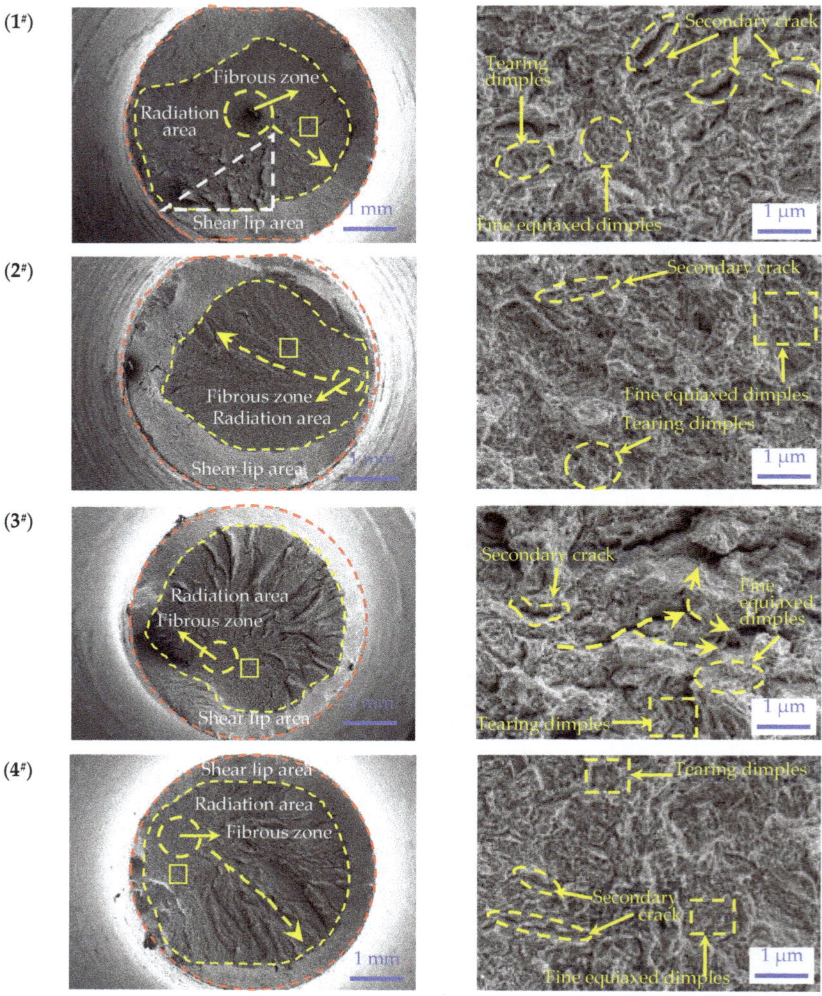

Figure 8. Micro and macro fracture morphologies of the steel samples.

When comparing the microstructures of the four steel samples' fractures, it was found that they all contained three kinds of microstructures, namely, secondary microcracks, tear dimples and fine equiaxed dimples. With the increase in alkalinity, the microstructure at the fracture surface of the samples changed greatly: the number and size of secondary microcracks gradually increased, there were high-density large and deep holes in the fracture of the 6[#] sample (these holes were likely caused by large hard inclusions), and there were not only large and deep holes but also many cleavage planes at the fracture of sample 7[#].

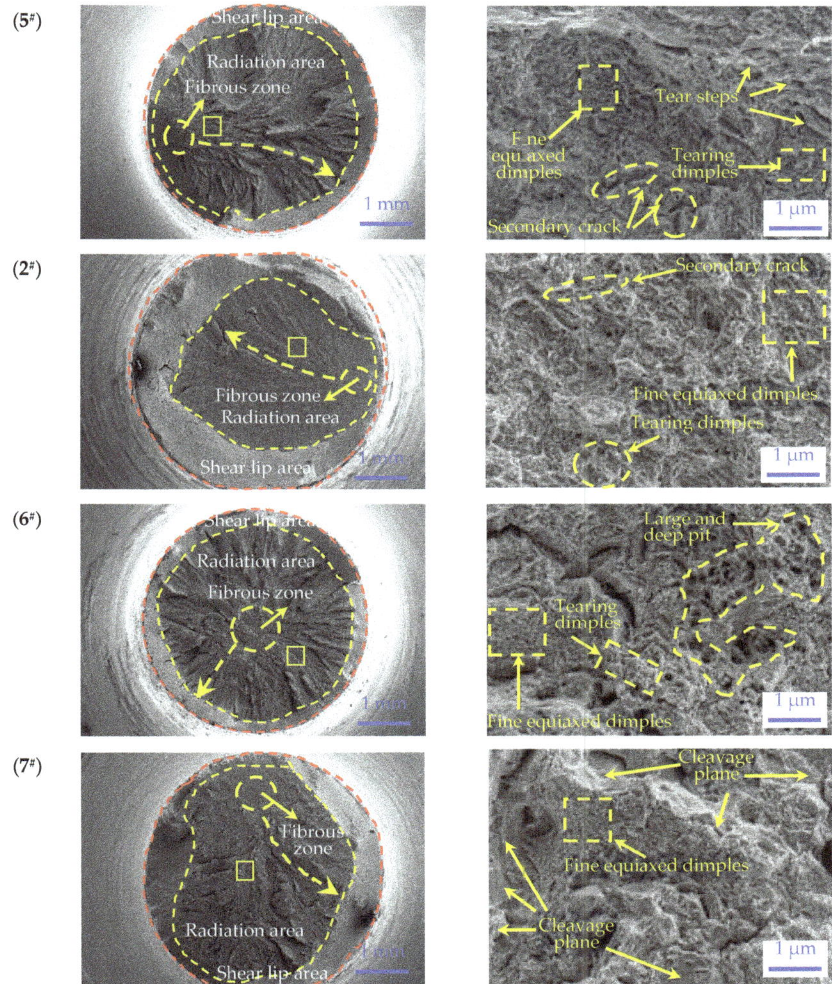

Figure 9. Micro and macro fracture morphologies of the steel samples.

3.5. Thermodynamic Calculations of Isoactivity Lines

The calculation of the component activity of the $CaO–SiO_2–Al_2O_3–MgO$ (10 wt.%) quaternary inclusion system was mainly completed by the phase diagram module in Factsage software. This part mainly calculated the isoactivity lines of CaO, SiO_2 and Al_2O_3 when the reaction reached equilibrium at 1873 K. The results are shown in Figure 10. It was obvious that the value of a_{SiO_2} decreased sharply from 0.7 to 0.02 as the value of R gradually increased. In contrast, the values of a_{CaO} and $a_{Al_2O_3}$ increased from 0.00001 to 0.01 and from 0 to 0.05, respectively. Within the low-melting-point zone, the value of a_{SiO_2} was one order of magnitude larger than that of a_{CaO}, which meant that the content of SiO_2 in the generated inclusions was significantly higher than that of CaO, which was consistent with the experimental results.

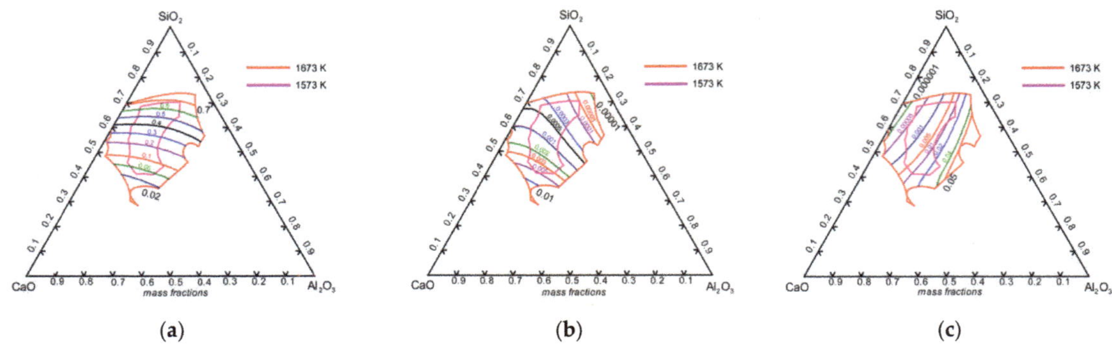

Figure 10. The isoactivity lines of SiO$_2$, CaO and Al$_2$O$_3$ at 1873 K: (**a**) SiO$_2$; (**b**) CaO; (**c**) Al$_2$O$_3$.

The contents of [O] and [Al] in steel have an important influence on the number, size and morphology of inclusions. Too high of a content of [O] will produce a large number of oxide inclusions and too high of a content of [Al] will produce brittle and hard inclusions with edges and corners, such as Al$_2$O$_3$ and MgO·Al$_2$O$_3$, which will seriously endanger the mechanical properties of spring steel. Therefore, the isooxygen (Iso-[O]) line and isoaluminum (Iso-[Al]) line were calculated to determine the ranges of [O] and [Al] contents where low-melting-point inclusions were formed.

The Equilib module and Phase Diagram module in Fact-sage software were used to calculate the Iso-[O] line and Iso-[Al] line when the 55SiCr molten steel was in equilibrium with CaO–SiO$_2$–Al$_2$O$_3$–MgO quaternary inclusions. The chemical composition of 55SiCr steel is shown in Table 1. The activity coefficients of C, Si, Mn, Cr, Ni, V, Al, O, S, P and other elements were mainly used in the calculation. The activity coefficients f_i of each element can be calculated according to Equation (1).

$$\lg f_i = \sum e_i^j [\%j] \tag{1}$$

where f_i is the activity coefficient of element i and e_i^j is the interaction coefficient of element j with i, as shown in Table 8.

Table 8. Interaction coefficients of elements in molten steel at 1873 K [25].

i \ j	Al	Si	Mn	O	P	S	C	Cr	V	Ni
Al	0.043 [26]	0.0056	0.0065	−1.867 [27]	0.0033	0.030	0.091	0.0120	/	−0.017
Si	0.058	0.1100	0.0020	−0.230 [28]	0.1100	0.056	0.180	0.0003	0.025	0.005
O	−3.900 [29]	−0.1310	−0.0210	−0.200	0.0700	0.133	0.450	0.0459	−0.300	0.006

(1) The isooxygen (Iso-[O]) line

The chemical reaction of [Si] and [O] in molten steel is shown in Equation (2) [25] and Equation (3) can be derived from it.

The activity value of SiO$_2$ in Figure 10 and the relevant data in Tables 1 and 6 were substituted into Equation (3) to calculate the isooxygen (Iso-[O]) line of 55SiCr molten steel and CaO–SiO$_2$–Al$_2$O$_3$–MgO quaternary inclusions in molten steel at 1873 K. The results are shown in Figure 11a. It was obvious that the oxygen content in the low-melting-point region near the calcline in the phase diagram was 0.0015 wt.%~0.0085 wt.%, with a large controllable range.

$$[Si] + 2[O] = SiO_2; \quad \Delta G\theta = -581900 + 221.8T \text{ J/mol} \tag{2}$$

$$[\%O] = \left(\frac{a_{SiO_2}}{K \times f_{Si} \times [\%Si] \times f_O^2}\right)^{1/2} \tag{3}$$

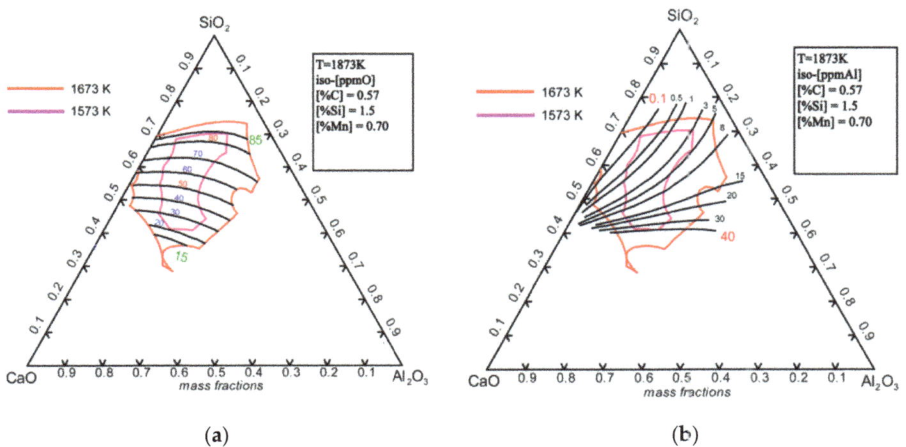

(a)   (b)

Figure 11. Isoactivity lines of the equilibrium between the molten spring steel 55SiCr and CaO–SiO₂–Al₂O₃–MgO inclusions at 1873 K. (a) Iso-[O] lines and, (b) iso-[Al].

(2)   The isoaluminum (Iso-[Al]) line

The chemical reaction of [Al] and [O] in molten steel is shown in Equation (4) [30] and Equation (5) can be derived from it.

The activity value of SiO₂ in Figure 10 and the relevant data in Tables 1 and 6 were substituted into Equation (5) to calculate the isooxygen (Iso-[Al]) line of 55SiCr molten steel and CaO–SiO₂–Al₂O₃–MgO quaternary inclusions in molten steel at 1873 K. The results are shown in Figure 11b. It was obvious that the oxygen content in the low-melting-point region near the calcline in the phase diagram was 0.00001 wt.%~0.0040 wt.%, with a small controllable range. Thus, the content of [Al] in steel needs to be strictly controlled at a very low level in order to obtain low-melting-point inclusions in the smelting process of low alkalinity refining slag combined with silicon deoxidation.

$$4[Al] + 3[SiO_2] = 2(Al_2O_3) + 3[Si]; \ \Delta G\theta = -658300 + 107.2T \ J/mol \tag{4}$$

$$[\%Al] = \left(\frac{a_{Al_2O_3}^2 \times f_{[Si]}^3 \times [\%Si]^3}{a_{SiO_2}^2 \times f_{[Al]}^4 \times K}\right)^{1/4} \tag{5}$$

In this study, the content of T.O in all steels was about 0.0020 wt.% and the content of all [Al] was about 0.0015 wt.%. Therefore, the inclusions in all of the group steels can be controlled in the low-melting-point region, which is consistent with the actual test results.

For the group made up of the 1#~4# steels, the compositions of inclusions were consistent with those of refining slag when the three phases of steel–slag–inclusion reached equilibrium. Therefore, the content of [Al]s in the molten steel and the Al₂O₃ content of inclusions in steel increased gradually as the Al₂O₃ content in the slag gradually increased.

For the group made up of the 5#, 2#, 6# and 7# steels, the activity of SiO₂ gradually decreased with the increasing alkalinity of the slag, which inhibited the progress of the reaction shown in Equation (4); as a result, the T.O content and the total number of inclusions in steel decreased gradually. In addition, the content of CaO in inclusions increased gradually with the basicity of slag gradually increasing due to the composition of inclusions being consistent with that of refining slag when the three phases of steel–slag–inclusion reached equilibrium.

3.6. Influence of Al_2O_3 Content and Basicity on the Viscosity and Structure of the Slag

Increasing the Al_2O_3 content and basicity in the refining slag had an obvious impact on the thermochemical and thermophysical properties of the slag, including their structure, viscosity and surface tension [31,32]. The effect of Al_2O_3 content and basicity on the viscosity of the slag at different temperatures was studied using a viscometer and the results are shown in Figure 12.

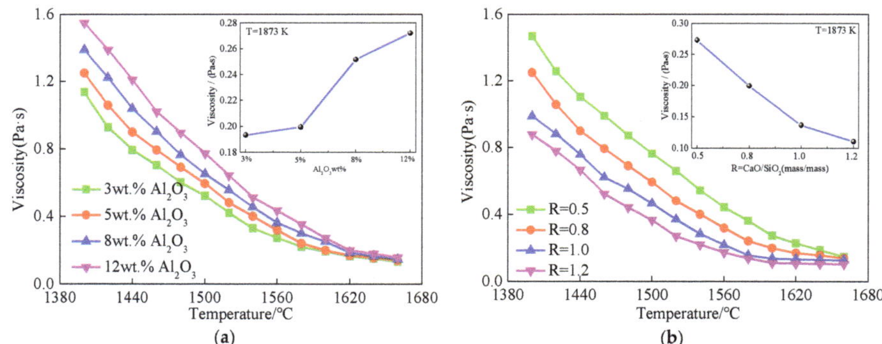

Figure 12. Effect of Al_2O_3 content and basicity in the $CaO–SiO_2–MgO–Al_2O_3$ slag on the viscosity of the slag at 1873 K. (**a**) Al_2O_3 content ranging from 3 wt.% to 12 wt.%; (**b**) basicity ranging from 0.5 to 1.2.

It was obvious that the viscosity of the slag increased gradually with the increase in the Al_2O_3 content. In contrast, the viscosity of the slag gradually decreased with the increase in basicity. In addition, the viscosity of the slag increased sharply as the temperature gently decreased. The change in slag viscosity was caused by the significant change in its microstructure.

The network structure of slag is shown in Figure 13. The viscosity of slag is often indirectly characterized by the degree of polymerization of the network structure, and the viscosity increases with the increase in the degree of polymerization.

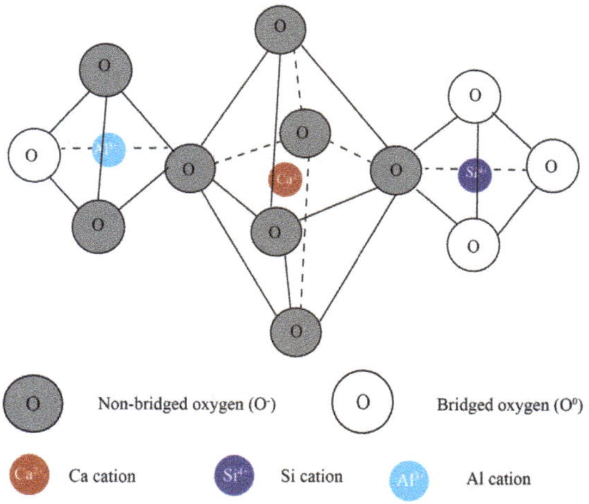

Figure 13. Schematic diagram of the molten slag structure.

In order to research the relationship between the structure and composition of molten slag, the molten slag quenched in water was further analyzed by using FTIR. Characteristic transmittance peaks in the FTIR spectra for silicate systems were observed in the wavenumber region between 1200 and 400 cm^{-1} [33]. In detail, the band group of Si-O bonds existed between 1030 and 750 cm^{-1}, which corresponds to NBO (non-bridging oxygen) of 1 to 4 [34]. The band group shown in the range of 750-630 cm^{-1} was related to the asymmetric stretching vibration of $[AlO_4]^{5-}$ tetrahedral units, and the trough near 500 cm^{-1} indicates Si-O-Al bending.

Figure 14 shows the effect of Al_2O_3 and basicity on the FTIR transmittance spectra of the CaO–SiO_2–Al_2O_3–10 wt% MgO slag. As can be seen in Figure 14a, the transmittance trough for the $[AlO_4]^{5-}$ tetrahedral stretching for wavenumbers of 750–630 cm^{-1} became deeper and more pronounced with the increase in Al_2O_3 content. This suggested the occurrence of the polymerization of complex aluminate structures using the tetrahedral $[AlO_4]^{5-}$ structural units with higher Al_2O_3. In addition, the Si-O-Al bending trough moved to higher wavenumbers, which suggested that the distance between Si/Al and O became shorter, i.e., the network structure was polymerized, as proposed by Badger's rule [35]. Moreover, the trough of the $[SiO]^{-4}$ tetrahedral band group experienced only a small change, which means that, in this experiment, Al_2O_3 content had little effect on the $[SiO]^{-4}$ tetrahedral structure. This showed that the amount of complex silicate structures increased. These results correlated well with the viscosity measurements.

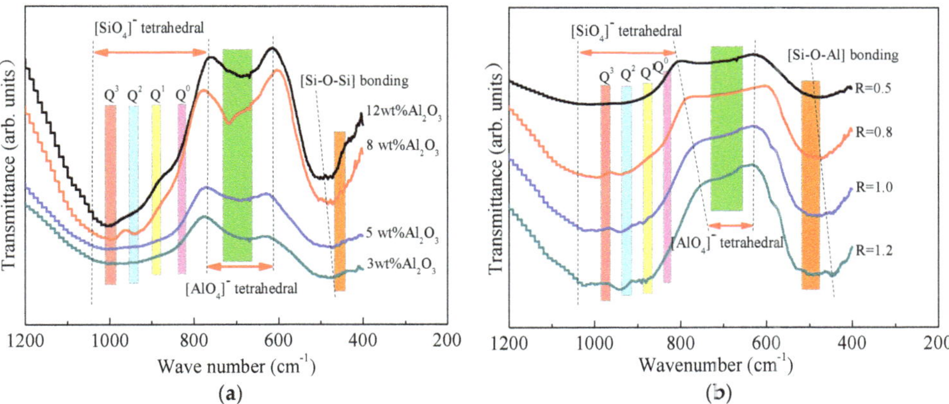

Figure 14. Effect of Al_2O_3 and basicity on the FTIR transmittance spectra of the CaO–SiO_2–Al_2O_3–10 wt.% MgO slag (**a**) at a fixed Al_2O_3 content and (**b**) at different basicities.

As can be seen from Figure 14b, the lower limit of the $[SiO_4]^-$ tetrahedral bands at about 1030-750 cm^{-1} shifted to a lower wavenumber from about 790 to 750 cm^{-1} with an increase in the basicity of the slag from 0.5 to 1.2. Furthermore, the broadening of the width of the $[SiO_4]^-$ tetrahedral bands suggested an increase in distance between the Si and O. This shows that the silicate network structures in the slag melt were depolymerized with an increase in the basicity of the slag. In addition, the trough of the $[AlO_4]^-$ tetrahedral bands at about 750–630cm^{-1} dampened with increasing basicity and almost disappeared at the basicity of 1.2 [36,37]. This indicated that the aluminate network structures in the slag melt were also depolymerized with increasing basicity. It seems reasonable to consider that the band groups observed at about 500 cm^{-1} were the Si-O-Al bending vibrations. It can be concluded that further depolymerization of a complex silicate and aluminate network structure occurred at a higher basicity. These results correlated well with the viscosity measurements.

3.7. Removal of Inclusions via the Adsorption of Refining Slag

The physical properties of slag can significantly affect its adsorption and removal of inclusions. The melting point of slag decreased with the increase in Al_2O_3 content, and the fluidity increased, which increased the adsorption capacity of slag to inclusions. In contrast, the viscosity of slag increased with the increase in Al_2O_3 content, which made it more difficult for inclusions to pass through the steel–slag interface. Therefore, the comprehensive effect caused the number of inclusions in the steel gradually decrease before it increased. In addition, the gradual increase in slag viscosity reduced the situation where it was entrapped in molten steel and became foreign inclusions, and thus, the proportion of large-sized inclusions in steel continued to decrease [38–40].

When the viscosity of slag decreased with the increase in slag basicity, it makes it easier for refining slag to be entrapped in molten steel and become foreign inclusions [38–40]. In addition, the fluidity will weaken sharply due to the melting point of slag increasing with the increase in CaO content in slag, resulting in a decrease in its adsorption capacity for inclusions [41]. Therefore, the comprehensive effect is that the percentage of the number of large inclusions in steel increased with the increase in slag basicity.

4. Conclusions

The effects of Al_2O_3 and basicity in the CaO–SiO_2–Al_2O_3–10 wt.% MgO system LF refining slag on inclusions removal in 55SiCr spring steel were investigated at 1873 K. The results can be summarized as follows:

(1) With the Al_2O_3 content increasing from 3 wt.% to 12 wt.%, the number percentage of inclusions with a diameter larger than 5 μm decreased sharply. Furthermore, the quality density of inclusions gradually increased simultaneously.

(2) In contrast, the opposite pattern was found as the basicity increased from 0.5 to 1.2.

(3) There were four kinds of typical inclusions in all steel samples, namely, CaO–SiO_2–Al_2O_3–MgO, CaO–SiO_2–Al_2O_3–MgO–MnS, MnS, and SiO_2. In addition, most of the compound oxide inclusions, especially for CaO–SiO_2–Al_2O_3–MgO inclusions, were concentrated in the low-melting-point region.

(4) The viscosity of the molten slag gradually increased with the content of Al_2O_3 increasing, which increased the degree of polymerization of the slag network structure, especially the $[AlO_4]^{5-}$ and [Si-O-Si] structures. In contrast, the viscosity of molten slag experienced the opposite pattern with the basicity of molten slag increasing. This was due to the fact that Ca^{2+} can significantly reduce the degree of polymerization of a slag network structure, especially the percentages of $[SiO_4]^{4-}$, $[AlO_4]^{5-}$ and [Si-O-Si] network structures.

Author Contributions: Conceptualization, Z.J., Y.L., H.H. and C.C.; methodology, Y.L., H.H., M.S. and C.C.; software, H.Y.; validation, Y.L., H.H. and C.C.; formal analysis, Y.L., H.H. and C.C.; investigation, Y.L., H.H. and C.C.; resources, H.H. and C.C.; data curation, Y.L., H.H. and C.C.; writing—original draft preparation, Y.L. and C.C.; writing—review and editing, Z.J.; visualization, M.S. and H.Y.; supervision, Z.J.; project administration, Z.J.; funding acquisition, Z.J. All authors read and agreed to the published version of the manuscript.

Funding: The authors are grateful for the support from the National Key Research and Development Program of China (grant no. 2016YFB0300105), the Transformation Project of Major Scientific and Technological Achievements in Shenyang (grant number Z17-5-003) and the Fundamental Research Funds for the Central Universities (N172507002).

Data Availability Statement: Not applicable.

Conflicts of Interest: On behalf of all authors, the corresponding author states that there is no conflict of interest.

References

1. Nam, W.J.; Choi, H.C. Effects of non-metallic inclusions on the fatigue life in bearing steels. *J. Korean Inst. Met. Mater.* **1993**, *31*, 526–535. (In Korean)
2. Kawakami, H.; Yamada, Y.; Ashida, S.; Shiwaku, K. Effect of chemical composition on sag resistanceof suspension spring. *SAE Trans.* **1982**, *91*, 464–470.
3. Furuya, Y.; Abe, T.; Matsuoka, S. Inclusion-controlled fatigue properties of 1800 MPa-class spring steels. *Metall. Mater. Trans. A.* **2004**, *35*, 3737–3744. [CrossRef]
4. Puff, R.; Barbieri, R. Effect of non-metallic inclusions on the fatigue strength of helical spring wire. *Eng. Fail. Anal.* **2014**, *44*, 441. [CrossRef]
5. Wang, Y.; Li, H.; Guo, J.; Wang, X.H.; Wang, W.J. Study on inclusions in ultra low oxygen 60SI2MNA spring steel during refining process. *Iron Steel* **2008**, *10*, 008.
6. Wang, Y.; Tang, H.; Wu, T.; Wu, G.; Li, J. Effect of acid-soluble aluminum on the evolution of non-metallic inclusions in spring steel. *Metall. Mater. Trans. B* **2017**, *48*, 943–955. [CrossRef]
7. Larsson, M.; Melander, A.; Nordgren, A. Effect of inclusions on fatigue behaviour of hardened spring steel. *Mater. Sci. Tech-lond.* **1993**, *9*, 235–245. [CrossRef]
8. Kawada, Y.; Kodama, S.J. A review on the effect of nonmetallic inclusions on the fatigue strength of steels. *Jpn Soc. Strength Fract. Mater.* **1971**, *6*, 1–11.
9. Suito, H.; Inoue, R. Thermodynamics on control of inclusions composition in ultra-clean steels. *ISIJ Int.* **1996**, *36*, 528–538. [CrossRef]
10. Hu, Y.M.; Wang, Q.; Hu, B.; Zhu, L.L.; Chen, W.M.; He, S.P. Application of high-basicity mould fluxes for continuous casting of large steel slabs. *Ironmak. Steelmak.* **2016**, *43*, 588–593.
11. Jiang, M.; Wang, X.H.; Chen, B.; Wang, W.J. Formation of $MgO \cdot Al_2O_3$ inclusions in high strength alloyed structural steel refined by $CaO-SiO_2-Al_2O_3-MgO$ Slag. *ISIJ Int.* **2008**, *48*, 885–890. [CrossRef]
12. Jiang, M.; Wang, X.H.; Chen, B.; Wang, W.J. Laboratory Study on Evolution Mechanisms of Non-metallic Inclusions in High Strength Alloyed Steel Refined by High Basicity Slag. *ISIJ Int.* **2010**, *50*, 95–104. [CrossRef]
13. Ma, M.J.; Bao, Y.P.; Wang, M.; Zhao, D.W. Influence of slag composition on bearing steel cleanness. *Ironmak. Steelmak.* **2014**, *41*, 26–30. [CrossRef]
14. Jiang, M.; Wang, X.H.; Wang, W.J. Study on refining slags targeting high cleanliness and lower melting temperature inclusions in Al killed steel. *Ironmak. Steelmak.* **2012**, *39*, 20–25. [CrossRef]
15. Zhang, J.; Wang, F.M.; Li, C.R. Thermodynamic analysis of the compositional control of inclusions in cutting-wire steel. *Int. J. Miner. Metall. Mater.* **2014**, *21*, 647–653. [CrossRef]
16. He, X.F.; Wang, X.H.; Chen, S.H.; Jiang, M.; Huang, F.X. Inclusion composition control in tyre cord steel by top slag refining. *Ironmak. Steelmak.* **2014**, *41*, 676–684. [CrossRef]
17. Yang, H.L.; Ye, J.S.; Wu, X.L.; Peng, Y.S.; Fang, Y.; Zhao, X. Optimum composition of $CaO-SiO_2-Al_2O_3-MgO$ slag for spring steel deoxidized by Si and Mn in production. *Metall. Mater. Trans. B.* **2016**, *47*, 1435–1444. [CrossRef]
18. Xin, C.P.; Yue, F.; Jiang, C.X.; Wu, Q.F. Study on control of inclusion compositions in tire cord steel by low basicity top slag. *High Temp Mat Pr-Isr.* **2016**, *35*, 47. [CrossRef]
19. Tang, H.Y.; Wang, Y.; Wu, G.H.; Lan, P.; Zhang, J.Q. Inclusion evolution in 50CrVA spring steel by optimization of refining slag. *J. Iron Steel Res. Int.* **2017**, *24*, 879–887. [CrossRef]
20. Zhou, L.; Wang, X.H.; Wang, J. Thermodynamic Analysis of Inclusions of Low Melting Point Area in Spring Steel. *J. Iron Steel Res. Int.* **2014**, *21*, 70–73. [CrossRef]
21. Wu, C.; Sun, Y.Q.; Luo, X.D.; Lu, X.Y. Influence of refining slag system with different alkalinity on inclusions in spring steel. *J. Wuhan U.* **2013**, *4*, 254–257.
22. Du, G.W.; Guo, H.J. A Study on Cleanliness of Spring Steel 55SiCr during 80t BOF-LF-RH-CC Steelmaking Process. *Spec. Steel* **2016**, *4*, 18–22.
23. Li, N.; Wang, W.; Liang, Q. Effect of hydrogen embrittlement and non-metallic inclusions on tensile fracture properties of 55CrSi spring steel. *Mater. Res. Express.* **2020**, *7*, 046520. [CrossRef]
24. Gao, Y.P.; Liang, X.B.; Deng, X.L. *GB/T 228.1-2010-Guide for implementation of Metallic Materials Tensile Testing Part 1: Room Temperature Test Method*; China Quality Inspection Press: Beijing, China, 2012; p. 53.
25. Warren, B.E.; Krutter, H.; Morningstar, O. Fourier analysis of X-Ray patterns of vitreous SiO_2 and B_2O_3. *J. Am. Ceram. Soc.* **1936**, *19*, 202–206. [CrossRef]
26. Ishii, F.; Ban, Y.S.; Hino, M. Thermodynamics of the deoxidation equilibrium of aluminum in liquid nickel and nickel-iron alloys. *ISIJ Int.* **1996**, *1*, 25–31. [CrossRef]
27. Satoh, N.; Taniguchi, T.; Mishima, S.; Oka, T.; Hino, M. Prediction of nonmetallic inclusion formation in Fe-40mass% Ni-5mass% Cr alloy production process. *Tetsu Hagane* **2009**, *12*, 827–836. [CrossRef]
28. Ohta, H.; Suito, H. Activities in $CaO-MgO-Al_2O_3$ slags and deoxidation equilibria of Al, Mg, and Ca. *ISIJ Int.* **1996**, *8*, 983–990. [CrossRef]
29. Cho, S.; Suito, H. Assessment of aluminum-oxygen equilibrium in liquid iron and activities in $CaO-Al_2O_3-SiO_2$ slags. *ISIJ Int.* **1994**, *2*, 177–185. [CrossRef]

30. Waseda, Y. Current Structural Information of Molten Slags by Means of a High Temperature X-Ray Diffraction. *Can. Metall. Q.* **1981**, *20*, 57–67. [CrossRef]
31. Sun, Y.Q.; Wang, H.; Zhang, Z.T. Understanding the relationship between structure and thermophysical properties of CaO-SiO$_2$-MgO-Al$_2$O$_3$ molten slags. *Metall. Mater. Trans. B.* **2018**, *49*, 677–687. [CrossRef]
32. Sohn, I.; Min, D.J. A review of the relationship between viscosity and the structure of calcium-silicate-based slags in ironmaking. *Steel Res. Int.* **2012**, *83*, 611–630. [CrossRef]
33. Kim, H.; Matsuura, H.; Tsukihashi, F.; Wang, W.L.; Min, D.J.; Sohn, I. Effect of Al$_2$O$_3$and CaO/SiO$_2$ on the viscosity of calcium-silicate–based slags containing 10 Mass Pct MgO. Metall. *Mater. Trans. B.* **2013**, *44*, 5–12. [CrossRef]
34. Park, J.H.; Min, D.J.; Song, H.S. FT-IR Spectroscopic Study on Structure of CaO-SiO$_2$ andCaO-SiO$_2$-CaF$_2$ Slags. *ISIJ Int.* **2002**, *42*, 344–351. [CrossRef]
35. Park, J.H.; Min, D.J.; Song, H.S. Structural investigation of CaO–Al$_2$O$_3$ and CaO–Al$_2$O$_3$–CaF$_2$ slags via fourier transform infrared spectra. *ISIJ Int.* **2002**, *42*, 38–43. [CrossRef]
36. Waseda, Y.; Toguri, J.M. The structure of molten binary silicate systems CaO-SiO$_2$ and MgO-SiO$_2$. *Mater. Trans.* **1977**, *8*, 563–568.
37. Lee, Y.S.; Min, D.J.; Jung, S.M.; Yi, S.H. Influence of Basicity and FeO Content on Viscosity of Blast Furnace Type Slags Containing FeO. *ISIJ Int.* **2004**, *44*, 1283–1290. [CrossRef]
38. Chen, C.Y.; Jiang, Z.H.; Li, Y.; Sun, M.; Wang, Q.; Chen, K.; Li, H.B. State of the art in the control of inclusions in spring steel for automobile—A review. *ISIJ Int.* **2020**, *4*, 617–627. [CrossRef]
39. Chen, C.Y.; Jiang, Z.H.; Li, Y.; Sun, M.; Li, H.B. Application of Alkali oxides in LF refining slag for enhancing inclusion removal in C96V saw wire steel. *ISIJ Int.* **2018**, *58*, 1232–1241. [CrossRef]
40. Chen, C.Y.; Jiang, Z.H.; Li, Y.; Sun, M.; Chen, K.; Wang, Q.; Li, H.B. Effect of Na$_2$O and Rb$_2$O on inclusion removal in C96V saw wire steels using low-basicity LF (Ladle Furnace) refining slags. *Metals* **2018**, *9*, 691. [CrossRef]
41. Mantovani, M.C.; Jr, L.R.M.; Cabral, E.F.; Silva, R.L.D.; Ramos, B.P. Interaction between molten steel and different kinds of MgO based tundish linings. *Ironmak. Steelmak.* **2013**, *5*, 319–325. [CrossRef]

Article

Experimental Analysis of Polycaprolactone High-Resolution Fused Deposition Manufacturing-Based Electric Field-Driven Jet Deposition

Yanpu Chao [1,*], Hao Yi [2,3,*], Fulai Cao [1], Shuai Lu [1] and Lianhui Ma [4]

[1] College of Mechatronics, Xuchang University, Xuchang 461000, China
[2] State Key Laboratory of High Performance Complex Manufacturing, Central South University, Changsha 410083, China
[3] State Key Laboratory of Mechanical Transmission, Chongqing University, Chongqing 400044, China
[4] SINOMACH Industrial Internet Research Co., Ltd., Zhengzhou 450007, China
* Correspondence: chaoyanpu@163.com (Y.C.); haoyi@cqu.edu.cn (H.Y.)

Abstract: Polycaprolactone (PCL) scaffolds have been widely used in biological manufacturing engineering. With the expansion of the PCL application field, the manufacture of high-resolution complex microstructure PCL scaffolds is becoming a technical challenge. In this paper, a novel PCL high-resolution fused deposition 3D printing based on electric field-driven (EFD) jet deposition is proposed to manufacture PCL porous scaffold structures. The process principle of continuous cone-jet printing mode was analyzed, and an experimental system was constructed based on an electric field driven jet to carry out PCL printing experiments. The experimental studies of PCL-fused deposition under different gas pressures, electric field voltages, motion velocities and deposition heights were carried out. Analysis of the experimental results shows that there is an effective range of deposition height (H) to realize stable jet printing when the applied voltage is constant. Under the stretching of electric field force and viscous drag force (F_D) with increasing movement velocities (V_s) at the same voltage and deposition height, the width of deposition lines was also gradually decreased. The width of the deposition line and the velocity of the deposition platform is approximately a quadratic curve. The bending phenomenon of deposition lines also gradually decreases with the increase of the movement velocities. According to the experiment results, a single layer linear grid structure was printed under the appropriate process parameters, with compact structure, uniform size and good straightness. The experimental results verify that the PCL porous scaffold structure can be accurately printed and manufactured.

Keywords: Polycaprolactone (PCL); scaffolds structure; Taylor cone; viscous drag force; single layer linear grid structure

Citation: Chao, Y.; Yi, H.; Cao, F.; Lu, S.; Ma, L. Experimental Analysis of Polycaprolactone High-Resolution Fused Deposition Manufacturing-Based Electric Field-Driven Jet Deposition. *Crystals* **2022**, *12*, 1660. https://doi.org/10.3390/cryst12111660

Academic Editor: Abdolhamid Akbarzadeh Shafaroudi

Received: 2 October 2022
Accepted: 15 November 2022
Published: 18 November 2022

Publisher's Note: MDPI stays neutral with regard to jurisdictional claims in published maps and institutional affiliations.

1. Introduction

The importance of micro (μ) and nano (η) scale manufacturing as research subjects has increased worldwide in both academia and industry [1]. The traditional micro/nano manufacturing technology mainly includes micro-electrical discharge machining (EDM) [2], nano-lithography [3], high-speed micro-milling, Lithographie, Galvanoformung and Abformung (LIGA) technology [4], micro/nano embossing [5], and other micro/nano manufacturing technologies. Such technologies are generally applied to fabricate simple 2D micro/nano geometries and are unable to manufacture complex and multi-material 3D structures on the micro/nano scale [6].

Micro/nano 3D printing technology is a new micro/nano manufacturing method based on the principle of additive manufacturing. It can directly print and form structures with minimum feature sizes, ranging from tens of nanometers to several microns [7].

At present, the micro-nano 3D printing technologies mainly include: two-photon polymerization (TPP) [8], electrohydrodynamic (EHD) jet printing [9], laser-induced forward transfer (LIFT) [10], electrochemical deposition [11], aerosol jet (AJ) deposition [12], projection microstereolithography (PμSL) [13], and focused electron beam (FEB) -induced deposition [14]. Micro/nano 3D printing technologies have been applied to micro actuators [15], micro-electro-mechanical systems, micro batteries/supercapacitors [16], microfluidic chips, micro/nano-optics [17], and multi-scale bio-active scaffolds [18].

With the development of biological tissue engineering, biological scaffolds are regarded as one of the most effective therapeutic methods in biomedicine [19]. Polycaprolactone (PCL) is a chemically synthesized, colorless crystalline solid polymer with a low molecular weight. PCL has good biocompatibility, biodegradability, drug permeability, high crystallization and a low melting point [20]. Biological cells can grow normally on their substrates and can be degraded into CO_2 and H_2O. PCL has been applied to sustained-release drug carriers, cell tissue culture scaffolds, degradable surgical sutures without removal, high-strength environmental protection film products, plasticizers for nanofiber spinning, low-temperature impact of plastic, medical modeling materials and toys, etc. [21–23].

In recent years, with the continuous expansion of the application field of PCL, the demand for the manufacture of high-resolution complex microstructure PCL scaffolds is increasing. At present, PCL structures can be processed by traditional injection molding, casting molding, extrusion molding, hot molding and other methods. However, the existing processing technology has poor dimensional precision and shape controllability, a high production cost and a long manufacturing cycle. There are some limitations to the forming of high-resolution and complex microstructures, so it is urgent to develop new high-resolution forming technology for PCL.

Additive manufacturing technology (ISO/ASTM 52900: 2021), which applies the additive shaping principle and thereby builds physical three-dimensional (3D) geometries through the successive addition of material [24], provides an entirely new solution for PCL processing, which is considered one of the top 12 disruptive technologies that will determine the future of the economy. Compared with the traditional material removal and cutting technology, additive manufacturing technology does not need tools, fixtures and molds, and can quickly and precisely manufacture parts with arbitrary complex shapes, which can significantly reduce the processing procedures, shorten the product opening cycle and reduce the cost input. Additive manufacturing is a powerful tool for the aerospace industry. According to Wohler's Report 2022, the global additive manufacturing market scale reached $15.244 billion in 2021, and additive manufacturing technology has been widely applied in aerospace, medical, automotive, consumer/electronic products, scientific research institutions, energy, government/military, construction and other fields [25,26]. At present, the additive manufacturing technologies used in tissue engineering mainly include fused deposition modeling (FDM) [27,28] and direct ink writing (DIW) [29], stereo lithography apparatus (SLA) [30], selected laser sintering [31], etc. The effects of fused filament fabrication (FFF) process parameters on mechanical and surface properties of Nylon 6/66 copolymer was analyzed [32]. The orientation effect in as-printed and as-sintered bending properties fabricated by metal-fused filament fabrication was researched [33]. A transient thermal finite-element analysis model of the fused filament fabrication process was built [34].

Jiao et al. [35] used FDM technology to fabricate nano-hyaluronic HA/PCL biological scaffolds. However, the larger fiber diameter (400 μm) makes it difficult for cells to attach, and the pore diameter is much larger than the cell diameter. Cells can only proliferate and differentiate along the path where the fiber is located, and the speed is slow, which cannot achieve the desired effect of cell culture. He et al. [36] used DIW technology to fabricate hydrogel biological scaffolds and studied the diffusion and deposition of hydrogels in different printing cycles. However, DIW technology is limited by the printing material and the printing line width (600 μm). The printing line width is thick, the cell adhesion is poor, the printed biological scaffolds collapse seriously, and the connectivity rate between pores is poor, which hinders the cell adhesion and the circulation of nutrients. Lu et al. [37] used SLA technology to fabricate calcium silicate bioceramic scaffolds with pores of different shapes. This technology has a high printing resolution (20 μm), a smooth sample surface and a fast printing speed, but the photoinitiator and residual resin may have a potential toxicity. Although the aforementioned printing technologies can achieve the rapid prototyping of biological scaffolds, these technologies also have problems such as low printing resolution, thick fiber diameter, large pore size, and toxicity, which cannot reach the microenvironment needed for cell survival.

In order to achieve the microenvironment needed for cell survival and meet the requirements of highly porous and high pore connectivity, Researchers also proposed novel printing technologies such as near-field electrospinning (NFES) [38,39] and melt electrospinning writing (MEW) [40–42]. By reducing the deposition height and controlling the stability of the jet, the micro and nano fibers were successfully deposited, and the biological scaffolds more consistent with the characteristic size of cells were prepared. He et al. [43] used NFES technology to fabricate PCL and hydroxyphosphate ash with different content of hydroxyapatite (HAP) composite biological scaffolds (the average pore size was 167 μm) and cell experiments were carried out. Eichholz et al. [44] used MEW technology to fabricate PCL biological scaffolds, conducted biocompatibility experiments, prepared biological scaffolds with different structures, and explored the influence of biological scaffold structure on cell morphology. Hryneevich et al. [45] used MEW technology to fabricate PCL biological scaffolds (pore size is 50 μm) and conducted biocompatibility experiments. The relationship between fiber diameter and printing speed and back pressure under the same voltage was studied. Kan et al. [46] used electrospinning technology to fabricate the Core-Shell PVA–PEG–SiO2@PVA–GO Fiber Mats, and the shell wall thickness and core were near 66 ± 18 nm and 173 ± 25 nm, respectively.

While the above electrospinning process can print scaffolds for the microenvironment in which cells live, the printing process needs to ensure that the nozzle and deposition substrate are conductive material, that the positive electrode of the high voltage needs to be connected to the nozzle, and that the negative electrode needs to be connected to the deposition substrate and an electric field forms between them. The distance between the nozzle and the substrate is limited to a very small range of dimensions. With the increase of the height of the printing structure, the voltage needs to continue to increase, with the forming height limit. On the other hand, due to the presence of residual charges in the stacking fibers, when the printed distance is very small, adjacent fibers will fuse together or have repulsive reactions, which cannot achieve accurate fiber deposition.

In order to solve the current technical problems, a novel high-resolution PCL-fused deposition 3D print based on electric field-driven (EFD) jet deposition [47–49] is proposed to manufacture PCL porous scaffold structures. The EFD technology, based on an electrostatic induction self-excited electric field, only needs the positive electrode of the high voltage power supply to be connected to the nozzle. The process principle of the continuous cone-jet printing mode was analyzed and PCL printing experiments were carried out. A single layer linear grid structure with a compact structure, uniform size and good straightness was printed.

2. Process Principle

EFD continuous cone-jet deposition technology is a novel, micro-scale 3D printing technology, which is based on an electrostatic induction self-excited electric field. The printing modes can be divided into a pulsed cone-jet mode and a continuous cone-jet mode. High viscosity materials can be efficiently printed using the continuous cone jet mode under the premise of meeting the required accuracy. A high-voltage DC is used as the driving signal in the continuous cone-jet mode. Figure 1 shows the principle of continuous cone-jet deposition based on a high-voltage-driven electric field.

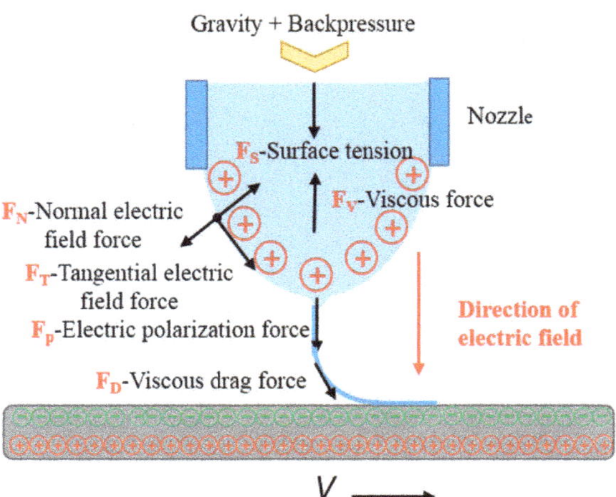

Figure 1. Schematic diagram of a continuous cone-jet deposition based on high-voltage electric field.

The forming principle is as follows: a meniscus shape of high viscosity liquid material is produced at the bottom of the nozzle under the action of gravity and air pressure. The conducting nozzle is connected to the positive electrode of the high-voltage DC power supply so that it has a high potential. The meniscus shape liquid surface has the positive charges.

When the Z moving platform drives the printing nozzle close to the deposition substrate, the positively charged nozzle will interact with the substrate. Electrostatic induction will be generated between the printing nozzle and the deposition substrate, resulting in charge redistribution inside the substrate. Negative charges will be attracted to the upper surface of the substrate facing the nozzle, while positive charges will be distributed on the lower surface of the substrate far away from the nozzle, as shown in Figure 1.

The change of charge position will change the original electric field distribution, and the electric field between the print nozzle and the target substrate will be enhanced by the negative charge attracted by the upper surface of the substrate. Under the combined effects of electric field force, surface tension, viscous force and air pressure, the curved liquid surface at the bottom of nozzle is gradually elongated to form a Taylor cone. Once the electric field force (including: F_N-normal electric field force, F_T-tangential electric field force, F_P-polarization electric field force) exceeds the liquid surface tension (F_S) and viscous force (F_V), the positively charged liquid is ejected from the top of the Taylor cone, forming a very fine cone jet.

When the continuous fine cone jet is deposited on the substrate, with the rapid movement of the deposition substrate, a viscous dragging force (F_D) is generated on the printing material, which further stretches and narrows the micro jet to form the microfilament. The diameter of the jet is usually less than one tenth of the nozzle diameter. In this process, the DC high-voltage power provides a continuous and stable electric field force for the printing melt to ensure the continuous jet injection. Combined with the back pressure, voltage and platform moving speed and other process parameters, the size effect of the microfiber is further reduced through the viscous dragging force, and the continuous and stable printing molding of higher resolution patterns is realized.

3. Experiment System

According to the above-mentioned principles, an experimental system was constructed based on an EFD continuous cone-jet deposition technology. The experimental system mainly included an air pump, a pressure-regulating valve, high-voltage power, a quartz crucible, an annular furnace, a crucible temperature control meter, a nozzle heating block, a nozzle temperature control meter, deposition substrate, an *XY*-axis movement platform, a

Z-axis movement platform, a high speed CCD, an image capture card, a stroboscope and an industrial personal computer.

Figure 2 shows a schematic diagram of the EFD jet deposition 3D-printed experimental system. The quartz crucible is installed in the annular furnace, and the temperature of 130 °C in the crucible can be precisely controlled by the temperature controller to meet the melting conditions of the PCL material in the crucible. The printing nozzle is installed at the bottom of the quartz crucible and is connected with the positive electrode of the high voltage power supply, which can achieve 0–30,000 V DC high voltage. The nozzle heating block is installed on the outer surface of the nozzle, and the nozzle temperature can be accurately controlled at 80 °C through the temperature controller, which can meet the temperature conditions of PCL molten jet injection. The annular heating furnace is fixed on the Z-axis moving platform, which can drive the nozzle to move up and down. The deposition substrate is mounted on the *XY*-axis movement platform, which can realize the precise control of the motion speed and trajectory in the *XY* direction. The air pump is communicated with the quartz crucible through the pressure regulating valve, which is used to accurately adjust the shape of the curved liquid surface at the nozzle and control the flow of extrusion PCL material. The high-speed camera CCD is connected with the industrial computer through the data acquisition card, which can monitor the deposition change process of the Taylor cone jet in real-time. Through analysis, the calculation and feedback control, accurate deposition and printing of the molten PCL continuous cone jet can be realized. According to the printed data of the 3D model, the printing layer by layer is superimposed, and the rapid manufacturing of the whole 3D structure is finally completed.

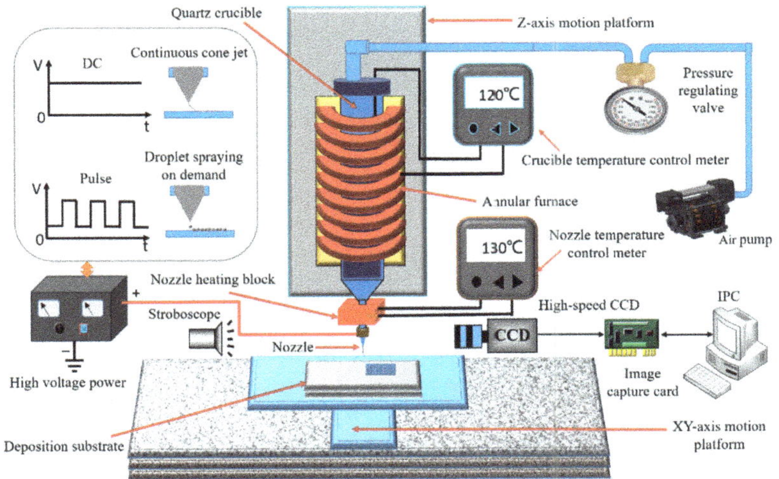

Figure 2. Schematic diagram of EFD jet deposition 3D printing experimental system.

4. Experiment Results and Discussion

4.1. Change of Extrusion Morphology of Fused PCL

In the HEF continuous cone-jet deposition printing process, in order to generate a micro-fine Taylor cone jet, it is necessary to ensure that the molten PCL material extruded from the bottom of the nozzle forms a stable menisolar surface under the action of air pressure. Polycaprolactone (PCL, CAS NO: 24980-41-4) was selected as experimental material, which is an organic polymer with the molecular formula $(C_6H_{10}O_2)$ N. By controlling the polymerization conditions, different molecular weights can be obtained. PCL is a white solid powder, non-toxic, with good biocompatibility, good organic polymer compatibility and good biodegradability. Its melting point is 65 °C, density: 1.145 g/mL at 26 °C, and after melting its liquid viscosity is 11.25 dL/Gm. Due to its high viscosity, under the action of its own gravity, the fused PCL material will not slide naturally from the nozzle to form the meniscus. Therefore, air pressure needs to be applied as back pressure. Due to the combination of gravity and air pressure, the molten, highly viscous PCL material is expelled from the bottom of the nozzle and forms a meniscus.

Figure 3 shows the extruding process of fused PCL material from the nozzle under air pressures. Figure 3a–d shows the process of the high viscosity PCL-fused material slowly ejecting from the bottom of the nozzle and finally forming the shape of the meniscus. As the air pressure continues to apply, the shape of the PCL-fused material will gradually grow. Figure 3e–h shows the high viscosity fused PCL material at the bottom of the nozzle gradually changes from the shape of the meniscus to a round sphere under the action of continuous pressure, indicating that the continuous extrusion molten PCL material needs to be deposited to maintain a dynamic equilibrium state.

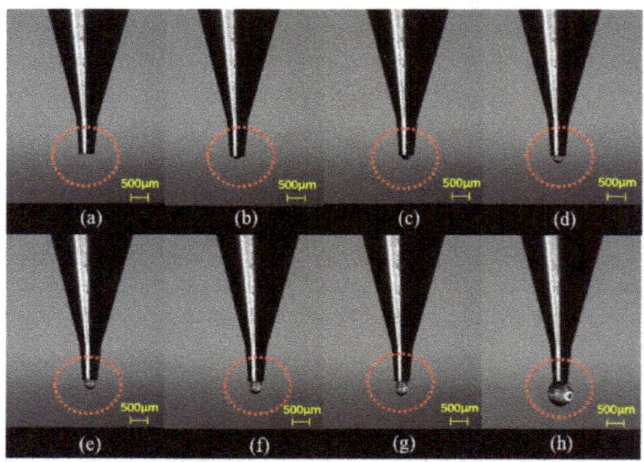

Figure 3. Extruding process of the fused PCL material under air pressures. (**a–d**) the forming process the shape of the meniscus; (**e–h**) the process of changing from the shape of the meniscus to a round sphere.

4.2. Continuous Cone Jet and Tensile Morphology

In this paper, the continuous cone-jet mode was studied. In order to reveal the influence of high-voltage electric field and viscous drag force on the morphology of the continuous cone jet, the jet drawing experiments were carried out at electric field voltages of 0 and 1600 V. Figure 4 shows that the high-viscosity PCL material was ejected from the bottom of the nozzle when the electric field voltage was 0 V. After the PCL jet was bonded to the deposited substrate, the shape of the jet was changed under the stretching of viscous drag force (F_D) through the upward movement of the nozzle. Figure 4a shows the meniscus initially formed at the bottom of the nozzle, keeping the pressure continuously acting and moving the nozzle downward so that the PCL material at the bottom bonded to the surface of the deposited substrate, as shown in Figure 4b. Then, the nozzle was moved upward, and the jet started to shrink under the viscous drag force (F_D) stretch, and the size became smaller. Figure 4c–i shows that the shape and size of the jet flow show obvious taper changes from the bottom of the nozzle to the deposited substrate. Continue to stretch upward, under the combined action of gravity and viscous drag force, the spindle shape appeared in the middle of the jet, as shown in Figure 4j–m. At higher stretching motions, the spindle shape in the middle of the jet remained unchanged, but the shape of the upper end of the jet was reduced more significantly, and the size became finer, and there was a tendency to fracture, as shown in Figure 4n–r.

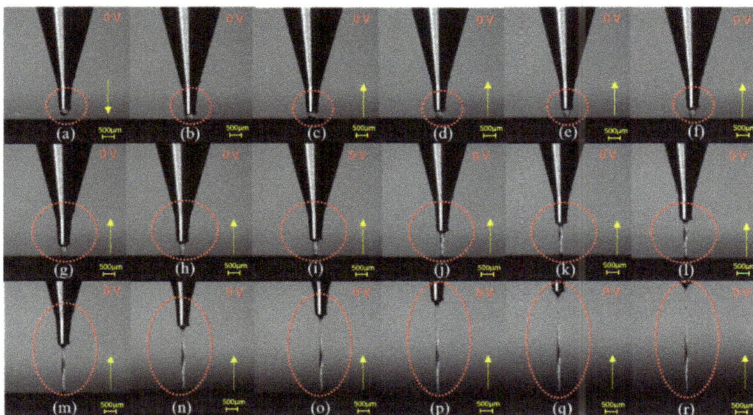

Figure 4. Vertical tensile morphology of jet at 0 V DC voltage. (**a**) the meniscus initially formed at the bottom of the nozzle; (**b**) the PCL material at the bottom was bonded to the surface of the deposited substrate; (**c–i**) the shape and size of the jet flow show obvious taper changes; (**j–m**) the spindle shape appeared in the middle of the jet; (**n–r**) the jet size became finer, and there was a tendency to fracture.

Figure 5 shows that the high-viscosity PCL material is ejected from the bottom of the nozzle, when the electric field voltage is 1600 V. After the PCL jet was bonded to the deposited substrate, the shape of the jet was changed under the stretching of viscous drag force (F_D) through the upward movement of the nozzle. Figure 5a shows the meniscus initially formed at the bottom of the nozzle, keeping the pressure continuously acting and moving the nozzle downward so that the PCL material at the bottom was bonded to the surface of the deposited substrate, as shown in Figure 5b. Then, the nozzle moved upward, and the continuously extruded jet started to change under the stretching of electric field force and viscous drag force (F_D), and the size and diameter of the jet became smaller. As shown in Figure 5c–f, from the bottom of the nozzle to the deposited substrate, the jet shape and size shows a weak taper change. As the upward stretching movement continued, Figure 5g–n shows that the overall diameter size of the jet gradually decreases, but the shape and size of a single jet is very uniform, without any taper change or spindle shape. At higher stretching motions, the overall diameter of the jet becomes smaller, and obvious diameter reduction occurs at the upper end of the jet near the nozzle, as shown in Figure 5o–r. Compared to the experimental results in Figure 4, it shows that after applying a high-voltage electric field, under the action of electric field force, the jet diameter can be guaranteed to be uniform and stable within a certain deposition height range that meets the printing accuracy requirements.

Figure 6 shows the formation of the Taylor cone jet and the horizontal deposition process of PCL material under continuous jet mode. Under the action of air pressure, the PCL-fused material was extruded from nozzle to form a meniscus. When the printing nozzle was close to the deposition substrate, the liquid at the tip of Taylor cone was sprayed under the action of electric field force, and a continuous cone jet was generated, as shown in Figure 6a. When the continuous jet was deposited on the Polyethylene terephthalate (PET) substrate, with the rapid movement of the nozzle, a viscous dragging force (F_D) was generated on the printing material, which further stretches the jet to form fine filaments, as shown in Figure 6b–f.

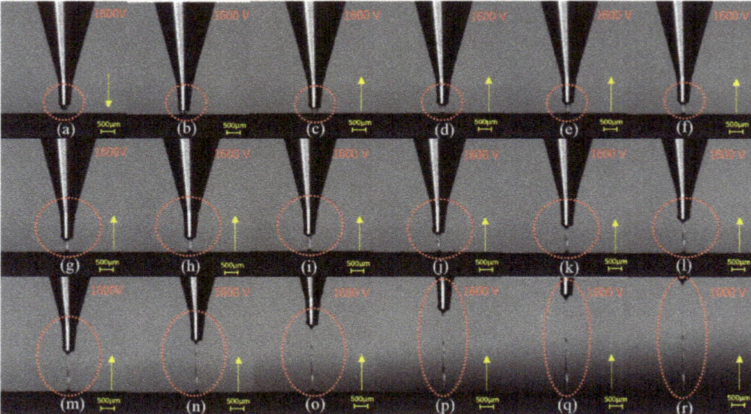

Figure 5. Vertical tensile morphology of jet at 1600 V DC voltage. (**a**) the meniscus initially formed at the bottom of the nozzle; (**b**) the PCL material at the bottom was bonded to the surface of the deposited substrate; (**c–f**) the nozzle moved upward, the jet shape and size shows a weak taper change; (**g–n**) the overall diameter size of the jet gradually decreases; (**o–r**) the obvious diameter reduction occurs at the upper end of the jet near the nozzle.

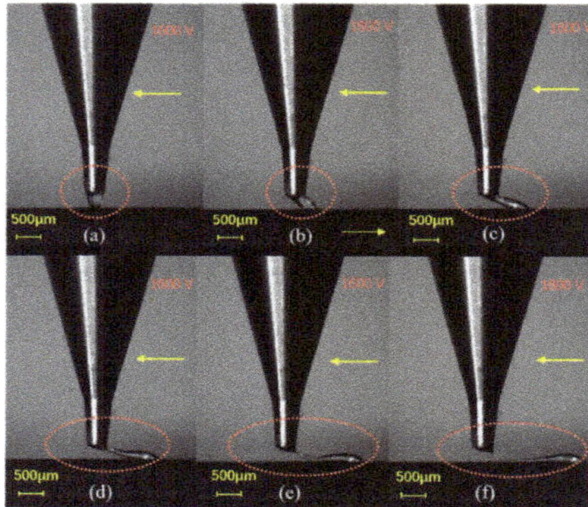

Figure 6. Continuous conical jet formation and horizontal deposition morphology. (**a**) a continuous cone jet was generated, (**b–f**) the process of cone jet is further stretched to form fine filaments under viscous dragging force (F_D).

4.3. Morphology of Jet at Different Deposition Heights

The deposition height (H) is the distance between the nozzle and the deposition substrate, which mainly affects the stability of the cone jet. Under a fixed voltage, when the H is too low, the discharge between the nozzle and the conductive substrate is easy to occur, the jet is prone to whipping, the deposited line is easy to bend, and the stable jet state is difficult to control. When H is too large, the electric field between the nozzle and the substrate becomes weak and the electric field force is small, and the viscous drag force in the jet is greater than the electric field force. The bending and sloshing of jet appeared on the deposition substrate, along with the accumulation phenomenon, which makes the width of the deposition line uneven. Only when H is suitable, the cone jet can achieve stable ejection, and the straightness and size uniformity of the deposition line can be guaranteed. Figure 7 shows the deposition experimental results of the continuous cone jet at different deposition heights. The initial value of H was set as 495.45 μm, as shown in Figure 7a. The deposition height was kept unchanged and the printing nozzle moved at a uniform speed in the horizontal direction. Figure 7b–e shows the process of forming fine fibers by the cone-jet stretching. Then, to keep the deposition substrate stationary, the nozzle begins to move upward at a uniform speed, as shown in Figure 7f–o. When the deposition height is 1.108 mm, stable jet deposition can be achieved by the cone jet, as shown in Figure 7i.

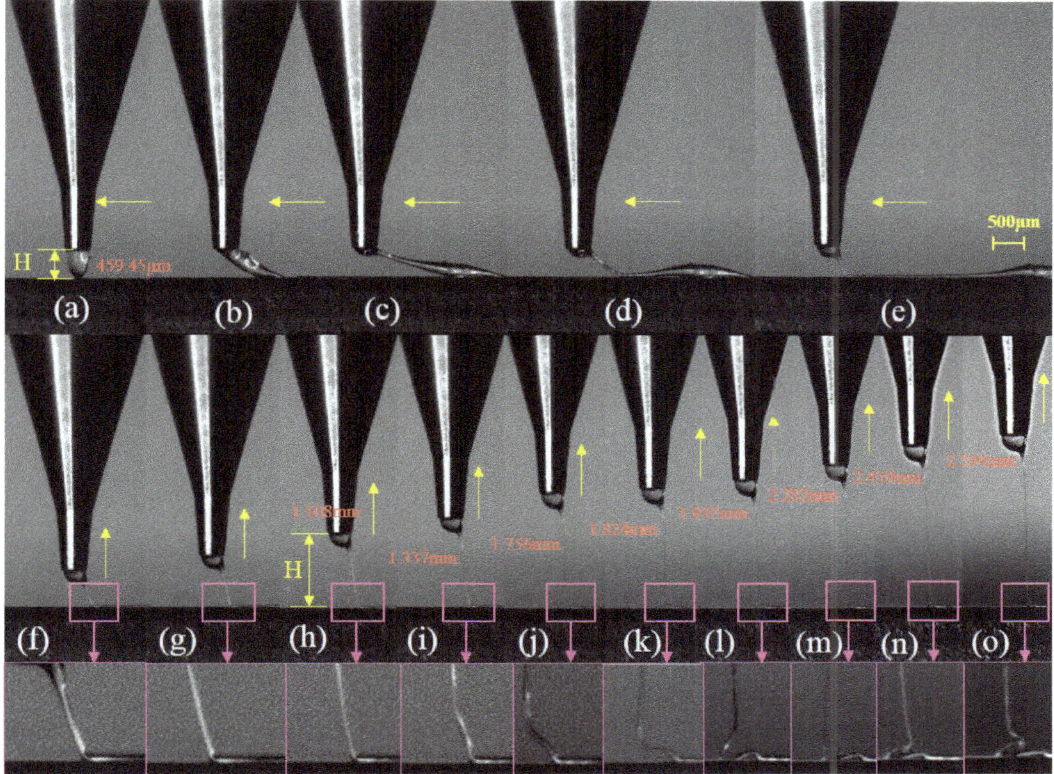

Figure 7. Deposition morphology of jet at different deposition heights. (**a**) The initial value of H was set as 495.45 µm; (**b–e**) the process of forming fine fibers by the cone-jet stretching; (**f–o**) the nozzle begins to move upward at a uniform speed.

As the deposition height gradually increased from 1.337 mm to 2.549 mm, it could be seen that the jet showed obvious bending, shaking and accumulating phenomena at the base deposition point, and the shape of the deposition line changed greatly. The experimental results show that when the applied voltage is 1600 V, the stable printing of the jet can be achieved in the deposition height of 495~1108 µm. Further analysis shows that there is a matching relationship between deposition height and voltage. When the applied voltage is constant, there is an effective range of deposition height to realize stable jet printing.

In order to further verify the matching relationship between deposition height and electric field voltage, the electric field voltage was increased to 1800 V. Figure 8 shows the photos of cone jet deposition of PCL materials at different deposition heights. The experimental results show that when *H* is 202.70 µm, 432.43 µm, 648.64 µm, 1162.16 µm and 1959.45 µm, the cone jet can achieve stable jet and deposition. There is no bending, sloshing and accumulating phenomenon at the base deposition point, and the deposition lines have good straightness and uniform size.

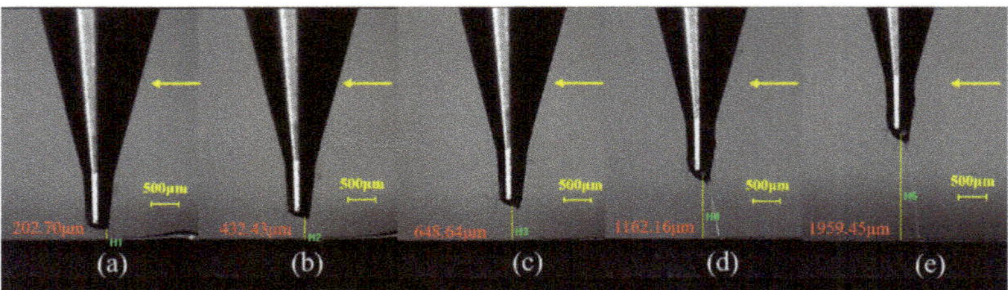

Figure 8. Stable deposition of continuous cone jet at an effective deposition height. (**a**) H is 202.70 μm; (**b**) H is 432.43 μm; (**c**) H is 648.64 μm; (**d**) H is 1162.16 μm; (**e**) H is 1959.45 μm.

4.4. Width of Deposition Lines

The velocity of deposition substrate is one of the important factors affecting the line width of printed graphics, which determines the deposition amount of material per unit time on substrate. The matching degree between velocity and ejection velocity affects the slant state of the jet flow, and then affects the line width of printing. The deposition state of the jet at different velocities is shown in Figure 9. When the moving velocity of F20 is low, the jet flow is approximately perpendicular to the deposition substrate. As the moving velocity gradually is increased to F50, the jet presents an obvious backward tilt state. When the moving velocity is set as F200, much larger than the jet velocity, the jet tilt is more obvious and the jet appears to be parallel to the deposition substrate.

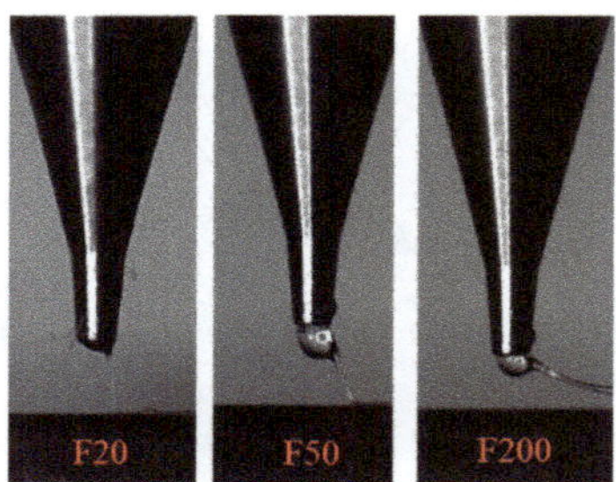

Figure 9. Deposition morphology of jet under different substrate velocities.

The experimental process parameters are shown in Table 1. The movement speeds of the deposition platform vs. were set as F10, F20, F30, F40, F50, F60, F70 and F100, respectively, in the experiment. Figure 10a shows the variation of deposition eight lines under different moving speed.

Table 1. The process parameters of printing deposition lines.

Process Parameters			
Printing Material	Deposition Substrate	Diameter of the Nozzle: D (μm)	Deposition Height: H (μm)
PCL (Polycaprolactone)	PET (Polyethylene terephthalate)	300	200
Electric field voltage: U (v)	Temperature of the crucible: T_c (°C)	Temperature of the nozzle: T_n (°C)	Air pressure: KPa
1600	130	80	15
Movement speed of the deposition platform: vs. (Plus/s)			

1	2	3	4	5	6	7	8
F10	F20	F30	F40	F50	F60	F70	F100

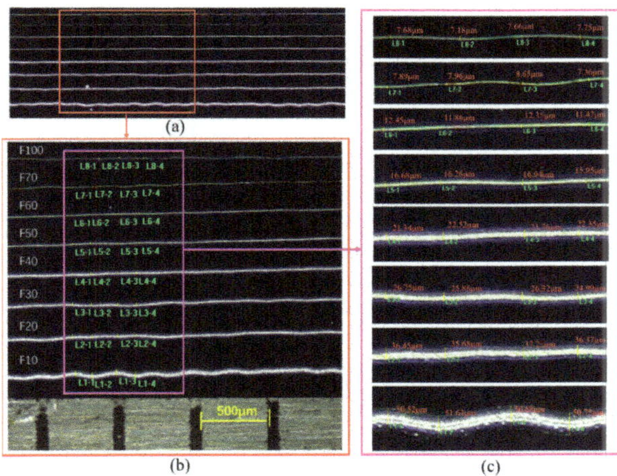

Figure 10. Variation of deposition line width under different moving speeds. (**a**) the variation of deposition eight lines under different moving speed; (**b**) four position points were selected for each deposition line; (**c**) the line width measured data of each deposition line.

For further analysis, four position points were selected for each deposition line as observation points for line width measurement, as shown in Figure 10b,c. The line width data of each deposition line were obtained by measurement and analysis, as shown in Table 2. The measurement results show that, with increasing V_s, the width of deposition lines also gradually decreased. The width of single deposition line was basically the same. The average widths of deposition lines at different movement velocities were 55.15 μm, 36.43 μm, 25.98 μm, 21.99 μm, 16.45 μm, 12.03 μm, 7.96 μm and 7.57 μm, respectively. The relationship between the width of the deposition line and the velocity of the deposition platform is approximately a quadratic curve, which was shown in Figure 11.

Table 2. The measurement data of depositing line.

		Line Width (μm)					Line Width (μm)		
		Average: 55.15					Average: 16.45		
F10	1.	54.52	3.	54.68	F50	1.	16.68	3.	16.26
	2.	55.63	4.	55.78		2.	16.94	4.	15.95
		Average: 36.43					Average: 12.03		
F20	1.	36.45	3.	37.23	F60	1.	12.45	3.	12.35
	2.	35.68	4.	36.37		2.	11.86	4.	11.47
		Average: 25.98					Average: 7.96		
F30	1.	26.75	3.	26.32	F70	1.	7.89	3.	8.65
	2.	25.86	4.	24.99		2.	7.96	4.	7.36
		Average: 21.99					Average: 7.57		
F40	1	21.34	3	21.76	F100	1	7.68	3	7.66
	2	22.52	4	22.35		2	7.18	4	7.75

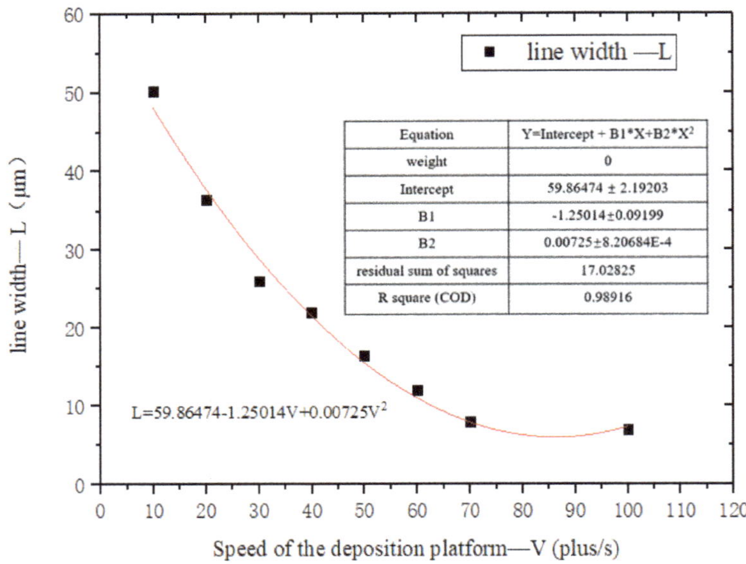

Figure 11. Conic relationship between line width and the velocity of deposition platform.

The analysis of the experimental results shows that the line width of the deposition line is related to the viscous drag force and the velocity of floor motion. After the PCL material is deposited on the substrate, due to the adhesion between the material and the substrate, a viscous dragging force (F_D) will be generated on the deposition jet, and the F_D will stretch the jet and make the jet evenly thin. Because the toughness of the PCL material is good, it will not be stretched and broken due to the increase of printing speed within a certain range.

When the movement velocities F10 is too low, on the one hand, the jet injection velocity is much higher than the deposition velocity, and the jet cannot flow back and accumulate deposition, resulting in a slight forward tilt of the jet. On the other hand, due to the small viscous drag force generated, its influence on the line width is insignificant, which eventually leads to the bending of the deposition line and the line width is thick. When the velocity of the deposition substrate gradually increases to be consistent with or greater than the jet velocity, the phenomenon of jet accumulation disappears, and the jet is vertical or backward inclined. Under the action of viscous dragging force, the deposition jet is stretched and tapered, leading to the tapering of the lines deposited on the substrate. Within a certain range, the higher the moving velocities, the finer the lines. However, when the velocity of the deposition substrate is too high, the breakage of deposition lines will occur. In this case, the pressure and voltage need to be readjusted to ensure the continuous formation of the jet to match the higher printing speed.

4.5. Straightness of Deposition Lines

The velocity of deposition substrate is one of the important factors affecting the straightness of the deposition line. The experimental process parameters are listed in Table 3, the vs. were set as F50, F100, F200, F250, F500, and F700, respectively, in the experiment. Figure 12 shows the morphology of deposition lines taken by optical microscope. The results show that the bending of the deposition line decreases gradually with the increase of the velocity. When the movement speed is F500 and F700, the straightness of deposition lines is very good. The reasons are as follows: on the one hand, the deposition height (H) is increased, the whipping effect of the jet is avoided, and the jet can achieve stable jet. On the other hand, as the velocity of deposition substrate increases, the jet deposition velocity matches the jet ejection velocity, and the jet accumulation phenomenon disappears. Under the action of the viscous drag force, the deposited jet was stretched and straightened, and the straightness of deposition lines and size uniformity was effectively improved, as shown in Figure 12e,f.

Table 3. The process parameters of printing deposition lines.

Process Parameters			
Printing Material	Deposition Substrate	Diameter of the Nozzle: D (μm)	Deposition Height: H (μm)
PCL (Poly-caprolactone)	PET (Polyethylene terephthalate)	300	1000
Electric field voltage: U (v)	Temperature of the crucible: T_c (°C)	Temperature of the nozzle: T_n (°C)	Air pressure: KPa
1800	130	80	15
Movement velocity of the deposition platform: vs. (Plus/s)			
1	2	3	4	5	6
F50	F100	F200	F250	F500	F700

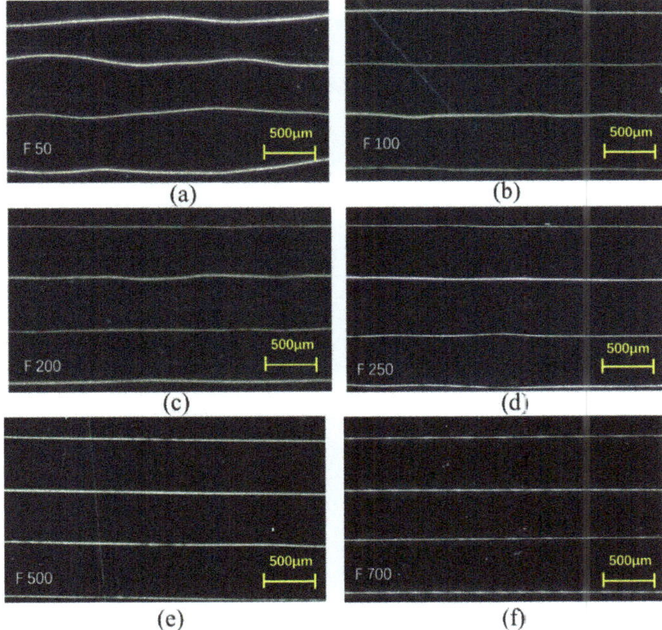

Figure 12. Variation of straightness of the deposition line under different velocities. (**a**) F50; (**b**) F100; (**c**) F200; (**d**) F250; (**e**) F500; (**f**) F700.

4.6. Deposition Single Layer Linear Grid

In order to further verify the feasibility of micro/nano 3D printing based on high-voltage electric field-driven jet deposition, on the basis of the above experimental research and analysis, the single layer linear grid was fabricated. The PLC was selected as printing material, and the PET film was used as a deposition substrate. The inner diameter of nozzle D was 300 μm, the air pressure in the crucible was set as 15 KPa, the temperature of the crucible T_c was set at 130 °C, and the temperature of the nozzle T_n was set at 80 °C, the motion speed of the deposition substrate was set to F700, and the applied voltage U was set to 1800 V.

Figure 13 shows the deposited results of single layer linear grid structure. The middle part and right-angle area were respectively selected for local amplification in Figure 12a. Figure 13b is an enlarged view of the middle area. It can be seen that the deposited PCL lines have a compact structure, uniform size and good straightness. Figure 13c is an enlarged view of the selected right-angle area. It can be seen that the deposited PCL lines did not deposit according to the set right angle path, but showed certain inclinations and arcs. The main reason is that the fused PCL material has certain viscoelastic properties, and there is a certain lag in the deposition process. The lag effect is obviously caused by the short printing distance of the adjacent lines at the end of the path. Five position points were selected for deposition line as observation points for line width measurement, as shown in Figure 13d. The line width measurement results of the five positions is: L1 = 10.25 μm, L2 = 10.35 μm, L3 = 10.67 μm, L4 = 10.73 μm, and L5 = 10.73 μm. The measurement results show that: the line width the deposition lines are consistent in size, the straightness of deposition lines is very good. The experimental results of deposition lines verify the feasibility of EFD continuous jet deposition PCL, which lays a technical foundation for the subsequent development of porous scaffold structure, micro lens array mold and cell culture template.

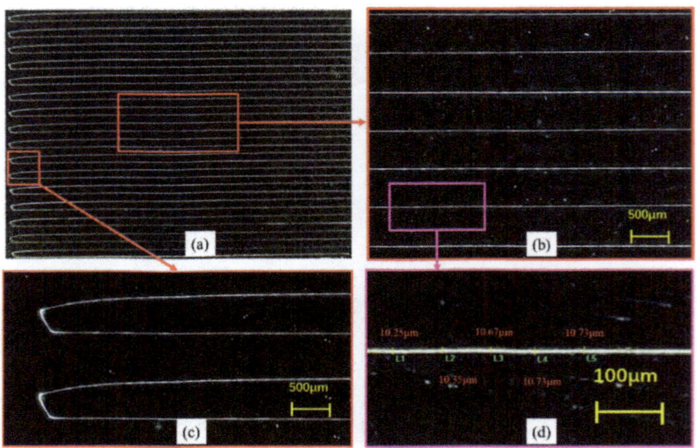

Figure 13. Deposition Single Layer Linear Grid. (**a**) the deposited single layer linear grid structure; (**b**) an enlarged view of the middle area; (**c**) an enlarged view of the selected right-angle area; (**d**) five position points were selected for deposition line as observation points for line width measurement.

5. Conclusions

(1) A novel PCL high-resolution fused deposition 3D Printing based on electric field-driven (*EFD*) jet deposition is proposed to manufacture PCL porous scaffold structures. The process principle of the continuous cone-jet printing mode was analyzed and an experimental system was constructed based on an *EFD* continuous cone-jet.

(2) *EFD* continuous cone-jet mode was studied, a Taylor cone-jet was generated under the action of the F_s, F_V, F_N, F_T and F_P. The Taylor cone-jet was further stretched by the viscous dragging force (F_D), the diameter of the jet is usually less than one-tenth of the nozzle diameter.

(3) There is an effective range of deposition height (*H*) to realize stable jet printing. Under the stretching of electric field force and viscous drag force (F_D), with the increasing of the movement velocities (V_s) the width of deposition lines was gradually decreased. The width of the deposition line and the velocity of the deposition platform is approximately a quadratic curve. The bending phenomenon of deposition lines also gradually decreases with the increase of the movement velocities.

(4)   A single layer linear grid structure was printed under the appropriate process parameters with a compact structure, uniform size and good straightness. The experimental results verify that the PCL porous scaffold structure can be accurately printed and manufactured.

Author Contributions: Y.C. conceived the idea and wrote the paper, H.Y. developed the theory and provided corrections. F.C. and S.L. performed the experiments and parameter optimization, L.M. analyzed the results and assisted with the editing of the paper. All authors have read and agreed to the published version of the manuscript.

Funding: This work was financially supported by the National Natural Science Foundation of China (grant nos. 51305128 and 52005059), Open Research Fund of State Key Laboratory of High Performance Complex Manufacturing, and Central South University (Kfkt2020-10), and Outstanding Young Backbone Teachers projects of Xu chang University.

Data Availability Statement: All the supplementary data to this article reported here can be made available on request by email.

Conflicts of Interest: The authors declare no conflict of interest.

References

1. Bodnariuk, M.; Melentiev, R. Bibliometric analysis of micro-nano manufacturing technologies. *Nanotechnol. Precis. Eng.* **2019**, *2*, 61–70. [CrossRef]
2. Kuo, C.L.; Yeh, T.H.; Nien, Y.P.; Chen, Y. Multi-objective optimization of edge quality and surface integrity when wire electrical discharge machining of polycrystalline diamonds in cutting tool manufacture. *J. Manuf. Process.* **2022**, *74*, 520–534. [CrossRef]
3. Sigita, G.; Aukse, N.; Edvinas, S.; Mangirdas, M.; Angels, S.; Jolita, O. Vegetable Oil-Based Thiol-Ene/Thiol-Epoxy Resins for Laser Direct Writing 3D Micro-/Nano-Lithography. *Polymers* **2021**, *13*, 872.
4. Hamid, H.M.A.; Celik-Butler, Z. A novel MEMS triboelectric energy harvester and sensor with a high vibrational operating frequency and wide bandwidth fabricated using UV-LIGA technique. *Sens. Actuators A Phys.* **2020**, *313*, 112175. [CrossRef]
5. Xu, J.; Su, Q.; Shan, D.B.; Guo, B. Sustainable micro-manufacturing of superhydrophobic surface on ultrafine-grained pure aluminum substrate combining micro-embossing and surface modification. *J. Clean. Prod.* **2019**, *232*, 705–712. [CrossRef]
6. Huang, Z.Y.; Shao, G.G.; Li, L.Q. Micro/nano functional devices fabricated by additive manufacturing. *Prog. Mater. Sci.* **2022**, *131*, 101020. [CrossRef]
7. Liu, P.P.; Guo, Y.W.; Chen, J.Y.; Yang, Y.B. A Low-Cost Electrochemical Metal 3D Printer Based on a Microfluidic System for Printing Mesoscale Objects. *Crystals* **2020**, *10*, 257. [CrossRef]
8. Pearre, B.W.; Michas, C.; Tsang, J.-M.; Gardner, T.J.; Otchy, T.M. Fast micron-scale 3D printing with a resonant-scanning two-photon microscope. *Addit. Manuf.* **2019**, *30*, 100887. [CrossRef]
9. Farjam, N.; Cho, T.H.; Dasgupta, N.P.; Barton, K. Subtractive patterning: High-resolution electrohydrodynamic jet printing with solvents. *Appl. Phys. Lett.* **2020**, *117*, 133702. [CrossRef]
10. Liang, P.; Shang, L.D.; Wang, Y.T.; Booth, M.J.; Li, B. Laser induced forward transfer isolating complex-shaped cell by beam shaping. *Biomed. Opt. Express* **2021**, *12*, 7024–7032. [CrossRef]
11. Wang, M.; Peng, Z.Y.; Huang, D.; Ning, Z.Q.; Chen, J.L.; Li, W.; Chen, J. Improving loading amount of nanodendrite array photo-electrodes on quantum dot sensitized solar cells by second electrochemical deposition. *Mater. Sci. Semicond. Process.* **2022**, *137*, 106219. [CrossRef]
12. Wilkinson, N.J.; Kay, R.W.; Harris, R.A. Electrohydrodynamic and aerosol jet printing for the copatterning of polydimethylsiloxane and graphene platelet inks. *Adv. Mater. Technol.* **2020**, *5*, 2000148. [CrossRef]
13. Ge, Q.; Li, Z.; Wang, Z.; Kowsari, K.; Zhang, W.; He, X. Projection micro stereolithography based 3D printing and its applications. *Int. J. Extrem. Manufac.* **2020**, *2*, 022004. [CrossRef]
14. Kumar Singh, A.; Choudhary, A.K. On the electrical characterization of focused ion/electron beam fabricated platinum and tungsten nano wires. *Mater. Today Proc.* **2020**, *28*, 127–130. [CrossRef]
15. Lee, Y.W.; Ceylan, H.; Yasa, I.C.; Kilic, U.; Sitti, M. 3D-printed multi-stimuli-responsive mobile micromachines. *ACS Appl. Mater. Interfaces* **2021**, *13*, 12759–12766. [CrossRef]
16. Lee, K.H.; Lee, S.S.; Ahn, D.B.; Lee, J.; Byun, D.; Lee, S.Y. Ultrahigh areal number density solid-state on-chip microsupercapacitors via electrohydrodynamic jet printing. *Sci. Adv.* **2020**, *6*, 1692. [CrossRef]
17. Udofia, E.N.; Zhou, W. 3D printed optics with a soft and stretchable optical material. *Addit. Manu.* **2020**, *31*, 100912.
18. Ding, A.; Lee, S.J.; Ayyagari, S.; Tang, R.; Cong, T.H.; Alsberg, E. 4D biofabrication via instantly generated graded hydrogel scaffolds. *Bioact. Mater.* **2022**, *7*, 324–332. [CrossRef]
19. Soundarya, S.P.; Menon, A.H.; Chandran, S.V. Bone tissue engineering: Scaffold preparation using chitosan and other biomaterials with different design and fabrication techniques. *Int. J. Biol. Macromol.* **2018**, *119*, 1228–1239. [CrossRef]

20. He, J.; Xia, P.; Li, D. Development of melt electrohydrodynamic 3D printing for Coplex microscale poly (s-caprolactone) scaffolds. *Biofabrication* **2016**, *8*, 035008. [CrossRef]

21. Qu, X.; Xia, P.; He, J.; Li, D. Microscale electrohydrodynamic printing of biomimetic PCL/nHA composite scaffolds for bone tissue engineering. *Mater. Lett.* **2016**, *185*, 554–557. [CrossRef]

22. Ovsianikov, A.; Khademhosseini, A.; Mironov, V. The Synergy of Scaffold-Based andScaffold-Free Tissue Engineering Strategies. *Trends Biotechnol.* **2018**, *36*, 348–357. [CrossRef] [PubMed]

23. Hong, X.Y.; Xiao, G.Q.; Zhang, Y.Z.; Zhou, J. Research on gradient additive remanufacturing of ultra-large hot forging die based on automatic wire arc additive manufacturing technology. *Int. J. Adv. Manuf. Technol.* **2021**, *116*, 2243–2254. [CrossRef]

24. Dou, Y.B.; Luo, J.; Qi, L.H.; Lian, H.C.; Huang, J.G. Drop-on-demand printing of recyclable circuits by partially embedding molten metal droplets in plastic substrates. *J. Mater. Process. Tech.* **2021**, *297*, 117268. [CrossRef]

25. Wohlers Report 2022, Analysis, Trends, Forecasts, 3D Printing and Additive Manufacturing State of the Industry. Available online: https://wohlersassociates.com/product/wohlers-report-2022/ (accessed on 20 October 2022).

26. Kollamaram, G.; Croker, D.M.; Walker, G.M.; Goyanes, A.; Basit, A.W.; Gaisford, S. Low temperature fused deposition modeling (FDM) 3D printing of thermolabile drugs. *Int. J. Pharm.* **2018**, *545*, 144–152. [CrossRef]

27. Corcione, C.E.; Gervaso, F.; Scalera, F.; Padmanabhan, S.K.; Madaghiele, M.; Montagna, F.; Sannion, A.; Licciulli, A.; Maffezzoli, A. Highly loaded hydroxyapatite microsphere/PLA porous scaffolds obtained by fused deposition modelling. *Ceram. Int.* **2019**, *452*, 2803–2810. [CrossRef]

28. Zuo, M.; Pan, N.; Liu, Q.; Ren, X.H.; Liu, Y.; Huang, T.S. Three-dimensionally printed polylactic acid/cellulose acetate scaffolds with antimicrobial effect. *RSC Adv.* **2020**, *10*, 2952–2958. [CrossRef]

29. Zhou, L.; Ramezani, I.H.; Sun, M. 3D printing of high-strength chitosan hydrogel scaffolds without any organic solvents. *Biomater. Sci.* **2020**, *8*, 5020–5028. [CrossRef]

30. Giustina, G.D.; Gandin, A.; Brigo, L.; Panciera, T.; Giulitti, S.; Sgarbossa, P.; D'Alessandro, D.; Trombi, L.; Danti, S.; Brusatin, G. Polysaccharide hydrogels for multiscale 3D printing of pullulan scaffolds. *Mater. Des.* **2019**, *165*, 107566. [CrossRef]

31. Meng, Z.; He, J.; Cai, Z.; Wang, F.; Zhang, J.; Wang, L.; Di, R.; Li., C. Design and additive manufacturing of flexible polycaprolactone scaffolds with highly-tunable mechanical properties for soft tissue engineering. *Mater. Des.* **2020**, *189*, 108508. [CrossRef]

32. Kaifur, R.; Abdullah, K.; Ranya, S.; Stuart, B. Fused filament fabrication of nylon 6/66 copolymer: Parametric study comparing full factorial and Taguchi design of experiments. *Rapid Prototyp. J.* **2022**, *28*, 1111–1128.

33. Chanun, S.; Anchalee, M. On the build orientation effect in as-printed and as-sintered bending properties of 17-4PH alloy fabricated by metal fused filament fabrication. *Rapid Prototyp. J.* **2022**, *28*, 1076–1085.

34. Chitralekha, N.; Kumar, G.P. Transient thermal finite-element analysis of fused filament fabrication process. *Rapid Prototyp. J.* **2022**, *28*, 1097–1110.

35. He, Y.; Yang, F.; Zhao, H.; Gao, Q.; Xia, B.; Fu, Z.J. Research on the printability of hydrogels in 3D bioprinting. *Sci. Rep.* **2016**, *6*, 29977. [CrossRef] [PubMed]

36. Lu, F.; Wu, R.; Shen, M.; Xie, L.; Gou, Z. Rational design of bioceramic scaffolds with tuning pore geometry bystereolithography: Microstructure evaluation and mechanical evolution. *J. Eur. Ceram. Soc.* **2021**, *41*, 1672–1682. [CrossRef]

37. Kolan, K.C.; Li, J.; Roberts, S.; Semon, J.A.; Park, J.; Day, D.E.; Ming, C.L. Near-field electrospinning of a polymer/bioactive glass composite to fabricate 3D biomimetic structures. *Int. J. Bioprinting* **2019**, *5*, 163. [CrossRef]

38. Park, Y.S.; Kim, J.; Oh, J.M.; Park, S.Y.; Cho, S.; Ko, H.; Cho, Y.K. Near-field electrospinning for three-dimensional stacked nanoarchitectures with high aspect ratios. *Nano Lett.* **2019**, *20*, 441–448. [CrossRef]

39. Robinson, T.M.; Hutmacher, D.W.; Dalton, P.D. The next frontier in melt electrospinning: Taming the jet. *Adv. Funct. Mater.* **2019**, *29*, 1904664. [CrossRef]

40. Ding, H.; Cao, K.; Zhang, F.; Bettcher, W.; Chang, R.C. A fundamental study of charge effects on melt electrowritten polymer fibers. *Mater. Des.* **2019**, *178*, 107857. [CrossRef]

41. Brown, T.D.; Dalto, P.D.; Hutmacher, D.W. Direct writing by way of melt electrospinning. *Adv. Mater.* **2011**, *23*, 5651–5657. [CrossRef]

42. He, F.L.; Li, D.W.; He, J.; Liu, Y.Y.; Ahmad, F.; Liu, Y.L.; Deng, X.; Ye, Y.J.; Yin, D.C. A novel layer-structured scaffold with large pore sizes suitable for 3D cell culture prepared by near-field electrospinning. *Mater. Sci. Eng. C* **2018**, *86*, 18–27. [CrossRef] [PubMed]

43. Eichholz, K.F.; Hoey, D.A. Mediating human stem cell behaviour via defined fibrous architectures by melt electrospinning writing. *Acta Biomater.* **2018**, *75*, 140–151. [CrossRef] [PubMed]

44. Hrynevich, A.; Elçi, B.Ş.; Haigh, J.N. Dimension based design of melt electrowritten scaffolds. *Small* **2018**, *14*, 1800232. [CrossRef] [PubMed]

45. Kan, Y.Y.; Bondareva, J.V.; Statnik, E.S.; Cvjetinovic, J.; Lipovskikh, S.; Abdurashitov, A.S.; Kirsanova, M.A.; Sukhorukhov, G.B.; Evlashin, A.A.; Salimon, A.I.; et al. Effect of Graphene Oxide and Nanosilica Modifications on Electrospun Core-Shell PVA–PEG–SiO$_2$PVA–GO Fiber Mats. *Nanomaterials* **2022**, *12*, 998. [CrossRef] [PubMed]

46. Tourlomousis, F.; Ding, H.; Kalyon, D.M.; Chang, R.C. Melt electrospinning writing process guided by a "Printability Number". *J. Manuf. Sci. Eng.* **2017**, *139*, 081004. [CrossRef]

47. Chao, Y.P.; Yi, H.; Cao, F.L.; Li, Y.H.; Cen, H.; Lu, S. Experimental Analysis of Wax Micro-Droplet 3D Printing Based on a High-Voltage Electric Field-Driven Jet Deposition Technology. *Crystals* **2022**, *12*, 277. [CrossRef]

48. Zhang, G.; Lan, H.; Qian, L.; Zhao, J.; Wang, F. A microscale 3D printing based on the electric-field-driven jet. *3D Print. Add. Manufact.* **2020**, *7*, 37–44. [CrossRef]
49. Wang, Z.; Zhang, G.M.; Huang, H.; Qian, L.; Liu, X.L.; Lan, H.B. The self-induced electric-field-driven jet printing for fabricating ultrafine silver grid transparent electrode. *Virtual Phys. Prototyp.* **2021**, *16*, 113–123. [CrossRef]

Article

Comparison of Dopant Incorporation and Near-Infrared Photoresponse for Se-Doped Silicon Fabricated by fs Laser and ps Laser Irradiation

Lingyan Du [1,2,*], Shiping Liu [1,2], Jie Yin [1,2], Shangzhen Pang [1,2] and Hao Yi [3]

1   School of Automation and Information Engineering, Sichuan University of Science and Engineering, Zigong 643000, China
2   Artificial Intelligence Key Laboratory of Sichuan Province, Yibin 644000, China
3   College of Mechanical Engineering, Chongqing University, Chongqing 400000, China
*   Correspondence: dulingyan927@163.com; Tel.: +86-13684308845

Abstract: Se-doped silicon films were fabricated by femtosecond (fs) laser and picosecond (ps) laser irradiating Si–Se bilayer film-coated silicon. The surface morphology, impurity distribution, crystal phase, and near-infrared photocurrent response of fs-laser-processed and ps-laser-processed Si are compared. With the same number of laser pulse irradiation, fs laser induces quasi-ordered micron-size columnar structures with some deeper gullies, and ps laser induces irregular nanoscale spherical particles with some cavities. Compared with the fs-laser-produced Se-doped layer, ps laser irradiation produces a Se-doped layer with better crystallinity and higher doping concentration, resulting in a higher photocurrent response for picosecond laser-processed Si in the near-infrared band. The changes brought about by ps laser processing facilitate the application of ultrafast laser-processed chalcogen-doped silicon for silicon-based integrated circuits.

Keywords: laser manufacturing; femtosecond laser; picosecond laser; Se doping silicon; near-infrared

Citation: Du, L.; Liu, S.; Yin, J.; Pang, S.; Yi, H. Comparison of Dopant Incorporation and Near-Infrared Photoresponse for Se-Doped Silicon Fabricated by fs Laser and ps Laser Irradiation. *Crystals* **2022**, *12*, 1589. https://doi.org/10.3390/cryst12111589

Academic Editor: Bertrand Poumellec

Received: 27 October 2022
Accepted: 7 November 2022
Published: 8 November 2022

Publisher's Note: MDPI stays neutral with regard to jurisdictional claims in published maps and institutional affiliations.

1. Introduction

After decades of vigorous development of microelectronics technology, up to now, silicon materials have gradually penetrated into the fields of optical detection, optical communication, solar cells, and so on, becoming the cornerstone of the semiconductor industry because of their mature technology and rich reserves. With the continuous expansion of the silicon application field, efforts to improve the properties of silicon have become essential. In the 21st century, micro/nanofabrication technology shines brilliantly in the fields of information science, aerospace, and engineering materials, promoting the realization of many new functional electronic and photonic devices [1]. Micro/nanostructured silicon has also emerged under this background, greatly improving the optical and electrical characteristics of silicon [2,3]. Due to the light trapping effect on the surface of micro/nanostructured silicon, the surface reflectance of single crystalline silicon can be reduced to about 10% [4]. In recent years, it has been reported that surface micro/nanostructured silicon can be obtained by different methods, such as alkaline etching [5,6], metal-assisted chemical etching [7,8], and ultrafast laser pulse processing [9–12]. In particular, impurity elements such as S, Se, and Te can be introduced into the silicon lattice when an ultrafast laser pulse interacts with silicon [10–12]. Theoretically [13] and experimentally [14], it has been demonstrated that chalcogen-doped (S, Se, Te) silicon can introduce an intermediate band energy level in the silicon band gap that is conducive to sub-bandgap absorption, further enhancing the effectiveness and photoelectric response of silicon-based photodetectors. The chalcogen-doped microstructured silicon thus breaks through the limitations of commercially available silicon-based optoelectronic devices, making silicon materials potentially useful in the field of infrared detectors [15], or is expected to improve the efficiency of silicon solar cells [16]. However, femtosecond laser irradiation induces sharp conical spikes on the silicon surface and leaves some lattice defects simultaneously, which are not conducive to the fabrication of optoelectronic devices for CMOS process. In addition, surface morphology, dopant distribution, and impurity energy levels will vary due to different laser processing parameters when the laser pulse interacts with silicon, thereby affecting photoelectric conversion efficiency.

In this paper, we explored the differences in the surface structure, impurity distribution, crystal phase, and near-infrared photocurrent response of Se-doped silicon prepared by fs laser and ps laser, respectively, and discussed the reasons for these differences based on the dopant incorporation mechanism of the laser processing of semiconductors. In this paper, we find ps laser induces less damage to the silicon lattice than the fs laser, and the selenium distribution on the surface of ps-laser-treated Si is more compact. Analysis of near-infrared photocurrent results revealed that a higher doping concentration contributes to a higher near-infrared photocurrent response.

2. Experimental

In the experiments, the Si–Se bilayer film-coated silicon samples were fabricated as follows. Firstly, N-type single-polished Si wafers ((100) orientation, 8–10 Ω cm, 520 μm thick) were cleaned by RCA standard cleaning procedure with each step for 15 min to remove organic and metallic contaminants [17]. A 100 nm-thick Se film was thermally evaporated onto the cleaned Si wafers. A 150 nm-thick Si film was then deposited on the surface of Se film by magnetic sputtering in order to protect some dopants from evaporating during laser processing [17]. As shown in Figure 1, the Si–Se-coated silicon sample was placed on sample stage exposed to the atmosphere and mounted on a computer-controlled three-dimensional translation stage, then scanned line by line using fs laser [17] and ps laser, respectively, in a raster manner. A Ti: sapphire amplifier laser (λ = 800 nm, τ = 100 fs, f = 1000 Hz) and a fiber ps laser (λ = 1064 nm, τ = 8.7 fs, f = 2000 kHz) were used successively to irradiate Si with an average of 200 pulses at normal incidence. According to the monitoring of Gaussian beam profile by a CCD camera, the fs laser pulses and ps laser pulses have full-width half maximum of 200 μm and 20 μm (FWHM of a Gaussian beam) after focused by convex lens, achieving an average fluence of 4.5 kJ/m^2 and 1.4 kJ/m^2 for each pulse, respectively. Average fluence ϕ can be calculated by the following formula $\phi = \frac{4P}{f \cdot \pi \cdot D^2}$, where P, f, and D represent the average power, laser repetition frequency, and diameter of the laser spot, respectively. In order to ensure the exposed Si surface receives uniform laser exposure, the distance between adjacent scanning lines during laser processing was designed to be 200 μm and 20 μm for fs laser and ps laser, respectively.

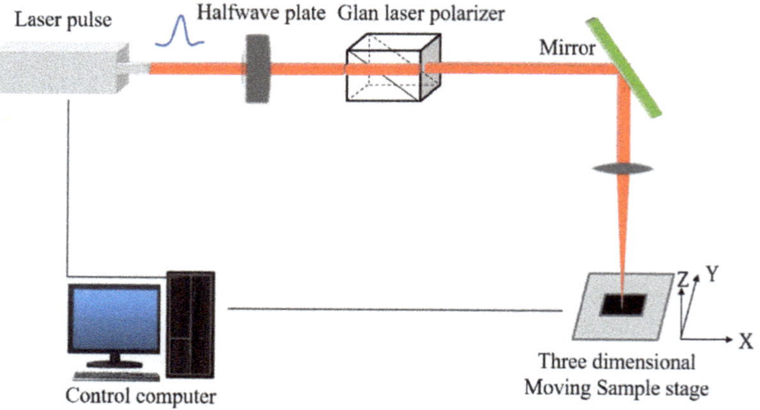

Figure 1. Schematic diagram of the experimental setup for laser processing.

After fs laser and ps laser treatment, the samples were washed in an ultrasonic bath of acetone for 10 min to remove any powder remaining on the surface. A Hitachi SU8010 field emission scanning electron microscope (FESEM) was used to investigate the sample morphology of structural modifications. Atom force microscope (AFM, Digital Instruments, Dimenson Icon) was used to analyze the surface roughness of the processed samples. Field scanning SEM, in combination with energy dispersive (X-ray) spectroscopy (EDX), was used to detect the selenium element and map the Se distribution. The crystallinity of the laser-processed samples was characterized using Thermo's DXR Raman microscope in backscattering geometry with 780 nm laser at a power of 5.1 mW. Finally, samples were annealed at 773 K for 1 h in a tube furnace under flowing nitrogen, then were used to fabricate Se-doped silicon N^+-N photodiode according to the method in [14]. Photoelectric response of the photodiodes in near-infrared band was studied using Fourier transform infrared spectroscopy (FTIR) combined with chopper and an external lock-in amplifier.

3. Experimental Results and Discussion

In order to reveal the influence of laser pulse parameters on the processing results of the microstructures, the fs-laser-treated and ps-laser-treated surface morphologies are shown in Figure 2 for comparison. It can be observed that quasi-ordered micron-size columnar structures, along with some deeper gullies, can be formed after fs laser processing of Si coated with Se film. For the case of ps laser processing, the surface morphologies are identified as cavities and irregular nanoscale spherical particles. Here, for a more detailed comparison, the surface morphologies of the fs-laser-treated and ps-laser-treated samples were further characterized by AFM to obtain a morphological comparison in detail. Figure 3 shows the AFM photographs obtained for the samples with a scale of 5×5 μm. The root mean square (rms) roughness for the fs-laser-treated sample and the ps-laser-treated sample are ~478 nm and ~122 nm, respectively. The maximum height of the fs-laser-formed columnar structure is about 3225 nm, and the maximum height of the ps-laser-formed nanospikes is about 889 nm.

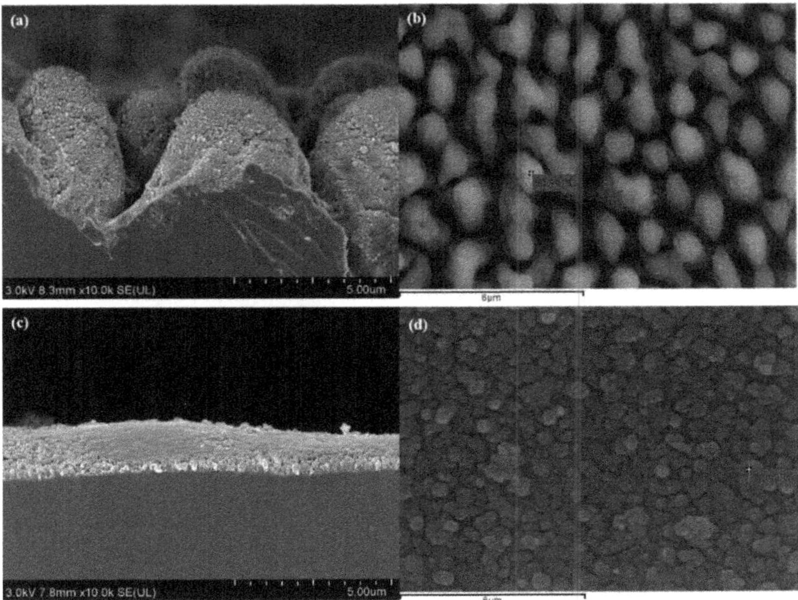

Figure 2. SEM images of Se-doped microstructure fabricated by (**a,b**) fs laser and (**c,d**) ps laser: (**a,c**) and (**b,d**) are the side and top view images. (The cross-section used for side image is obtained by brittle fracture of liquid nitrogen).

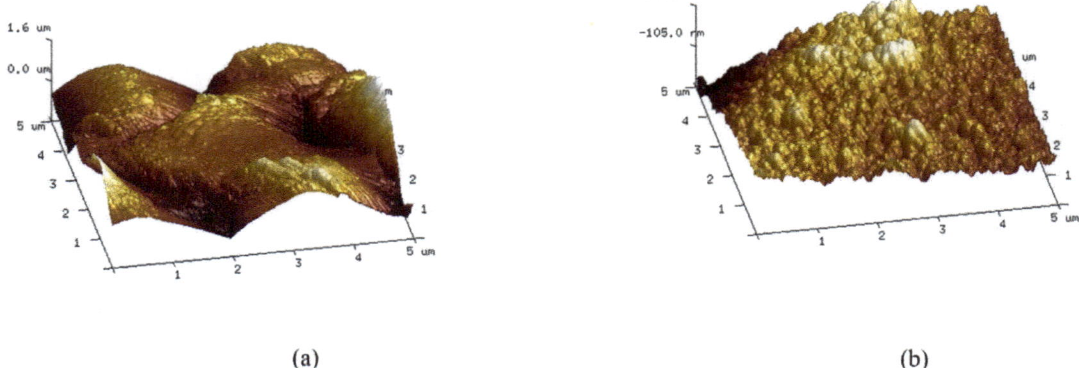

Figure 3. Large−scale (5 × 5 μm) AFM images for (**a**) fs−laser-treated Si and (**b**) ps−laser-treated Si.

With the same number of incident laser pulses, the large variation in surface morphologies obtained involves different pulse duration, laser fluence and wavelength. Melting, ablation, and microstructure self-organization formation take place successively during the laser interaction with the silicon [18]. With higher average fluence and wider spot size in diameter, fs laser deposit more energy to the sample surface, resulting heat diffusion into the processed-area is wider and deeper. The difference in energy density and spot diameter causes a difference in the volume of the ablation material. Fs laser ablate away larger volume of silicon to form the resulting 3–4 μm tall columnar structures, however, ps laser can only ablate away the silicon at the shallower surface to form the resulting nano-scale spherical particles. A series of processes of energy deposition, thermal melting and molten layer redepositon can be completed for fs laser treatment on a timescale faster than the ps laser treatment [19]. As a result, fs laser processing of silicon has the characteristics of higher spatial resolution and less thermal damage to the processed-area periphery [3]. The dependence of machining characteristics on pulse width can explain why the surface morphologies processed by fs laser are more regular and smoother than those processed by ps laser.

For a demonstration of the incorporation of selenium atoms during laser pulse irradiating silicon in the presence of dopant precursors, SEM-EDX (Figure 4) of the convex position in micro/nanostructures was taken in spot mode in the same positions marked by the pointer in Figures 2b and 3d. In order to highlight the selenium peak, the insets in Figure 4 are enlarged to show the EDX counts between 0.8 KeV to 1.6 KeV. Obviously, the presence of selenium is confirmed. Six repeated measurements of different points and EDX quantitative analysis show that the selenium concentration of the fs-laser-processed sample is higher than that of the ps-laser-processed sample, and both of them are close to the resolution limit of EDX on the order of 1%. Then, SEM-EDX of the two laser-treated samples was taken in surface scanning mode for the areas in Figure 5a,e to investigate the distribution of selenium. The maps of oxygen, silicon, and selenium for fs-laser-treated silicon and ps-laser-treated silicon are shown in Figures 5b–d and 5f–h, respectively.

Figure 5b,f shows that oxygen is present on the sample surface, and the distribution of oxygen is approximately complementary to that of silicon, indicating the micro/nanostructures of the two samples are mainly composed of SiO_x, which reveals that oxidation occurs during laser pulse processing. Figure 5d,h for the Se maps shows a difference in selenium distribution between the fs-laser-treated and ps-laser-treated silicon surfaces. Obviously, the selenium distribution on the surface of ps-laser-treated Si is more compact, indicating a higher doping concentration of selenium in ps-laser-processed silicon. This could be ascribed to fs lasers ablation of a larger volume of silicon, leaving deep and wide micron-scale gullies between the cylinder structures. Previous studies have shown that chalcogen dopants can be incorporated into the topmost 300 nm of silicon by laser irradiation on Si coated with chalcogen thin film [20]. It is considered that impurity atoms are mainly distributed on the surface of the convex cylinder structure, and there is almost no selenium distributed in the gully between the cylinders. However, EDX detection in spot mode showed a lower dopant concentration of ps-laser-treated silicon than that of fs-laser-treated silicon. Dopants can be introduced into silicon using laser pulse irradiating as an evaporated thin film, and the hyper-doping mechanism is that molten liquid can contain more dopants than the solid phase in equilibrium [3]. Due to the higher average fluence of the fs laser pulse, the fs laser pulse accumulatively deposits more energy than the ps laser pulse after the same number of pulse radiations. The more energy is deposited, the deeper the heat diffuses into the processing area and converts it into a molten layer. The molten layer enables dopants to diffuse in and traps more dopants in silicon, resulting in higher Se atom counts in the microregion.

Then, for further investigation of the crystalline properties of the fs-laser-processed and ps-laser-processed regions, Raman analysis was conducted with a spectral resolution of less than 2 cm^{-1}. Figure 6 shows the Raman spectra for fs-laser-processed silicon (solid blue line) and ps-laser-processed silicon (pink dashed line). It can be observed from Figure 6 that there is only one sharp spectral peak for the ps-treated sample at 520 cm^{-1}, which is evidence of the single-crystal phase corresponding to the original diamond structure (Si-I). However, for the fs-laser-treated silicon, besides the sharp Raman peak at 520 cm^{-1}, the Raman spectra show broadband peaks centered at 470 cm^{-1}, which is characteristic of amorphous silicon (α-Si) [21]. The simultaneous appearance of two peaks at 520 cm^{-1} and t 470 cm^{-1} demonstrates that the mixed phase silicon of Si-I and α-Si exists in the fs-laser-treated sample. The formation of amorphous silicon could be attributed to the structural instability caused by laser pulse radiation and silicon doping beyond the solubility limit [22]. Compared with ps laser pulses, the fs laser pulses generate large peak electric fields [3], which break the bonds of the atoms to electrons and are more likely to excite the instability of the lattice structure. In addition, the fs laser pulse interacts with silicon on a timescale shorter than the electron–phonon coupling relaxation time [3]. Most of the laser energy is absorbed by electrons and leaves the ions completely "cold", resulting in a change in the lattice structure. Moreover, the EDX counts in Figure 4 show a higher dopant concentration in the microregions of fs-treated Si, meaning a greater probability of lattice damage. This indicates that the doped layer produced by ps laser radiation has better crystallinity than that produced by fs laser radiation.

The above results show that the ps laser produces a Se-doped layer with a higher dopant concentration than the fs-laser-produced Se-doped layer over a larger area, albeit at a lower dopant concentration in the microregion. Chalcogenide doping concentration will affect the sub-bandgap absorption while increasing the doping concentration is beneficial to obtain more sub-bandgap absorption [3], thereby improving device performance. To compare the optoelectronic properties of fs-laser-produced Se-doped Si and ps-laser-produced Se-doped Si, we tested the photocurrent response spectra of N^{+}-N junction photodiodes fabricated from these two materials in the near-infrared band from 1.1 to 2.5 µm, as shown in Figure 7. Before fabricating the diodes, both the two materials were annealed at 773 K for 1 h to ensure optimal rectification characteristics and eliminate the influence of the amorphous silicon structure. Hall measurements (Lake Shore 8400) show that the bulk electron concentration of fs-laser-treated Si and ps-laser-treated Si are 3.57×10^{14} cm^{-3} and 1.49×10^{15} cm^{-3}, respectively, and both are higher than that of the substrate with 2.8×10^{13} cm^{-3}. As a result, Figure 7 shows that the photocurrent intensities of the two photodiodes over the range 1.1–1.2 µm are basically the same, and the absorption of photons in this range is independent of selenium doping and depends on the band gap of intrinsic silicon. For the wavelength from 1.2 to 2.5 µm, the photocurrent of the ps-laser-treated silicon diode is greater than that of the fs-laser-treated diode; this can be ascribed to a higher dopant concentration of ps-treated Si with more sub-band absorption.

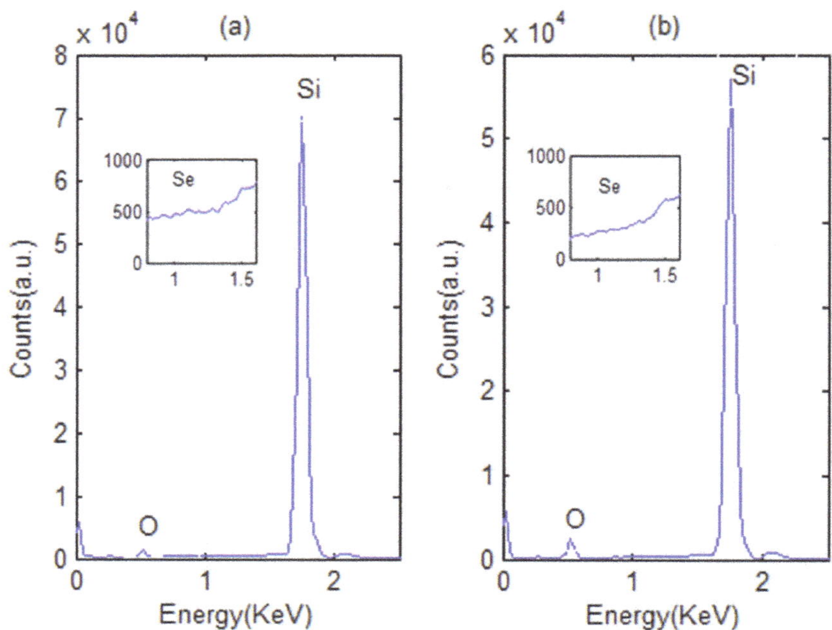

Figure 4. EDX spectral taken in black silicon microstructure (marked by the pointer in Figure 2) for (**a**) fs-laser-treated Si and (**b**) ps-laser-treated Si.

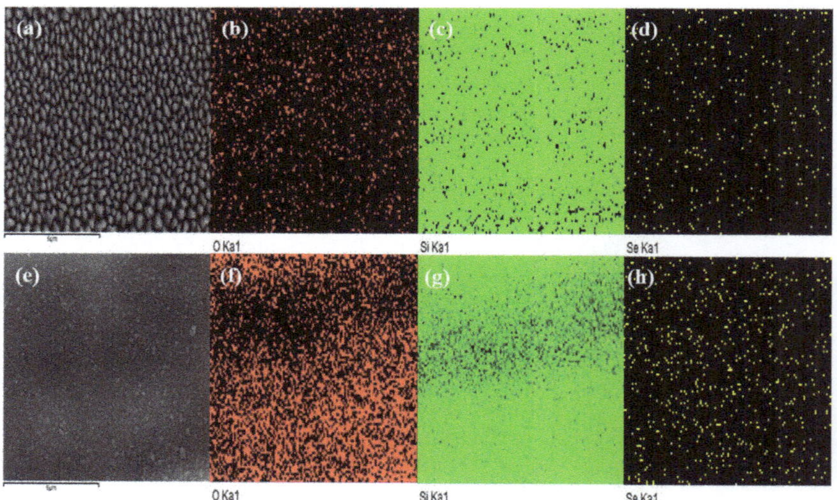

Figure 5. EFSEM maps of oxygen, silicon, and selenium for (**a**–**d**) fs-laser-treated Si and (**e**–**h**) ps-laser-treated Si.

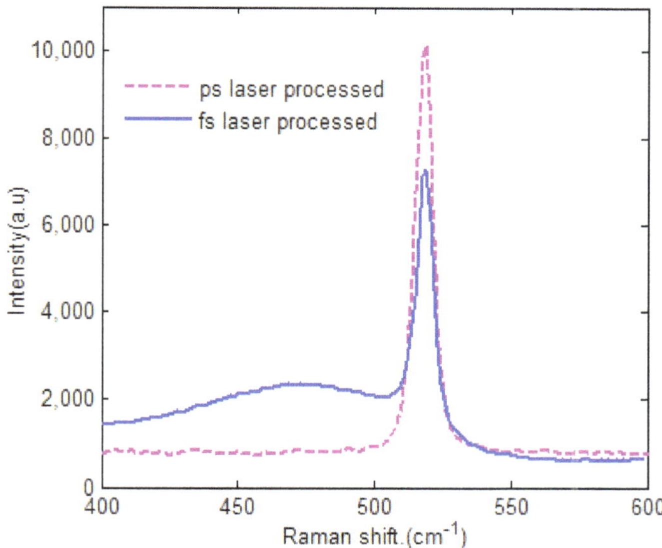

Figure 6. Raman spectra of fs−laser−processed silicon and ps−laser−processed silicon.

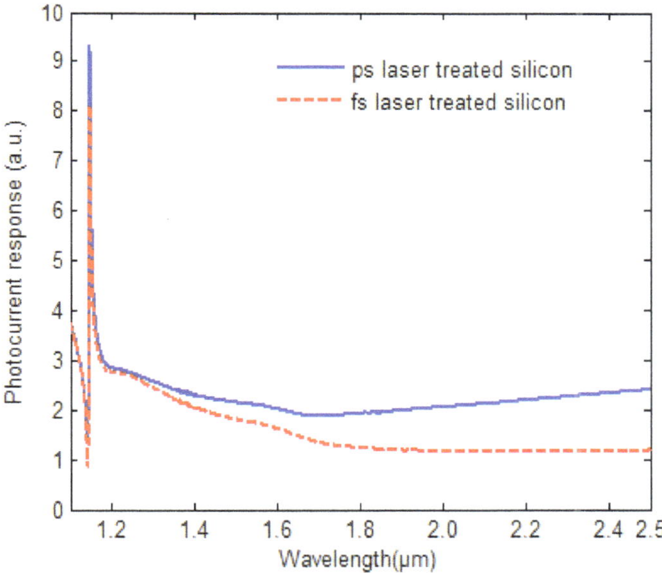

Figure 7. The photocurrent response at near-infrared wavelength for fs-laser-treated silicon and ps-laser-treated silicon.

4. Conclusions

In summary, the same number of laser pulses were used to irradiate silicon coated with Si–Se bilayer films. The fs laser induces quasi-ordered micron-size columnar structures, and the ps laser induces irregular nanoscale spherical particles on the silicon surface. Both fs laser processing and ps laser processing can introduce dopants into the target silicon. Raman measurements indicated that ps doping produces doped layers with better crystallinity compared with fs-produced doped layers. EDX analysis taken in spot mode showed that the dopant concentration of fs-laser-processed Si is higher than that of ps-laser-processed Si, but the EDX map scanned in surface mode showed that impurities are more densely distributed in doped silicon produced by the ps laser than the fs laser. Generally, a higher doping concentration of ps-laser-processed Si makes it have a stronger photocurrent in the near-infrared band. Our results suggest that ps doping may be able to compensate for the uneven distribution of impurities caused by fs laser processing and increase the doping concentration, which will improve the performance of silicon-based photodetectors.

Author Contributions: Experimental design, L.D.; methodology, L.D. and J.Y.; data curation and analysis, J.Y. and S.L.; writing—original draft preparation, L.D.; writing—review and editing, S.P.; supervision, H.Y. All authors have read and agreed to the published version of the manuscript.

Funding: This research was funded by the Science and Technology Support Program of Sichuan Province (No. 2021YFSY0049) and the National Natural Science Foundation of China (No. 61421002).

Data Availability Statement: Not applicable.

Acknowledgments: Thanks for the support of instruments from Chongqing University.

Conflicts of Interest: The authors declare that they do not have any commercial or associative interest that represents a conflict of interest in connection with the work submitted.

References

1.   Zhou, W.; Li, R.; Qi, Q.; Yang, Y.; Wang, X.; Dai, S.; Song, B.; Xu, T.; Zhang, P. Fabrication of Fresnel zone plate in chalcogenide glass and fiber end with femtosecond laser direct writing. *Infrared Phys. Technol.* **2022**, *120*, 104004. [CrossRef]

2.   Her, T.H.; Finlay, R.J.; Wu, C.; Deliwala, S.D.; Mazur, E. Microstructuring of silicon with femtosecond laser pulses. *Appl. Phys. Lett.* **1998**, *73*, 1673–1675. [CrossRef]

3.   Phillips, K.C.; Gandhi, H.H.; Mazur, E.; Sundaram, S.K. Ultrafast laser processing of materials: A review. *Adv. Opt. Photonics* **2015**, *7*, 685–712. [CrossRef]

4.   Hsu, C.H.; Liu, S.M.; Wu, W.Y.; Cho, Y.S.; Huang, P.H.; Huang, C.J.; Lien, S.Y.; Zhu, W.Z. Nanostructured pyramidal black silicon with ultralow reflectance and high passivation. *Arab. J. Chem.* **2020**, *13*, 8239–8247. [CrossRef]

5.   Ma, L.L.; Zhou, Y.C.; Jiang, N. Wide-band "black silicon" based on porous silicon. *App. Phys. Lett.* **2006**, *88*, 171907. [CrossRef]

6.   Lv, H.; Shen, H.; Jiang, Y.; Gao, C.; Zhao, H.; Yuan, J. Porous-pyramids structured silicon surface with low reflectance over a broad band by electrochemical etching. *Appl. Surf. Sci.* **2012**, *258*, 5451–5454. [CrossRef]

7.   Oh, J.; Yuan, H.C.; Branz, H.M. An 18.2% efficient black-silicon solar cell achieved through control of carrier recombination in nanostructures. *Nat. Nanotechnol.* **2012**, *7*, 743–748. [CrossRef]

8.   Noor, N.A.; Mohamad, S.K.; Hamil, S.S.; Devarajan, M.; Pakhuruddin, M.Z. Effects of etching time towards broadband absorption enhancement in black silicon fabricated by silver-assisted chemical etching. *Optik* **2018**, *176*, 586–592. [CrossRef]

9.   Liu, X.; Ma, S.; Zhu, S.; Zhao, Y.; Ning, X.; Zhao, L.; Zhuang, J. Light stimulated and regulated gas sensing ability for ammonia using sulfur-hyperdoped silicon. *Sens. Actuators B: Chem.* **2019**, *291*, 345–353. [CrossRef]

10.   Smith, M.J.; Sher, M.J.; Franta, B.; Lin, Y.T.; Mazur, E.; Gradečak, S. Improving dopant incorporation during femtosecond-laser doping of Si with a Se thin-film dopant precursor. *Appl. Phys. A* **2014**, *114*, 1009–1016. [CrossRef]

11.   Wang, X.; Huang, Y.; Liu, D.; Wang, B.; Zhu, H. Fabrication of Tellurium Doped Silicon Detector by Femtosecond Laser and Excimer Laser. *Chin. J. Lasers* **2013**, *40*, 1688–1691.

12.   Li, X.; Chang, L.; Qiu, R.; Wen, C.; Li, Z.; Hu, S. Microstructuring and doping of silicon with nanosecond laser pulses. *Appl. Surf. Sci.* **2012**, *258*, 8002–8007. [CrossRef]

13.   Sánchez, K.; Aguilera, I.; Palacios, P.; Wahnón, P. Formation of a reliable intermediate band in Si heavily coimplanted with chalcogens (S, Se, Te) and group III elements (B, Al). *Phys. Rev. B* **2010**, *82*, 165201. [CrossRef]

14.   Du, L.; Yin, J.; Wen, Y.; Zeng, W. Possible excited states in Si:Se and Si:Te prepared by femtosecond-laser irradiation of Si coated with Se or Te film. *Infrared Phys. Technol.* **2020**, *104*, 103150. [CrossRef]

15.   Said, A.J.; Recht, D.; Sullivan, J.T.; Warrender, J.M.; Buonassisi, T.; Persans, P.D.; Aziz, M.J. Extended infrared photoresponse and gain in chalcogen-supersatured silicon photodiodes. *Appl. Phys. Lett.* **2011**, *99*, 073503. [CrossRef]

16.   SiOnyx Inc. Black Silicon Enhanced Thin Film Silicon Photovoltaic Devices. 2010. Available online: https://www.osti.gov/bridge/servlets/purl/984305-ly0wxh (accessed on 31 July 2010).

17.   Du, L.; Wu, Z.; Su, Y.; Li, R.; Tang, F.; Li, S.; Zhang, T.; Jiang, Y. Se doping of silicon with Si/Se bilayer films prepared by femtosecond-laser irradiation. *Mater. Sci. Semicond. Process.* **2016**, *54*, 51–56. [CrossRef]

18.   Sundaram, S.K.; Mazur, E. Inducing and probing non-thermal transitions in semiconductors using femtosecond laser pulses. *Nat. Mater.* **2002**, *1*, 217–224. [CrossRef]

19. Buividas, R.; Mikutis, M.; Kudrius, T.; Greičius, A.; Šlekys, G.; Juodkazis, S. Femtosecond laser processing—A new enabling technology. *Lith. J. Phys.* **2012**, *52*, 301–311. [CrossRef]
20. Tabbal, M.; Kim, T.; Woolf, D.N.; Shin, B.; Aziz, M.J. Fabrication and sub-band-gap absorption of single-crystal Si supersaturated with Se by pulsed laser mixing. *Appl. Phys. A* **2010**, *98*, 589–594. [CrossRef]
21. Wang, X.C.; Zheng, H.Y.; Tan, C.W.; Wang, F.; Yu, H.Y.; Pey, K.L. Femtosecond laser induced surface nanostructuring and simultaneous crystallization of amorphous thin silicon film. *Opt. Express* **2010**, *18*, 19379–19383. [CrossRef]
22. Fabbri, F.; Lin, Y.T.; Bertoni, G.; Rossi, F.; Smith, M.J.; Gradečak, S.; Mazur, E.; Salviati, G. Origin of the visible emission of black silicon microstructures. *Appl. Phys. Lett.* **2015**, *107*, 021907. [CrossRef]

Review

A Review on Ultrafast-Laser Power Bed Fusion Technology

Yuxiang Wu, Yongxiong Chen *, Lingchao Kong, Zhiyuan Jing and Xiubing Liang

Defense Innovation Institute, Academy of Military Science, Beijing 100072, China
* Correspondence: famon1599@163.com

Abstract: Additive manufacturing of metals by employing continuous wave and short pulse lasers completely changes the way of modern industrial production. But the ultrafast laser has the superiority to short pulse laser and continuous wave laser in additive manufacturing. It has higher peak power, small thermal effect, high machining accuracy and low damage threshold. It can effectively perform additive manufacturing for special materials and improve the mechanical properties of parts. This article reviews the mechanism of the interaction between ultrafast laser and metal materials to rule the manufacturing processes. The current application of ultrafast laser on forming and manufacturing special materials, including refractory metals, transparent materials, composite materials and high thermal conductivity materials are also discussed. Among the review, the shortcomings and challenges of the current experimental methods are discussed as well. Finally, suggestions are provided for the industrial application of ultrashort pulse laser in the field of additive manufacturing in the future.

Keywords: ultrafast laser; selective laser melting; special material manufacturing; additive manufacturing; powder bed fusion

Citation: Wu, Y.; Chen, Y.; Kong, L.; Jing, Z.; Liang, X. A Review on Ultrafast-Laser Power Bed Fusion Technology. *Crystals* **2022**, *12*, 1480. https://doi.org/10.3390/cryst12101480

Academic Editors: Huajun Cao, Menglin Liu, Le Jia and Hao Yi

Received: 26 September 2022
Accepted: 17 October 2022
Published: 18 October 2022

Publisher's Note: MDPI stays neutral with regard to jurisdictional claims in published maps and institutional affiliations.

1. Introduction

In the past few decades, additive manufacturing (AM) technology has received widespread attention due to its ability to create three-dimensional free structures of almost any geometric shape. Selective laser melting (SLM, also termed Laser Powder Bed Fusion) is one of the most common and attractive AM technologies that directly forms three-dimensional complex components according to a digital model by adding material layer by layer [1]. Although continuous wave (CW) laser and short pulse laser have been widely used in the field of laser machining, there are still some limitations and disadvantages. For example: some special materials with high melting point or high reflectivity continue to pose challenges and are difficult to process [2]; some micro engines must be manufactured with remarkable precision at the micro level [3]. A more effective method to reach very high temperatures while causing minimal heat affected zone (HAZ) is needed to process such materials.

With the development of laser technology, researchers can use mode locking to generate pulses of femtosecond (fs) or picosecond (ps) duration with higher peak intensities compared to CW lasers or long-pulsed lasers. For the extension of the application field of the femtosecond laser processing, several efforts including various theoretical and experimental studies have already been made so far. It has been demonstrated that as the pulse width decreases from nanosecond to femtosecond, the heat affected area of the metal solids is suppressed and the processing quality can be significantly improved [4–6]. During ultrafast pulse laser interaction with materials, laser allow for rapid energy delivery into the material (~ps), which is significantly faster than plasma expansion time (from ns to μs). Heat accumulation combined with high peak powers of ultrafast pulse laser increases the local temperature rapidly rise to temperatures as high as 6000 °C, while the surrounding area is still cold. Benefiting from ultrashort pulse width and smaller energy input to the materials, the usage of ultrafast laser in high quality manufacturing processes can reduce heat dissipation and thus reduce HAZ in the surrounding environment. This highly intensively pulse with a duration of less than picoseconds can modify or ablate materials, creating precise features on micro- and nano-scales in a highly localized and controlled manner [7]. Thus the use of ultrafast pulse lasers as a manufacturing tool is gaining significant momentum across industries [8,9]. Though the applications of ultrafast lasers in laser AM have not been extensively studied yet, they have been widely used for various material processing such as drilling, micro-structuring and glass welding [10–14] with its unique characteristics of ultrashort pulse width and extremely high peak intensity. Compared with CW laser and short pulse laser, ultrafast pulse laser can increase part resolution, possess rapid cooling rate, minish residual stresses, reduce oxidation effect and decrease substrate damage [7]. In this regard, ultrafast lasers are well suited for lifting the shortcomings of the existing laser AM techniques, especially in situations where high melting point materials are involved or extreme precision and reproducibility are required.

While femtosecond laser-based AM is still in its infancy, this review is an attempt to highlight some of the fs laser-based research endeavors, identify the intended applications and outline prospects of the fs laser AM. We summarize the research status of ultrashort pulse laser manufacturing. Namely this article discusses the process of the interaction between ultrashort pulse laser and materials, and reviews current research status of ultrashort pulse laser SLM process in four special processing fields: refractory metals, transparent material, multi-material layered structure and high thermal conductivity materials. Among these, the influence of laser power, scan rate and other process parameters on the forming quality of the parts are analyzed, and the research hotspots and existing problems in the current ultrashort pulse laser forming manufacturing process are also put forward, aiming to provide a practical and effective reference for the research of SLM process.

2. Theoretical Research on the Interaction between Ultrashort Pulse Laser and Solid Materials

In SLM, melting takes place due to the absorption of the thermal energy resulting from the laser–matter interaction. Consider specific heat capacity and latent heat vary from one material to another, inappropriate laser parameters may affect the final product's quality. From the theoretical point of view the ultrashort time of excitation allows the separation of the involved processes as excitation, melting, and material removal. Compared with continuous laser and short pulse laser, ultrashort pulse laser have two notable characteristics: (1) extremely short pulse time, even on the femtosecond scale; (2) generating extremely high levels of 10^{13}–10^{14} W/cm^2 energy density. In some cases, the provision of insufficient energy affects the material build quality, leading to detrimental structural impacts. Extremely high laser power can result in excessive evaporation of molten material, even generates deep grooves in the center of the wall [15]. While ultrafast lasers are distinguished mainly by three factors: extremely short pulse duration, high intensity and a broad spectral bandwidth [16]. Through theoretical studies, researchers can explain the advantages of ultrafast lasers, guide process parameters and provide guidance for future research.

2.1. Mechanisms of Ultrafast Laser–Matter Interaction

In a single laser–materials interactions, fs lasers allow for rapid energy delivery into the material (~ps) [17], the whole process of heat transfer between ultrafast laser and solid material can be divided into two steps [18]. The first step starts with the deposition of laser energy into the material, including the spatial and temporal distribution of energy. The second step is the process in which the energy deposited in the material is conducted inside the material through free electrons or lattice vibrations after irritation. Since most metals are opaque to visible light, the laser energy is deposited in the penetration depth range at the beginning of the process and then gradually transferred to the thermal diffusion depth under conventional heat transfer equations.

The interaction mechanism between the femtosecond laser and the material can be obtained as shown in Figure 1 [19]. It can be found that the pulse width of the ultrashort pulse laser is less than the electron-grain relaxation time (10–100 ps), so when the ultrashort pulse laser interacts with the material, only the free electrons on the top of the material can absorb the energy of the ultrashort pulse laser electrons in the conduction band absorb photons and gain higher energy. At this time, the electrons are heated to a high temperature as shown in Figure 1a, this excitation of the solid takes place during the laser pulse, and the lattice system is still at an ambient temperature as shown in Figure 1b. Due to the interaction of laser and matter, only the process of electrons being stimulated to absorb and storing energy occurs, which fundamentally avoids the transfer and conversion of energy, as well as the existence and thermal diffusion of thermal energy [20]. Therefore, the heat transfer mechanism between the ultrashort pulse laser and the material is different from the traditional heat transfer theory (Fourier law). After the laser pulse depositing energy, the nonequilibrium of temperatures drives the energy exchange between electrons and phonons. Thus, the thermodynamic equilibrium between the electronic system and the crystal lattice can be reached after multiple electron-phonon relaxation times (in the range of picoseconds) as shown in Figure 1c. Since the electron–phonon coupling strength depends on the properties of the conduction band electrons, the characteristic relaxation time of the temperature equilibration leads to a huge difference between different materials. Through this thermal conduction mechanism, the strength of the electron-phonon coupling determines the energy loss of the ultrashort pulse laser into the material and thermal damage zone [21]. The literature [22] pointed out that the electron-phonon coupling coefficient of the material has a decisive effect on the energy transfer. Corkum [23] also pointed out that the heat transfer range is determined by the electro-phonon coupling coefficient. So, the physical properties of the material and parameters such as the pulse time of the laser will significantly affect the energy absorption of the powder and the shape of the molten pool, and ultimately affect the performance of the products [24].

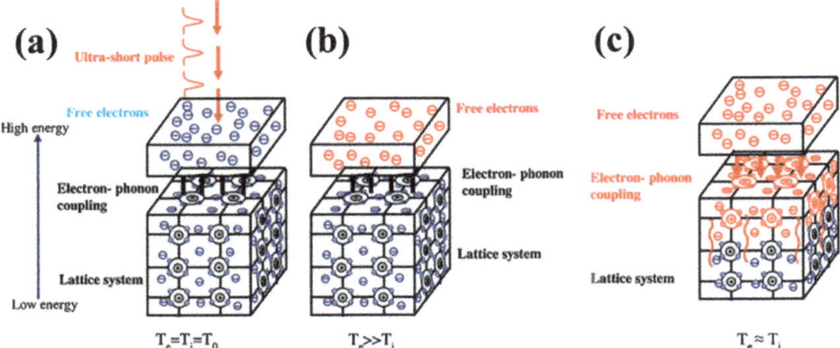

Figure 1. Mechanism of interaction between ultrashort pulse laser and materials (T_e-electron temperature, T_i-lattice temperature, T_0-ambient temperature). (**a**) absorption. (**b**) Heating. (**c**) Energy transfer [19].

Extremely narrow pulse duration of femtosecond lasers allows extremely small energies to reach extremely high powers compared to CW or short pulse laser. Since most of the electron energy transferers to the crystal lattice through the electron-phonon coupling process, with almost no thermal diffusion, so that the thermal damage of the metal is limited to a small laser focus area [25,26]. While the CW and short pulse laser is different. The electron energy diffuses in the metal for a longer time, resulting in lower precision of CW and short pulse laser processing and higher HAZ.

Based on different material and laser parameters, a series of phenomena which shows a very strong threshold effect begin to occur after laser irritation. An overview of typical involved phenomena in the considered parameter range is given in Figure 2 which shows pathways of the material from excitation to ablation, depending on the relevant timescale and intensity ranges [27]. When the laser energy exceeds the melting threshold, the material surface exhibits heterogeneous melting and gradually stabilizes; while exceed ablation threshold, the relaxation after homogeneous melting of the surface results in the expansion of the surface, followed by the generation of tensile stress waves, and when the tensile stress is large enough the matter is going to move, which is also called spallation; laser energy further reaches the evaporation threshold, the material will evaporate and even become plasma, the phase explosion state is reached. Thus, selecting a suitable processing parameters window based on the mechanism of action is necessary, it can also be used to microscopically analyze the non-equilibrium phase transition process of the interaction between ultrashort pulse laser and metal.

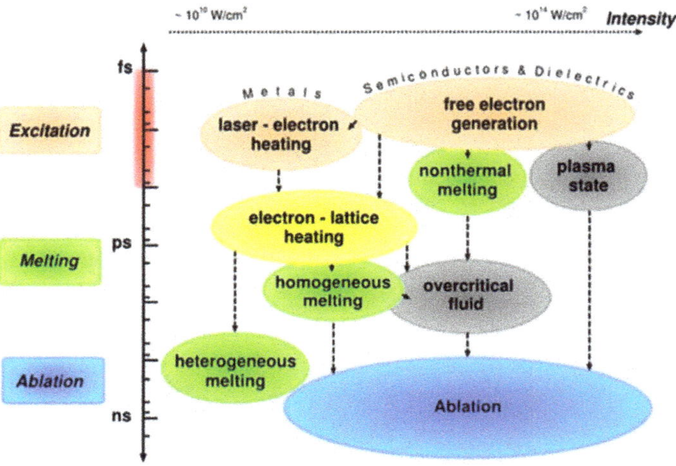

Figure 2. Typical timescales and intensity ranges of several phenomena and processes occurring during and after irradiation of a solid with an ultrashort laser pulse of about 100 fs duration [27].

2.2. Numercial Simulation of Ultrafast Laser–Matter Interaction

The process of interaction between the ultrafast laser and materials is very complex, the time and space involved are very wide and the existing monitoring methods are almost impossible to visually characterize these microscopic phenomena during the printing process. As a result, it is a reliable solution to study the interaction details by numerical models and provide parameter windows for experiments. With the deepening of research on the interaction between laser and solid materials, the theory of interaction between ultrashort pulse laser and solid materials has been further developed. Among them, some new theoretical models have been proposed, such as two temperature model(TTM) [28], molecular dynamics model(MD) [29], fluid Dynamics Model (FD) [30] and so on. The TTM is used to describe the process of fs laser-matter interaction: when a metal is irradiated by an ultrashort pulse laser, its electrons and photons absorb the laser energy within the ultrashort pulse duration, and then electron-phonon interactions assign energy to the lattice, which is described as:

$$\begin{cases} C_e \frac{\partial T_e}{\partial t} = \frac{\partial}{\partial x}\left(\kappa_e(T_e, T_l)\frac{\partial T_e}{\partial x}\right) - G(T_e - T_l) + I(z,t) \\ C_l \frac{\partial T_l}{\partial t} = \frac{\partial}{\partial x}\left(\kappa_l(T_l)\frac{\partial T_l}{\partial x}\right) + G(T_e - T_l) \end{cases} \tag{1}$$

where C is the heat capacity, κ is the thermal conductivity and G is the electron-phonon coupling factor, and their subscripts e and l are electron and lattice, and $I(z,t)$ is the laser heating source term. TTM can well describe the process of electron absorption of laser energy, energy transfer between electrons and phonons, etc., but TTM cannot be used to observe the deformation of ultrashort pulse laser radiation targets. MD simulation can calculate the motion trajectories and interactions of atoms and molecules. It is an effective tool to analyze the thermodynamic behavior of femtosecond laser ablation of metal materials from a microscopic perspective. MD can simulate the processes including melting, ablation, spalling and microstructural evolution of metals under ultrashort pulsed laser radiation. The FD can be used to simulate dynamic processes such as phase transitions and metastable states during the interaction between femtosecond lasers and metals based on the conservation of mass, momentum and energy. But the FD methods oversimplify the process of laser absorption. In addition, FD at the microscale is computationally challenging, as it requires considering micro time steps, which increases the computational costs. These three models have their own advantages and disadvantages, and they are generally used in combination with each other. They allow for realizing a better understanding underlying physical interactions of the processes and their effects on the quality of the manufactured part.

When the laser energy is below the threshold, it is not enough to melt the powder. Above this threshold, a homogeneous melting track could be generated. A further increase of the pulse energy would result in the blasting effect at the center of the track, where the particles in the focus were ablated/blasted away by single pulses without significant melting of the remaining particles [31–33]. Due to the strong threshold effect of ultrafast laser–matter interaction, numerical modelling is usually used to predicted the influence of laser parameters [34]. Literature [35] systematically investigated the damage and ablation thresholds of gold under each irradiation condition from both experimentally and theoretically analysis. Xie [36] studied the melting and disintegration behaviors of Cu thin films by femtosecond lasers with different pulse widths in the range of 35–500 fs based on TTM-MD. Literature [37] simplified TTM-MD and figured out the ablation threshold which matches well with the ablation experiments of nickel. And the predicted ablation depth for copper and aluminum subjected to an fs laser was also calculated in literature [38]. Qiu [39] concluded that a pulse repetition rate in the range of 0.1–1 GHz aids in maximizing ablation and reducing melt pool generation.

Apart from the influence of laser parameters, the structure evolution and its generation mechanism were also discussed. The ultrafast laser induced atomic structure transformation of Au nanoparticles(NPs) were studied [40], they found that when the temperature of Au NPs was higher than the melting point, complete melting occurs. Then, Au NPs cooled down, and a dynamic stress peak occurs. The stacking faults forms at the edges owing to the increased dynamic stress during the solidification process. The ultrafast laser processing of single-crystal Cu was also investigated, and the atomic structure evolution mechanism was revealed [41]. In the study, three different layers named recast layer, high density dislocation layer, and unaffected layer were found. And the melting phenomena of Al film irradiated by the femtosecond laser was also studied, Meng [42] concluded that a rapid melting stage dominated by homogeneous melting occurs within the first 2 ps and it is followed by the slow melting stage dominated by heterogeneous melting within 20 ps. However, it is still a daunting challenge to understand the whole interaction between laser and materials more detail during the whole fs-SLM due to the Limitations of computing resources [43], which involves solid-liquid-gas three phase transition, and includes a very wide range of length and time scale. Although it is nearly impossible to understand the clear microscopic characteristics only based on the numerical simulation methods, the already appeared simulations were previously conducted to explore the phase transition process and the SLM molten pool behavior by simplifying certain conditions. Literature [44] clarified the formation mechanism of pores, spatter, and denudation zones in CW-SLM by MD, so that certain remedial measurements can be taken; Also, literature [45] studied the grain evolution process to guide the specification of CW-SLM processing parameters according to the same way. Using numerical methods to clarify the formation of internal structural defects, process parameters can be optimized to achieve microstructure control of components.

Through the above analysis, it is clear that after the use of ultra-short laser pulses as the heating source of AM, small HAZ, high power and so on can be obtained. According to the differences in physical properties of different materials, it is necessary to select suitable process parameters. But more efficient calculation methods are needed. MD is difficult to simulate large-scale and long-term processing due to the limitation of computing power, and it is difficult for FD to reflect the real femtosecond laser action on the time scale. The current conditions can only do targeted research on certain issues under specific conditions or simplifications, there is still a long way to go to fully guide for industrial production. But femtosecond laser-based AM has shown great potentials due to its extraordinary characteristics.

3. Specialty Materials Manufacturing and Analysis

Over the past periods, some breakthroughs have been made on ultrashort pulse laser AM [46]. Due to its small thermal effect and high peak power, it has different characteristics from continuous laser and short laser. Therefore, it can be used in some fields to overcome the defects of continuous laser and short laser forming manufacturing or improve product performance [47]. Based on previous researches, different laser parameters will lead to different melting modes [15], therefore, it is necessary to study the products obtained with different laser parameters. Since different types of materials present different challenges and requirements, this review is presented in subsections based on the base metal considered for fabrication. Some of the most recent developments in fs-based AM are discussed below.

3.1. Refractory Metal

In special manufacturing fields such as aerospace, there is an urgent need for high performance materials with high temperature resistance and complex structures, such as tungsten and some ceramics [2,48–50]. Those parts like tungsten are usually manufactured by traditional powder metallurgy methods [49,51], but these refractory/hard materials have the characteristics of high melting point (>3000 °C) and high thermal conductivity (>100 (W/mK)), which requires extremely high energy to be melted, making it difficult using traditional additive manufacturing technologies.

SLM has obvious advantages in the preparation of metal parts with complex internal and external structures, and provides a promising choice for tungsten processing. In SLM, in order to achieve densification, it is necessary to form a continuous melting track. If the melt channel breaks, spheroidization will occurs, which may reduce the final density of the finished product [52]. If the tungsten melt solidifies too fast to completely diffuse to the cooled substrate during the CW-SLM process, the spheroidization effect will occur. However, due to the high melting point, high thermal conductivity, and oxidation tendency of tungsten, it is particularly difficult to form a continuous track in the CW-SLM process. Many studies on the CW-SLM of tungsten show that tungsten has a strong tendency to spheroidize, and the density of the final product increases with the increase of laser power [53,54]. Compared with CW laser and short laser, the extremely narrow pulse duration of femtosecond laser pulses allows most of the electron energy to be transferred to the crystal lattice through the electron-phonon coupling process. Since there is almost no thermal diffusion, and the thermal damage of the metal is limited to a small amount. Using ultrashort pulse laser AM methods, researchers have realized processing and manufacturing of some special parts that cannot be achieved by traditional manufacturing methods, such as gears with complex geometries [55].

Due to higher power and faster cooling rate, ultrashort pulse laser AM not only reduces the difficulty of forming refractory metals, but also improves the performance of the product by more fully melting and refining grains. Shuang Bai [56] used an SLM experimental device to manufacture tungsten products with different laser pulse widths. The cross-sectional SEM images of the product under different process conditions is shown in Figure 3. It does show that the fs laser-based AM generates less and smaller cracks and defects as compared with those AM parts made with short pulse or CW lasers. And the grain shape and size of the products produced by femtosecond laser and CW laser are also discussed. The results show that the hardness of the products processed by femtosecond laser is better than that of short pulse laser and CW laser, as listed in Table 1. Through the comparison of the process-structure-performance relationship, it can be found that the tungsten samples obtained by the femtosecond laser AM have achieved better results in terms of microstructure and mechanical properties. And through function fitting, it is concluded that the shorter the laser pulse width, the higher the density of the tungsten product and the lower the porosity can be obtained. This is mainly because the melting characteristics (such as melting pool, heating rate, and possible cooling rate) induced by higher peak power of fs laser are much higher in rate and confined (melting pool) than short lasers due to the higher laser intensity and less HAZ.

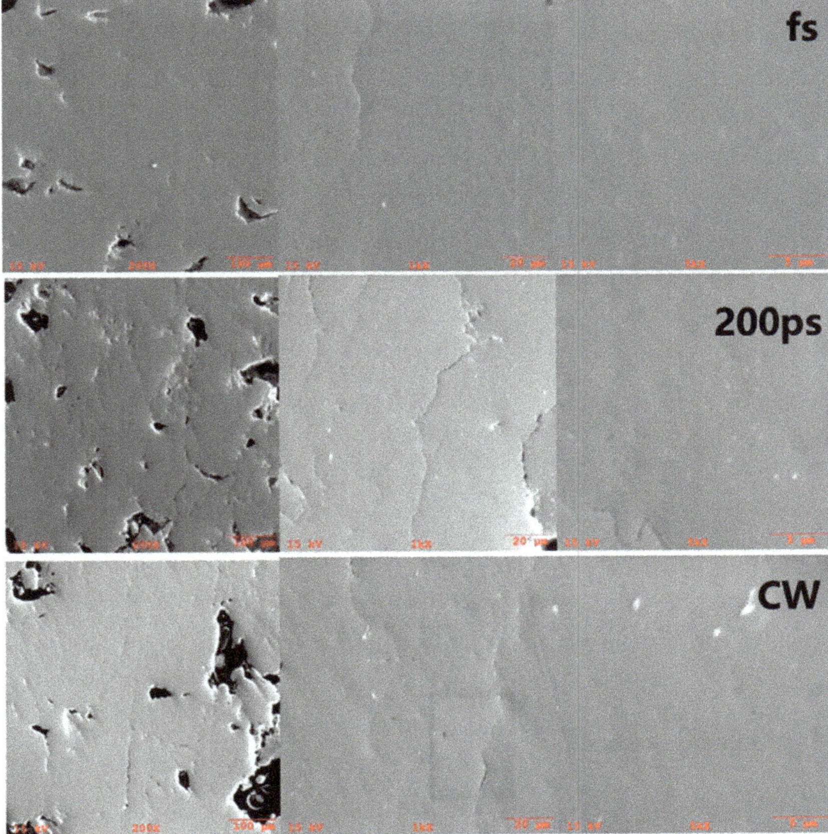

Figure 3. Cross section of tungsten sample, from top to bottom, made by fs,200 ps and CW; from left to right, SEM image of ×200, ×1000, ×5000 [56].

Table 1. Summary of the hardness testing results for different samples [56].

Hardness (HRC)	Fs	20 ps	200 ps	CW	Tungsten Substrate
Top	45.4	44.1	42.4	44.7	44.9
Cross section	47.7	41.8	45.1	44.9	45.8

This work revealed that the fs AM samples had lower cracks and defects compared to the other laser printed samples, which proved that the microstructure or grain size of AM parts can be manipulated or modified via laser parameters. For refractory metals, ultrashort laser have presented great application prospects [57]. However, pores and cracks are still unavoidable, which means that improving process parameters has the potential to further improvements in product performance. In addition to improving the scanning strategy, adding toughness elements to improve the whole product toughness is also a feasible route.

3.2. Transparent Material

Transparent and fragile glass are widely used in semiconductor, energy, communications, biomedicine and aerospace industries due to its high hardness, high chemical and thermal stability, and high transparency of the visible spectrum, making it as a kind of key material in modern society [58]. Due to its high hardness and brittleness, glass is very difficult to be processed via traditional technologies. Moreover, most laser sources have high transmittance to penetrate glass, and nonlinear optical materials have a low energy absorption rate for low power, which makes it difficult to process completely [59]. Compared with CW laser and short laser, the ultrashort pulse laser has the characteristics of high energy peak and low HAZ. When the time between two continuous pulses is much shorter than the characteristic time of thermal diffusion, heat accumulation will cause temperature constant increase. The melting and resolidification mechanism produced by the interaction time between the ultrashort irradiation area and the laser radiation surpasses the traditional thermal equilibrium process. This extremely short solidification time can prevent the segregation of different material components in the molten pool. Therefore, the distribution of different species is more uniform, and the resulting microstructure supports higher mechanical strength, while also having better optical properties [60,61]. Based on the interaction of ultrashort pulse laser nonlinear absorption effect and thermal accumulation effect, the use of ultrashort pulse laser has great potential for nonlinear optical material processing [62].

Literature [33] used a laser pulse with 500 fs pulse width and a center wavelength of 515 nm to melt glass in selected areas to realize the production of porous bulk glass parts, and the structure thickness is less than 30 μm. At repetition rates below the MHz-regime, no formation of molten tracks could be achieved due to insufficient heat accumulation within the powder. A further rise in pulse energy leads to a regime, where glass particles in the focus were ablated/blasted away by single pulses without significant melting of the remaining glass particles. So, it is very important to choose a suitable machining window when consider the interaction of nonlinear absorption and heat accumulation effects. A stable structure exhibiting fully molten tracks with powder sintered to the melt tracks was achieved by adjusting the laser parameters to repetition rate: 20 MHz, pulse energy: 0.55 mJ and scanning velocity: 20 mm/s, corresponding to a line energy of about 575 J/m, as shown in Figure 4a. By using ultrashort pulse induced melting, fabrication of solid cuboids and complex gear wheel structures could be realized, as shown in Figure 4b. Due to the high viscosity of glass, if there are no bubbles wrapped in the glass matrix, it cannot be partially melted. This behavior prevents the evolution of the powder structure to the transparent glass structure during laser processing.

Yttrium stabilized zirconium dioxide (YSZ) has the characteristics of brittleness, light weight, and the tendency of fine particles to agglomerate [63]. Traditional manufacturing method are hardly used to realize to the manufacture of parts with high porosity and special geometric shapes (flat or tubular). Jian Liu [64] used femtosecond laser for laser AM of YSZ, and discussed the influence of laser power and other parameters on the morphology and structure of molten parts. It is shown that increasing the laser power to 131 W and the scanning speed to 300 mm/s can obtain high-quality fully dense thin-layer melting parts with a total area of 20 × 60 mm. The measured micro-hardness is 18.84 GPa and uniform with the density of larger than 99%, which is much higher than the hardness of industrial granulated YSZ solid disk (fuel material of 13.66 GPa).

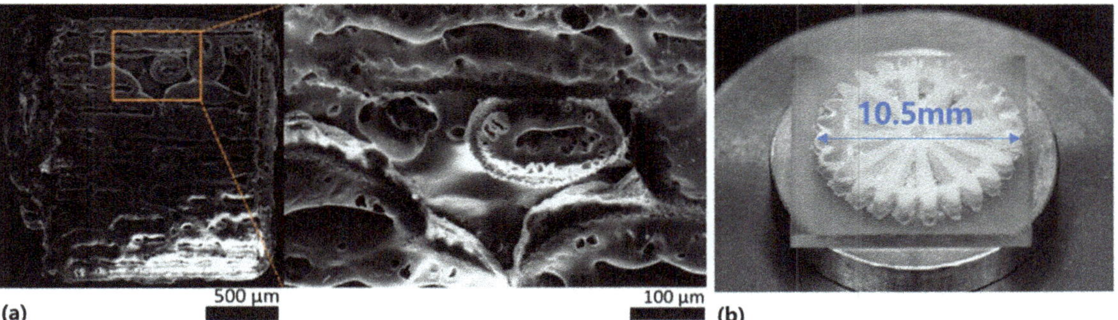

Figure 4. (**a**) SEM image (top view) of a fabricated cubic sample, exhibiting molten tracks and gas inclusions due to the high viscosity of glass. (**b**) Fabricated gear wheel with dimensions of 10.5 mm in diameter and 1.5 mm height [33].

The above results show that the use of ultrashort pulse laser AM has great potential in the semiconductor and energy industries, and its product performance is significantly improved as compared with the traditional manufacturing methods with a very high local resolution. Ultrashort pulse laser AM is a potential method in the manufacture of composite glass devices, based on the characteristics such as low thermal expansion, chemical inert behavior, or electrical and thermal insulation of optical functional devices. But the current research on the preparation of brittle materials such as transparent and fragile glass by AM methods is still in the exploratory stage, while the samples show a relatively rough surface and porous appearance. The fabricated elements exhibited a porous morphology due to the low-dense source material in combination with the high viscosity of glass. Further improvements regarding the overall density of glass AM parts may be realized with the help of spherical powder grains that support the processing of layers with thicknesses significantly below 50 µm. In actual operations, some problems such as thermal ablation caused by the accumulation of slag and the secondary action of laser light still hinder laser processing from moving towards a more precise processing level, so processing transparent and brittle material powder is still a big challenge in industrial practice [65].

3.3. Multi-Material Layered Structure

Using the advantages of 3D printing, this technology can realize product manufacturing with multiple materials, complex geometries and additional functions [66]. One of the important application scenarios is the AM of composite materials. Due to the layer-by-layer slice manufacturing method, there is no weld that causes stress concentration. This unique combination of multiple metals is better than traditional processes. Since these two materials both start in powder form, a variety of metals that are difficult to combine with traditional methods can be combined more easily by SLM [67]. Because it can control the production material in a single manufacturing operation, which can change performance, such as the functionally graded material manufactured can be locally improved by depositing metal or ceramic materials at a specific location in this way. When these operating processes are selected reasonably, structures and products that have never been seen before can be created [68].

SLM has the advantages that traditional methods are difficult to replace, however, there are still some problems of SLM need to be solved. At present, due to the shortcomings of poor surface finish, low productivity, poor quality control, unrepeatability, limited component size, and limited range of printable materials, only a few AM processes are used in modern manufacturing [69]. For metal materials, various alloys have been proposed [16,70,71]. However, in actual production, the mismatch of material with different cooling rate or thermal expansion coefficient will cause many problems, such as peeling of different layers, thermal effects caused by high power, ablation of low melting point materials on the base layer, and different types problems caused by the formation and distribution of crystal images. The low thermal effect of ultrashort pulse laser can avoid some of the above-mentioned problems. Compared with continuous laser, ultrashort pulse lasers provide the use of heat accumulation that depends on the repetition frequency to control the thermal diffusion in the focal area. Moreover, the extremely short curing time can prevent the segregation of different material components in the molten pool. As a result, the distribution of different materials is more uniform, and the resulting microstructure has higher mechanical strength [15]. And high cooling rate and crucible-free processing help reduce pollution, expand solid solubility and fine microstructure.

Bai [72] used a femtosecond laser to study the laser AM of a multi-material multilayer structure. By adding a thin layer of yttrium-stabilized zirconia (YSZ) and a Ni-YSZ layer, a solid oxide fuel cell (SOFC) was formed. Then a layer of lanthanum strontium manganese (LSM) is added to form a basic three-layer battery, the structure is shown in Figure 5a. The system evaluates the performance of a single fuel cell unit with the best density and porosity parameters such as laser power, scanning speed, and scanning mode. Figure 5b is a photo of the SOFC cells with consistent and uniform three-layered cells. Figure 5c shows that the high strength of Ni–YSZ supporting anode can be achieved with controlled porosity at properly adjusted laser power and scan speed. Through ultra-short pulse, the complex procedures of the traditional process are simplified, and the mechanical performance and product accuracy are improved at the same time.

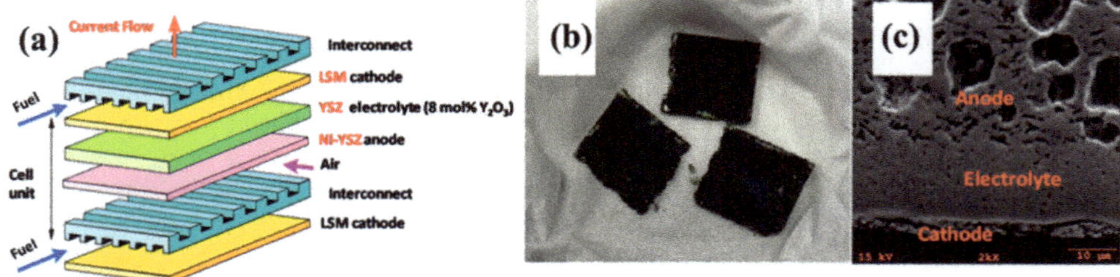

Figure 5. (**a**) Solid oxide fuel cell (SOFC) design architecture. A single cell unit include cathode (LSM), electrolyte (YSZ), and anode (Ni–YSZ), air is supplied to the cathode surface whereas fuel is supplied to the anode surface. (**b**) Complete three-layered cell samples (30 × 30 mm). (**c**) cross-section SEM image of the interfaces for a complete cell sample [72].

To successfully fabricating multi-material structures, the hatching space was discussed. The melting effect of YSZ on Ni–YSZ as a function of hatching space was investigated and summarized in Figure 6. When the hatching space is too small, continuous ripple lines in the narrow space can be found. This resulted in more cracks propagating in the cross direction. Consequently, the porosity is large (as shown in Figure 6a,b. When the hatching space is too large, melted lines cannot be formed continuously, resulting in unmelt regions and fractured structure with large porosity (as shown in Figure 6g,h. At the hatching space of 20 μm, large uniformly melted regions with less cracks were found, as shown in Figure 6c–f.

For the first time, a three layered structure was built and demonstrated via AM, and more materials are also conceivable. But the interface strength and uniformity of the YSZ electrolyte thickness, reduction of the Ni–YSZ surface roughness and the LAM integration of the interconnect still need to improve. Meanwhile, connecting two different materials at the same time may bring about not only the advantage of each component, but also some undesired properties, Thus, careful and overall consideration should be taken before combining. Moreover, there are still no small challenges for the hybrid manufacturing of materials with completely different physical properties such as metal-polymer-ceramics.

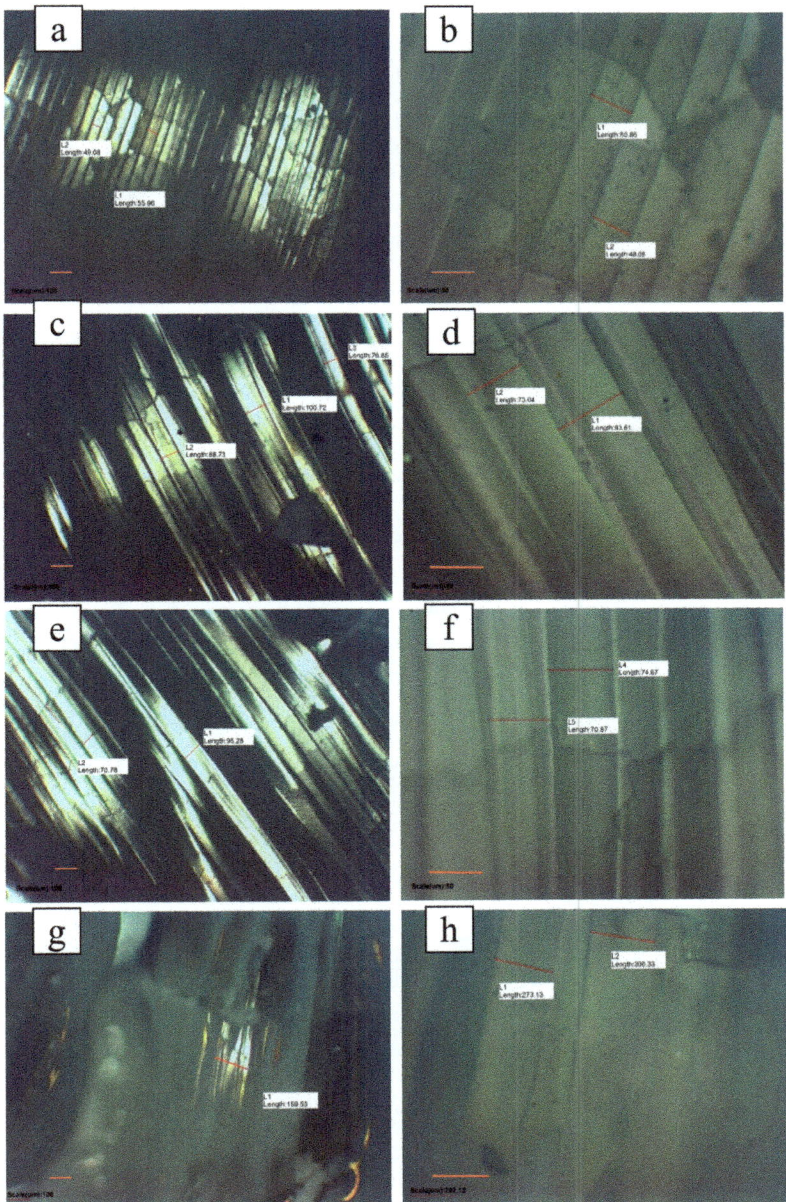

Figure 6. Microscopic images (different scales) of the YSZ layer printed on the Ni–YSZ anode substrates. (**a,b**) Hatching space of 15 μm, scanning speed of 80 mm/s, and target power of 78 W. (**c,d**) hatching space of 20 μm, scanning speed of 100 mm/s, and target power of 78 W. (**e,f**) Hatching space of 20 μm, scanning speed of 80 mm/s, and target power of 78 W. (**g,h**) Hatching space of 25 μm, scanning speed of 100 mm/s, and target power of 78 W [72].

3.4. High Thermal Conductivity Material

Al-40 Si has special hardness and mechanical strength. However, the achievable cooling rate in the traditional casting process may lead to obvious micro and macro segregation and the formation of brittle δ-AlLi phase [73]. Different from traditional manufacturing processes, SLM results in finer primary Si crystals and more uniform Si distribution due to a faster cooling rate. Compared with CW laser, the pulses of ultrashort pulse laser are shorter, resulting in a molten pool width of less than 100 μm, faster heat extraction from the molten pool, finer structure, and better mechanical properties. For this reason, literature [15] studied the influence of pulse width and pulse energy on the size and shape of the Al-40 Si molten pool in the SLM process. In Figure 7a the top view of fused walls with a total height of 4 mm using a scan speed of 200 mm/s and a layer height of 15 μm were shown for different pulse modes in comparison to CW. With increasing pulse duration, the melt pool width increased combined with a reduced uniformity along the scanned path. It was found that the use of 500 fs laser pulses can significantly reduce the porosity, and at higher pulse energy and shorter pulse duration, the relative density of the resulting bulk sample increases while being significantly higher than that of CW as shown in Figure 7b. Therefore, the ultrashort pulse laser pulse shows a huge heat source potential in the high-precision laser powder bed smelting of Al-Si alloys [74].

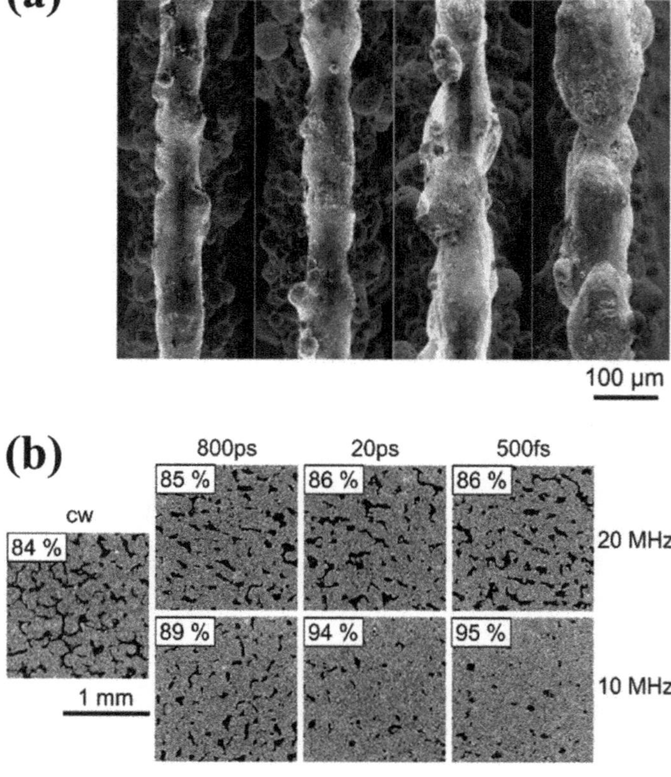

Figure 7. (a) SEM images of walls (top view) produced at different pulse width. (b)The corresponding relative densities of the samples produced using different pulse durations and repetition rates obtained by X-ray CT [15].

Copper as a metal with high reflectivity and high thermal conductivity (400 W/(mK)) further promotes the dissipation of heat, hinders the temperature rise and affects the shape of the molten pool [75]. And through the heat accumulation at the repetition frequency, the heat diffusion in the focal area is controlled. The ultra-short melting and resolidification process avoids the segregation of crystal images, so that the distribution of different materials is uniform and the final mechanical properties are improved. Literature [76,77] studies the effects of different pulse energies, different repetition frequencies and scanning speeds on the performances of the structure, as shown in Figure 8. This study shows that for a repetition frequency of 200 kHz, solid interconnection with the substrate cannot be achieved. After increasing the pulse frequency to the MHz range, this behavior changed. The 20 MHz processing window was determined by the surface sintering behavior under pulse energy below 0.75 µJ and the ablation effect of pulse energy above 1.5 µJ. Therefore, at 20 MHz and a pulse energy of about 1 µJ, a rough processing window can be identified. On the one hand, it shows the firm connection between the first floor and the building platform, and on the other hand, it involves the firm melting in the layer system. Assuming a spot diameter of 35 µm, this corresponds to a peak flux range of 0.04 to 0.06 J/cm^2. When the line energy exceeds 100 J/m, the HAZ is greatly expanded, which will result in a decrease in processing resolution and geometric accuracy.

Figure 8. (**a**)Typical sample of thin-wall structures with diameter of 3 mm processed with different laser parameter sets. (**b**) SEM image of a single wall with a thickness below 100 µm, which is marked by the dashed lines. A pulse energy of 1.0 µJ and scan velocity 666 mm/s was applied [64].

As mentioned above, the ultra-high thermal conductivity counteracts the induced melting process, allowing the heat to quickly dissipate into the bulk. Basic research has revealed the importance of using high pulse repetition rates for heat accumulation in the MHz range, demonstrating the advantages of ultrashort pulse laser over continuous laser. Furthermore, in the process of SLM high thermal conductivity materials, using its ultra-short melting and resolidification process to improve melting and reduce residual stress has great potential.

Finally, we summarized the most suitable ultrashort pulse laser processing parameters and their material properties in the above-mentioned literature of the special materials, hoping to provide guidance for subsequent industrial production, as shown in Table 2.

Table 2. The suitable ultrashort pulse laser processing parameters and material properties.

Material	Power (W)	Pulse Repetition Frequency (MHZ)	Energy (µJ)	Pulse Width (fs)	Scan Speed (mm/s)	Peak Power (MW)	Melting Point (°C)	Thermal Conductivity (W/(m × k))
rhenium [78]	50	1	22.5	400	20	56.25	3250	
Tungsten [79]	50	1	50	400	25	125	3422	174
iron [64]	50	80	0.625	350	50	1.79	1538	173
Borosilicateglass [33]	11	20	0.55	900	20	0.575		0.8–1
YSZ [64]	131	80	1.638	800	300	2.05		1.8
YSZ layer on substrate [72]	78	80	0.975	800	100	1.22		
Al-Si alloy [12]	25	20	1.25	500		2.5		121–151
copper [76]	20	20	1	500	666	2	1084	400

4. Summary and Conclusions

Laser additive manufacturing has become a commanding height of future technology and industry due to its innate technical characteristics. The majority of current research efforts considering the SLM process has been carried out with CW and short laser, while ultrafast lasers exhibit better performance. This article discusses the interaction process between ultrashort pulse laser and material, and summarizes the ultrashort pulse laser forming manufacturing in the current ultrashort pulse laser forming process of four special processing fields. The process of interaction between ultrafast laser and materials reaches from the femtosecond to the nanosecond scale, i.e. from the initial energy absorption to the final material removal. Comparing fs laser sintered parts with CW laser and short laser AM parts, it was found that the defects present in fs laser AM samples were fewer compared to other types of lasers. Further fs laser produced greater hardness and higher density parts as compared to CW laser sintered parts. SLM of pure borosilicate glass parts by using ultrashort laser pulses was achieved. A three-layered structure, as a basic cell unit of SOFC, was built and demonstrated via fs AM. It showed that the density and porosity can be well controlled during the fs AM process, and that the basic SOFC cell sample functions. CW and short pulse laser failed to achieve copper parts at the micron or submicron levels because of HAZs, the use of fs laser clearly alleviated the limitation. Based on the research status, the following conclusions suggestions are required to be further considered, before the ultrashort pulse laser forming manufacturing can be practically applied into industrial use:

(1) The fs laser-based melting process enhances the material processing library of traditional techniques based on CW or short pulse lasers by adding full control over the processing parameters. In particular, challenging materials such as copper, YSZ and tungsten have been reported to produce standard functional parts that meet the requirements of many applications. In addition to the current material, materials with high melting points or brittleness that are difficult to manufacture by traditional methods, such as high-entropy alloys and ceramics, are also worth a try. The potential benefits of fs-AM can be exploited in a wide range of applications due to its high power and low HAZ. Guiding experiments and production through numerical simulation is the focus of future research. The rapid development and application of big data and large-scale parallel computing provide possibilities for process modeling and analysis. With the development of artificial intelligence technology, in the future, through data collection, data processing, machine learning and neural network methods to obtain a reliable process-structure-performance relationship for the SLM process is also an effective way to shorten the calculation time and improve the accuracy of the calculation.

(2) At present, systematic experiments are mainly carried out on laser power, scanning speed, scanning method, etc. to obtain the process parameters for the best target performance. The parameters involved are numerous, as they are related to materials, laser and the process itself, all of which can affect the performance and quality of the processed parts. This method is time-consuming and labor intensive, and the acquisition of ideal parameters depends on the setting of windows and gradients, which may not suitable as global optimal results. It may only be the local optimum in the set parameters, which also shows that the performance of manufactured parts has the potential for further improvement. As tungsten parts have been successfully fabricated with fs laser sources, microcracks are still prevalent. The next step is to continue to strengthen the process optimization research of fs SLM, especially the intelligent process optimization based on computer numerical calculation technology. While the rapid heating and cooling associated with ultrafast laser processing helps to homogenize the composition and prevent segregation, it also creates residual stress and affects the performance of the fabricated part, and SLM-based products still suffer from surface finish defects, so post-processing is needed. At the same time, expensive powder preparation and other costs are also factors that must be considered for the use of this technique.

(3) Although the current rapid development of AM, there are still problems in actual experimental operations, such as limited size of printed samples and long printing time. At the same time, there are problems such as complex and expensive powder preparation. It is also necessary to speed up the research and development of additive manufacturing equipment. At present, many of fs-SLM are based on the transformation and upgrading of traditional CW laser SLM equipment. There are many limitations, and there is still a lack of market-oriented mature printing equipment sales. The next step is to accelerate the research in this area, especially the research and development of high-performance lasers and optical path systems suitable for the fs SLM process.

Author Contributions: Conceptualization, formal analysis, original draft preparation, writing, Y.W.; review and editing, L.K. and Y.C.; investigation, Z.J. and X.L. All authors have read and agreed to the published version of the manuscript.

Funding: This research received no external funding.

Data Availability Statement: Not applicable.

References

1. Cao, L. Mesoscopic-scale numerical investigation including the influence of scanning strategy on selective laser melting process. *Comput. Mater. Sci.* **2021**, *189*, 110263. [CrossRef]
2. Lonergan, J.; Fahrenholtz, W.; Hilmas, G. Zirconium Diboride with High Thermal Conductivity. *J. Am. Ceram. Soc.* **2014**, *97*, 1689–1691. [CrossRef]
3. Gieseke, M.; Senz, V.; Vehse, M.; Fiedler, S.; Irsig, R.; Hustedt, M.; Sternberg, K.; Noelke, C.; Kaierle, S.; Wesling, V.; et al. Additive Manufacturing of Drug Delivery Systems. *Biomed. Technik. Biomed. Eng.* **2012**, *57*, 398–401. [CrossRef]
4. Chichkov, B.N.; Momma, C.; Nolte, S.; von Alvensleben, F.; Tünnermann, A. Femtosecond, picosecond and nanosecond laser ablation of solids. *Appl. Phys. A* **1996**, *63*, 109–115. [CrossRef]
5. Noël, S.; Hermann, J.; Itina, T. Investigation of nanoparticle generation during femtosecond laser ablation of metals. *Appl. Surf. Sci.* **2007**, *253*, 6310–6315. [CrossRef]
6. Chowdhury, I.; Xu, X. Heat Transfer in Femtosecond Laser Processing of Metal. *Numer. Heat Transf. Part A-Appl.* **2003**, *44*, 219–232. [CrossRef]
7. Lei, S.; Zhao, X.; Yu, X.; Hu, A.; Vukelic, S.; Jun, M.; Joe, H.-E.; Yao, Y.; Shin, Y. Ultrafast Laser Applications in Manufacturing Processes: A State of the Art Review. *J. Manuf. Sci. Eng.* **2020**, *142*, 031005. [CrossRef]
8. Birnbaum, M. Semiconductor Surface Damage Produced by Ruby Lasers. *J. Appl. Phys.* **1965**, *36*, 3688–3689. [CrossRef]

9. Hodgson, N.; Laha, M.; Lee, T.; Haloui, H.; Heming, S.; Steinkopff, A. Industrial Ultrafast Lasers Systems, Processing Fundamentals, and Applications. In Proceedings of the 2019 Conference on Lasers and Electro-Optics (CLEO), San Jose, CA, USA, 5–10 May 2019.

10. Xiao, K.; Li, M.; Li, M.; Dai, R.; Hou, Z.; Qiao, J. Femtosecond laser ablation of AZ31 magnesium alloy under high repetition frequencies. *Appl. Surf. Sci.* **2022**, *594*, 153406. [CrossRef]

11. Chen, C.; Zhang, F.; Zhang, Y.; Xiong, X.; Ju, B.-F.; Cui, H.; Chen, Y.-L. Single-pulse femtosecond laser ablation of monocrystalline silicon: A modeling and experimental study. *Appl. Surf. Sci.* **2022**, *576*, 151722. [CrossRef]

12. Roth, G.-L.; Rung, S.; Hellmann, R. Welding of transparent polymers using femtosecond laser. *Appl. Phys. A-Mater. Sci. Process.* **2016**, *122*, 86. [CrossRef]

13. Tamaki, T.; Watanabe, W.; Nishii, J.; Itoh, K. Welding of transparent materials using femtosecond laser pulses. *Jpn. J. Appl. Phys. Part 2-Lett. Express Lett.* **2005**, *44*, L687–L689. [CrossRef]

14. Li, X.; Jing, F.; Fan, J.; Yong, P.; Wang, K. Parametric investigation of femtosecond laser welding on alumina using Taguchi technique. *Optik* **2021**, *225*, 165891. [CrossRef]

15. Ullsperger, T.; Liu, D.; Yürekli, B.; Matthäus, G.; Schade, L.; Seyfarth, B.; Kohl, H.; Ramm, R.; Rettenmayr, M.; Nolte, S. Ultra-short pulsed laser powder bed fusion of Al-Si alloys: Impact of pulse duration and energy in comparison to continuous wave excitation. *Addit. Manuf.* **2021**, *46*, 102085. [CrossRef]

16. Aversa, A.; Marchese, G.; Saboori, A.; Bassini, E.; Manfredi, D.; Biamino, S.; Ugues, D.; Fino, P.; Lombardi, M. New Aluminum Alloys Specifically Designed for Laser Powder Bed Fusion: A Review. *Materials* **2019**, *12*, 1007. [CrossRef]

17. Raciukaitis, G. Ultra-Short Pulse Lasers for Microfabrication: A Review. *IEEE J. Sel. Top. Quantum Electron.* **2021**, *27*, 1–12. [CrossRef]

18. von der Linde, D.; Sokolowski-Tinten, K.; Bialkowski, J. Laser–solid interaction in the femtosecond time regime. *Appl. Surf. Sci.* **1997**, *109–110*, 1–10. [CrossRef]

19. Kabotu, K.; Samuel, G.L.; Shunmugam, M.s. Theoretical and experimental Investigations of Ultra-short Pulse Laser Interaction on Ti6Al4V alloy. *J. Mater. Process. Technol.* **2018**, *263*, 266–275. [CrossRef]

20. Wu, B.; Shin, Y. A simple model for high fluence ultra-short pulsed laser metal ablation. *Appl. Surf. Sci.* **2007**, *253*, 4079–4084. [CrossRef]

21. Wellershoff, S.; Hohlfeld, J.; Güdde, J.; Matthias, E. The role of electron-phonon coupling in femtosecond laser damage of metals. *Appl. Phys. A* **1999**, *69*, S99–S107. [CrossRef]

22. Li, X.; Guan, Y. Theoretical fundamentals of short pulse laser–metal interaction: A review. *Nanotechnol. Precis. Eng.* **2020**, *3*, 105–125. [CrossRef]

23. Corkum, P.B.; Brunel, F.; Sherman, N.K.; Srinivasan-Rao, T. Thermal Response of Metals to Ultrashort-Pulse Laser Excitation. *Phys. Rev. Lett.* **1988**, *61*, 2886–2889. [CrossRef] [PubMed]

24. Gusarov, A.V.; Yadroitsev, I.; Bertrand, P.; Smurov, I. Model of Radiation and Heat Transfer in Laser-Powder Interaction Zone at Selective Laser Melting. *J. Heat Transf.* **2009**, *131*, 072101. [CrossRef]

25. Le Harzic, R.; Breitling, D.; Weikert, M.; Sommer, S.; Föhl, C.; Valette, S.; Donnet, C.; Audouard, E.; Dausinger, F. Pulse width and energy influence on laser micromachining of metals in a range of 100 fs to 5 ps. *Appl. Surf. Sci.* **2005**, *249*, 322–331. [CrossRef]

26. Yang, J.; Zhao, Y.; Zhu, X. Theoretical studies of ultrafast ablation of metal targets dominated by phase explosion. *Appl. Phys. A* **2007**, *89*, 571–578. [CrossRef]

27. Rethfeld, B.; Ivanov, D.S.; Garcia, M.E.; Anisimov, S.I. Modelling ultrafast laser ablation. *J. Phys. D: Appl. Phys.* **2017**, *50*, 193001. [CrossRef]

28. Anisimov, S.I.; Relhfeld, B. On the theory of ultrashort laser pulse interaction with a metal. *Nauk Fiz* **1997**, *61*, 1642–1655.

29. Wang, X.L.; Wu, H.; Chang, Y.X.; Zhu, W.H.; Chen, Z.Y.; Lu, P.X. Influence of thermophysical parameters by femtosecond laser ablation of metals: Molecular dynamics simulation. *Guangzi Xuebao/Acta Photonica Sin.* **2009**, *38*, 3052–3056.

30. Iartsev, B.; Vichev, I.; Tsygvintsev, I.; Sidelnikov, Y.; Krivokorytov, M.; Medvedev, V. On experimental and numerical study of the dynamics of a liquid metal jet hit by a laser pulse. *Exp. Fluids* **2020**, *61*, 119. [CrossRef]

31. Zhu, G.; Xu, Z.; Jin, Y.; Chen, X.; Yang, L.; Xu, J.; Shan, D.; Chen, Y.; Guo, B. Mechanism and application of laser cleaning: A review. *Opt. Lasers Eng.* **2022**, *157*, 107130. [CrossRef]

32. Kerse, C.; Kalaycioglu, H.; Elahi, P.; Cetin, B.; Kesim, D.K.; Akcaalan, O.; Yavas, S.; Asik, M.D.; Oktem, B.; Hoogland, H.; et al. Ablation-cooled material removal with ultrafast bursts of pulses. *NATURE* **2016**, *537*, 84–88. [CrossRef] [PubMed]

33. Seyfarth, B.; Schade, L.; Ullsperger, T.; Matthäus, G.; Tünnermann, A.; Nolte, S. Selective Laser Melting of Borosilicate Glass Using Ultrashort Laser Pulses. Available online: https://www.spiedigitallibrary.org/conference-proceedings-of-spie/10523/105230C/Selective-laser-melting-of-glass-using-ultrashort-laser-pulses/10.1117/12.2289614.short?SSO=1 (accessed on 13 October 2022).

34. Stuart, B.C.; Feit, M.D.; Herman, S.; Rubenchik, A.M.; Shore, B.W.; Perry, M.D. Nanosecond-to-femtosecond laser-induced breakdown in dielectrics. *PHYSICAL REVIEW B* **1996**, *53*, 1749–1761. [CrossRef] [PubMed]

35. Lizunov, S.A.; Bulgakov, A.V.; Campbell, E.E.B.; Bulgakova, N.M. Melting of gold by ultrashort laser pulses: Advanced two-temperature modeling and comparison with surface damage experiments. *Appl. Phys. A* **2022**, *128*, 602. [CrossRef]

36. Xie, J.; Yan, J.; Zhu, D. Atomic simulation of irradiation of Cu film using femtosecond laser with different pulse durations. *J. Laser Appl.* **2020**, *32*, 022016. [CrossRef]

37. Zhou, Y.; Wu, D.; Luo, G.; Hu, Y.; Qin, Y. Efficient modeling of metal ablation irradiated by femtosecond laser via simplified two-temperature model coupling molecular dynamics. *J. Manuf. Process.* **2022**, *77*, 783–793. [CrossRef]

38. Colombier, J.P.; Combis, P.; Bonneau, F.; Le Harzic, R.; Audouard, E. Hydrodynamic simulations of metal ablation by femtosecond laser irradiation. *Phys. Rev. B* **2005**, *71*, 165406. [CrossRef]

39. Qiu, T.Q.; Tien, C.L. Heat-Transfer Mechanisms during Short-Pulse Laser-Heating of Metals. *J. Heat Transf. -Trans. Same* **1993**, *115*, 835–841. [CrossRef]

40. Zhu, D.; Yan, J.; Xie, J.; Liang, Z.; Bai, H. Ultrafast Laser-Induced Atomic Structure Transformation of Au Nanoparticles with Improved Surface Activity. *ACS Nano* **2021**, *15*, 13140–13147. [CrossRef]

41. Xie, J.; Yan, J.; Zhu, D.; He, G. Atomic-Level Insight into the Formation of Subsurface Dislocation Layer and Its Effect on Mechanical Properties During Ultrafast Laser Micro/Nano Fabrication. *Adv. Funct. Mater.* **2022**, *32*, 2108802. [CrossRef]

42. Meng, Y.; Ji, P.; Jiang, L.; Lin, G.; Guo, J. Spatiotemporal insights into the femtosecond laser homogeneous and heterogeneous melting aluminum by atomistic-continuum modeling. *Appl. Phys. A* **2022**, *128*, 520. [CrossRef]

43. Panwisawas, C.; Qiu, C.; Anderson, M.J.; Sovani, Y.; Turner, R.P.; Attallah, M.M.; Brooks, J.W.; Basoalto, H.C. Mesoscale modelling of selective laser melting: Thermal fluid dynamics and microstructural evolution. *Comput. Mater. Sci.* **2017**, *126*, 479–490. [CrossRef]

44. Khairallah, S.; Anderson, A.; Rubenchik, A.; King, W. Laser powder-bed fusion additive manufacturing: Physics of complex melt flow and formation mechanisms of pores, spatter, and denudation zones. *Acta Mater.* **2016**, *108*, 36–45. [CrossRef]

45. Yang, M.; Wang, L.; Yan, W. Phase-field modeling of grain evolution in additive manufacturing with addition of reinforcing particles. *Addit. Manuf.* **2021**, *47*, 102286. [CrossRef]

46. Kruth, J.-P.; Froyen, L.; Vaerenbergh, J.; Mercelis, P.; Rombouts, M.; Lauwers, B. Selective laser melting of iron-based powder. *J. Mater. Process. Technol.* **2004**, *149*, 616–622. [CrossRef]

47. Lassner, E. Tungsten: Properties, Chemistry, Technology of the Element, Alloys, and Chemical Compounds. 2000. Available online: https://onlinelibrary.wiley.com/doi/10.1002/14356007.a27_229 (accessed on 13 October 2022).

48. Duriagina, Z.; Kulyk, V.; Kovbasiuk, T.; Vasyliv, B.; Kostryzhev, A. Synthesis of Functional Surface Layers on Stainless Steels by Laser Alloying. *Metals* **2021**, *11*, 434. [CrossRef]

49. Zhou, Z.; Pintsuk, G.; Linke, J.; Hirai, T.; Rödig, M.; Ma, Y.; Ge, C. Transient high heat load tests on pure ultra-fine grained tungsten fabricated by resistance sintering under ultra-high pressure. *Fusion Eng. Des.* **2010**, *85*, 115–121. [CrossRef]

50. Ma, J.; Zhang, J.; Liu, W.; Shen, Z. Suppressing pore-boundary separation during spark plasma sintering of tungsten. *J. Nucl. Mater.* **2013**, *438*, 199–203. [CrossRef]

51. Wang, D.; Yu, C.; Zhou, X.; Ma, J.; Liu, W.; Shen, Z. Dense Pure Tungsten Fabricated by Selective Laser Melting. *Appl. Sci.* **2017**, *7*, 430. [CrossRef]

52. Ebert, R.; Ullmann, F.; Hildebrandt, D.; Schille, J.; Hartwig, L.; Klötzer, S.; Streek, A.; Exner, H. Laser Processing of Tungsten Powder with Femtosecond Laser Radiation. *JLMN-J. Laser Micro/Nanoeng.* **2012**, *7*, 38–43. [CrossRef]

53. Zhang, D.; Cai, Q.; Liu, J. Formation of Nanocrystalline Tungsten by Selective Laser Melting of Tungsten Powder. *Mater. Manuf. Process.* **2012**, *27*, 1267–1270. [CrossRef]

54. Zhou, X.; Liu, X.; Zhang, D.; Shen, Z.; Liu, W. Balling phenomena in selective laser melted tungsten. *J. Mater. Process. Technol.* **2015**, *222*, 33–42. [CrossRef]

55. Bai, S.; Liu, J.; Yang, P.; Huang, H.; Yang, L.-M. Femtosecond fiber laser additive manufacturing of tungsten. *Laser 3d Manuf. III* **2016**, *9738*, 96–105.

56. Bai, S.; Yang, L.; Liu, J. Manipulation of microstructure in laser additive manufacturing. *Appl. Phys. A* **2016**, *122*, 495. [CrossRef]

57. Zhang, H.; Zhao, Y.; Huang, S.; Zhu, S.; Wang, F. Manufacturing and Analysis of High-Performance Refractory High-Entropy Alloy via Selective Laser Melting (SLM). *Materials* **2019**, *12*, 720. [CrossRef] [PubMed]

58. Komatsu, T.; Honma, T. Laser patterning and growth mechanism of orientation designed crystals in oxide glasses: A review. *J. Solid State Chem.* **2019**, *275*, 210–222. [CrossRef]

59. Minghui, H.; Koji, S.; Yongfeng, L.; Katsumi, M.; Tow Chong, C. Optical Diagnostics in Laser-Induced Plasma-Assisted Ablation of Fused Quartz. Available online: https://www.spiedigitallibrary.org/conference-proceedings-of-spie/4088/0000/Optical-diagnostics-in-laser-induced-plasma-assisted-ablation-of-fused/10.1117/12.405763.short?SSO=1 (accessed on 13 October 2022).

60. Xia, Y.; Jing, X.; Zhang, D.; Wang, F.; Jaffery, S.; Li, H. A comparative study of direct laser ablation and laser-induced plasma-assisted ablation on glass surface. *Infrared Phys. Technol.* **2021**, *115*, 103737. [CrossRef]

61. Seyfarth, B.; Schade, L.; Matthäus, G.; Ullsperger, T.; Heidler, N.; Hilpert, E.; Nolte, S. Laser powder bed fusion of glass: A comparative study between CO2 lasers and ultrashort laser pulses. *Laser 3D Manuf. VII* **2020**, *11271*, 87–92.

62. Liu, J.; Deng, C.; Bai, S. Glass surface metal deposition with high-power femtosecond fiber laser. *Appl. Phys. A* **2016**, *122*, 1064. [CrossRef]

63. Huang, W.L.; Zhu, Q.; Xie, Z. Gel-cast anode substrates for solid oxide fuel cells. *J. Power Sources* **2006**, *162*, 464–468. [CrossRef]

64. Liu, J.; Bai, S. Femtosecond laser additive manufacturing of YSZ. *Appl. Phys. A* **2017**, *123*, 293. [CrossRef]

65. Zhang, Z.; Yang, Z.; Wang, C.; Zhang, Q.; Zheng, S.; Xu, W. Mechanisms of femtosecond laser ablation of Ni3Al: Molecular dynamics study. *Opt. Laser Technol.* **2021**, *133*, 106505. [CrossRef]

66. Huang, S.; Liu, P.; Mokasdar, A.; Liang, H. Additive manufacturing and its societal impact: A literature review. *Int. J. Adv. Manuf. Technol.* **2012**, *67*, 1191–1203. [CrossRef]

67. Bandyopadhyay, A.; Heer, B. Additive manufacturing of multi-material structures. *Mater. Sci. Eng. R Rep.* **2018**, *129*, 1–16. [CrossRef]

68. Cui, J.; Ren, L.; Mai, J.; Zheng, P.; Zhang, L. 3D Printing in the Context of Cloud Manufacturing. *Robot. Comput. -Integr. Manuf.* **2022**, *74*, 102256. [CrossRef]

69. Heemsbergen, L.; Daly, A.; Lu, J.; Birtchnell, T. 3D-printed Futures of Manufacturing, Social Change and Technological Innovation in China and Singapore: The Ghost of a Massless Future? *Sci. Technol. Soc.* **2019**, *24*, 254–270. [CrossRef]

70. Ullsperger, T.; Matthäus, G.; Kaden, L.; Engelhardt, H.; Rettenmayr, M.; Risse, S.; Tünnermann, A.; Nolte, S. Selective laser melting of hypereutectic Al-Si40-powder using ultra-short laser pulses. *Appl. Phys. A* **2017**, *123*, 798. [CrossRef]

71. Gorsse, S.; Hutchinson, C.; Goune, M.; Banerjee, R. Additive manufacturing of metals: A brief review of the characteristic microstructures and properties of steels, Ti-6Al-4V and high-entropy alloys. *Sci. Technol. Adv. Mater.* **2017**, *18*, 584–610. [CrossRef]

72. Bai, S.; Liu, J. Femtosecond Laser Additive Manufacturing of Multi-Material Layered Structures. *Appl. Sci.* **2020**, *10*, 979. [CrossRef]

73. Dursun, T.; Soutis, C. Recent developments in advanced aircraft aluminium alloys. *Mater. Des.* **2014**, *56*, 862–871. [CrossRef]

74. Grigoriev, S.; Tarasova, T.; Gvozdeva, G. Optimization of Parameters of Laser Surfacing of Alloys of the Al–Si System. *Met. Sci. Heat Treat.* **2016**, *57*, 589–595. [CrossRef]

75. Cheng, C.; Chen, J. Femtosecond laser sintering of copper nanoparticles. *Appl. Phys. A* **2016**, *122*, 289. [CrossRef]

76. Kaden, L.; Matthäus, G.; Ullsperger, T.; Engelhardt, H.; Rettenmayr, M.; Tünnermann, A.; Nolte, S. Selective laser melting of copper using ultrashort laser pulses. *Appl. Phys. A* **2017**, *123*, 596. [CrossRef]

77. Kaden, L.; Seyfarth, B.; Ullsperger, T.; Matthäus, G.; Nolte, S. Selective Laser Melting of Copper Using Ultrashort Laser Pulses at Different Wavelengths. 2018, p. 41. Available online: https://www.semanticscholar.org/paper/Selective-laser-melting-of-copper-using-ultrashort-Kaden-Matth%C3%A4us/073040f72798f8047864da3ee38cd1769c61319f#citing-papers (accessed on 13 October 2022). [CrossRef]

78. Nie, B.; Huang, H.; Bai, S.; Liu, J. Femtosecond laser melting and resolidifying of high-temperature powder materials. *Appl. Phys. A* **2015**, *118*, 37–41. [CrossRef]

79. Nie, B.; Yang, L.; Huang, H.; Bai, S.; Wan, P.; Liu, J. Femtosecond laser additive manufacturing of iron and tungsten parts. *Appl. Phys. A* **2015**, *119*, 1075–1080. [CrossRef]

Article

Tribological Performance of Microcrystalline Diamond (MCD) and Nanocrystalline Diamond (NCD) Coating in Dry and Seawater Environment

Hui Zhang [1,2], Hui Song [2,*], Ming Pang [3,*], Guoyong Yang [2], Fengqin Ji [3,4], Nan Jiang [2,*] and Kazuhito Nishimura [5]

[1] College of Transportation Science and Engineering, Civil Aviation University of China, Tianjin 300300, China
[2] Key Laboratory of Marine Materials and Related Technologies, Zhejiang Key Laboratory of Marine Materials and Protective Technologies, Ningbo Institute of Materials Technology and Engineering (NIMTE), Chinese Academy of Sciences, Ningbo 315201, China
[3] College of Aeronautical Engineering, Civil Aviation University of China, Tianjin 300300, China
[4] State Key Laboratory of Solid Lubrication, Lanzhou Institute of Chemical Physics, Chinese Academy of Sciences, Lanzhou 730000, China
[5] Advanced Nano-Processing Engineering Lab, Mechanical Engineering, Kogakuin University, Tokyo 192-0015, Japan
* Correspondence: songhui@nimte.ac.cn (H.S.); pangming1980@126.com (M.P.); jiangnan@nimte.ac.cn (N.J.)

Abstract: In the present study, the tribological properties of diverse crystalline diamond coating with micro (MCD) and nanometer (NCD) sizes, fabricated by the microwave plasma chemical vapor deposition (MPCVD) method, are systematically investigated in dry and seawater environments, respectively. Owing to the SiO_2 lubricating film with extraordinary hydrophilicity performance by a tribochemical reaction, the average friction coefficient (COF) and wear rate of NCD coating under seawater decreased by 37.8% and 26.5%, respectively, comparing with in dry conditions. Furthermore, graphite would be generated with the increment of surface roughness. Graphite transformed from the diamond under high contact pressure. Thus, with the synergism between SiO_2 lubricating film with extraordinary hydrophilicity performance and graphite, the corresponding COF and wear rate of MCD would be further decreased by up to 64.1% and 39.5%. Meanwhile, various characterizations on morphology, spectra, and tribological performance of the deposited diamond coating were conducted to explore the in-depth mechanism of the enhanced tribological performance of our NCD and MCD coatings in the extreme under seawater working conditions. We envision this work would provide significant insights into the wear behavior of diamond coatings in seawater and broaden their applications in protective coatings for marine science.

Keywords: diamond coating; tribological performance; seawater environment; lubricating film

Citation: Zhang, H.; Song, H.; Pang, M.; Yang, G.; Ji, F.; Jiang, N.; Nishimura, K. Tribological Performance of Microcrystalline Diamond (MCD) and Nanocrystalline Diamond (NCD) Coating in Dry and Seawater Environment. *Crystals* 2022, 12, 1345. https://doi.org/10.3390/cryst12101345

Academic Editor: Hao Yi

Received: 30 August 2022
Accepted: 19 September 2022
Published: 23 September 2022

Publisher's Note: MDPI stays neutral with regard to jurisdictional claims in published maps and institutional affiliations.

1. Introduction

Chemical vapor deposition (CVD) polycrystalline diamond coatings have excellent mechanical properties, high thermal conductivity, and superior chemical inertness [1–3], which are widely applied in various fields [4–8]. Specially, polycrystalline diamond coatings as surface protection coatings possess a low coefficient of friction (COF), outstanding wear resistance, and high chemical stability, which shows great promise for applications in multifactor complex environment [9–11]. Moreover, polycrystalline diamond coatings with different morphologies and structures can be prepared by choosing the different deposition conditions of the CVD process such as the ratio of the reactive gas, the temperature, and the pressure of the substrate [12–15]. For example, the MCD coating deposited by CVD has excellent wear resistance and can be used in many fields, but its high surface roughness may lead to a high friction coefficient during the process of wear; the NCD coating has the characteristics of continuous film surface smoothness and other characteristics, which have good anti-friction performance during the process of wear [16]. Currently, scientists have also conducted many in-depth studies on friction and wear properties of MCD and NCD coatings in some typical environments. Lei et al. [17] found that the friction coefficient of MCD and NCD after dry sliding was stable in the range of 0.053–0.062, whereas COF decreased to the range of 0.023–0.025 under water lubrication; De Barros Bouchet M I et al. [18] found that the ultra-low friction of NCD in the presence of water vapor was associated with OH and H passivation of the sliding surface, hydrogen-passivated surfaces were generated by dissociative adsorption of hydrogen molecules and were more effective than OH-terminated surfaces in further reducing friction. Cui et al. [19] investigated the tribological properties of fabricated polished MCD (MCD-p) films using a ball-plate reciprocating friction test with Si_3N_4 ceramic balls as counterparts under water lubrication, the grown MCD films, NCD films, and Si_3N_4 plates were used as comparison samples. MCD, NCD, and MCD-p films showed similar stable COF states after running in, which were 0.036, 0.032, and 0.035, respectively, and the wear rate of Si_3N_4 on MCD-p specimens was 2–3 orders of magnitude lower than the wear rate of sliding on grown MCD or NCD specimens. Feng et al. [20] studied that the friction coefficient of a diamond sliding in an ultra-high vacuum (UHV) (-4×10^{-9} Torr) was between 0.6 and 1.0, which was about ten times that was measured in air. However, in molecular gases at low pressure (-1×10^{-5} Torr), only oxygen leads to a significant reduction in the friction coefficient. Perle M et al. [6] deposited MCD coatings by a hot filament chemical vapor deposition process (HFCVD) and studied the friction properties under various atmospheres (oxygen, argon, nitrogen, ambient air); the displayed value of the coefficient of friction was in the range of 0.1 to 0.4. Lin et al. [21] investigated the frictional behavior of MCD and ultrananocrystalline diamond (UNCD) films at high temperatures, the results demonstrated that the wear of diamond film at 25–500 °C was mainly caused by the strong adhesion between diamond and substrate, and the wear resistance of UNCD film was significantly reduced by oxygen corrosion above 600 °C, whereas the MCD film still maintained a certain wear resistance at 700 °C. Furthermore, we have conducted a summary of current research into the friction and corrosion of several surface protective coatings in seawater. Wang et al. [22–24] have investigated the friction and corrosion mechanisms of Cr/GLC multilayered coating, 316 L stainless steel, and Al/Ti co-doped diamond-like carbon films in seawater. Ou et al. [25] have reported the superhard yet tough Ti-C-N coatings that showed excellent seawater-lubricating performance with an extremely low friction coefficient of 0.03 and a mild wear track in 3.5 wt% NaCl aqueous solution. Nonetheless, the current research on the friction and wear of diamond as a protective coating mainly focuses on the friction and wear behavior and related friction mechanism in a dry friction environment, water environment, even high-temperature environment, vacuum, low-pressure environment, and various atmospheric environment, and the friction and corrosion of diamond coatings have not studied in seawater. In fact, diamond has great application advantages as an anti-corrosion and wear-resistant material due to its high chemical stability. With the rapid development of marine equipment and marine technology, there is a high demand for surface protection materials of parts with high bearing capacity and sea-related friction, which requires that the protective material not only has good load-bearing capacity, but also maintains good lubricating properties under the boundary and fluid lubrication in the presence of seawater in order to ensure a satisfactory service life under severe working conditions [24–27]. However, the research on the tribological behavior of a diamond coating in a seawater environment has not been involved, and there is no research on the wear mechanism of common MCD and NCD coatings in a seawater environment. Considering the practical service condition will face a complex friction situation, in this study, both the tribological performance and corresponding friction mechanism of a typical diamond film in dry and seawater conditions have been studied exhaustively.

Bearing this aspect in mind, in this study, MCD and NCD coatings were prepared by MPCVD, we compared the friction and wear behavior of MCD and NCD coatings in a dry friction environment and simulated a seawater environment by carrying the friction test on a reciprocating ball-disk friction tester. Subsequently, we have systematically analyzed the surface morphology, chemical composition changes, and phase evolution caused by sliding action in the worn areas, and thus revealed the frictional wear mechanism of different diamond coatings in a seawater environment.

2. Materials and Methods

2.1. Preparation of Diamond Coatings

In this experiment, we employed microwave plasma chemical vapor deposition (MPCVD, HMPS-2050, Provided by Ningbo Institute of Materials Technology and Engineering, Chinese Academy of Sciences, China) to grow the diamond coating. Single-side polished N (100) Si wafers (Bought from Zhejiang Lijing Silicon Material Co., Ltd., Jiaxing, China) whose thickness was 2 mm and an area of 2×2 cm^2 were used as substrates. Before the deposition process, the substrate was sequentially placed in acetone (Purchased from Sinopharm Chemical Reagent Co., Ltd., Shanghai, China) and alcohol (Bought from TAICANG XINTAI ALCOHOL CO., LTD., Suzhou, China) and sonicated for 5 min each. Afterward, two-step pretreatment was introduced on the substrates during the deposition to improve the nucleation density of diamond coatings. Firstly, the substrate specimens were sanded with sandpaper (2000 grit, Bought from SUISUN CO., LTD., Qingdao, China). Secondly, as shown in Figure 1a, the substrate specimens were seeded by ultrasonic oscillation with 200 nm diamond powder (Purchased from Zhongyuan Super Abrasives Co., Ltd., Zhengzhou, China) mixed with anhydrous ethanol solution in the ultrasonic device (KQ-100DE, KUNSHAN ULTRASONIC INSTRUMENT CO., LTD., Kunshan, China) for 10 min. After the seed crystal was completed, the surface was cleaned again by ultrasonic cleaned with anhydrous ethanol for 2 min. Thereafter, the treated substrate was dried with nitrogen gas and set aside in a dry place. Subsequently, as can be seen from Figure 1b, a microwave plasma chemical vapor deposition (MPCVD) with a rated power of 5 KW apparatus was applied for the MCD and NCD coatings deposition. The preparation parameters of MCD and NCD coatings are shown in Table 1. The MCD and NCD coatings were deposited with a ratio of CH$_4$/H$_2$ = 12/400 and a ratio of CH$_4$/H$_2$ = 20/400, respectively. A high gas pressure of 12 KPa was used during the MCD coatings deposition, and a low gas pressure of 9 KPa was used for the NCD coating process. In addition, an infrared thermometer (CHINO, IR-AH, Chino, Japan) was used to observe the deposition process of microwave chemical vapor deposition and measure the substrate temperature during the experiment. The substrate temperature of the MCD coatings was maintained at 900 ± 20 °C during the deposition growth, whereas the substrate temperature of the NCD coatings was kept at 670 ± 20 °C. The deposition time of MCD coating was 4 h and the deposition time of NCD coating was 9 h.

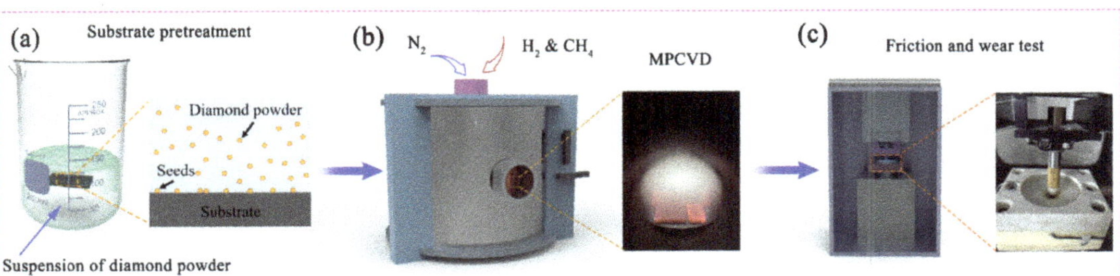

Figure 1. (**a**) Pretreatment of the substrate; (**b**) Preparation of diamond coating; (**c**) Testing of tribological properties.

Table 1. Synthesis parameters of MCD and NCD coatings.

	CH$_4$/sccm	H$_2$/sccm	N$_2$/sccm	Pressure/KPa	Power/KW	Substrate Temperature/°C
MCD	12	400	–	12	4.4	900 ± 20
NCD	20	400	20	9	3	670 ± 20

2.2. Friction and Wear Test

The tribological performance of the diamond coatings was carried out on ball-to-disk friction and wear tester (Rtec, Rtec Instruments, San Jose, CA, USA) configuration in reciprocating sliding mode. The test condition was at room temperature under a relative humidity of 50~80%. Then, the tribotests of MCD and NCD coatings were performed in the simulated sea environment (3.5 wt.% NaCl solution) and dry environment, respectively. SiC balls (Φ6 mm) were used as friction pairs after ultrasonic cleaned with ethanol and dried in air. The applied load was fixed at 15 N and the average sliding velocity of 20 mm/s with a stroke length of 4 mm for all tests, the test period was 5 h, and the friction coefficient of diamond coatings was recorded automatically by the friction tester. In order to reduce errors, every test was conducted at least three times. The schematic diagram of the tribological test is shown in Figure 1c

2.3. Characterization of Diamond Coatings

In order to obtain the precise surface morphology and structural information of as-prepared coatings, a scanning electron microscopy (SEM, Regulus8230, HITAGHI, Tokyo, Japan) at an accelerated voltage of 10 kV was used as characterization method to analyze the structural information of diamond coatings. The Raman spectrum of diamond coatings was measured by Raman spectrum (Raman, LabRAM Odyssey, HORIBA FRANCE SAS, Kyoto, Japan) with a wavelength of 532 nm. Atomic mechanics microscopy (AFM, Dimension 3100, Veeco, San Jose, CA, USA) was used to calculate the average roughness of MCD and NCD coatings. The cross-sectional transmission electron microscopy (TEM) sample was prepared on worn scars deposited at 960 °C using focused ion beam microscopy (FIB, Helios 5 CX, Thermo Fisher Scientific, Waltham, MA, USA). The TEM sample was cut along the direction perpendicular to the worn scars sliding. The microstructure of diamond and EDS analysis of lubricating film of the cross-sections of worn scars after the friction test were obtained by high-resolution transmission electron microscopy (HRTEM, Talos F200X, Thermo Fisher Scientific, Waltham, MA, USA) at 200 kV detailly. X-ray diffraction (XRD, D8 Advance, BRUKER, Madison, WI, USA) with a scanning rate of 8°/min was employed to determine the crystal structures of diamond coatings in the 10°–90° range (2θ). The contact angle of a seawater droplet on the diamond surface was measured by a contact angle measurement apparatus (DCAT21, Ningbo Jinmao Import and Export Co., Ltd., Ningbo, China) at a relative humidity of 50~80%. The chemical composition of MCD and NCD coatings before and after the friction test was determined using X-ray photoelectron spectroscopy (XPS, AXIS SUPRA, Kratos, Manchester, UK). A Revetest Scratch Test System (CSM Revetest, Ningbo Jinmao Import & Export Co., Ningbo, China) was conducted to test the adhesion of diamond coatings. A diamond tip with an applied normal load was increased from 0 N to 80 N, a scratching speed of 1 mm/min, and a scratching length was 5 mm. In order to reduce errors, all experiments were repeated at least three times.

3. Results

3.1. Characterization of As-Deposited Diamond Coatings

The surface and cross-sectional morphologies of as-deposited MCD and NCD coatings are observed by SEM, which are shown in Figure 2. Figure 2a shows the MCD coating has well-faceted crystallites exhibiting a rough surface, and the crystalline grain size of the MCD coating is about 10 μm. Unlike MCD coating, Figure 2d shows the NCD coating that was composed of small diamond particles which exhibited a smoother surface. As shown in Figure 2e, grains on the NCD coating surface clustered and formed a typical cauliflower appearance, which could be clearly observed by SEM. Additionally, it can be clearly seen that continuous diamond coatings were deposited on the substrates, and the thicknesses of MCD and NCD coatings were 7.78 μm and 7.65 μm, respectively. The similar thicknesses of diamond coatings can eliminate the influence of the thickness on the properties of tribological performance. The size of NCD coatings cannot be distinguished individually from the SEM image due to the small grain size. We further analyzed the XRD pattern of the diamond coating surface as shown in Figure 2g. The average grain size of NCD can be determined using the (111) diffraction peak in the XRD pattern [28]. The diamond (111) peak of the NCD coating in Figure 2g was located at 2θ = 44.68°, according to Scherrer formula, and the average grain size of NCD coating was calculated at half of the maximum value of the diamond (111) peak [29,30]; therefore, the average grain size was calculated to be 20.8 nm. Figure 2h,i showed the Raman spectra of MCD and NCD coating, respectively. In the Raman spectrogram of the diamond coating, it can be clearly seen that there was a distinct peak at 1332 cm^{-1} all over, which proved the presence of sp^3 diamond phase. Furthermore, in the Raman spectrum of NCD coating, in addition to the characteristic peak of diamond out of 1332 cm^{-1}, there were also peaks of Raman shift at 1192.23 cm^{-1}, 1469.35 cm^{-1}, and 1556.27 cm^{-1}, respectively. The peaks 1192.23 cm^{-1} and 1469.35 cm^{-1} indicated the presence of the trans-polyacetylene (t-PA) phase, which was characteristic of nanoscale diamond particles [3,31–33], corresponding to the peak at 1556.27 cm^{-1} that was ascribed to graphite. This was attributed to the higher CH_4/H_2 concentration ratio and lower reaction gas pressure in the NCD coating deposition, and this raised the level of sp^2 impurities in the membrane [30]. However, in Raman spectroscopy, the sensitivity of graphite to Raman signals was tens of times higher than that of diamond [17]; therefore, it can be concluded that the diamond phase dominated in both coatings.

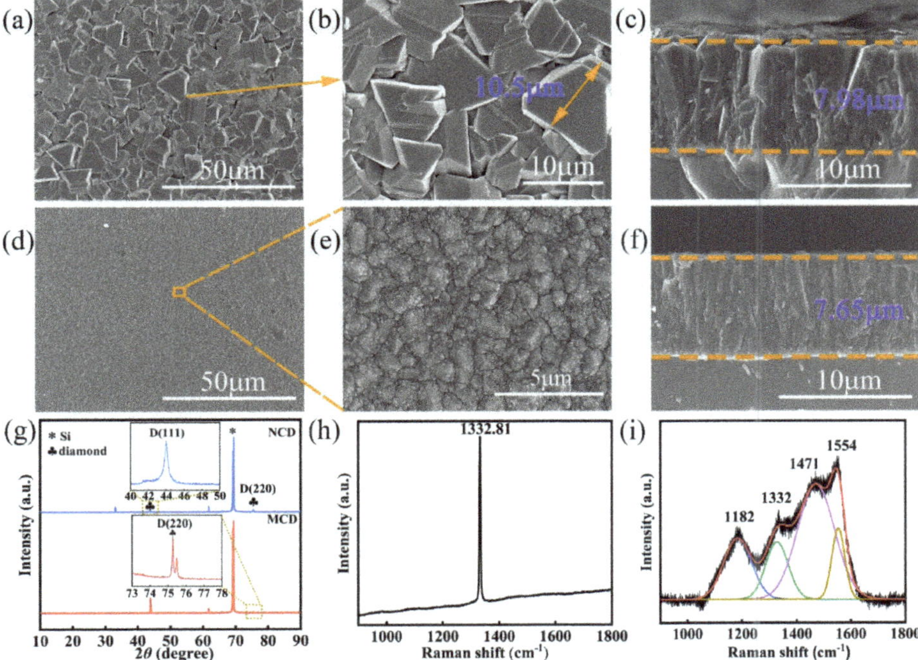

Figure 2. The surface and cross-sectional morphologies of diamond coatings: (**a–c**) MCD; (**d–f**) NCD; (**g**) The XRD pattern of diamond coatings; (**h**) The Raman spectrum of MCD coatings; (**i**) The Raman spectrum of NCD coatings.

The AFM topographies of different diamond films were acquired by AFM with a scanning range of 10 μm × 10 μm. The average roughness (Ra) of the coating was calculated by the software NanoScope Analysis. As illustrated in Figure 3a, the surface Ra of the MCD coating with obvious crystal spikes was calculated to be 248 nm. Compared with the MCD coating, the surface morphology of the NCD coating was relatively smooth, and the surface Ra of the NCD coating was calculated to be 51.9 nm. Figure 3b shows the wetting angle of seawater on MCD and NCD coatings. A surface is hydrophobic when its static wetting angle θ is >90° and hydrophilic when θ is <90° [34]. The wetting angle of seawater on NCD coating was 82.26°, whereas the MCD coating was 94.48°. Therefore, the NCD coating exhibited hydrophilic properties and the MCD coating exhibited hydrophobic properties. Figure 3c showed the scratch morphology of the MCD and NCD coatings as observed by the scratch test system. The first critical load when the film was peeled from the substrate can be used to assess the adhesion properties of the film [35]. The location of the MCD coating and the NCD coating peeled off from the substrate corresponded to loads of 73.1 N and 47.2 N, respectively. The more detailly Scratch morphologies of diamond are shown in Figure S1. This indicated that the adhesion of the MCD coating was better than NCD coating, which was consistent with the other works reported [36]. Furthermore, compared with other reports [36,37], our prepared MCD and NCD coatings have good adhesion properties.

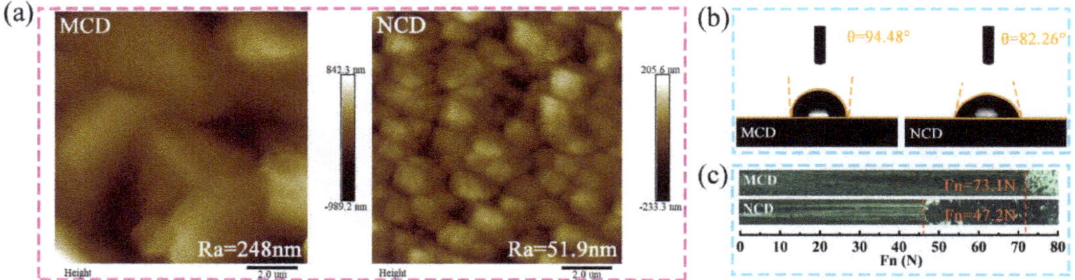

Figure 3. (**a**) The AFM three-dimensional morphologies of diamond coatings; (**b**) The wetting angle of the sea water on the diamond coatings; (**c**) Scratch morphologies of diamond coatings.

3.2. Tribological Performance of MCD and NCD Coatings

The COF of MCD and NCD coatings under different test conditions are given in Figure 4a,b, respectively. The results showed that the friction coefficient of diamond coating in the seawater environment was lower than in the dry friction environment. In Figure 4a,b, the friction coefficient has occurred obvious changes under the seawater environment. We compared the change in the average coefficient of friction for the first ten minutes and the last ten minutes of the wear process. For the MCD coating, in the dry environment, the coefficient of friction has changed from 0.3303 to 0.1509, and the friction coefficient of NCD coating has changed from 0.3343 to 0.0875. Furthermore, the coefficient of friction has changed from 0.1389 to 0.0360, and the friction coefficient of NCD coating has changed from 0.3304 to 0.0358 in seawater. This was contributed to by the seawater which can provide a certain thickness of the liquid film for the friction pairs in seawater lubrication conditions [17]; therefore, the roughness of the diamond coating surface was greatly reduced during the wear process and the actual contact area of the friction pair was minimized, thus reducing the coefficient of friction. In addition, the rough structure of the hydrophobic surface of the MCD coating may have also improved the load-bearing capacity of the friction pair, and reduced the friction coefficient. During hydrodynamic lubrication, the hydrodynamic effect was generated to achieve the synergistic effect of lubrication and anti-wear [38]. As shown in Figure 4a,b, the friction coefficient of MCD and NCD coatings showed a similar pattern of variation: the coefficient of friction rose rapidly during the initial period, which was due to the abrasive wear of ceramic balls rubbing on the diamond coating and the plowing of microconvex bodies on the surface. Subsequently, all friction coefficients dropped rapidly to a steady state after a short break-in period [39]. Furthermore, compared with NCD coatings, MCD coatings reached the stabilization phase of COF in the seawater environment more quickly. In general, the diamond coatings have a lower COF in a seawater environment than in a dry friction environment, which may be related to the fluid lubrication effect in the liquid environment. In particular, as shown in Figure 4c, the average friction coefficient of MCD under dry friction environment was 0.1310, whereas the average friction coefficient of NCD coating was 0.1001. The MCD coatings have a higher COF in a dry friction environment due to the relatively rougher surface profile of MCD coatings resulting in a higher coefficient of friction. However, the friction coefficient in seawater environment showed an overall lower trend, the average friction coefficients of MCD and NCD coatings under seawater were 0.0402 and 0.0672, respectively. Compared with dry friction, the average friction coefficients of MCD and NCD coatings were reduced by 64.1% and 37.8%, respectively, and the wear rate was decreased by 39.5% and 26.5%, respectively. Figure 4c gives the wear rates of different diamond coatings under different test conditions. The formula for calculating the wear rate of a coating is as follows [32]:

$$W = \frac{V}{FL} \tag{1}$$

Herein, W represents the wear rate and the unit is mm^3/Nm, V indicates the wear volume and the unit is μm^3, the unit of normal load (F) is N and the unit of wear scar length (L) is m. From Figure 4b, it can be seen that the wear rate of MCD coating was higher than that of NCD coating. Meanwhile, the wear rate of diamond coatings in the seawater environment was lower than those in the dry friction environment, which showed that the wear resistance of the diamond coating was better under the seawater environment.

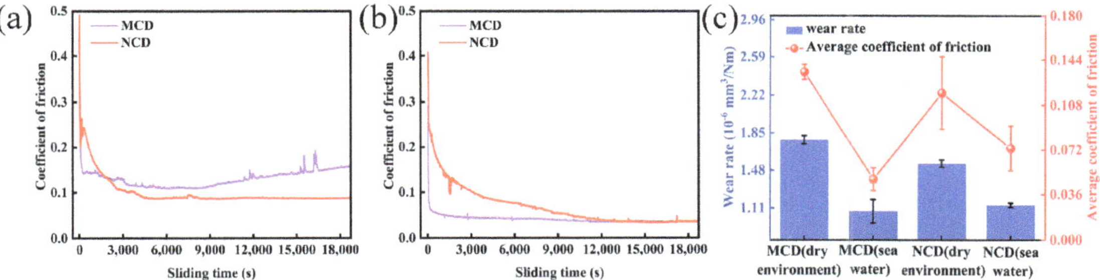

Figure 4. (**a**) Friction coefficient of diamond coatings in the dry environment; (**b**) Friction coefficient of diamond coatings in the seawater environment; (**c**) Average wear rate and average friction coefficient of diamond coatings.

3.3. Friction and Wear Analysis of Diamond Coatings in Dry Environment

In order to further investigate the corresponding wear mechanism, the worn morphology of different types of diamond film in every test condition was given in Figure 5. As shown in Figure 5a–h, in a dry friction environment, both MCD and NCD have more severe wear surfaces compared with a seawater environment, and it can be seen that the worn surface has obvious abrasive grains and scratches, and the rough surface generation and destruction may be closely related to their main friction mechanism under a dry friction environment. For MCD coating, obvious wear debris can be observed on the surface of the worn scars, as the surface of SiC balls was subjected to the sharp part of the diamond particles of the MCD coating with high shear stress during the process of wear. Moreover, the diamond particles also shattered during the process of wear, both SiC and diamond debris adhered to the diamond particles on the coating surface, and the counterpart ball surface also showed friction marks corresponding to the surface of the abrasion marks. The surface was continuously smoothed so that the friction coefficient was relatively smooth after the running in period. However, most of the wear debris came from SiC balls which can be observed by EDS; the EDS analysis of the wear debris revealed debris consisting mainly of Si and O elements, the Si obtained from the SiC balls, and the O elements possibly associated with oxygen in the air (Figure S2, Supporting Information). The high hardness of the wear debris makes the surface of the SiC ball scratched and discontinuous, and the rough surface will fail rapidly due to scraping during repeated friction. For NCD coating, in Figure 5e,f, the worn scars of NCD coating was cleaner due to its smooth surface, as there was less wear debris left and the size was smaller. Its excellent dry friction performance could be contributed to its smoother surface, which should be closely related to the size of the grain scale. In general, the surface of the worn scars was significantly rougher and had deeper scratches under dry friction conditions; whereas, in the seawater environment, the width of the worn scars increased but the surface of the worn scars was smoother with only a slight scratch.

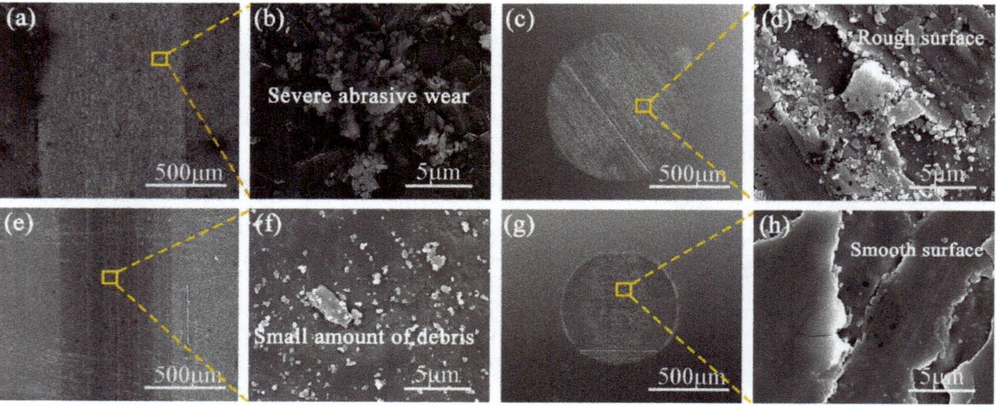

Figure 5. (**a,b**) The typical worn scars of MCD coatings sliding against SiC balls in the dry environment; (**c,d**) The typical worn scars of SiC sliding against MCD coatings in the dry environment; (**e,f**) The typical worn scars of NCD coatings sliding against SiC balls in the dry environment; (**g,h**) The typical worn scars of SiC sliding against NCD coatings in the dry environment.

Figure 6a,b show the Raman spectrum of diamond coating before and after friction in dry environment. It can be observed that the diamond peaks in the Raman spectrum show a right shift after the friction test, which was related to residual stress caused by the presence of non-diamond components and structural defects, such as rubbing process dislocations or impurities [17,39]. Moreover, the graphitic phase in the Raman spectrum after test increased compared with that before rubbing, which indicated that the diamond to graphite phase in the coating was transformed due to local pressure and shear stress during the process of wear [40]. Figure 6c–f are the XPS spectra of the worn scar surface before and after dry friction. It can be seen from the full spectra that the coating surface contained a higher oxygen concentration before friction, which was caused by the exposure of the coating to the atmospheric environment. After the dry friction test, the oxygen concentration on the coating surface increased, and the oxygen concentration of the NCD coating was higher than the MCD coating. In Figure 6d–f, C 1s spectra shows peaks at a bonding energy of 283.80 eV, 284.79 eV, and 286.42 eV before and after friction of the NCD coating, the peak at 283.80 eV corresponded to sp^2, and the peak at 284.79 eV represented sp^3. Furthermore, the peak at 286.42 eV represented CH_2-O/C=O which maybe originated from the friction pairs contacted with air during test. In contrast, the C 1s spectrum of the MCD coating before rubbing showed the two peaks at bonding energy of 284.78 eV and 286.50 eV, the peak at 284.78 eV corresponded to sp^3 and the peak at 286.50 eV represented

CH$_2$-O/C=O. After friction, there was an additional peak of MCD coating at 284.38 eV referred to sp^2 [39,41,42], which was because, during recurrent friction, the process of local pressure and shear stress and the heat generated by the relative motion may have caused sp^3 to sp^2 structural transformation which occurs at the friction contacts [43]. The graphite phase transition occurred by repeatedly applying compression (before ball contact) and tension (after ball contact) to the subsurface region of the wear track, and the formation of sp^2 bonds was easier to shear, resulting in a lower coefficient of friction [32,40]. In addition, it can be seen that the ratio of sp^2/sp^3 in the MCD and NCD coatings increased compared with those before friction, which indicated that the graphite phase increased after rubbing. This was consistent with the Raman spectroscopy results. Meanwhile, the sp^2 concentration of NCD coating was higher than MCD coating, which was known for low friction due to easy in-plane shearing [44,45]. This may be the reason for the better tribological properties of NCD under dry friction conditions.

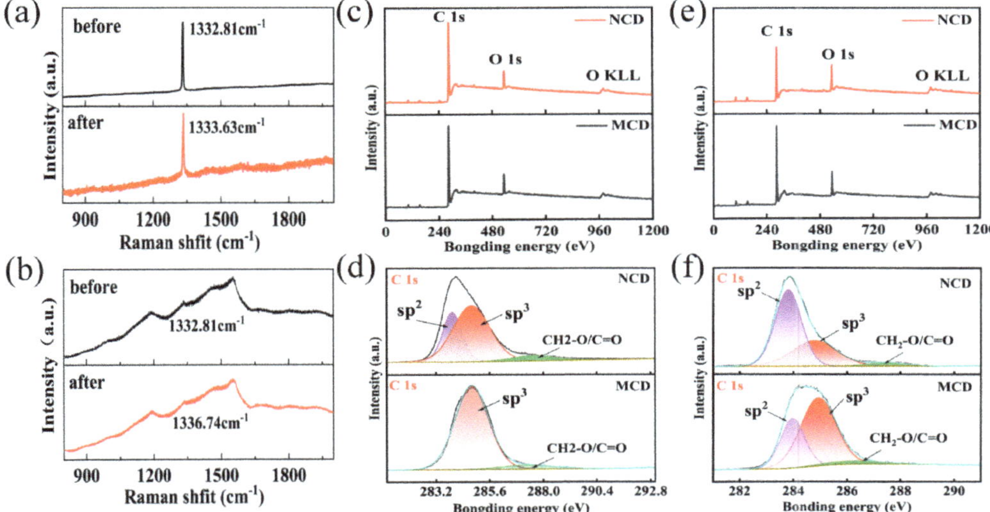

Figure 6. (**a**) The Raman spectrum of MCD coating before and afetr test; (**b**) The Raman spectrum of NCD coating before and afetr test; (**c,d**) The XPS spectra of the surface of the worn track before test; (**e,f**) The XPS spectra of the surface of the worn track after test under dry environment.

From the above analysis results, the wear mechanism of typical diamond coatings in the dry environment was mainly abrasive wear. The COF of the diamond coating was larger due to the three-body wear of the surface. Due to its rough surface, the MCD coating generated large wear debris during repeated wear and remained in the wear track, resulting in a high friction coefficient and serious wear. For NCD coating, the surface was smoother and produced less wear debris. From the XPS analysis results, the sp^2/sp^3 of the NCD coating surface was higher, indicating that the NCD surface produced more graphite, which reduced the friction coefficient. Therefore, the NCD coating has relatively excellent dry friction performance, which was consistent with the relevant literature reports [38,39,44,45].

3.4. The Influence of Seawater Environment on Friction Behavior of Diamond Coatings

To further investigate the wear mechanism of diamond coatings in a seawater environment, the wear morphology of MCD and NCD in a seawater environment is given in Figure 7a–h. As shown in Figure 7, the surface of the worn scars changed significantly compared to the dry friction environment, and the worn surface became smooth with only slight scratches. A friction-induced lubricating film has been observed on the worn surface of diamond film, which was distinguished from the morphology of coating in a dry environment clearly. This indicated that the friction mechanism in a seawater environment was significantly different from that in a dry friction environment, and the formation of a lubricating film on the diamond surface may be related to the frictional chemical reaction in the seawater environment. As shown in Figure 7i,j, both MCD and NCD coatings produce a lubricating film on the surface. It can be observed that the thickness of the lubricating film of NCD can reach about 50 nm, whereas the lubricating film of MCD coating can reach about 200 nm. The MCD coating produced a thicker film. The EDS further confirm the composition of the cross-sectional morphologies of lubricating film. The main components of lubricating film were O and Si elements, where the O element may be adsorbed by atoms or molecules (e.g., H_2O, hydrogen and hydroxyl), the element Si was due to the SiC on the grinding ball in the friction process to produce grinding chips, a part of these abrasive chips in the friction will be expelled from the friction interface. The other part will be sandwiched between the two friction surfaces, and the lubricating film formed may be composed of SiO_2. In fact, under an applied load of 15 N, the degree of wear in these two environments was not significant for either MCD or NCD coatings, which may be related to the higher bonding force between the diamond and the substrate. In addition, the Na^+ and Cl^- in the seawater environment do not affect the surface of the diamond coating, indicating that the diamond coating has good anti-wear and anti-corrosion properties. Furthermore, more details about corrosion resistance were revealed, which were shown in Figure S3, and the corresponding analyses were exhibited in supporting information.

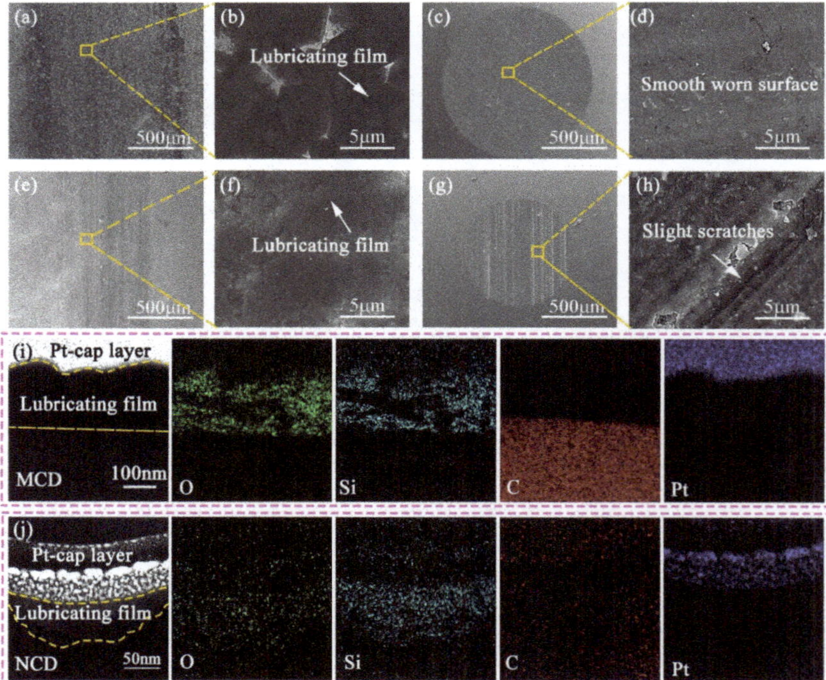

Figure 7. (**a,b**) The worn scars of MCD coatings sliding against SiC balls in the seawater environment; (**c,d**) The worn scars of SiC sliding against MCD coatings in the seawater environment; (**e,f**) The worn scars of NCD coatings sliding against SiC balls in the seawater environment; (**g,h**) The worn scars of SiC sliding against NCD coatings in the seawater environment; (**i**) The EDS mapping of the transfer film formed on the frictional zones of MCD diamond coatings; (**j**) The EDS mapping of the transfer film formed on the frictional zones of NCD diamond coatings.

Moreover, the worn scars of SiC balls sliding on NCD coating were all smaller than those on MCD coating under the same friction conditions, whereas SiC balls sliding on the MCD coating exhibited larger worn scars. These different worn profiles should be closely related to the different scales of diamond coating performance. The worn scars of SiC balls sliding on the MCD coating in the seawater environment were relatively large compared with the dry friction environment, which showed that the wear resistance of diamond coating was better under the seawater environment.

The X-ray photoelectron spectroscopy (XPS) spectra of C 1s, O 1s, and Si 2p on the worn surface under the sea water environment are given in Figure 8. The overall distribution of elements was shown in the Figure 8a. The two peaks of 99 eV and 150 eV corresponded to the Si 2p and Si 2s peaks, respectively. The spectra of C 1s was shown in Figure 8b. After friction and wear in the seawater environment, the surface of the worn scars of the MCD coating showed an increase in several components with a bonding energy of 283.89 eV, 285.34 eV, and 287.66 eV, respectively. The peak at 283.89 eV represented to sp^2; in addition, the peak of 285.34 eV represented the group of $C-COO/CH_3/CH_2$-O and the peak of 287.66 eV was corresponded to CH_2-O group, whereas the peak of sp^3 was located at the bonding energy of 284.78 eV. However, the surface of the worn scars of the NCD coating showed only three peaks at 283.45 eV, 284.80 eV, and 287.30 eV, respectively. The peak at 283.45 eV corresponded to sp^2, the peak at 284.80 eV represented sp^3, and the peak at 287.30 eV represented CH_2-O. It showed that there were various oxygen fragments and OH groups on the surface of the diamond coating after friction, which was due to the fact that the friction surface of the MCD coating was exposed to seawater in the seawater environment. Seawater contained a large amount of O and H elements. The spectra of Si 2p is shown in Figure 8c. On the worn scars of MCD and NCD coatings, the Si 2p spectra showed SiO_2 at a bonding energy of 103.36 eV [39,41,42], and the Si 2p spectra showed that more SiO_2 was generated on the surface of the MCD coating, which was consistent with the EDS results. The generated SiO_2 might have reduced the coefficient of friction during the friction process.

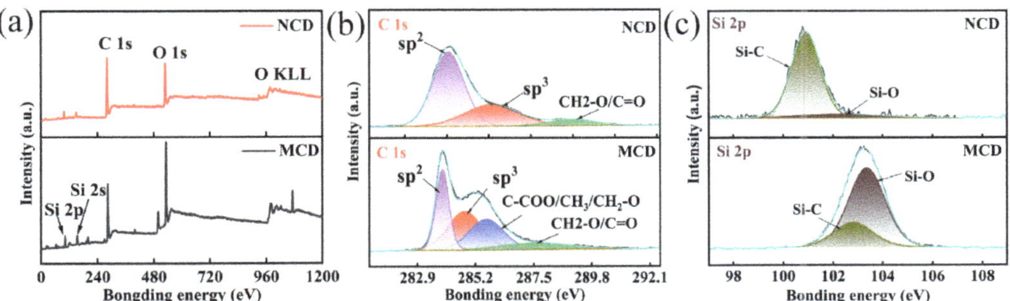

Figure 8. (**a**) The overall distribution of elements of the worn track in the seawater environment; (**b**) The C 1s spectra of the worn track in the seawater environment; (**c**) The Si 2p spectra of the worn track in the seawater environment.

In order to disclose the distinct tribological performance of above-mentioned coating, the subtle worn morphology was revealed in Figure 9, which shows the obtained TEM image of the wear scarred surface of the MCD and NCD coating and the corresponding SAED pattern. The TEM image in Figure 9a demonstrates the lubricating film on the surface of the MCD coating as well as the graphitization image close to the diamond surface. The diamond (111) lattice, graphite, and lubricating film were clearly separated by a dashed boundary marked in the HRTEM (high-resolution transmission electron microscope) image. The Fourier-transformed (FT) diffractogram image of Figure 9a is shown in the lower left inset, which indicated that both diamond (several of the diffraction points originated from diamond particles were marked by Dia) and nano-graphite (diffraction ring of graphite was marked by G) were contained in the figure [46]. More details of the graphite and diamond area were unveiled with the zoom-in HRTEM image in Figures 9b and 9c, respectively. As presented in Figure 9b, the fringes of the lattice plane of graphite were clear, and it can be observed that the interplanar spacing of its G (200) orientation is 0.34 nm [47]; the FT diffractogram images of the graphite lattice structure inside the corresponding areas is exhibited in Figure 9(b1). Figure 9c showed that the fringes of the diamond lattice plane are clear, and the (111) lattice plane can be observed with a lattice spacing of 0.205 nm [48], which was the same structure as the crystalline diamond. The selected area electron diffraction image (SAED) of Figure 9c is shown in Figure 9(c1), which can be clearly observed in the diamond orientation of the (111) and (200) diffraction rings. The SAED of the diamond region provides a clearer pattern indicating the high quality and crystallinity of the diamond. The TEM image in Figure 9d shows the lubricating film on the surface of the NCD coating. As can be seen, compared with the MCD coating, only a lubricating film was formed on the surface of the NCD coating without the transformation of diamond to graphite phase. Under HRTEM investigation in Figure 9e, the (111) lattice plane can be observed with a lattice spacing of 0.206 nm. The FT diffractogram images of the diamond lattice structure inside the corresponding areas is shown in Figure 9(e1), which also can be clearly observed in the diamond orientation of the (111) and (200) diffraction rings.

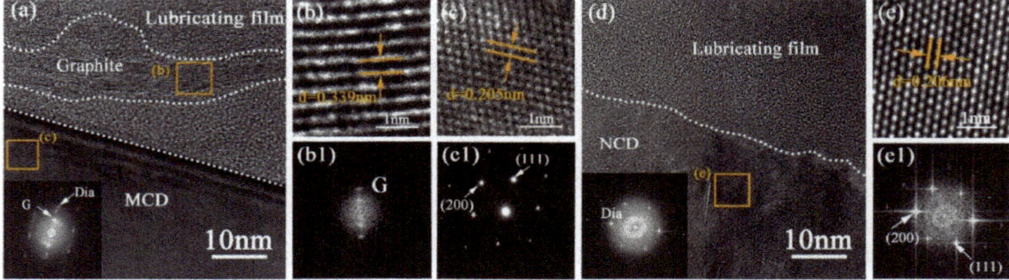

Figure 9. (a) The TEM image of the wear-scarred surface of the MCD coating; (b) HRTEM image of the graphite area of (a); (b1) the FT diffractogram images of (b); (c) HRTEM image of the diamond area of (a); (c1) the FT diffractogram images of (c); (d) The TEM image of the wear-scarred surface of the MCD coating; (e) HRTEM image of the diamond area of (d); (e1) the FT diffractogram images of (e).

From the above characterization results, the MCD coating and NCD coating have the effect of storing seawater due to their rough surfaces, so that a layer of water film was formed between the two worn surfaces, which hindered the direct contact between the two worn surfaces, resulting in a certain hydrodynamic effect for the low friction coefficient of the diamond coating in the seawater environment. In addition, the SiC wear debris generated during the friction process has a tribochemical reaction with water. The reaction process as follows [49–52]:

$$Si\text{-}C + H_2O \rightarrow Si\text{-}OH + C\text{-}H, \; Si\text{-}OH + Si\text{-}OH \rightarrow O\text{-}Si\text{-}O + H_2O$$

Therefore, the SiO_2 that was generated by a tribochemical reaction which formed a lubricating film on the surfaces of both the MCD coating and the NCD coating; SiO_2 lubricant films with high hydrophilicity retain water at the sliding interface. Retention of water at the sliding interface resulted in high load capacity and stable friction [52]. In addition, as the MCD coating has a rougher surface, a part of its surface area experienced a rather high local contact pressure, where the silica layer was destroyed by the contact force, and the asperity to the SiC ball surface and the diamond coating are sheared and destroyed, leaving the contact surface flat. Frictional heat is concentrated in the contact zone and the local temperature rises, leading to the transformation of sub-stable sp^3 bonds into stable sp^2 bonds and graphitization of the contact area [53]. According to literature reports, in the case of graphite, its layered structure yields low friction only when the π–π* orbitals were separated by intercalated donors or acceptors. Water acts as a Lewis base (a weak donor for the high electron density of the π cloud), reducing the electron-hole attraction between the basal planes; therefore, the effect of water on graphite is to reduce its friction [17].

Based on the above analysis of the research results, we propose a feasible friction mechanism as shown in Figure 10. The surface of MCD coating in the dry friction environment was rough and the size of the generated wear debris was large, which led to severe abrasive wear during the repeated process of wear. In contrast, the surface of NCD coating was smoother, and the wear debris was smaller; in addition, more graphite was generated during the repeated process of wear, which made the NCD coating have better tribological properties in a dry environment. For the seawater environment, the excellent tribological properties of diamond coatings were attributed to the SiO$_2$ lubricant film generated by the tribochemical reaction, as is illustrated in Figure 10. Due to the micro-scale surface roughness of diamond, it was able to store water and wear debris between its grains, so that wear debris did not adhere between two wear surfaces and abrasive wear occurred. On the other hand, the SiO$_2$ generated by the tribochemical reaction also formed a lubricating film on the diamond surface, which was hydrophilic and therefore retained water at the sliding interface, both of which resulted in low friction and low wear of the diamond coating in a seawater environment. Furthermore, the MCD coating has a rougher surface, and the high local contact pressure allows the SiO$_2$ layer to break down. In these areas, the diamond phase produced a graphite phase change at high contact pressures and graphite also produced low frictional effects in water. Thus, for MCD coating in a seawater environment, the wear mechanism was a mixed lubrication state, with SiO$_2$ particles playing a major role. Although the graphite phase was formed at the top of the diamond surface asperity, the SiO$_2$ particles were also always generated at the wear interface and are always supplied to the lubricated interface, giving it a stable low friction state [54].

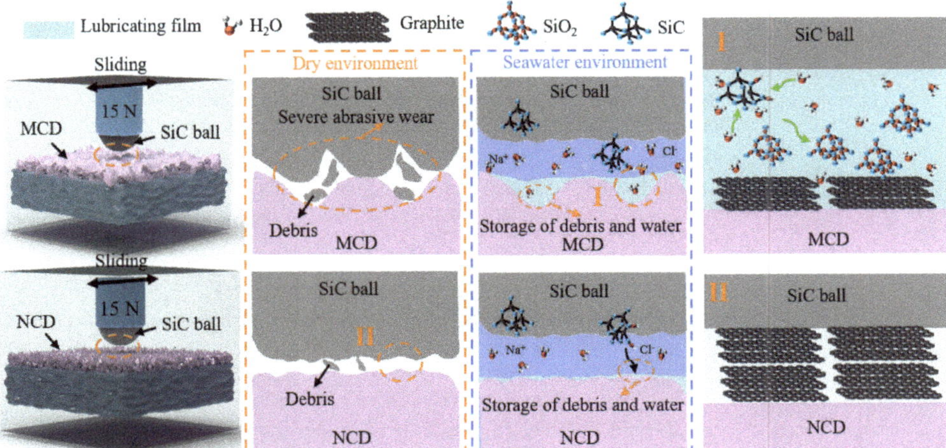

Figure 10. Schematic illustration of wear mechanism in dry friction and seawater environment.

4. Conclusions

In this study, MCD and NCD coatings were deposited by MPCVD, and the friction and wear behavior of the two coatings on diamond coatings in dry and seawater environments was investigated. This work provides significant insights into the sea water wear behavior of diamond coatings and the application of marine protective coatings. The main conclusions are as follows:

(1) Continuous and similar thicknesses of MCD and NCD coatings were successfully deposited by MPCVD under different deposition parameters, which eliminates the great influence of the thicknesses on the properties of tribological performance. Both MCD and NCD coatings have good bonding properties.

(2) The main wear mechanism of MCD coating and NCD coating in a dry friction environment is abrasive wear. The surface of the wear track of the NCD coating is smooth and the wear debris generated is small. Additionally, more sp^2 and graphite phases are generated in the wear track due to frictional heat, which makes the NCD coating have better tribological performance in a dry friction environment.

(3) Compared with the dry friction environment, the average COF of MCD and NCD coatings in the seawater environment are reduced 64.1% and 37.8%, respectively, and the wear rate is reduced by 39.5% and 26.5%, respectively. The main wear mechanism of the diamond coating is the SiO_2 lubricating film generated by the tribochemical reaction. Moreover, the MCD coating is a synergistic lubrication effect consisting of graphite phase and SiO_2 layer, resulting in a low average coefficient of friction and low wear.

(4) Based on the discussion on the friction mechanism of MCD and NCD coatings on diamond coatings in a dry environment and seawater environment, the diamond coating has good tribological properties and corrosion resistance in a seawater environment, which provides guidance on protective coatings on the surface of sea-related friction parts.

Supplementary Materials: The following supporting information can be downloaded at: https://www.mdpi.com/article/10.3390/cryst12101345/s1, Figure S1: (a) Scratch morphologies of MCD coatings in detail; (b) Scratch morphologies of NCD coatings in detail; Figure S2: (a) The EDS analysis of debris of MCD coating; (b) The EDS analysis of debris of NCD coating; Figure S3: (a) Nyquist plot for MCD coating obtained in a 3.5 wt.% NaCl solution; (b) Nyquist plot for NCD coating obtained in a 3.5 wt.% NaCl solution. (c) Bode plots for MCD and NCD coating were obtained in a 3.5 wt.% NaCl solution. References [55,56] are cited in the Supplementary Materials.

Author Contributions: Investigation, H.Z.; data curation, H.Z.; writing—original draft, H.Z.; writing—review and editing, H.S.; conceptualization, H.S. and M.P.; methodology, H.S. and M.P.; formal analysis, M.P.; supervision, N.J., F.J., G.Y., and K.N.; investigation, N.J., F.J., G.Y., and K.N. All authors have read and agreed to the published version of the manuscript.

Funding: The authors gratefully acknowledge the financial supports from the Open Project of the State Key Laboratory of Solid Lubrication (Grant No. LSL-2102), the Youth Fund of the Chinese Academy of Sciences (No. JCPYJJ-22030) and the Project of the Chinese Academy of Sciences (ZDKYYQ20200001).

Data Availability Statement: Not applicable.

Conflicts of Interest: The authors declare that they have no known competing financial interest or personal relationship that could have appeared to influence the work reported in this paper.

References

1. Gicquel, A.; Hassouni, K.; Silva, F.; Achard, J. CVD diamond films: From growth to applications. *Curr. Appl. Phys.* **2001**, *1*, 479–496. [CrossRef]
2. Wang, X.; Shen, X.; Sun, F.; Shen, B. Tribological properties of MCD films synthesized using different carbon sources when sliding against stainless steel. *Tribol. Lett.* **2016**, *61*, 1–16. [CrossRef]
3. Gaydaychuk, A.; Linnik, S. Tribological and mechanical properties of diamond films synthesized with high methane concentration. *Int. J. Refract. Met. Hard Mater.* **2019**, *85*, 105057. [CrossRef]
4. Wang, X.; Zhang, J.; Sun, F.; Zhang, T.; Shen, B. Investigations on the fabrication and erosion behavior of the composite diamond coated nozzles. *Wear* **2013**, *304*, 126–137. [CrossRef]
5. Chandran, M.; Kumaran, C.R.; Dumpala, R.; Shanmugam, P.; Natarajan, R.; Bhattacharya, S.S.; Ramachandra Rao, M.S. Nanocrystalline diamond coatings on the interior of WC–Co dies for drawing carbon steel tubes: Enhancement of tube properties. *Diamond Relat. Mater.* **2014**, *50*, 33–37. [CrossRef]

6. Perle, M.; Bareiss, C.; Rosiwal, S.M.; Singer, R.F. Generation and oxidation of wear debris in dry running tests of diamond coated SiC bearings. *Diamond Relat. Mater.* **2006**, *15*, 749–753. [CrossRef]

7. Sochr, J.; Švorc, Ľ.; Rievaj, M.; Bustin, D. Electrochemical determination of adrenaline in human urine using a boron-doped diamond film electrode. *Diamond Relat. Mater.* **2014**, *43*, 5–11. [CrossRef]

8. Chattopadhyay, A.; Sarangi, S.K.; Chattopadhyay, A.K. Effect of negative dc substrate bias on morphology and adhesion of diamond coating synthesised on carbide turning tools by modified HFCVD method. *Appl. Surf. Sci.* **2008**, *255*, 1661–1671. [CrossRef]

9. Miyake, S.; Shindo, T.; Miyake, M. Friction properties of surface-modified polished chemical-vapor-deposited diamond films under boundary lubrication with water and poly-alpha olefin. *Tribol. Int.* **2016**, *102*, 287–296. [CrossRef]

10. Ashcheulov, P.; Škoda, R.; Škarohlíd, J.; Taylor, A.; Fekete, L.; Fendrych, F.; Vega, R.; Shao, L.; Kalvoda, L.; Vratislav, S.; et al. Thin polycrystalline diamond films protecting zirconium alloys surfaces: From technology to layer analysis and application in nuclear facilities. *Appl. Surf. Sci.* **2015**, *359*, 621–628. [CrossRef]

11. Wang, X.; Shen, X.; Gao, J.; Wang, X.; Shen, X.; Gao, J.; Sun, F. Consecutive deposition of amorphous SiO2 interlayer and diamond film on graphite by chemical vapor deposition. *Carbon* **2017**, *117*, 126–136. [CrossRef]

12. Din, S.H.; Shah, M.A.; Sheikh, N.A.; Mursaleen Butt, M. CVD Diamond. *Trans. Indian Inst. Met.* **2019**, *72*, 1–9. [CrossRef]

13. Yang, Q.; Yang, S.; Li, Y.S.; Lu, X.; Hirose, A. NEXAFS characterization of nanocrystalline diamond thin films synthesized with high methane concentrations. *Diamond Relat. Mater.* **2007**, *16*, 730–734. [CrossRef]

14. Fabisiak, K.; Torz-Piotrowska, R.; Staryga, E.; Szybowicz, M.; Paprocki, K.; Banaszak, A.; Popielarski, P. The influence of working gas on CVD diamond quality. *Mater. Sci. Eng. C* **2012**, *177*, 1352–1357. [CrossRef]

15. Salgueiredo, E.; Amaral, M.; Neto, M.A.; Fernandes, A.J.S.; Oliveira, F.J.; Silva, R.F. HFCVD diamond deposition parameters optimized by a Taguchi Matrix. *Vacuum* **2011**, *85*, 701–704. [CrossRef]

16. Shabani, M.; Abreu, C.S.; Gomes, J.R.; Silva, R.F.; Oliveira, F.J. Effect of relative humidity and temperature on the tribology of multilayer micro/nanocrystalline CVD diamond coatings. *Diamond Relat. Mater.* **2017**, *73*, 190–198. [CrossRef]

17. Lei, X.; Shen, B.; Chen, S.; Wang, L.; Sun, F. Tribological behavior between micro-and nano-crystalline diamond films under dry sliding and water lubrication. *Tribol. Int.* **2014**, *69*, 118–127. [CrossRef]

18. De Barros Bouchet, M.I.; Zilibotti, G.; Matta, C.; Clelia Righi, M.; Vandenbulcke, L.; Vacher, B.; Martin, J.M. Friction of diamond in the presence of water vapor and hydrogen gas. Coupling gas-phase lubrication and first-principles studies. *J. Phys. Chem. C* **2012**, *116*, 6966–6972. [CrossRef]

19. Cui, Y.; He, Y.; Ji, C.; Lin, B.; Zhang, D. Anti-wear performance of polished microcrystalline diamond films sliding against Si3N4 under water lubrication. *Surf. Rev. Lett.* **2020**, *27*, 2050008. [CrossRef]

20. Feng, Z.; Tzeng, Y.; Field, J.E. Friction of diamond on diamond in ultra-high vacuum and low-pressure environments. *J. Phys. D Appl. Phys.* **1992**, *25*, 1418. [CrossRef]

21. Lin, Q.; Chen, S.; Ji, Z.; Huang, Z.; Zhang, Z.; Shen, B. High-temperature wear behavior of micro-and ultrananocrystalline diamond films against titanium alloy. *Surf. Coat. Technol.* **2021**, *422*, 127537. [CrossRef]

22. Liu, Y.; Li, S.; Li, H.; Ma, G.; Sun, L.; Guo, P.; Ke, P.; Lee, K.R.; Wang, A. Controllable defect engineering to enhance the corrosion resistance of Cr/GLC multilayered coating for deep-sea applications. *Corros. Sci.* **2022**, *199*, 110175. [CrossRef]

23. Liu, Y.; Liu, L.; Li, S.; Wang, R.; Guo, P.; Wang, A.; Ke, P. Accelerated deterioration mechanism of 316L stainless steel in NaCl solution under the intermittent tribocorrosion process. *J. Mater. Sci. Technol.* **2022**, *121*, 67–79. [CrossRef]

24. Xu, X.; Guo, P.; Zuo, X.; Sun, L.; Li, X.; Lee, K.R.; Wang, A. Understanding the effect of Al/Ti ratio on the tribocorrosion performance of Al/Ti co-doped diamond-like carbon films for marine applications. *Surf. Coat. Technol.* **2020**, *402*, 126347. [CrossRef]

25. Ou, Y.; Wang, H.; Hua, Q.; Liao, B.; Ouyang, X. Tribocorrosion behaviors of superhard yet tough Ti-CN ceramic coatings. *Surf. Coat. Technol.* **2022**, *439*, 128448. [CrossRef]

26. Dong, M.; Zhu, Y.; Xu, L.; Ren, X.; Ma, F.; Mao, F.; Li, J.; Wang, L. Tribocorrosion performance of nano-layered coating in artificial seawater. *Appl. Surf. Sci.* **2019**, *487*, 647–654. [CrossRef]

27. Wang, X.; Jiang, Y.; Wang, Y.; Ye, C.; Du, C. Probing the tribocorrosion behaviors of three nickel-based superalloys in sodium chloride solution. *Tribol. Int.* **2022**, *172*, 107581. [CrossRef]

28. Wang, H.; Wang, C.; Wang, X.; Sun, F. Effects of carbon concentration and gas pressure with hydrogen-rich gas chemistry on synthesis and characterizations of HFCVD diamond films on WC-Co substrates. *Surf. Coat. Technol.* **2021**, *409*, 126839. [CrossRef]

29. Patterson, A.L. The Scherrer formula for X-ray particle size determination. *Phys. Rev.* **1939**, *56*, 978. [CrossRef]

30. Wang, H.; Song, X.; Wang, X.; Sun, F. Fabrication, tribological properties and cutting performances of high-quality multilayer graded MCD/NCD/UNCD coated PCB end mills. *Diamond Relat. Mater.* **2021**, *118*, 108505. [CrossRef]

31. Klauser, F.; Steinmüller-Nethl, D.; Kaindl, R.; Bertel, E.; Memmel, N. Raman studies of nano-and ultra-nanocrystalline diamond films grown by hot-filament CVD. *Chem. Vap. Depos.* **2010**, *16*, 127–135. [CrossRef]

32. Sharma, N.; Kumar, N.; Dhara, S.; Dash, S.; Bahuguna, A.; Kamruddin, M.; Tyagi, A.K.; Raj, B. Tribological properties of ultra nanocrystalline diamond film-effect of sliding counterbodies. *Tribol. Int.* **2012**, *53*, 167–178. [CrossRef]

33. Kumar, N.; Kozakov, A.T.; Sankaran, K.J.; Sidashov, A.V.; Lin, I.N. Controlled atmosphere dependent tribological properties of thermally annealed ultrananocrystalline diamond films. *Diamond Relat. Mater.* **2019**, *97*, 107437. [CrossRef]

34. Law, K.Y. Definitions for hydrophilicity, hydrophobicity, and superhydrophobicity: Getting the basics right. *J. Phys. Chem. Lett.* **2014**, *5*, 686–688. [CrossRef]

35. Song, H.; Ji, L.; Li, H.; Liu, X.; Zhou, H.; Liu, L.; Chen, J. Interface design for aC: H film with super long wear life in high vacuum environment. *Tribol. Int.* **2016**, *95*, 298–305. [CrossRef]

36. Yan, G.; Wu, Y.; Cristea, D.; Liu, L.; Tierean, M.; Wang, Y.; Lu, F.; Wang, H.; Yuan, Z.; Munteanu, D.; et al. Mechanical properties and wear behavior of multi-layer diamond films deposited by hot-filament chemical vapor deposition. *Appl. Surf. Sci.* **2019**, *494*, 401–411. [CrossRef]

37. Buijnsters, J.G.; Shankar, P.; Van Enckevort, W.J.P.; Schermer, J.J.; Ter Meulen, J.J. Adhesion analysis of polycrystalline diamond films on molybdenum by means of scratch, indentation and sand abrasion testing. *Thin Solid Films* **2005**, *474*, 186–196. [CrossRef]

38. Guo, Y.; Zhu, Y.; Zhang, X.; Luo, B. Effects of Superhydrophobic Surface on Tribological Properties:Mechanism, Status and Prospects. *Prog. Chem.* **2020**, *32*, 320. [CrossRef]

39. Wang, C.; Wang, X.; Sun, F. Tribological behavior and cutting performance of monolayer, bilayer and multilayer diamond coated milling tools in machining of zirconia ceramics. *Surf. Coat. Technol.* **2018**, *353*, 49–57. [CrossRef]

40. Voevodin, A.A.; Phelps, A.W.; Zabinski, J.S.; Donley, M.S. Friction induced phase transfor-mation of pulsed laser deposited diamond-like carbon. *Diamond Relat. Mater.* **1996**, *5*, 1264–1269. [CrossRef]

41. Kumar, N.; Sankaran, K.J.; Kozakov, A.T.; Sidashov, A.V.; Nicolskii, A.V.; Haenen, K.; Kolesnikov, V.I. Surface and bulk phase analysis of the tribolayer of nanocrystalline diamond films sliding against steel balls. *Diamond Relat. Mater.* **2019**, *97*, 107472. [CrossRef]

42. Huang, K.; Hu, X.; Xu, H.; Shen, Y.; Khomich, A. The oxidization behavior and mechanical properties of ultrananocrystalline diamond films at high temperature annealing. *Appl. Surf. Sci.* **2014**, *317*, 11–18. [CrossRef]

43. Erdemir, A.; Fenske, G.R.; Krauss, A.R.; Gruen, D.M.; McCauley, T.; Csencsits, R.T. Tribological properties of nanocrystalline diamond films. *Surf. Coat. Technol.* **1999**, *120*, 565–572. [CrossRef]

44. Abreu, C.S.; Oliveira, F.J.; Belmonte, M.; Fernandes, A.J.S.; Gomes, J.R.; Silva, R.F. CVD diamond coated silicon nitride self-mated systems: Tribological behaviour under high loads. *Tribol. Lett.* **2006**, *21*, 141–151. [CrossRef]

45. Kumar, N.; Panda, K.; Dash, S.; Popov, C.; Reithmaier, J.P.; Panigrahi, B.K.; Tyagi, A.K. Baldev Raj. Tribological properties of nanocrystalline diamond films deposited by hot filament chemical vapor deposition. *AIP Adv.* **2012**, *2*, 032164. [CrossRef]

46. Chen, C.; Fan, D.; Xu, H.; Jiang, M.; Li, X.; Lu, S.; Ke, C.; Hu, X. Monoatomic tantalum induces ordinary-pressure phase transition from graphite to n-type diamond. *Carbon* **2022**, *196*, 466–473. [CrossRef]

47. Shen, B.; Ji, Z.; Lin, Q.; Gong, P.; Xuan, N.; Chen, S.; Liu, H.; Huang, Z.; Xiao, T.; Sun, Z. Graphenization of Diamond. *Chem. Mater.* **2022**, *34*, 3941–3947. [CrossRef]

48. Lu, Y.; Man, W.; Wang, B.; Rosenkranz, A.; Yang, M.; Yang, K.; Yi, J.; Song, H.; Li, H.; Jiang, N. (100) oriented diamond film prepared on amorphous carbon buffer layer containing nano-crystalline diamond grains. *Surf. Coat. Technol.* **2020**, *385*, 125368. [CrossRef]

49. Gates, R.S.; Hsu, S.M. Tribochemistry between water and Si3N4 and SiC: Induction time analysis. *Tribol. Lett.* **2004**, *17*, 399–407. [CrossRef]

50. Ootani, Y.; Xu, J.; Adachi, K.; Kubo, M. First-principles molecular dynamics study of silicon-based ceramics: Different tribo-chemical reaction mechanisms during the running-in period of silicon nitride and silicon carbide. *J. Phys. Chem. C* **2020**, *124*, 20079–20089. [CrossRef]

51. Wang, Y.; Yukinori, K.; Koike, R.; Ootani, Y.; Adachi, K.; Kubo, M. Selective Wear Behaviors of a Water-Lubricating SiC Surface under Rotating-Contact Conditions Revealed by Large-Scale Reactive Molecular Dynamics Simulations. *J. Phys. Chem. C* **2021**, *125*, 14957–14964. [CrossRef]

52. Kasuya, M.; Hino, M.; Yamada, H.; Mizukami, M.; Mori, H.; Kajita, S.; Ohmori, T.; Suzuki, A.; Kurihara, K. Characterization of Water Confined between Silica Surfaces Using the Resonance Shear Measurement. *J. Phys. Chem. C* **2013**, *117*, 13540–13546. [CrossRef]

53. Li, J.; Yu, X.; Zhang, Z.; Zhao, Z. Exploring a diamond film to improve wear resistance of the hydraulic drilling impactor. *Surf. Coat. Technol.* **2019**, *360*, 297–306. [CrossRef]

54. Ootani, Y.; Xu, J.; Nakamura, F.; Kawaura, M.; Uehara, S.; Kanda, K.; Wang, Y.; Ozawa, N.; Adachi, K.; Kubo, M. Three Tribolayers Self-Generated from SiC Individually Work for Reducing Friction in Different Contact Pressures. *J. Phys. Chem. C* **2022**, *126*, 2728–2736. [CrossRef]

55. Ponthiaux, P.; Wenger, F.; Drees, D.; Celis, J.P. Electrochemical techniques for studying tribocorrosion processes. *Wear* **2004**, *256*, 459–468. [CrossRef]

56. Namus, R.; Rainforth, W.M. The influence of cathodic potentials on the surface oxide layer status and tribocorrosion behaviour of Ti6Al4V and CoCrMo alloys in simulated body fluid. *Biotribology* **2022**, *30*, 100212. [CrossRef]

Communication

Nanomechanical Characterization of High-Velocity Oxygen-Fuel NiCoCrAlYCe Coating

Feifei Zhou [1,*], Donghui Guo [1], Baosheng Xu [1,*], Yiguang Wang [1] and You Wang [2]

1   Institute of Advanced Structure Technology, Beijing Institute of Technology, Beijing 100081, China
2   Department of Materials Science, School of Materials Science and Engineering, Harbin Institute of Technology, Harbin 150001, China
*   Correspondence: zhoufeifei@bit.edu.cn (F.Z.); xubsh@bit.edu.cn (B.X.)

Abstract: MCrAlY (M = Ni or/and Co) coatings have played an indispensable role in the high-temperature protection system for key components of aero-engines due to their excellent high-temperature oxidation and hot corrosion resistance. Nanoindentation is a useful and highly efficient method for characterizing the nanomechanical properties of materials. The rich information reflecting materials can be gained by load-displacement curves. In addition to common parameters such as elastic modulus and nanohardness, the indentation work and creep property at room temperature can also be extracted. Herein, nanomechanical properties of NiCoCrAlYCe coatings using high-velocity oxygen-fuel (HVOF) spraying were investigated systematically by nanoindentation. The microstructure of as-sprayed NiCoCrAlYCe coatings present mono-modal distribution. Results of nanoindentation reveal that the elastic modulus and nanohardness of NiCoCrAlYCe coatings are 121.08 ± 10.04 GPa and 6.09 ± 0.86 Gpa, respectively. Furthermore, the indentation work of coatings was also characterized. The elastic indentation work is 10.322 ± 0.721 nJ, and the plastic indentation work is 22.665 ± 1.702 nJ. The ratio of the plastic work to the total work of deformation during indentation is 0.687 ± 0.024, which can predict excellent wear resistance for NiCoCrAlYCe coatings. Meanwhile, the strain rate sensitivity determined by nanoindentation is 0.007 ± 0.001 at room temperature. These results can provide prediction of erosion resistance for MCrAlY coatings.

Keywords: microstructure; nanomechanical properties; HVOF spraying; NiCoCrAlYCe coatings; indentation work; strain rate sensitivity

Citation: Zhou, F.; Guo, D.; Xu, B.; Wang, Y.; Wang, Y. Nanomechanical Characterization of High-Velocity Oxygen-Fuel NiCoCrAlYCe Coating. *Crystals* **2022**, *12*, 1246. https://doi.org/10.3390/cryst12091246

Academic Editors: Hao Yi, Huajun Cao, Menglin Liu and Le Jia

Received: 11 August 2022
Accepted: 31 August 2022
Published: 2 September 2022

Publisher's Note: MDPI stays neutral with regard to jurisdictional claims in published maps and institutional affiliations.

1. Introduction

MCrAlY coatings, as a high-temperature protective coating of superalloys (bond layer of thermal barrier coatings and overlay coatings), have been widely used in aerospace, shipping, energy, and other industries due to the excellent resistance to high-temperature oxidation and thermal corrosion [1,2]. The evolution of bond coats has proceeded from diffusion coatings, including aluminum and platinum aluminum, MCrAlY (M = Ni or/and Co) coatings to thermal barrier coatings. As for MCrAlY coatings, the rich-Ni and rich-Co provide resistance to oxidation and hot corrosion, respectively. Al and Cr are used to promote the formation of oxide film, and Y is used to improve the adhesion of oxide film [3]. Furthermore, MCrAlY coatings can also relieve thermal mismatches between the substrate and top layer [4]. However, with the increasing work temperature of hot aero-engine components, MCrAlY coatings are subjected to complex interactions, such as thermal stress and mechanical stress [5,6]. Therefore, to improve the efficiency of engines and prolong the service life of hot components of engines, it is urgent to improve the comprehensive properties of MCrAlY coatings.

Currently, there are many methods by which to prepare MCrAlY coatings, such as arc spraying [7], electro-plating [8], plasma spraying [9], flame spraying [10], and so on. Among them, high-velocity oxygen-fuel (HVOF) spraying is one of the most commonly used industrial thermal spraying technologies due to its high flame velocity and relatively low temperature, which is suitable for spraying MCrAlY powders [11–14]. Karaoglanli et al. [15] compared the oxidation behavior of CoNiCrAlY coatings fabricated by atmospheric plasma spraying (APS), supersonic atmospheric plasma spraying (SAPS), HVOF, and detonation gun methods. The results indicated that the HVOF-CoNiCrAlY coatings had a better oxidation resistance. Srivastava et al. [16] fabricated an $Al_{1.4}Co_{2.1}Cr_{0.7}Ni_{2.45}Si_{0.2}Ti_{0.14}$ high-entropy alloy (HEA) as a bond coat for the TBC system on a Ni-based superalloy by HVOF spraying, and showed that coatings containing HEA had more outstanding high-temperature properties compared with MCrAlY. Praveen et al. [17] investigated the erosion resistance of the $NiCrSiB-Al_2O_3$ coating on AISI304 stainless steel by HVOF thermal spraying, and indicated that the $NiCrSiB-Al_2O_3$ coating had a ductile erosion behavior. Rajendran et al. [18] prepared the WC-10Ni-5Cr coatings on 35 Mo Cr steel by HVOF process, and implied that the oxygen flow rate had a larger effect on the porosity and microhardness of coatings. Sacriste et al. [19] prepared MCrAlY coatings by twin wire arc spray, and found that the NiCrAlY coatings arc sprayed by using nitrogen had a lower roughness due to the lower oxide content compared to air atomization. In addition, the composition and structure of MCrAlY coatings determine chemical and mechanical properties of coatings to a certain degree [20]. To ensure that the hot components of engine can work stably at high temperature for a long time, some researchers optimize the composition of MCrAlY alloy. Zakeri et al. [21] investigated the high-temperature oxidation behavior of the $NiCoCrAlY-CeO_2$ coatings fabricated by HVOF spraying and found that the oxidation rate of the nano-CeO_2 modified coatings decreased by 87% compared to the NiCoCrAlY coatings. Yu et al. [22] studied the oxidation behaviors of the NiCrAlY alloy with different silicon content, and indicated that doping silicon was beneficial to improve the oxidation resistance of coatings. Yang et al. [23,24] demonstrated that the addition of Pt in NiCoCrAlY coatings promoted the diffusion of Al, which was conducive to selective oxidation and the formation of highly protective Al_2O_3 film. In short, doping alloying elements or nanoparticles is an effective way to improve the oxidation resistance of MCrAlY coatings. As for MCrAlY coatings, except for high-temperature oxidation and corrosion, it is well known that the mechanical properties of coatings are also one of the important parameters by which to evaluate the service life. Hence, studying the mechanical properties of coatings, especially the nanomechanical properties, has vital practical significance for the improvement of MCrAlY coatings system.

According to our previous studies, we designed the NiCoCrAlYCe coatings by HVOF spraying, and indicated that the NiCoCrAlYCe coatings exhibited a good bonding strength and thermal shock resistance [25,26]. The structural properties of multicomponent coatings have been investigated in previous studies, such as the composition, uniformity of the distribution of elements, and phase composition [26]. Nevertheless, the nanomechanical properties of NiCoCrAlYCe coatings are not analyzed systematically. In this work, the NiCoCrAlYCe coatings were fabricated by HVOF spraying, and the nanomechanical properties of NiCoCrAlYCe coatings were discussed in detail.

2. Experimental

2.1. Material Preparation

The substrates are K417G nickel-based superalloy ($\Phi 25.4 \times 6$ mm^2) and feedstocks are customized Ni-20Co-19.75Cr-11.4Al-0.68Y-0.79Ce (wt.%) powders (NiCoCrAlYCe, Institute of Metal Research, Chinese Academy of Sciences, Shenyang, China) according to composition requirements. Before spraying, the pretreatment of the substrates' surface is necessary, including ultrasonic cleaning in the acetone solvent for 10 min and sand blasting with parameters of 46 mesh brown corundum and 0.4 MPa pressure. The average surface roughness determined by the 3D profiler of sand-blasted substrates is 4.01 μm. Subsequently, NiCoCrAlYCe coatings with the thickness of 200 ± 30 μm were deposited onto substrates by high-velocity oxygen fuel spraying (HVOF, JP-5000, Praxair, Inc., Danbury, CT, USA). The main parameters are set in Table 1.

Table 1. Main parameters for HVOF-sprayed NiCoCrAlYCe coatings.

Flow rate of kerosene (L/h)	18.9
Flow rate of N_2 (L/min)	10.4
Flow rate of O_2 (L/min)	873.2
Feeding rate (g/min)	70
Spray distance (mm)	360

2.2. Characterization Technology

The microstructure of NiCoCrAlYCe feedstocks and as-sprayed coatings was characterized by scanning electron microscope (SEM, Nova NanoSEM 430, FEI Company, Eindhoven, The Netherlands). The nanoindentation test (Anton-Paar, Graz, Austria) was employed on the polished cross-section of NiCoCrAlYCe coatings to analyse the elastic–plastic properties. In order to reduce the error, 20 indentation points are used. The maximum loading force of nanoindentation is 98 mN with 10 s dwelling time and the loading rate is 3.27 mN/s. To ensure the reproducibility of results, twenty samples of each configuration were analysed.

3. Results and Discussion

Figure 1a shows the surface morphology of NiCoCrAlYCe feedstocks. It is clear that feedstocks are spherical and smooth. The particle size ranges from 10 μm to 45 μm and this meet the demands of HVOF spraying. In the case of thermal-sprayed coatings, the flowability of feedstocks is necessary. The spherical and smooth NiCoCrAlYCe feedstocks can ensure the flowability. In addition, whether the microstructure inside the powder is dense or not has a great influence on the wear resistance of as-sprayed coatings. Figure 1b shows the cross-section morphology of feedstocks. It is significant that there is a dense microstructure inside the feedstocks. The feedstocks with dense microstructure can obtain dense coatings after suitable spraying parameters and the wear resistance of coatings will be improved in turn.

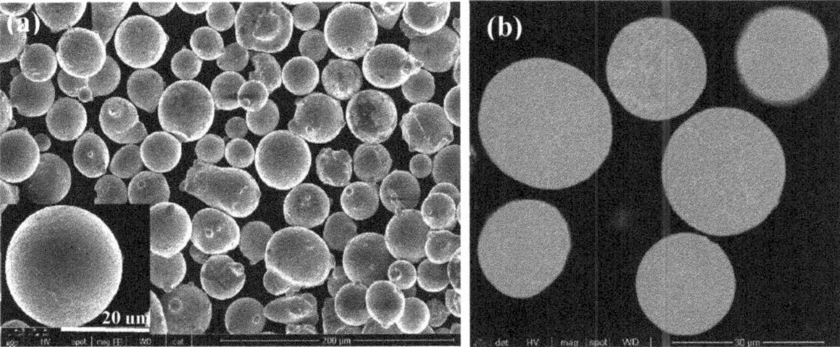

Figure 1. Surface morphology (**a**) and cross-section morphology (**b**) of NiCoCrAlYCe feedstocks. The inset in (**a**) is magnified image of surface morphology.

Figure 2a shows surface morphology of as-sprayed NiCoCrAlYCe coatings. It can be observed that the melting state of feedstocks is good and only a small amount of unmelted particles exists in the coatings. This indicate that NiCoCrAlYCe coatings can be regarded as a fully melted zones, i.e., mono-modal distribution. The cross-section image further demonstrates the coatings are compact, as displayed in Figure 2b.

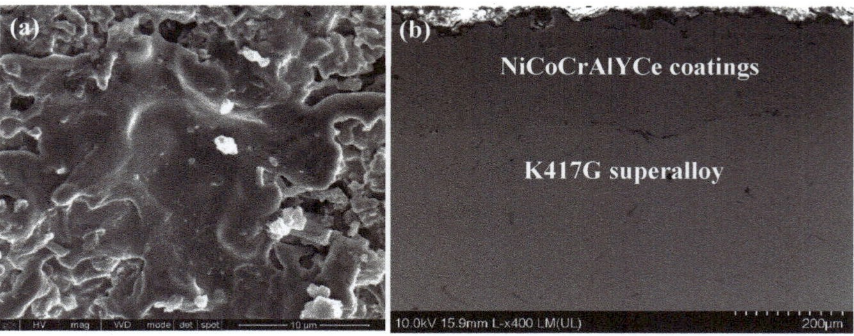

Figure 2. Surface (**a**) and cross section (**b**) images of as-sprayed NiCoCrAlYCe coatings.

As for the NiCoCrAlYCe coatings, nanomechanical properties are vital. Figure 3a shows load-displacement curves measured by nanoindentation of NiCoCrAlYCe coatings. It can be seen that these curves are relatively close and have good repeatability. Based on characteristics of load-displacement curves, it can be inferred that coatings are mainly composed of molten regions. The elastic modulus (E) and nanohardness (H) are 121.08 ± 10.04 GPa and 6.09 ± 0.86 GPa, respectively. Further Weibull analysis of E and H for NiCoCrAlYCe coatings is shown in Figure 3b. There is only a single line for the Weibull plots of elastic modulus or nanohardness, and this demonstrates the mono-modal microstructure in NiCoCrAlYCe coatings.

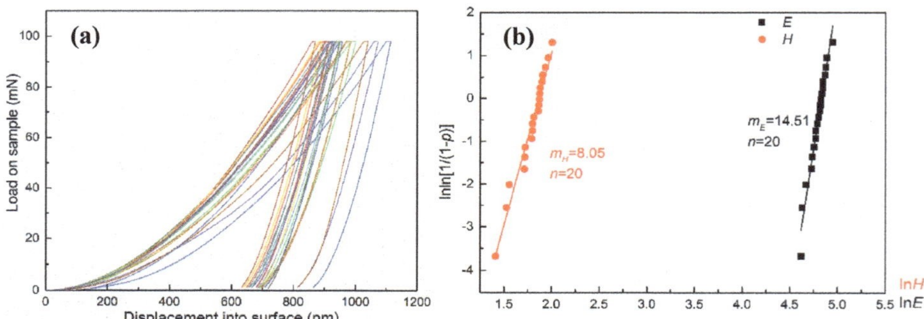

Figure 3. Load-displacement curves (**a**) and Weibull distribution plots of E and H (**b**) of NiCoCrAlYCe coatings.

In addition to the nanohardness and elastic modulus, the indentation work is also a crucial parameter by which to evaluate the elastic–plastic property of coatings. The indentation work can be obtained from load-displacement curves in nanoindentation. The total indentation work (W_t) consists of elastic work (W_e) and plastic work (W_p). The equation can be written as follows [27]:

$$W_t = W_e + W_p \tag{1}$$

where W_t is the total indentation work, W_e is the elastic work, and W_p is the plastic work. The total indentation work can be calculated by the following equation [27]:

$$W_t = \int_{0}^{h_{\max}} F(h)\,\mathrm{d}h \tag{2}$$

where h_{max} is the maximum indentation depth during loading, and $F(h)$ is the function of load with indentation depth during the loading curve. We have

$$W_e = \int_{h_f}^{h_{max}} P(h)\mathrm{d}h \tag{3}$$

where h_{max} is the maximum indentation depth during loading, $P(h)$ is the function of load with indentation depth during the unloading curve, and h_f is the remaining indentation depth [27]. According to Equations (1)–(3), the indentation work of NiCoCrAlYCe coatings is displayed in Figure 4a. The elastic indentation work ranges from 8.551 nJ to 11.969 nJ, and the plastic indentation work ranges from 19.960 nJ to 26.169 nJ. Figure 4a also shows that the plastic work of the coating is sensitive to the microstructure, whereas the elastic work is the opposite. The wear resistance of coatings can be evaluated by a new parameter, the microhardness dissipation parameter (MDP). The MDP is defined as the ratio of plastic indentation work to total indentation work, as shown in the following equation [27],

$$MDP = W_p / W_t \tag{4}$$

where MDP is the microhardness dissipation parameter, W_p is the plastic work, and W_t is the total indentation work.

The higher the MDP value of the coating, the better the wear resistance. As shown in Figure 4b, the calculated MDP of NiCoCrAlYCe coatings is 0.687 ± 0.024. This result suggests that NiCoCrAlYCe coatings have the ability to dissipate most of the deformation energy (about 69%), predicting excellent wear resistance under the erosion or abrasive wear condition. Furthermore, there are no cracks near the indentation impression displayed in Figure 4c.

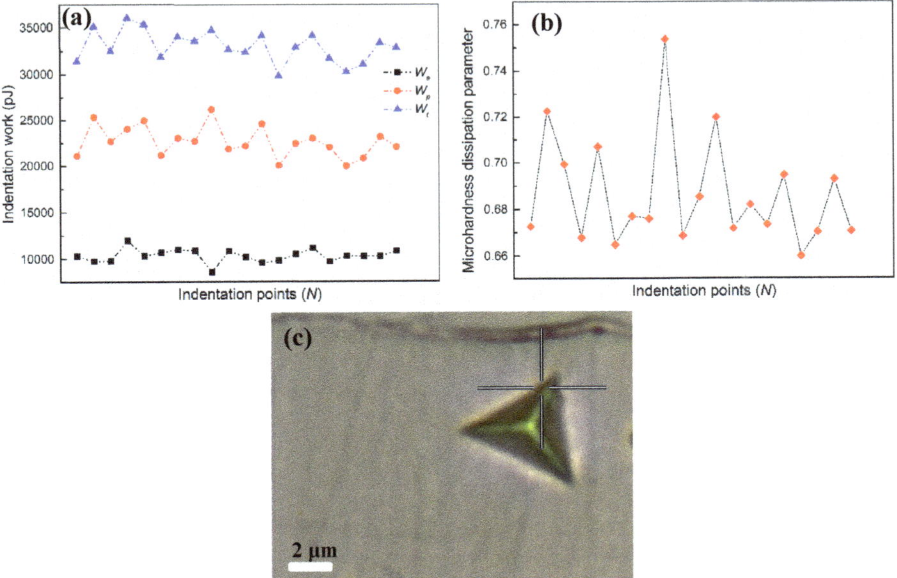

Figure 4. Indentation work (**a**), MDP (**b**), and an indentation impression at 10 gf load (**c**) of NiCoCrAlYCe coatings.

Furthermore, it is obvious that there is a platform, namely creep process, in the load displacement curve (Figure 3a). The stable creep stage can be described by the following equation [28,29],

$$\frac{1}{h}\frac{dh}{dt} = K\sigma^{\frac{1}{m}} \tag{5}$$

$$\sigma = C_1\frac{P_{max}}{h^2}, \tag{6}$$

where h is the indentation depth, t is time, σ is the flow stress, P_{max} is the maximum load (10 gf), K and C_1 are constant, and m is the strain rate sensitivity. Based on Equations (5) and (6), the integral result is written as follows [30]:

$$h = A(t - t_c)^{\frac{m}{2}}, \tag{7}$$

where h is the indentation depth, t is time, t_c is the constant, and m is the strain rate sensitivity. In the process of creep behavior, the indentation depth (h) versus holding time can be fitted by using Equation (7). The fitting curves are shown in Figure 5a (different colored lines represent randomly selected indentation points), and the fitting results are very good ($R^2 > 99\%$). The strain rate sensitivity of NiCoCrAlYCe coatings at room temperature is 0.007 ± 0.001 (Figure 5b). As for high-temperature protection coatings, creep property is very important. This indentation method can provide a new way to calculate the strain rate sensitivity.

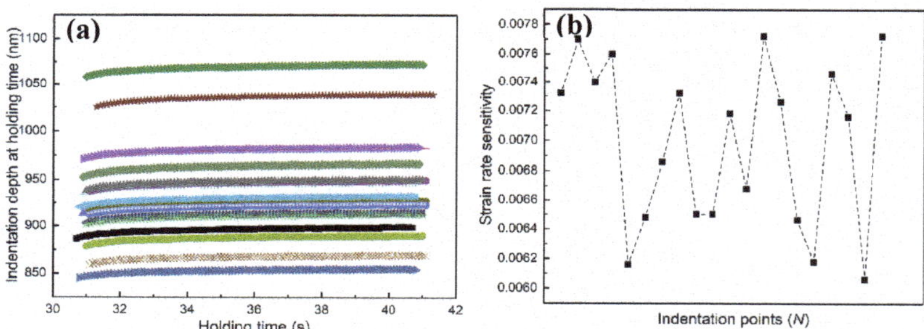

Figure 5. Indentation depth as a function of holding time (**a**) and the calculated strain rate sensitivity (**b**) of NiCoCrAlYCe coatings.

Based on the above results, NiCoCrAlYCe coatings exhibit outstanding nanomechanical properties due to the mono-modal microstructure of NiCoCrAlYCe coatings. Furthermore, compared to nanostructured $La_2(Zr_{0.75}Ce_{0.25})_2O_7$ thermal barrier coatings, the elastic modulus and nanohardness of NiCoCrAlYCe coatings is dramatically improved [28]. In addition, the ratio of the plastic work to the total work of deformation during indentation is larger than that of $La_2(Zr_{0.75}Ce_{0.25})_2O_7$ coatings, which can predict excellent wear resistance for NiCoCrAlYCe coatings.

4. Conclusions

In this work, NiCoCrAlYCe coatings were fabricated by HVOF spraying by using spherical and dense feedstocks. The microstructure of NiCoCrAlYCe coatings was characterized by SEM. Moreover, the nanomechanical properties of coatings were investigated by the nanoindentation method. Some meaningful conclusions can be drawn as follows:

(1) The nanohardness and elastic modulus of NiCoCrAlYCe coatings are 6.09 ± 0.86 GPa and 121.08 ± 10.04 GPa, respectively.

(2) The Weibull distribution reveals the mono-modal microstructure in NiCoCrAlYCe coatings.

(3) The elastic indentation work is 10.322 ± 0.721 nJ, and the plastic indentation work is 22.665 ± 1.702 nJ. The microstructure of NiCoCrAlYCe coatings has little effect on the elastic work, whereas it has a great influence on the plastic work.

(4) It can be predicted that NiCoCrAlYCe coatings have a good wear resistance based on the value of MDP (0.687 ± 0.024).

(5) The strain rate sensitivity of NiCoCrAlYCe coatings is 0.007 ± 0.001 under room temperature.

Author Contributions: Conceptualization, F.Z. and B.X.; methodology, Y.W. (Yiguang Wang) and Y.W. (You Wang); formal analysis, F.Z.; investigation, D.G.; writing—original draft preparation, F.Z. and D.G.; writing—review and editing, F.Z.; funding acquisition, B.X. All authors have read and agreed to the published version of the manuscript.

Funding: This research was funded by the Beijing Institute of Technology Research Fund Program for Young Scholars, National Science and Technology Major Project (2017-VI-0020-0093) and National Natural Science Foundation of China (12090031, 11602125).

Data Availability Statement: Results presented in this paper are not available publicly at this time but may be obtained from the authors upon reasonable request.

Conflicts of Interest: The authors declare no conflict of interest.

References

1. Ghadami, F.; Aghdam, A.S.R.; Ghadami, S. Microstructural characteristics and oxidation behavior of the modified MCrAlX coatings: A critical review. *Vacuum* **2020**, *185*, 109980. [CrossRef]
2. Padture, N.P.; Gell, M.; Jordan, E.H. Thermal Barrier Coatings for Gas-Turbine Engine Applications. *Science* **2002**, *296*, 280–284. [CrossRef] [PubMed]
3. Chen, Y.; Zhao, X.; Xiao, P. Effect of surface curvature on oxidation of a MCrAlY coating. *Corros. Sci.* **2019**, *163*, 108256. [CrossRef]
4. Jackson, G.A.; Sun, W.; McCartney, D.G. The influence of microstructure on the ductile to brittle transition and fracture be-haviour of HVOF NiCoCrAlY coatings determined via small punch tensile testing. *Mater. Sci. Eng. A* **2019**, *754*, 479–490. [CrossRef]
5. Salam, S.; Hou, P.; Zhang, Y.-D.; Wang, H.-F.; Zhang, C.; Yang, Z.-G. Compositional effects on the high-temperature oxidation lifetime of MCrAlY type coating alloys. *Corros. Sci.* **2015**, *95*, 143–151. [CrossRef]
6. Meng, G.-H.; Liu, H.; Liu, M.-J.; Xu, T.; Yang, G.-J.; Li, C.-X.; Li, C.-J. Highly oxidation resistant MCrAlY bond coats prepared by heat treatment under low oxygen content. *Surf. Coat. Technol.* **2019**, *368*, 192–201. [CrossRef]
7. Xie, S.; Lin, S.; Shi, Q.; Wang, W.; Song, C.; Xu, W.; Dai, M. A study on the mechanical and thermal shock properties of MCrAlY coating prepared by arc ion plating. *Surf. Coat. Technol.* **2021**, *413*, 127092. [CrossRef]
8. Li, Z.; Qian, S.; Wang, W. Characterization and oxidation behavior of NiCoCrAlY coating fabricated by electrophoretic deposition and vacuum heat treatment. *Appl. Surf. Sci.* **2010**, *257*, 4616–4620. [CrossRef]
9. Meng, G.-H.; Liu, H.; Liu, M.-J.; Xu, T.; Yang, G.-J.; Li, C.-X.; Li, C.-J. Large-grain α-Al$_2$O$_3$ enabling ultra-high oxidation-resistant MCrAlY bond coats by surface pre-agglomeration treatment. *Corros. Sci.* **2019**, *163*, 108275. [CrossRef]
10. Zakeri, A.; Bahmani, E.; Aghdam, A.S.R. Impact of MCrAlY feedstock powder modification by high-energy ball milling on the microstructure and high-temperature oxidation performance of HVOF-sprayed coatings. *Surf. Coat. Technol.* **2020**, *395*, 125935. [CrossRef]
11. Liu, J.; Wang, J.; Memon, H.; Fu, Y.; Barman, T.; Choi, K.-S.; Hou, X. Hydrophobic/icephobic coatings based on thermal sprayed metallic layers with subsequent surface functionalization. *Surf. Coat. Technol.* **2018**, *357*, 267–272. [CrossRef]
12. Han, Y.J.; Zhu, Z.Y.; Zhang, B.S.; Chu, Y.J.; Zhang, Y.; Fan, J.K. Effects of process parameters of vacuum preoxidation on themicrostructural evolution of CoCrAlY coating deposited by HVOF. *J. Alloys Compd.* **2018**, *735*, 547–559. [CrossRef]
13. Mora-García, A.; Ruiz-Luna, H.; Mosbacher, M.; Popp, R.; Schulz, U.; Glatzel, U.; Muñoz-Saldaña, J. Microstructural analysis of Ta-containing NiCoCrAlY bond coats deposited by HVOF on different Ni-based superalloys. *Surf. Coat. Technol.* **2018**, *354*, 214–225. [CrossRef]
14. Sun, X.; Chen, S.; Wang, Y.; Pan, Z.; Wang, L. Mechanical Properties and Thermal Shock Resistance of HVOF Sprayed NiCrAlY Coatings Without and With Nano Ceria. *J. Therm. Spray Technol.* **2012**, *21*, 818–824. [CrossRef]
15. Karaoglanlia, A.C.; Ozgurluka, Y.; Doleker, K.M. Comparison of microstructure and oxidation behavior of CoNiCrAlY coatings produced by APS, SSAPS, D-gun, HVOF and CGDS techniques. *Vacuum* **2020**, *180*, 109609. [CrossRef]
16. Srivastava, M.; Jadhavb, M.S.; Chethana; Chakradhara, R.P.S.; Singh, S. Investigation of HVOF sprayed novel Al$_{1.4}$Co$_{2.1}$Cr$_{0.7}$Ni$_{2.45}$Si$_{0.2}$Ti$_{0.14}$HEA coating as bond coat material in TBC system. *J. Alloys Compd.* **2022**, *924*, 166388. [CrossRef]
17. Praveen, A.S.; Arjunan, A. Parametric optimisation of high-velocity oxy-fuel nickel-chromium-silicon-boron and alumini-um-oxide coating to improve erosion wear resistance. *Mater. Res. Express* **2019**, *6*, 096560. [CrossRef]

18. Rajendran, P.R.; Duraisamy, T.; Seshadri, R.C.; Mohankumar, A.; Ranganathan, S.; Balachandran, G.; Murugan, K.; Renjith, L. Optimisation of HVOF Spray Process Parameters to Achieve Minimum Porosity and Maximum Hardness in WC-10Ni-5Cr Coatings. *Coat.* **2022**, *12*, 339. [CrossRef]
19. Sacriste, D.; Goubot, N.; Dhers, J.; Ducos, M.; Vardelle, A. An Evaluation of the Electric Arc Spray and (HPPS) Processes for the Manufacturing of High Power Plasma Spraying MCrAlY Coatings. *J. Therm. Spray Technol.* **2001**, *10*, 352–358. [CrossRef]
20. Feizabadi, A.; Doolabi, M.S.; Sadrnezhaad, S.; Rezaei, M. Cyclic oxidation characteristics of HVOF thermal-sprayed NiCoCrAlY and CoNiCrAlY coatings at 1000 °C. *J. Alloys Compd.* **2018**, *746*, 509–519. [CrossRef]
21. Zakeri, A.; Bahmani, E.; Aghdam, A.S.R.; Saeedi, B.; Bai, M. A study on the effect of nano-CeO_2 dispersion on the characteristics of thermally-grown oxide (TGO) formed on NiCoCrAlY powders and coatings during isothermal oxidation. *J. Alloys Compd.* **2020**, *835*, 155319. [CrossRef]
22. Yu, M.; Cui, T.; Zhou, D.; Li, R.; Pu, J.; Li, C. Improved oxidation and hot corrosion resistance of the NiSiAlY alloy at 750 °C. *Mater. Today Commun.* **2021**, *29*, 102939. [CrossRef]
23. Yang, Y.F.; Yao, H.R.; Bao, Z.B.; Ren, P.; Li, W. Modification of NiCoCrAlY with Pt: Part I. Effect of Pt depositing location and cyclic oxidetion performance. *J. Mater. Sci. Technol.* **2019**, *35*, 341–349. [CrossRef]
24. Yu, C.T.; Liu, H.; Jiang, C.Y.; Bao, Z.B.; Zhu, S.L.; Wang, F.H. Modification of NiCoCrAlY with Pt: Part II. Application in TBC with pure metastable tetragonal (t′) phase YSZ and thermal cycling behavior. *J. Mater. Sci. Technol.* **2019**, *35*, 350–359. [CrossRef]
25. Zhou, F.; Zhang, Z.; Liu, S.; Wang, L.; Jia, J.; Wang, Y.; Gong, X.; Gou, J.; Deng, C.; Liu, M. Effect of heat treatment and synergistic rare-earth modified NiCrAlY on bonding strength of nanostructured 8YSZ coatings. *Appl. Surf. Sci.* **2019**, *480*, 636–645. [CrossRef]
26. Zhou, F.; Wang, Y.; Liu, M.; Deng, C.; Zhang, X.; Wang, Y. Acoustic emission monitoring of the tensile behavior of a HVOF-sprayed NiCoCrAlYCe coating. *Appl. Surf. Sci.* **2019**, *504*, 144400. [CrossRef]
27. Zhou, F.; Xu, L.; Deng, C.; Song, J.; Wang, Y.; Liu, M. Nanomechanical characterization of nanostructured $La_2(Zr_{0.75}Ce_{0.25})_2O_7$ thermal barrier coatings by nanoindentation. *Appl. Surf. Sci.* **2019**, *505*, 144585. [CrossRef]
28. Li, C.; Ding, J.; Zhu, F.; Yin, J.; Wang, Z.; Zhao, Y.; Kou, S. Indentation creep behavior of Fe-based amorphous coatings fabricated by high velocity Oxy-fuel. *J. Non-Cryst. Solids* **2018**, *503–504*, 62–68. [CrossRef]
29. Babu, P.S.; Jha, R.; Guzman, M.; Sundararajan, G.; Agarwal, A. Indentation creep behavior of cold sprayed aluminum amorphous/nano-crystalline coatings. *Mater. Sci. Eng. A* **2016**, *658*, 415–421. [CrossRef]
30. Dean, J.; Campbell, J.; Aldrich-Smith, G.; Clyne, T. A critical assessment of the "stable indenter velocity" method for obtaining the creep stress exponent from indentation data. *Acta Mater.* **2014**, *80*, 56–66. [CrossRef]

Article

The Effect of Scanning Strategies on FeCrAl Nuclear Thin-Wall Cladding Manufacturing Accuracy by Laser Powder Bed Fusion

Fusheng Cao [†], Haitian Zhang [†], Hang Zhou, Yu Han, Sai Li, Yang Ran, Jiawei Zhang, Kai Miao *, Zhongliang Lu * and Dichen Li

State Key Laboratory for Manufacturing Systems Engineering, Xi'an Jiaotong University, Xi'an 710049, China
* Correspondence: kaimiao@xjtu.edu.cn (K.M.); zllu@xjtu.edu.cn (Z.L.)
† These authors contributed equally to this work.

Abstract: FeCrAl alloy has been proposed as an alternative material for accident-tolerant fuel (ATF) cladding for nuclear reactors. Thin-wall cladding can be rapidly fabricated by laser powder bed fusion (LPBF). In this paper, a finite element model is established to simulate the transient temperature fields of the cladding under two different laser scanning strategies (linear scanning and ring scanning). In linear scanning simulations, bidirectional scanning, compared with unidirectional scanning, had a smaller temperature gradient along the radial direction. In the ring scanning simulation, the maximum temperature gradually increased and then became stable with the increase of layers. Then, FeCrAl thin-wall cladding with a wall thickness of 0.14 mm was fabricated by LPBF. FeCrAl cladding using the ring scanning strategy had a smaller roughness value (Ra = 4.061 µm). Ring scanning had better accuracy than bidirectional scanning for FeCrAl thin-wall cladding with a wall thickness below 0.4 mm. Therefore, compared with the bidirectional scanning, the ring scanning is more suitable for the high-accuracy manufacturing of FeCrAl thin-wall cladding.

Keywords: FeCrAl alloy; thin-wall cladding; LPBF; simulation; accuracy

Citation: Cao, F.; Zhang, H.; Zhou, H.; Han, Y.; Li, S.; Ran, Y.; Zhang, J.; Miao, K.; Lu, Z.; Li, D. The Effect of Scanning Strategies on FeCrAl Nuclear Thin-Wall Cladding Manufacturing Accuracy by Laser Powder Bed Fusion. *Crystals* **2022**, *12*, 1197. https://doi.org/10.3390/cryst12091197

Academic Editor: Umberto Prisco

Received: 21 July 2022
Accepted: 21 August 2022
Published: 25 August 2022

Publisher's Note: MDPI stays neutral with regard to jurisdictional claims in published maps and institutional affiliations.

1. Introduction

The safety accident at the Fukushima Daiichi nuclear power plant in 2011 highlighted a lack of fault tolerance among traditional nuclear fuel element [1]. Zirconium-based alloys are often used as nuclear fuel cladding and structural components in conventional light water reactors (LWR). However, in high-temperature steam environments, the aggressive oxidation and high heat generation of zirconium-based alloys may significantly increase the risk of hydrogen-induced explosions [2]. Therefore, the development of advanced accident-tolerant fuel (ATF) with greater safety margins has become a major focus [3,4]. At present, FeCrAl alloy is a promising ATF cladding material to replace zirconium [5]. Because of the formation of an alumina film on the surface, FeCrAl alloys have unrivalled oxidation resistance in air and steam at temperatures above 1000 °C, with a melting point of approximately 1500 °C [6–8]. They are being investigated as an option for ATF cladding.

The traditional cladding manufacturing method is generally rolling forming, which requires a series of complicated process flows, has low processing efficiency, and encounters difficulties in manufacturing complex parts. Due to the larger thermal neutron absorption cross-section of FeCrAl [9], the cladding thickness is more strictly controlled than that of conventional Zr alloys, which put forward higher requirements for the fabrication of cladding. Laser powder bed fusion (LPBF) is a promising method for the rapid fabrication of FeCrAl nuclear fuel cladding. Compared with the traditional fabrication process, LPBF has many advantages [10–12]. LPBF method is an additive manufacturing process; it can produce complex three-dimensional structures. In addition, the LPBF process has the characteristics of a fast forming speed, high dimensional accuracy and customization [13,14].

In the LPBF process, metal powders are rapidly melted and solidified by the laser heat source. The transient temperature field plays an important role in the process and has a direct influence on the deformation and residual stress [15]. However, the processing parameters, such as laser power, scanning speed and scanning strategy, greatly affect the temperature field distribution, and then affect the final part quality [16]. Therefore, in recent years, many scholars have simulated the forming process of LPBF through the finite element method to predict the quality and performance of the final formed parts.

Roberts et al. [17] adopted a new simulation technology, Element Birth and Death, to simulate the three-dimensional temperature field of a multilayer powder bed. The results show that there was a slight increasing temperature accumulation in the parts and the base plate as the number of layers increases. Hussein et al. [18] developed a nonlinear transient model based on sequential coupled thermo-field analysis using ANSYS APDL and simulated the temperature field and stress field of monolayers 316 L stainless steel LPBF. It was found that the width and depth of the molten pool decreased with the increase of scanning speed. Meanwhile, the length of the molten pool increased due to the limitation of the cooling rate. Parry et al. [19] studied the influence of scanning strategy on residual stress by using finite element analysis. It was found that there is no significant numerical difference between the two scanning strategies (bidirectional and unidirectional), but their distribution differs due to the thermal history difference.

Some researchers have studied the manufacturing of FeCrAl alloys by LPBF. Walk et al. [20] fabricated the simple thin-wall FeCrAl structure. They researched the effects of laser power and scanning speed on the fine oxide particle distribution. Boegelein et al. [21] prepared PM2000 (ODS-FeCrAl) samples by LPBF. Fine dispersion of Y-containing precipitates was observed in the manufactured thin-wall component. However, although these studies obtained the FeCrAl thin-walled structure, they did not study the effect of scanning strategy on it.

This paper aims to explore the effect of the scanning strategy on forming accuracy and quality of FeCrAl alloy cladding by LPBF. Since the cladding is tubular, the ring scanning strategy matches its shape better than other methods. Meanwhile, linear scanning is a widely used scanning strategy. Therefore, this paper will focus on these two scanning strategies. The temperature distribution of linear scanning and ring scanning strategies were studied by numerical simulation. Herein, numerical simulations were operated by Ansys software using the finite element method (FEM). At the same time, the test and analysis of the FeCrAl cladding were studied with white light interferometry and electron backscatter diffraction (EBSD). The results of this paper can provide an effective way to prepare nuclear thin-wall cladding by LPBF.

2. Model for LPBF Process

Firstly, a numerical model based on the finite element method is established by ANSYS 19.2 software (ANSYS, Inc., Canonsburg, PA, USA). Figure 1 illustrates the LPBF process. The laser beam illuminates the surface of the powder bed in a preset path. The metal powder rises in temperature and melts beyond its melting point. Subsequently, the molten pool is rapidly cooled and solidified. Repeat this step, layer by layer, to complete the parts. Therefore, heat transfer must be taken into consideration in LPBF. The FeCrAl alloy used in this study is a new type of second-generation nuclear FeCrAl alloy and many of its thermal properties parameters are not complete data. C35M is a kind of FeCrAl alloy, and its chemical composition is close to that of the FeCrAl used in this paper, as shown in Table 1 [22]. So, the simulation in this paper uses C35M data.

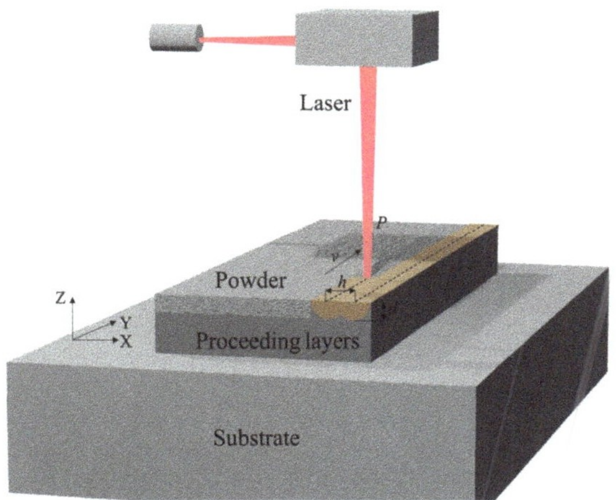

Figure 1. LPBF process.

Table 1. Chemical composition of FeCrAl alloy and C35M (in wt.%).

IDs	Fe	Mo	Al	Cr	C	O	N	S
FeCrAl	81.27	1.98	3.92	12.82	<0.018	<0.008	<0.074	<0.003
C35M	80.15	2	4.5	13	-	-	-	-

2.1. Thermal Models

In LPBF, the thermal transfer is affected by inherent properties such as density, heat capacity and thermal conductivity. The distribution of the temperature field satisfies the differential equation of three-dimensional heat conduction, which can be expressed as [18]:

$$\rho c \frac{\partial T}{\partial t} = \frac{\partial}{\partial x}\left(k\frac{\partial T}{\partial x}\right) + \frac{\partial}{\partial y}\left(k\frac{\partial T}{\partial y}\right) + \frac{\partial}{\partial z}\left(k\frac{\partial T}{\partial z}\right) + Q \tag{1}$$

where ρ is the material density, c is the specific heat capacity, T is the temperature, t is the time, k is the thermal conductivity, and Q is heat generated per volume within the component.

The initial temperature was set to the preheating temperature. The initial temperature was 25 °C at 0 s [18];

$$T(x,y,z,t)|_{t=0} = T_0(x,y,z) \in D \tag{2}$$

where T_0 is the preheating temperature and D denotes the manufactured area.

The coefficients of thermal conductivity, density and specific heat capacity change with temperature in the LPBF process. The values of the above three temperature-dependent properties are listed in Table 2 [23].

Table 2. Thermal physical of FeCrAl.

Temperature (K)	270	400	600	800	1000	1200	1400	1600	1800
Thermal conductivity (W/m K)	12.5	14.3	17.0	19.5	21.9	24.0	26.1	28.0	29.7
Specific heat (J/kg K)	423.3	513.8	616.8	791.4	742.9	709.4	741.3	855.9	1082.5
Density (kg/m³)	7.10	7.06	7.01	6.94	6.84	6.72	6.59	6.43	6.26

2.2. Heat Source Model

The laser energy distribution in LPBF was the Gaussian distribution, which was mathematically presented as [24]:

$$q(r) = \frac{2\alpha P}{\pi \omega^2} \cdot \exp\left(-\frac{2r^2}{\omega^2}\right) \tag{3}$$

where $q(r)$ is the input heat flux; α is the absorption rate of powder, which is 70%; ω is the radius of laser spot, which is 0.5 mm; P is the laser power and r is the radial distance from the center of the laser beam.

2.3. Finite Element Model

The 3D finite element model of the LPBF is shown in Figure 2a. The gray part represents the substrate, and the yellow part represents the power bed. In the simulation, the model is simplified to reduce the amount of calculation. Set the size of the substrate to 30 mm × 30 mm × 2 mm. The inner diameter of the cladding is 10 mm, the wall thickness is 1 mm. The thickness of a single powder layer was 30 μm, and three powder layers were set in the model. The mesh dimensions of cladding are 100 μm × 100 μm × 30 μm, and the mesh dimensions of the substrate are 500 μm × 500 μm × 500 μm. The three-dimensional simulation model is meshed into 38,644 nodes and 21,600 elements in total.

The scanning strategies of the laser beam are unidirectional, bidirectional and employ ring scanning, as shown in Figure 2b. Because the FeCrAl cladding has a ring structure, many scanning paths need to be defined when linear scanning is adopted. This simulation is very complicated. Therefore, parts of the ring model are used for simplification, as shown in Figure 2c. Three paths are defined to simulate the local linear scanning of the ring. According to the scanning spacing of 0.05 mm, the effective width of the three paths is 0.15 mm. The central angle of the arc length in the local model is only about 1.7 °, which is simplified as a rectangle (1 mm × 0.15 mm). The simplified substrate size is 2 mm × 1 mm × 0.25 mm. So, the three-dimensional simulation model of simplified model is meshed into 21,750 nodes and 4240 elements.

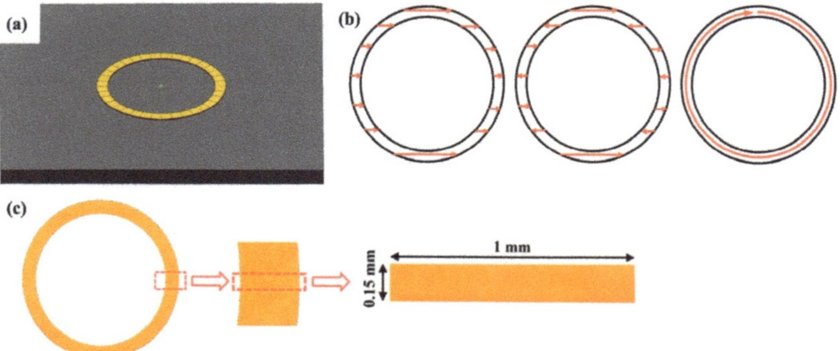

Figure 2. (**a**) 3D finite element model; (**b**) unidirectional, bidirectional and ring scanning strategies; (**c**) parts of the ring model.

In the present model, some assumptions are made as follows [25]:

(1) The whole FeCrAl powder bed is assumed to be a continuous and uniform medium and the heat transfer between the powder pores is not considered.
(2) The shrinkage of the powder bed is negligible for simplifying the calculation.
(3) The surface of the molten pool is assumed to be flat without respect to evaporation and capillary flow.

(4) The moving laser heat source is modeled as the Gaussian distribution and is input directly on the surface of the powder layer.

2.4. Boundary Conditions

In this model, the temperature of the air and the heat transfer coefficient between the object and the air are specified. The overall initial temperature of the substrate is set to a different preheating temperature or the normal temperature (25 °C), and the bottom of the substrate is set to be insulated. The vertical contact side is static air, and its convective coefficient is set to 5 W/K m^2 [26]. In the LPBF process, the powders are densely spread on the substrate to form powder layers. The flow rate of air on the powder layer and substrate surface is 0. In addition, there must be radiation heat transfer between the powder layer, the substrate and the surrounding environment, but the heat exchanged by radiation heat transfer is far less than that by convection heat transfer. As such, only convection heat transfer needs to be considered, which can be expressed as [27],

$$-\kappa \frac{\partial T}{\partial z} = h_i(T - T_E) \tag{4}$$

where h_i is the thermal convection coefficient; κ is the thermal conductivity; T is temperature and T_E is the environment temperature.

Table 3 lists the simulation parameters of this paper.

Table 3. Simulation parameters.

Parameter	Unit	Value
Laser power	W	200
Scan speed	mm/s	500
Hatching spaces	mm	0.05
Layer thickness	mm	0.03
Initial temperature	°C	25
Thermal conductivity	W/m K	Table 2
Specific heat	J/kg K	Table 2
Density	kg/m^3	Table 2
Absorption	%	70
Radius of laser spot	mm	0.5
Convective heat transfer coefficient	W/K m^2	5

3. Materials and Experiment

Self-made gas-atomized FeCrAl alloy powders were used as raw material. Table 1 shows the chemical composition of FeCrAl alloy powder. The micromorphology of the FeCrAl alloy powders is shown in Figure 3a and the particle size distribution is shown in Figure 3b.

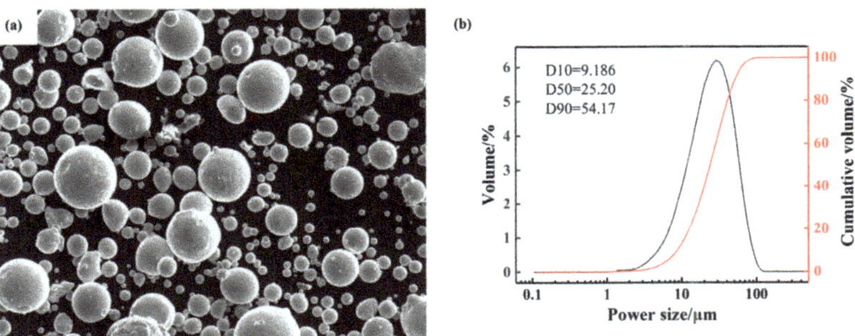

Figure 3. (a) The micromorphology of FeCrAl alloy powders and (b) the particle size distribution.

Laser powder bed fusion equipment (the home-built SLM-M150) was used to fabricate FeCrAl alloy specimens. The specimens were fabricated using various combinations of laser powers (200 W), scanning speeds (500 mm/s), hatching spaces (0.05 mm) and layer thickness (0.03 mm). Bidirectional scanning and ring scanning strategy were used to manufacture thin-wall cladding, respectively.

The surface roughness of cladding was measured by white light interferometry (1000 WLI). A scanning electron microscope (SU3500) was used for microstructure analysis.

4. Results and Discussion

4.1. Linear Scanning Simulation

In multi-pass linear scanning, remelting zones are formed between the weld passes to achieve metallurgical bonding. Unidirectional scanning and bidirectional scanning methods are selected to conduct simulation respectively. The simulation curves of maximum temperature are shown in Figure 4a. The simple change of scanning mode has little influence on the trend of maximum temperature but has a great influence on the whole temperature field. In the process of unidirectional scanning, each laser track is not continuous. Therefore, the maximum temperature of the molten pool is stable at a constant value. About bidirectional scanning, the laser beam is continuous at the ends. Heat accumulates gradually, so the molten pool temperature continues to rise. The maximum temperature of bidirectional scanning is higher than that of unidirectional scanning. Figure 4b gives the temperature field distribution of different linear scanning at the same time. The temperature gradient along the axis of bidirectional scanning is obviously smaller than that of unidirectional scanning. The bidirectional scanning has a wide heat distribution. This is related to the thermal history of the laser scanning path. This helps to reduce axial stress and deformation of the cladding.

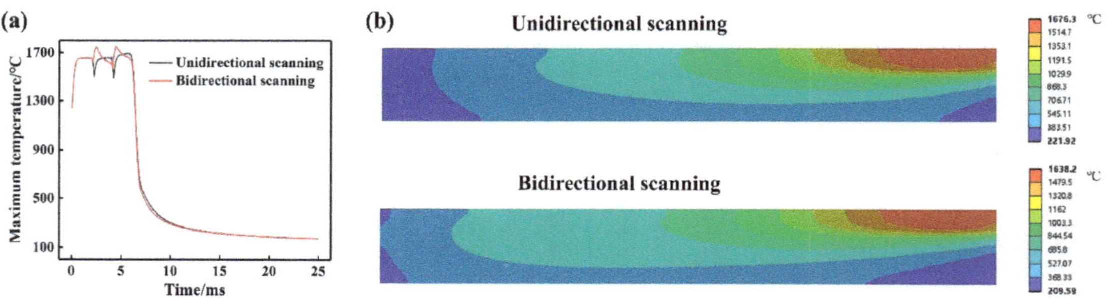

Figure 4. (a) The simulation curves of maximum temperature under different linear scanning; (b) temperature field distribution of different linear scanning at the same time.

4.2. Ring Scanning Simulation

Figure 5a shows the maximum temperature variation curve of three layer at the ring scanning strategy. Figure 5b shows the temperature distribution of the first layer at six moments. After the laser scanning starts operation, the maximum temperature of the molten pool rose rapidly to above the melting point and fluctuated within a small temperature range until the scanning was completed. At the end of the first layer ($t = 0.07$ s), the laser spot returned near the starting point, and the residual temperature at the starting point increased the maximum temperature, as shown in Figure 5b. Then the laser heat source disappeared, and the maximum temperature decreased sharply, and gradually decreased to near the preheating temperature.

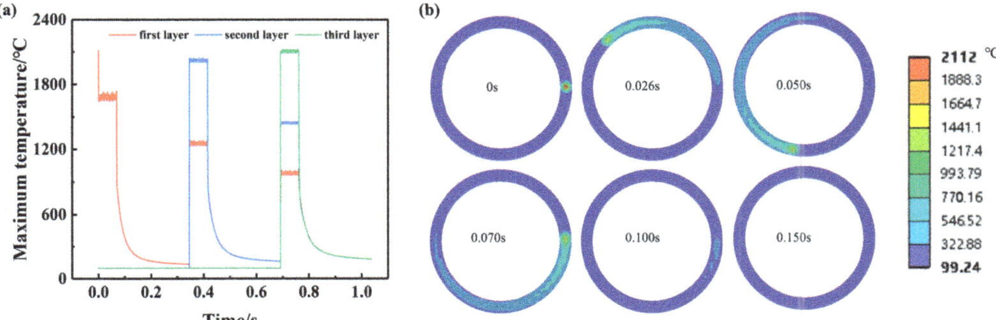

Figure 5. (**a**) Maximum temperature curve of ring scanning, (**b**) temperature field distribution of the first layer of the ring scanning.

As shown in Figure 5a, because of the short interval between layers and significant heat accumulation, the maximum temperature of the third layer was significantly higher than that of the first layer. However, the temperature increases of the third layer compared with the second layer were significantly less than that of the second layer compared with the first layer. The maximum temperature tends to stabilize as the number of layers increases. This indicates that the temperature gradient decreases as the number of layers increases. The grain size is larger when the higher layer is formed. When scanning the third layer, it can be seen that the maximum temperature of the third layer is much higher than that of the second layer. There is a large temperature gradient between layers along the building direction. As the printing continues, the formed layer will still have a temperature fluctuation, so the bottom layer will go through several thermal cycles in the whole process, which will make the bottom grain also have a trend of coarsening.

Figure 6 shows the four points selected on the cladding model, and their temperature changes are shown in Figure 6b. There was almost no obvious difference between point B and point C. It can be seen that the inner and outer temperature changes of ring scanning were little affected. The maximum temperature difference between point A and D was about 530 °C, and that between point A and point C was about 120 °C.

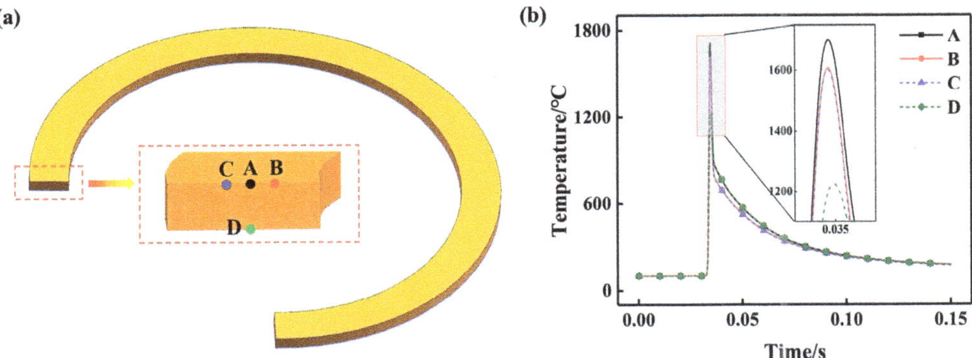

Figure 6. The temperature changes with time at different points on the longitudinal section: (**a**) the distance from point A to point B is 0.1 mm, the distance from point A to point C is 0.1 mm, the distance from point A to point D is 0.03 mm, and (**b**) temperature curve over time.

4.3. Forming Accuracy

Due to the particularity of thin-wall cladding and ring shape of thin-wall cladding, ring scanning is one of the best candidates. In order to compare the advantages and disadvantages of the two scanning methods in forming FeCrAl thin-wall cladding, bidirectional and ring scanning were used respectively, as shown in Figure 7. The inner diameter of the design cladding model is 10 mm, and the wall thickness is 0.05, 0.1, 0.2, 0.4 and 0.8 mm, respectively. Figure 7a shows the actual samples used for the ring scanning and the measured wall thicknesses were 0.14, 0.18, 0.30, 0.54, and 0.96 mm, respectively. Figure 7b gives actual samples used for the bidirectional scanning and the measured wall thickness were 0.37, 0.47, 0.65 and 1.03 mm, respectively. When the wall thickness of the model is 0.05 mm, FeCrAl cladding cannot be fabricated by bidirectional scanning. It can be seen the thickness of forming cladding is greater than that of the designed model. The thinnest thickness that can be formed by bidirectional scanning is 0.37 mm and the thinnest thickness can reach 0.14 mm by ring scanning. So, the range of bidirectional scanning meeting the wall thickness requirement is small and ring scanning can fully meet the requirements of wall thickness.

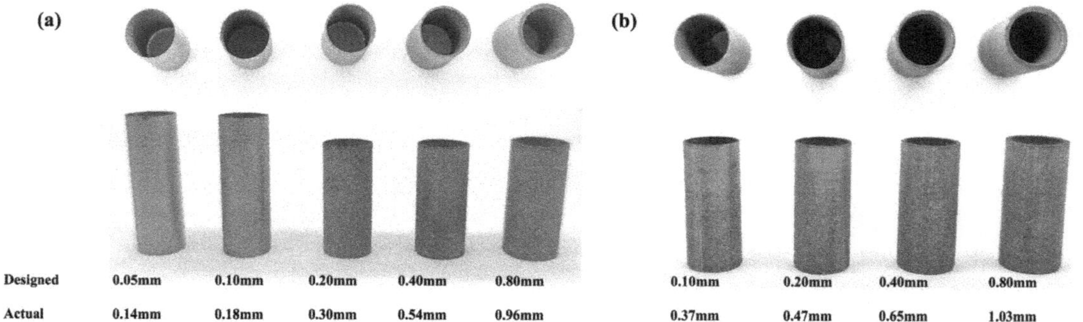

Figure 7. FeCrAl cladding of different thickness in LPBF, (**a**) ring scanning and (**b**) bidirectional scanning.

White light interferometry was used to collect the outer three-dimensional shape of the FeCrAl cladding by ring scanning and bidirectional scanning respectively, as shown in Figure 8. The cladding which used ring scanning had a smaller roughness value (Ra = 4.061 μm) than the cladding using bidirectional scanning (Ra = 7.946 μm). Figure 8a illustrates that the outer part of the cladding was evenly distributed with obvious small burrs, but the outer surface was still smooth. These burrs were caused by the flow of the ring scanning pool into the gaps between the powders on both sides and can be removed by subsequent sandblasting or surface finishing. Figure 8b gives the morphology of the outer part of the cladding using bidirectional scanning. The outer surface of the cladding was uneven and had many burrs. This was because the end point and start point of each scan line of bidirectional scanning were on the outer surface, which made the surface quality of the cladding significantly worse than that of the ring scanning. Therefore, in order to control the wall thickness within 0.4 mm and obtain better surface quality, the ring scanning was obviously better than the bidirectional scanning.

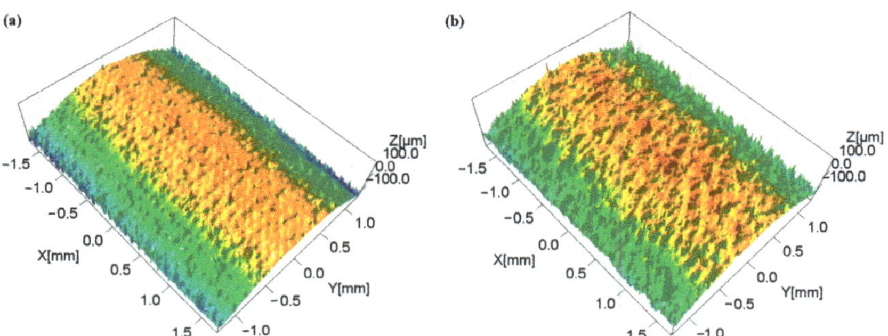

Figure 8. The outer three-dimensional shape of FeCrAl cladding in LPBF, (**a**) ring scanning and (**b**) bidirectional scanning.

4.4. Microstructure

FeCrAl cladding of 50 mm high and 1 mm wall thickness manufactured using a ring-type scanning strategy was selected. The top (45–50 mm) and bottom (0–5 mm) of the cladding were cut, respectively. The vertical and cross-sections at the top and bottom of the sample were characterized by EBSD characterization.

The microstructure of the top and bottom vertical sections of the FeCrAl cladding is shown in Figure 9. The grains at the bottom were more concentrated in <001> and <101> direction, while in <111> direction grains were fewer. The grain orientation at the top was distributed in <001>, <101> and <111> directions without obvious optimization of grain direction. Meanwhile, Figure 9c,d give the pole figure of the top and bottom vertical sections of the cladding, respectively. There was no strong texture in either. Figure 10 illustrates the microstructure of the top and bottom cross-sections of the FeCrAl cladding. There was no obvious optimization of grain orientation in the two sections. The orientation of grain was distributed in <001>, <101> and <111> directions. Figure 10c,d are polar diagrams of cross-sections at the top and bottom of the cladding, respectively. The two cross-sections also show no strong texture. Figure 10b shows that the orientation of grains in the middle region were very different from those on both sides. The grains in the middle region were concentrated in <100> and <111> directions, and the grains on both sides were concentrated in <001> direction. The heat dissipation conditions of different parts of the ring were different and the orientation of different parts of the grain was different, and so the optimization of the orientation of the whole grain was not obvious. However, the metal powders around the cladding forming had preheating and heat transfer qualities. The temperature was obviously higher than the top of the air; the vertical direction was still the main direction of heat dissipation. The temperature gradient in the building direction was still significantly greater than that in the horizontal direction, and the grains were still columnar.

Figure 11 shows the grain size distribution of the vertical and cross sections at the top and bottom of the cladding. The average grain size at the bottom of the vertical section was 434.82 μm^2, and the average grain size at the top was 613.63 μm^2. The average grain size at the bottom of the cross section was 132.44 μm^2, and the average grain size at the top was 149.41 μm^2. The grains at the top were significantly larger than those at the bottom. With the increase in the number of processing layers, the temperature gradient at the beginning of the condensation of the top molten pool was much smaller than that at the beginning of the condensation of the bottom molten pool when the heat accumulation tends to be stable.

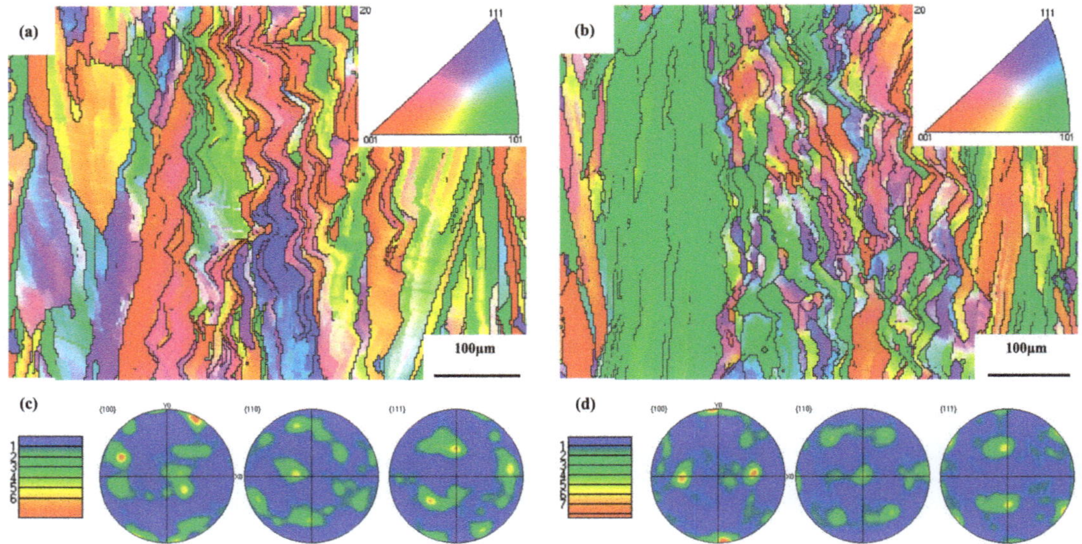

Figure 9. EBSD analysis of vertical sections of top (**a,c**) and bottom (**b,d**) of thin-wall cladding, (**c,d**) pole figure of (**a,b**).

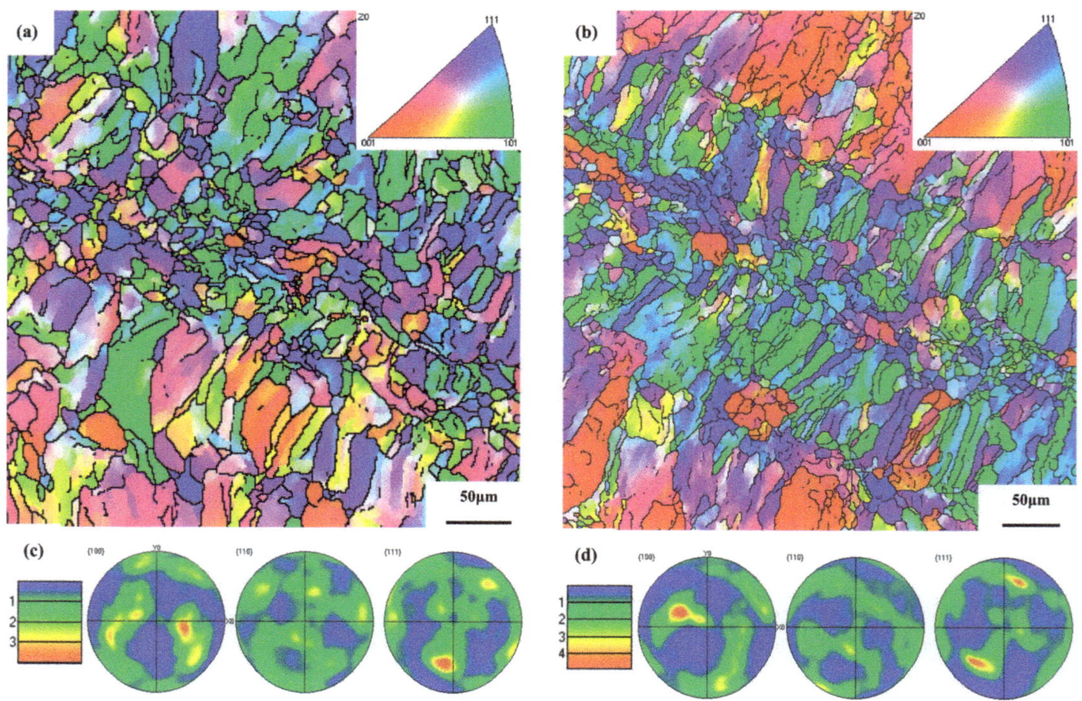

Figure 10. EBSD analysis of cross-sections of top (**a,c**) and bottom (**b,d**) of thin-wall cladding, (**c,d**) pole figure of (**a,b**).

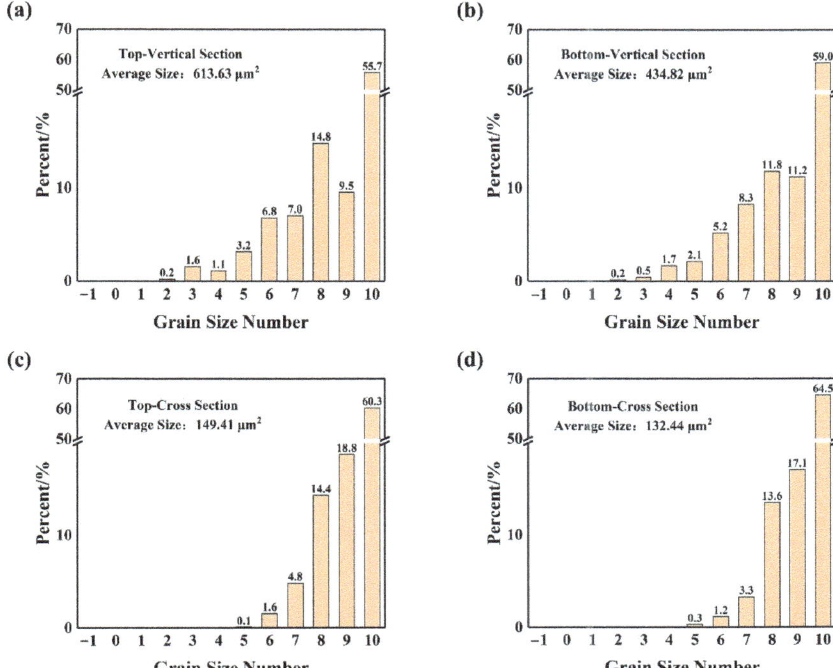

Figure 11. Grain size distribution of vertical sections (**a**,**b**) and cross-sections (**c**,**d**) of FeCrAl cladding at the top region (**a**,**c**) and the bottom region (**b**,**d**).

5. Conclusions

In this study, thin-wall FeCrAl claddings are fabricated by laser powder bed fusion. The effect of the scanning strategy was studied by simulation and experiment. The main conclusions are as follows:

(1) Linear scanning simulation shows that the maximum temperature of unidirectional scanning is stable. The maximum temperature of bidirectional scanning is higher than that of unidirectional scanning. The bidirectional scanning has a smaller temperature gradient along the radial direction. It can help to improve the dimensional accuracy of the FeCrAl cladding.

(2) Ring scanning simulation gives the maximum temperature fluctuated within a very small temperature range after the laser scanning began. The maximum temperature gradually increases and then tends to be stable with the increase of layers. The temperature gradient is greatest along the building direction.

(3) Thin-wall cladding samples were fabricated by LPBF. The thinnest shell thickness is 0.37 mm in the bidirectional scanning strategy, and 0.14 mm in ring scanning strategy. The cladding used for the ring scanning has a smaller roughness value (Ra = 4.061 μm) than the cladding used for bidirectional scanning (Ra = 7.946 μm). When the thickness of thin-walled cladding is less than 0.4 mm, the manufacturing accuracy of ring scanning is better than that of bidirectional scanning.

(4) The grain size of the top and bottom of the cladding was analyzed. The grain of the cladding is columnar. The grain size at the top is larger than that at the bottom due to thermal accumulation.

Author Contributions: Conceptualization, F.C. and H.Z. (Haitian Zhang); methodology, F.C. and H.Z. (Haitian Zhang); software, H.Z. (Haitian Zhang) and H.Z. (Hang Zhou); validation, F.C. and H.Z. (Hang Zhou); formal analysis, Y.H. and S.L.; investigation, F.C., Y.R. and J.Z.; resources, F.C., K.M. and Z.L.; data curation, Y.H.; writing—original draft preparation, F.C., H.Z. (Haitian Zhang) and H.Z. (Hang Zhou); writing—review and editing, H.Z. (Hang Zhou), K.M. and Z.L.; visualization, F.C.; supervision, D.L.; project administration, Z.L.; funding acquisition, Z.L. All authors have read and agreed to the published version of the manuscript.

Funding: This work was funded by the Key project of Nuclear Safety and advanced Nuclear Technology, grant number 2019YFB1901002.

Institutional Review Board Statement: Not applicable.

Informed Consent Statement: Not applicable.

Data Availability Statement: Not applicable.

Conflicts of Interest: The authors declare no conflict of interest.

References

1. Chen, P.; Qiu, B.; Wu, J.; Zheng, M.; Gao, S.; Zhang, K.; Zhou, Y.; Wu, Y. A comparative study of in-pile behaviors of FeCrAl cladding under normal and accident conditions with updated FROBA-ATF code. *Nucl. Eng. Des.* **2021**, *371*, 110889. [CrossRef]
2. Tang, C.; Jianu, A.; Steinbrueck, M.; Grosse, M.; Weisenburger, A.; Seifert, H.J. Influence of composition and heating schedules on compatibility of FeCrAl alloys with high-temperature steam. *J. Nucl. Mater.* **2018**, *511*, 496–507. [CrossRef]
3. Duan, Z.; Yang, H.; Satoh, Y.; Murakami, K.; Kano, S.; Zhao, Z.; Shen, J.; Abe, H. Current status of materials development of nuclear fuel cladding tubes for light water reactors. *Nucl. Eng. Des.* **2017**, *316*, 131–150. [CrossRef]
4. Park, D.J.; Kim, H.G.; Jung, Y.I.; Park, J.H.; Yang, J.H.; Koo, Y.H. Microstructure and mechanical behavior of Zr substrates coated with FeCrAl and Mo by cold-spraying. *J. Nucl. Mater.* **2018**, *504*, 261–266. [CrossRef]
5. Chang, K.; Meng, F.; Ge, F.; Zhao, G.; Du, S.; Huang, F. Theory-guided bottom-up design of the FeCrAl alloys as accident tolerant fuel cladding materials. *J. Nucl. Mater.* **2019**, *516*, 63–72. [CrossRef]
6. Field, K.G.; Yamamoto, Y.; Pint, B.A.; Gussev, M.N.; Terrani, K.A. Accident tolerant FeCrAl fuel cladding: Current status towards commercialization. In Proceedings of the 18th International Conference on Environmental Degradation of Materials in Nuclear Power Systems–Water Reactors, Portland, OR, USA, 13–17 August 2019; pp. 1381–1389.
7. Park, D.J.; Kim, H.G.; Park, J.Y.; Jung, Y.I.; Park, J.H.; Koo, Y.H. A study of the oxidation of FeCrAl alloy in pressurized water and high-temperature steam environment. *Corros. Sci.* **2015**, *94*, 459–465. [CrossRef]
8. Gupta, V.K.; Larsen, M.; Rebak, R.B. Utilizing FeCrAl oxidation resistance properties in water, air and steam for accident tolerant fuel cladding. *ECS Trans.* **2018**, *85*, 3. [CrossRef]
9. Alrwashdeh, M.; Alameri, S.A. Preliminary neutronic analysis of alternative cladding materials for APR-1400 fuel assembly. *Nucl. Eng. Des.* **2021**, *384*, 111486. [CrossRef]
10. Bartkowiak, K.; Ullrich, S.; Frick, T.; Schmidt, M. New developments of laser processing aluminium alloys via additive manufacturing technique. *Phys. Procedia* **2011**, *12*, 393–401. [CrossRef]
11. Amato, K.; Gaytan, S.; Murr, L.E.; Martinez, E.; Shindo, P.; Hernandez, J.; Collins, S.; Medina, F. Microstructures and mechanical behavior of Inconel 718 fabricated by selective laser melting. *Acta Mater.* **2012**, *60*, 2229–2239. [CrossRef]
12. Thijs, L.; Kempen, K.; Kruth, J.-P.; Van Humbeeck, J. Fine-structured aluminium products with controllable texture by selective laser melting of pre-alloyed AlSi10Mg powder. *Acta Mater.* **2013**, *61*, 1809–1819. [CrossRef]
13. Nagahari, T.; Nagoya, T.; Kakehi, K.; Sato, N.; Nakano, S. Microstructure and creep properties of Ni-Base Superalloy IN718 built up by selective laser melting in a vacuum environment. *Metals* **2020**, *10*, 362. [CrossRef]
14. Xiao, Z.; Yang, Y.; Wang, D.; Song, C.; Bai, Y. Structural optimization design for antenna bracket manufactured by selective laser melting. *Rapid Prototyp. J.* **2018**, *24*, 539–547. [CrossRef]
15. Mohanty, S.; Hattel, J. Cellular scanning strategy for selective laser melting: Capturing thermal trends with a low-fidelity, pseudo-analytical model. *Math. Probl. Eng.* **2014**, *2014*, 715058. [CrossRef]
16. He, K.; Zhao, X. 3D thermal finite element analysis of the SLM 316L parts with microstructural correlations. *Complexity* **2018**, *2018*, 6910187. [CrossRef]
17. Roberts, I.A.; Wang, C.; Esterlein, R.; Stanford, M.; Mynors, D. A three-dimensional finite element analysis of the temperature field during laser melting of metal powders in additive layer manufacturing. *Int. J. Mach. Tools Manuf.* **2009**, *49*, 916–923. [CrossRef]
18. Hussein, A.; Hao, L.; Yan, C.; Everson, R. Finite element simulation of the temperature and stress fields in single layers built without-support in selective laser melting. *Mater. Des.* **2013**, *52*, 638–647. [CrossRef]
19. Parry, L.; Ashcroft, I.; Wildman, R.D. Understanding the effect of laser scan strategy on residual stress in selective laser melting through thermo-mechanical simulation. *Addit. Manuf.* **2016**, *12*, 1–15. [CrossRef]
20. Walker, J.C.; Berggreen, K.M.; Jones, A.R.; Sutcliffe, C.J. Fabrication of Fe–Cr–Al oxide dispersion strengthened PM2000 alloy using selective laser melting. *Adv. Eng. Mater.* **2009**, *11*, 541–546. [CrossRef]

21.	Boegelein, T.; Louvis, E.; Dawson, K.; Tatlock, G.J.; Jones, A.R. Characterisation of a complex thin walled structure fabricated by selective laser melting using a ferritic oxide dispersion strengthened steel. *Mater. Charact.* **2016**, *112*, 30–40. [CrossRef]
22.	Yamamoto, Y.; Pint, B.A.; Terrani, K.A.; Field, K.G.; Yang, Y.; Snead, L.L. Development and property evaluation of nuclear grade wrought FeCrAl fuel cladding for light water reactors. *J. Nucl. Mater.* **2015**, *467*, 703–716. [CrossRef]
23.	Field, K.G.; Snead, M.A.; Yamamoto, Y.; Terrani, K.A. *Handbook on the Material Properties of FeCrAl Alloys for Nuclear Power Production Applications*; Nuclear Technology Research and Development; Oak Ridge National Laboratory: Oak Ridge, TN, USA, 2017.
24.	Yin, J.; Zhu, H.; Ke, L.; Lei, W.; Dai, C.; Zuo, D. Simulation of temperature distribution in single metallic powder layer for laser micro-sintering. *Comput. Mater. Sci.* **2012**, *53*, 333–339. [CrossRef]
25.	Huang, W.; Zhang, Y. Finite element simulation of thermal behavior in single-track multiple-layers thin wall without-support during selective laser melting. *J. Manuf. Processes* **2019**, *42*, 139–148. [CrossRef]
26.	Kosky, P.; Balmer, R.T.; Keat, W.D.; Wise, G. *Exploring Engineering: An Introduction to Engineering and Design*, 5th ed.; Academic Press: Cambridge, MA, USA, 2021; pp. 317–340.
27.	Duan, W.; Yin, Y.; Zhou, J. Temperature field simulations during selective laser melting process based on fully threaded tree. *Chin. Foundry* **2017**, *14*, 405–411. [CrossRef]

crystals

Article

Experimental Investigation and Prediction of Mechanical Properties in a Fused Deposition Modeling Process

Amanuel Diriba Tura [1], Hirpa G. Lemu [2,*] and Hana Beyene Mamo [1]

[1] Faculty of Mechanical Engineering, Jimma University, MVJ4+R95 Jimma, Ethiopia; diriba.amanuel@ju.edu.et (A.D.T.); hanicho2102@gmail.com (H.B.M.)
[2] Faculty of Science and Technology, University of Stavanger, N-4036 Stavanger, Norway
* Correspondence: hirpa.g.lemu@uis.no

Abstract: Additive manufacturing, also known as three-dimensional printing, is a computer-controlled advanced manufacturing process that produces three-dimensional items by depositing materials directly from a computer-aided design model, usually in layers. Due to its capacity to manufacture complicated objects utilizing a wide range of materials with outstanding mechanical qualities, fused deposition modeling is one of the most commonly used additive manufacturing technologies. For printing high-quality components with appropriate mechanical qualities, such as tensile strength and flexural strength, the selection of adequate processing parameters is critical. Experimentally, the influence of process parameters such as the raster angle, printing orientation, air gap, raster width, and layer height on the tensile strength of fused deposition modeling printed items was examined in this work. Through analysis of variance, the impact of each parameter was measured and rated. The system's response was predicted using an adaptive neuro-fuzzy technique and an artificial neural network. In Minitab software, the Box-Behnken response surface experimental design was used to generate 46 experimental trials, which were then printed using acrylonitrile butadiene styrene polymer materials on a three-dimensional forge dreamer II fused deposition modelling printing machine. The results revealed that the raster angle, air gap, and raster width had significant impacts on the tensile strength. The adaptive neuro-fuzzy approach and artificial neural network predicted tensile strength accurately with an average percentage error of 0.0163 percent and 1.6437 percent, respectively. According to the findings, the model and experimental data are in good agreement.

Keywords: fused deposition modeling; mechanical properties; tensile strength; adaptive neuron-fuzzy methods; artificial neural network

Citation: Tura, A.D.; Lemu, H.G.; Mamo, H.B. Experimental Investigation and Prediction of Mechanical Properties in a Fused Deposition Modeling Process. *Crystals* **2022**, *12*, 844. https://doi.org/10.3390/cryst12060844

Academic Editors: Hao Yi, Huajun Cao, Menglin Liu and Le Jia

Received: 15 May 2022
Accepted: 13 June 2022
Published: 15 June 2022

Publisher's Note: MDPI stays neutral with regard to jurisdictional claims in published maps and institutional affiliations.

1. Introduction

Additive manufacturing is a novel technology that uses a layer-based production method to create a product straight from a computer-aided design model (CAD) model. Fused deposition modeling (FDM) is one of several three-dimensional (3D) printing techniques that employ flexible thermoplastic filament injected through a heated nozzle to build components. The thermoplastics and reinforced thermoplastic materials that can be printed with FDM include acrylonitrile butadiene styrene (ABS), polylactic acid (PLA), polycarbonate, unfilled polyetherimide (PEI), Polyether ether ketone (PEEK), Polyethylene terephthalate glycol (PETG), and fiber-reinforced thermoplastics. FDM-produced components are increasingly displacing conventional components in a variety of industries, including the automotive, aviation, and medical sectors [1–5]. The process variables and their settings have a substantial impact on the mechanical qualities of FDM-printed components. As a result, enhancing the mechanical qualities of printed components requires analyzing the effects of input factors and anticipating results by using adequate settings [6–9].

Several studies and predicted models of the impact of printing settings on the mechanical qualities of FDM components have been conducted by various scholars using various approaches such as the adaptive neuro-fuzzy technique (ANFIS), artificial neural network (ANN), response surface method (RSM), analysis of variance (ANOVA), group method for data handling (GMDH), and differential evolution (DE). According to Zhou, et al. [10], the infill density and printing pattern, for example, had a substantial impact on the tensile strength of polylactic acid FDM-printed components. Gebisa and Lemu [11] investigated the impact of process factors such as air gap, raster size, raster angle, contour quantity, and contour width on the tensile characteristics of components manufactured using the FDM technique and ULTEM 9085 polymeric material. According to their research, the raster angle has a significant effect on tensile properties. Byberg, et al. [12] looked at how layer alignment and build direction influenced the mechanical qualities of ULTEM 9085 thermoplastic. According to their results, the layer orientations and construction directions had a huge impact on the mechanical characteristics of the polymer.

The impact of principal directions, inclination angle, and air separation on the flexural characteristics of ULTEM 9085 fabricated by using the FDM technique with both solid and sparse building methods was examined by Motaparti, et al. [13]. Their study revealed that vertical (edge) designs had a greater elastic yield point than horizontal ones. According to Gebisa and Lemu [14], raster angle and width have a significant impact on the ULTEM 9085 polymer's flexible pavements. The impact of process parameters (layer height and printing speed) on the mechanical characteristics of 3D-printed ABS composite was investigated by Christiyan, et al. [15]. They revealed that the material's optimal tensile and flexural strength were achieved while using a low production speed and a minimal layer height. Hsueh, et al. [16] studied how FDM process factors impacted the mechanical characteristics of PLA and PETG materials (printing temperatures and rates). The results reveal that when the printing temperature increases, the PLA and PETG materials' mechanical characteristics (tension, compression, and bending) are enhanced. Furthermore, when manufacturing speed rises, the PLA material's mechanical behavior improves while the mechanical features of the PETG material deteriorate. According to Enemuoh, et al. [17], infill density has a significant impact on the tensile properties of the FDM component, followed by layer, speed, and infill patterns. Hsueh, et al. [18] investigated the impact of printing temperature and infill rate on the mechanical characteristics of FDM-printed PLA components using tensile and Shore D hardness tests. Raising the infill proportion or printing temperature, according to their results, could dramatically improve the material's longitudinal elastic modulus, ultimate strength, elasticity, and Shore hardness.

Patil, et al. [19] evaluated the tensile and flexural strength of FDM-printed PLA components using experimental testing and finite element analysis. Using 33 trials and result data, Manoharan, et al. employed ANN to develop mathematical models for predicting the tensile properties of FDM-made PLA components. The observed tensile strength results were compared to the anticipated values using the RSM, ANN, and ANOVA findings. Pazhamannil and Govindan [20] used an artificial neural network to estimate the tensile properties of FDM-printed items at various nozzle temperatures, layer thicknesses, and infill rates. Rayegani and Onwubolu [21] used DE and the GMDH to predict and optimize the link between FDM component strength properties and operating parameters (part alignment, inclination angle, raster size, and air gap). Srinivasan, et al. [22] employed response surface methodology to predict and optimize the impact of process factors on tensile properties of FDM-produced ABS components.

According to the literature, pre-processing settings have a considerable effect on the mechanical features of FDM-produced parts. It was also crucial to examine the combined influence of FDM settings on the mechanical features of produced components. As a result, five crucial pre-processing parameters (raster angle, printing orientation, air gap, raster width, layer height) were chosen as inputs for the current study: The tensile strength characteristics (UTS) were selected as the output response. Furthermore, the application of ANFIS and ANN was used to predict the response output, validated with experimental results.

2. Materials and Methods

2.1. 3D Printer and Materials

A Flash Forge Guider II 3D printer was used to create the specimens in this study. The build envelope of the printer measures $280 \times 250 \times 300$ mm^3 and can generate components with 0.2 mm accuracy. The 3D printer's characteristics are shown in Table 1. The study employed ABS printing materials since it is a strong thermoplastic and a typical FDM material. ABS is best suited for parts that require strength and flexibility, such as car components or household appliances. ABS is synthesized using three monomers: acrylonitrile, butadiene, and styrene, in an emulsion or continuous mass process. Acrylonitrile is a chemically resistant and thermally stable synthetic monomer made from propylene and ammonia that is used to make ABS. Butadiene is a by-product of ethylene synthesis from steam crackers that gives ABS polymer hardness and impact strength. ABS plastic gets its stiffness and processability from styrene, which is made by dehydrogenating ethyl benzene. Acrylonitrile butadiene styrene has the chemical formula $C_8H_8 \cdot C_4H_6 \cdot C_3H_3N$ [23–26]. ABS has a low melting point, which enables it to be easily used in 3D printing. It is very resistant to physical impacts and chemical corrosion, which allows the finished plastic to withstand heavy use and adverse environmental conditions [27,28]. The mechanical properties of the printing materials are shown in Table 2.

Table 1. The specification of the Flash Forge Guider II 3D printer.

Name	Guider II
Number of extruder	1
Print technology	Fused deposition modeling (FDM)
Build volume	$280 \times 250 \times 300$ mm
Layer resolution	0.05–0.4 mm
Build accuracy	$\pm$0.2 mm
Positioning accuracy	Z axis 0.0025 mm; XY axis 0.011 mm
Filament diameter	1.75 mm ($\pm$0.07)
Nozzle diameter	0.4 mm
Nozzle temperature	210–250 °C
Platform temperature	0–120 °C
Print speed	10–200 mm/s

Table 2. The specification of the ABS printing materials.

Properties	Specification
Material	ABS
Color	White
Wire diameter	1.75 $\pm$ 0.05 mm
Recommended printing temperature	230–250 °C
Recommended printing speed	30–90 mm/s
Extrusion temperature (°C)	190–210
Heated bed temperature (°C)	80
Density (g/cm^3)	1.04

2.2. Experimental Design

According to the literature, the mechanical characteristics of FDM-produced components are highly impacted by process parameters. It was also vital to examine the combined effects of FDM parameters on the mechanical characteristics of the produced components. As a result, five crucial process parameters were chosen as inputs for this research: raster angle, orientation angle, air gap, raster width, and layer height with three levels. The values of each element were adjusted according to machine manufacturer recommendations. Table 3 lists the process parameters and their ranges that were explored during this study. The other parameters were retained at their default settings. A total of 46 experiments were utilized, based on the Box–Behnken response surface experimental design, which depends on the number of input variables, and their levels as shown in Table 4.

Table 3. Process parameters and their range for experiments.

S.No.	Process Parameters	Units	Levels		
			Low (−1)	Medium (0)	High (+1)
1	Raster angle		0	30	60
2	Printing orientation		0	30	60
3	Air gap	mm	0	0.003	0.006
4	Raster width	mm	0.4064	0.4464	0.4864
5	Layer height	mm	0.14	0.185	0.23

Table 4. Box—Behnken response surface experimental design matrix and measured responses.

Run Order	Std Order	Raster Angle (°)	Printing Orientation (°)	Air Gap (mm)	Raster Width (mm)	Layer Height (mm)	Tensile Strength (MPa)
8	1	0	0	0.003	0.4464	0.185	27.835
2	2	60	0	0.003	0.4464	0.185	15.585
34	3	0	60	0.003	0.4464	0.185	10.248
30	4	60	60	0.003	0.4464	0.185	16.198
45	5	30	30	0	0.4064	0.185	20.608
13	6	30	30	0.006	0.4064	0.185	13.283
1	7	30	30	0	0.4864	0.185	19.983
14	8	30	30	0.006	0.4864	0.185	19.758
31	9	30	0	0.003	0.4464	0.14	19.642
17	10	30	60	0.003	0.4464	0.14	10.854
40	11	30	0	0.003	0.4464	0.23	23.129
10	12	30	60	0.003	0.4464	0.23	14.942
16	13	0	30	0	0.4464	0.185	23.804
15	14	60	30	0	0.4464	0.185	16.654
46	15	0	30	0.006	0.4464	0.185	16.029
9	16	60	30	0.006	0.4464	0.185	16.879
33	17	30	30	0.003	0.4064	0.14	12.835
41	18	30	30	0.003	0.4864	0.14	17.810
37	19	30	30	0.003	0.4064	0.23	18.673
43	20	30	30	0.003	0.4864	0.23	19.548
22	21	30	0	0	0.4464	0.185	29.415
27	22	30	60	0	0.4464	0.185	11.027
38	23	30	0	0.006	0.4464	0.185	15.740
18	24	30	60	0.006	0.4464	0.185	17.152
19	25	0	30	0.003	0.4064	0.185	19.639
4	26	60	30	0.003	0.4064	0.185	12.592
21	27	0	30	0.003	0.4864	0.185	18.604
36	28	60	30	0.003	0.4864	0.185	19.404
29	29	30	30	0	0.4464	0.14	19.285
12	30	30	30	0.006	0.4464	0.14	12.960
42	31	30	30	0	0.4464	0.23	20.523
7	32	30	30	0.006	0.4464	0.23	19.298
11	33	0	30	0.003	0.4464	0.14	17.006
20	34	60	30	0.003	0.4464	0.14	13.506
3	35	0	30	0.003	0.4464	0.23	20.443
25	36	60	30	0.003	0.4464	0.23	17.644
23	37	30	0	0.003	0.4064	0.185	18.239
44	38	30	60	0.003	0.4064	0.185	13.902
5	39	30	0	0.003	0.4864	0.185	25.314

Table 4. *Cont.*

Run Order	Std Order	Raster Angle (°)	Printing Orientation (°)	Air Gap (mm)	Raster Width (mm)	Layer Height (mm)	Tensile Strength (MPa)
28	40	30	60	0.003	0.4864	0.185	12.677
6	41	30	30	0.003	0.4464	0.185	15.210
24	42	30	30	0.003	0.4464	0.185	15.211
35	43	30	30	0.003	0.4464	0.185	15.200
32	44	30	30	0.003	0.4464	0.185	15.220
26	45	30	30	0.003	0.4464	0.185	15.210
39	46	30	30	0.003	0.4464	0.185	15.200

2.3. Specimen Fabrication

The test specimen was 3D modeled using CATIA V5 software, as per the criteria. The stereo lithography (STL) format is used to save the CAD file, which is then transferred to the slicer for separation into the needed number of layers. Using flash print slicing software, the printing settings are also included. The slicer then transforms the STL file to a G-code file, which the printers use to begin layer-by-layer fabrication of the specimen. The tensile test specimens were made according to the American Society for Testing and Materials (ASTM) D638-I standard. The length, width, and thickness of ASTM D638-I are 165 × 13 × 3.2 mm as shown in Figure 1.

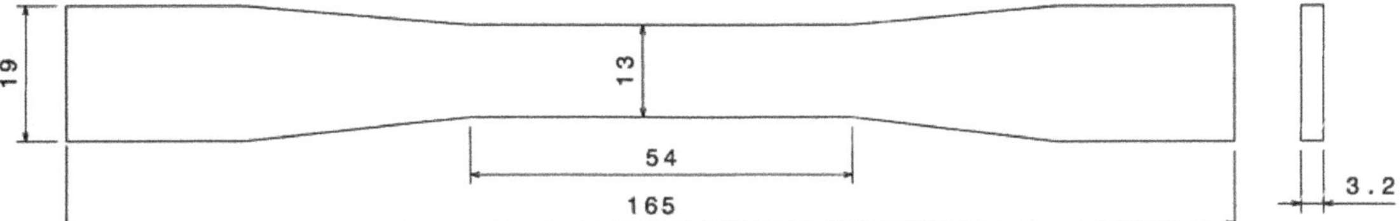

Figure 1. The ASTM D638-I tensile test sample.

2.4. Experimental Procedure

The universal testing equipment was used to assess the tensile strength of ABS specimens that had been conditioned according to the ASTM D638 standard as shown in Figure 2. The top grip moved at a continuous rate of 2.5 mm/min with a maximum load of 100 KN and a 5 Hz signal sampling rate. The built-in program recorded the elongation and force load data. When the specimen elongates or fractures by more than 2.5 percent, the ASTM D638 test is completed. The tensile strength is calculated from the breaking load and is divided by the initial cross section area according to Equation (1). In this study, 46 experimental tests were conducted with two replications, and no differences between the two replications were observed (presented in Table 4).

$$Tensile\ strength\ (UTS) = \frac{Breaking\ load(Pf)}{orignal\ cross\ sectional\ area(Ao)} \tag{1}$$

Figure 2. Universal testing machine with testing specimen.

2.5. Adaptive Neuro-Fuzzy Modeling

The architecture and learning approach of neuro-fuzzy techniques (also known as an adaptive neuro-fuzzy model) was initially developed by Jang [29]. ANFIS is a powerful approach that integrates neural networks' optimization and learning skills with the reasoning capabilities of fuzzy logic linguistic IF–THEN rules, which are made up of membership functions (MF). It combines the benefits of fuzzy logic with artificial neural networks. The ANFIS model is especially beneficial when data are inconsistent or nonlinear and established methodologies fail or are too difficult to apply with greater precision [30–34].

In this study, the ANFIS model was stimulated using the fuzzy inference system concept as a five-layered neural network. The parameters of the ANFIS model are listed in Table 5. MATLAB R2019a software was used to run the simulations. The first and last levels of the ANFIS structure and layer indicate the respective input variables (raster angle, orientation angle, air gap, raster width, and layer height) and output variable (tensile strength), as shown in Figure 3. In the second layer, the model employed first-Sugeno inference systems, which turn input parameters into membership values using membership functions called fuzzification. The model output is then deduced using a set of logical rules in the third layer.

Table 5. Details of the ANFIS model used in this study.

ANFIS Information	Takagi–Sugeno–Kang
Number of MFs	3 3 3 3 3
MF type (Input)	Trapmf
Output MF types	Constant
Optimization method	Hybrid
Error tolerance	1×10^{-7}
Epochs	100
FIS generation	Grid partitioning (GP)
Data points	32
Number of fuzzy rules	243

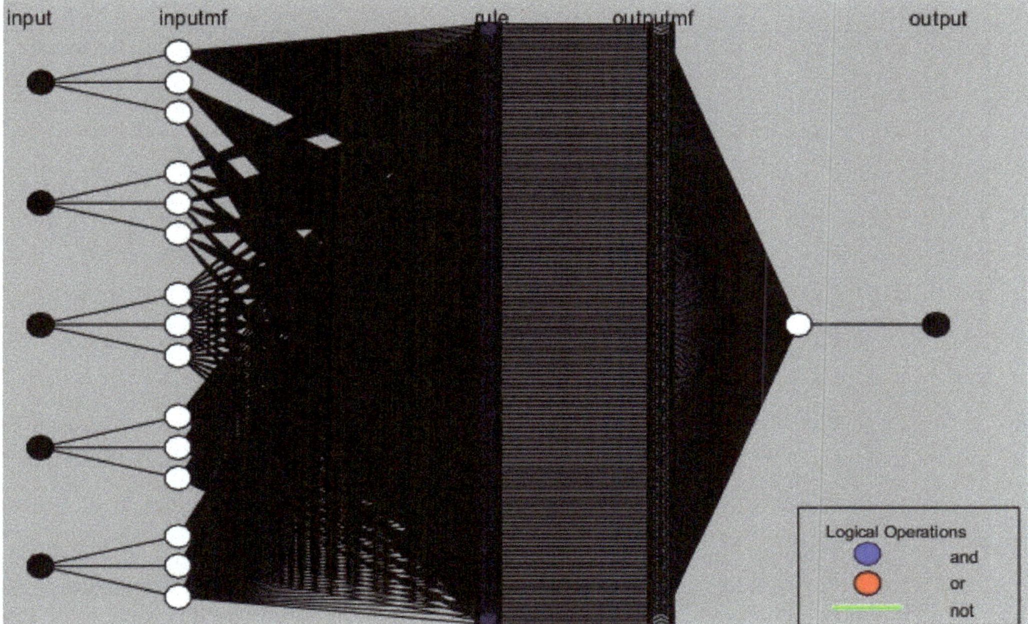

Figure 3. ANFIS multilayer architecture.

2.6. Artificial Neural Network Modeling

Artificial neural networks (ANNs) are computational models based on the organic neural networks in the human brain. An ANN is used to mimic non-linear situations and predict output values using training data. Input and output layers, as well as multiple hidden layer neurons, comprise an ANN structural network [35–37]. ANN uses samples of data rather than the whole data set available in the system for quick prediction, saving money and time. ANNs can be easily replaced by existing data analysis systems [38].

The ANN model was used to train and assess the 3D printing data models. An input layer with five inputs, a hidden layer with feed-forward conditions added, and one output layer were created using MATLAB as shown in Figure 4. The experimental data from Table 4 were used to train the network. The parameters of the ANN model are listed in Table 6. Figure 4 depicts the ANN architecture and learning variables used in this investigation.

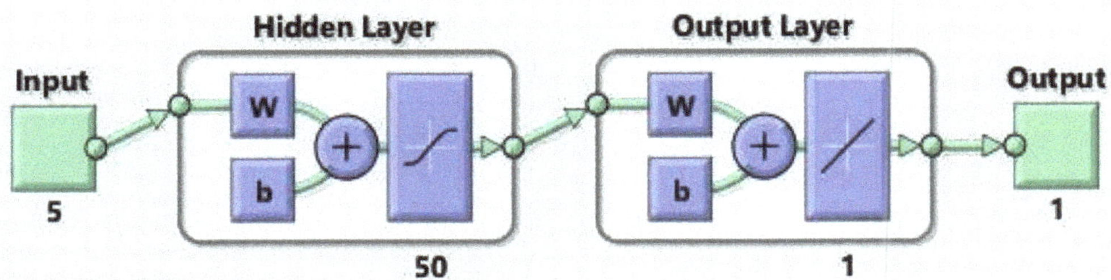

Figure 4. Structure of the neural network.

Table 6. Learning parameters designated for the ANN.

Type of Network	Feed-Forward Neural Network
Training function	Train Levenberg–Marquardt (LM) algorithm
Adaption Learning function	LEARNGD (Gradient descent)
Performance function	Mean square error
Network topology	5-50-1-1
Transfer function	TANSIG
Number of Hidden Layers	1
Number of Hidden Neurons	50
Training method	Back-propagation
Number of Epochs	1000

3. Results and Discussion

3.1. Effect of Process Parameters on Tensile Strength

Analysis of variance (ANOVA) was used to examine the results for tensile characteristics in order to explore the major factors that influence the quality measures. Factors with a very modest probability (Prob. > F-value less than 0.05) are considered significant in the ANOVA table, whereas factors with a probability (Prob. > F-value) larger than 0.1 are considered inconsequential. Furthermore, larger F-values and lower p-values have a greater impact on the performance characteristics derived from designed process parameters. Table 7 shows the ANOVA of each process parameter with respect to the ultimate tensile strength of ABS parts, where A is the raster angle, B is the printing orientation, C is the air gap, D is the raster width, and E is the layer height. According to the ANOVA table (Table 7), printing orientation, layer height, raster angle, and layer height combinations, the air gap and layer height combinations are insignificant factors that impact the ultimate tensile strength of ABS components. Since the p-values of the raster angle, printing orientation, raster width, air gap, and their combinations are less than 0.05, they have a significant impact on the UTS of ABS-printed parts.

Table 7. ANOVA for ultimate tensile strength of ABS parts.

Source	DF	Adj. SS	Adj. MS	F-Value	p-Value	
Regression	20	772.083	38.6042	133.88	0.000	
A	1	30.224	30.2237	104.82	0.000	Significant
B	1	1.175	1.175	4.07	0.054	Insignificant
C	1	39.473	39.4735	136.89	0.000	Significant
D	1	8.424	8.4238	29.21	0.000	Significant
E	1	0.643	0.643	2.23	0.148	Insignificant
A*A	1	11.292	11.2923	39.16	0.000	Significant
B*B	1	11.127	11.1274	38.59	0.000	Significant
C*C	1	35.055	35.0547	121.57	0.000	Significant
D*D	1	12.655	12.6547	43.89	0.000	Significant
E*E	1	5.761	5.7614	19.98	0.000	Significant
A*B	1	82.81	82.81	287.19	0.000	Significant
A*C	1	16	16	55.49	0.000	Significant
A*D	1	15.603	15.6025	54.11	0.000	Significant
A*E	1	0.122	0.1225	0.42	0.52	Insignificant
B*C	1	98.01	98.01	339.9	0.000	Significant
B*D	1	17.222	17.2225	59.73	0.000	Significant
B*E	1	0.09	0.09	0.31	0.581	Insignificant
C*D	1	12.603	12.6025	43.71	0.000	Significant
C*E	1	6.503	6.5025	22.55	0.000	Significant
D*E	1	4.203	4.2025	14.57	0.001	Significant
Error	25	7.209	0.2883			
Lack-of-Fit	20	7.209	0.3604			
Pure Error	5	0.000	0.0001			
Total	45	779.292				

Figure 5 illustrates the main effects plot for UTS, which shows how UTS vary depending on the inputs. The mean UTS value started to decrease with increasing raster angle, printing orientation, and air gap, as shown in this main effect graph. The UTS also improved as the raster width and layer height started escalating. According to the main effects plot shown in Figure 5, a lower raster angle, printing orientation, air gap, and higher raster width and layer height would result in enhanced UTS.

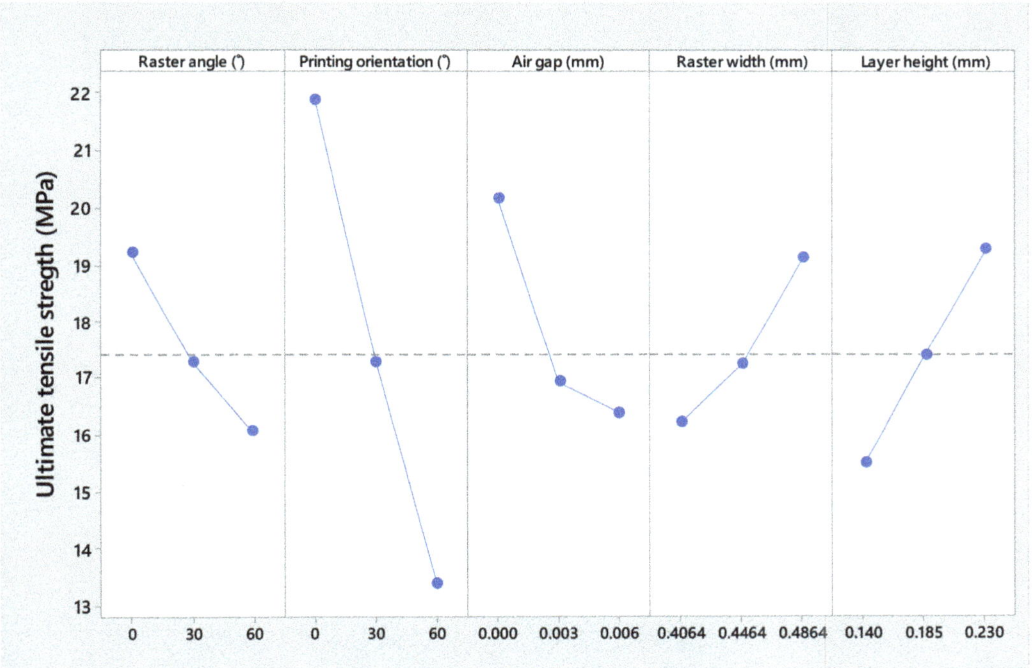

Figure 5. Main effect plots of UTS for means with all process parameters.

Figures 6–8 show 3D and contour graphs depicting the impact of both of the input factors on the ultimate tensile strength. In these figures, blue-colored zones represent very low and low levels, green zones reflect low–medium, medium, and high–medium levels, while yellow zones represent high and very high volumes. The effect of raster angle (input 1) and printing orientation (input 2) on the UTS is shown in Figure 6. It shows that lower raster angles and printing orientation led to increased UTS values. Figure 7 depicts the influence of the air gap (input 3) and raster width (input 4) on the UTS in a similar way, showing that reduced air gaps and larger raster widths yielded higher UTS values. These interaction graphs also show that raster width had a moderate influence on UTS. The effects of raster width (input 4) and layer height (input 5) on the UTS are also shown in Figure 8. Higher UTS values were achieved when the raster width and layer height were increased.

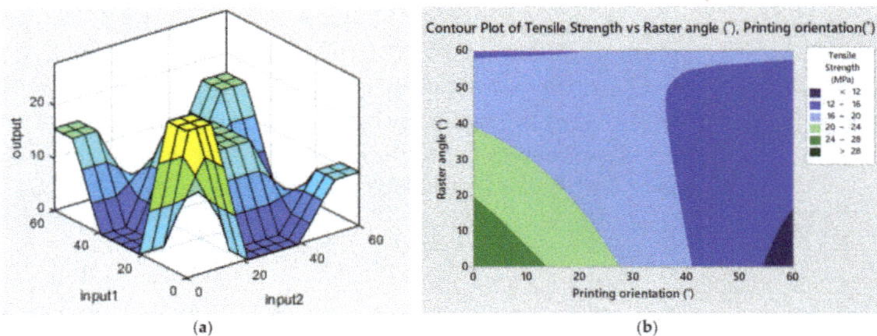

(a)   (b)

Figure 6. 3D plot (**a**) and controur plot (**b**) of tensile strength vs. raster angle and printing orientation.

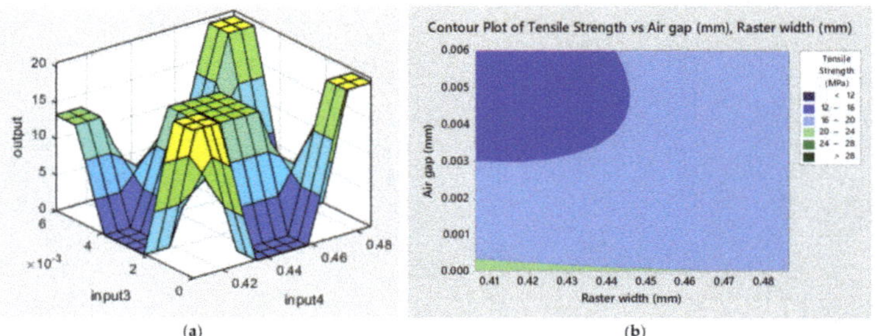

(a)   (b)

Figure 7. 3D plot (**a**) and contour plot (**b**) of tensile strength vs. air gap and raster width.

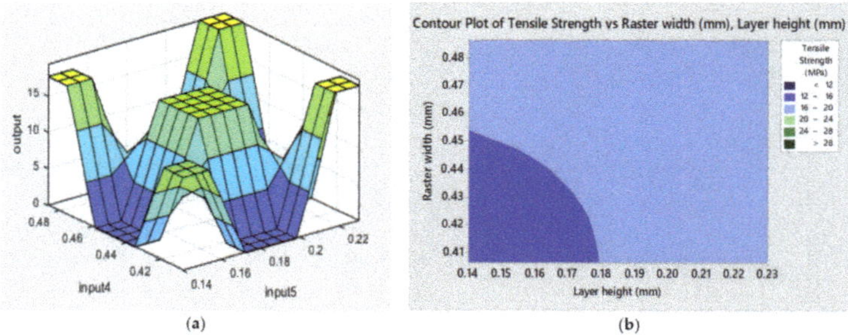

(a)   (b)

Figure 8. 3D plot (**a**) and contour plot (**b**) of tensile strength vs. raster width and layer height.

3.2. Results of ANFIS Modeling

For the prediction of the tensile strength of ABS-printed components using FDM, an ANFIS model was created utilizing the data gained from experimental measurements. A number of factors must be adjusted to identify the best model architecture for the ANFIS model. To develop a robust model with sufficient predictive performance, parameters such as the number of rules, type, and number of MFs, and logical operators must be modified. Based on the data in Table 4, the results of the proposed ANFIS model were obtained using the MATLAB R2019a program, and the predicted data are shown in Table 6. Figure 9 shows the two hundred and forty-four IF–THEN membership functions (MF) rules used to train the ANFIS model in this study. It also illustrates that the predicted results of UTS were 15.2 MPa when the raster angle (input 1) was 30 degrees, the printing orientation (input 2) was 30 degrees, the air gap (input 3) was 0.4466 mm, the raster width (input 4) was 0.185 mm, and the layer height (input 5) was 100 mm/s. The suggested model was valid after evaluating the checking data with the derived FIS model, with a minimal checking RMSE of 0.002500, which is less than 0.3080 % error. As shown in Figure 10, the R^2 value between predicted and targeted values for ANFIS tensile strength demonstrated a high level of accuracy, with $R^2 = 0.9999$, indicating that the model is valid.

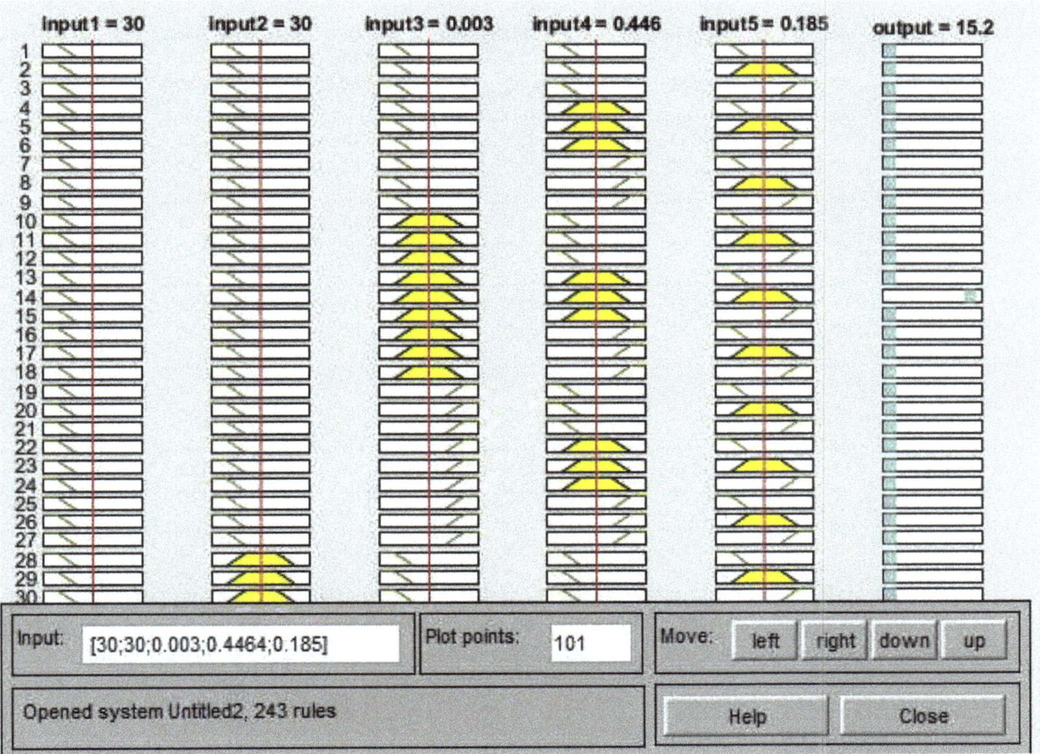

Figure 9. Representation of rules for the generated FIS.

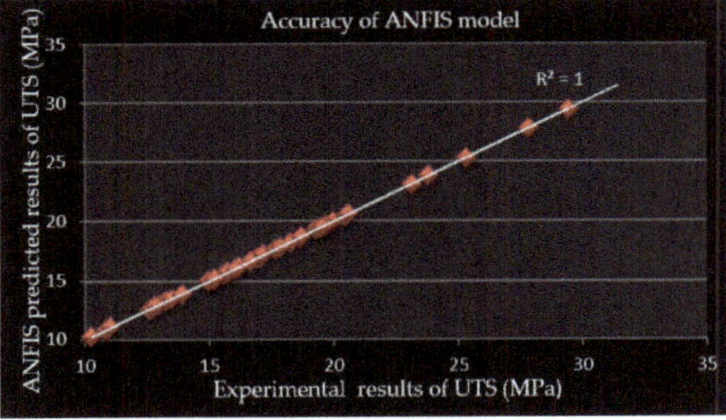

Figure 10. Regression results of the ANFIS model.

3.3. Results of ANN Modeling

When the error, or the difference between the expected and predicted output, is less than a defined upper bound, or when the number of epochs exceeds a specified threshold, the ANN stops training. A score near 1 implies a strong connection, while a value close to 0 shows a random relation. Figure 11 illustrates the 12-iteration regression graphs developed by artificial neural networks. The regression plots obtained reveal that for training, testing, validation, and total data, the values were 0.98228, 0.85711, 0.92749, and 0.917, respectively, suggesting the best fitness after repeated training. This indicates that the ANN model's anticipated outcomes appear to be in line with the experimental data. The ANN model worked adequately, as shown in Table 8, with an average percentage error of 1.6437 from the experimental results, demonstrating its potential for future usage.

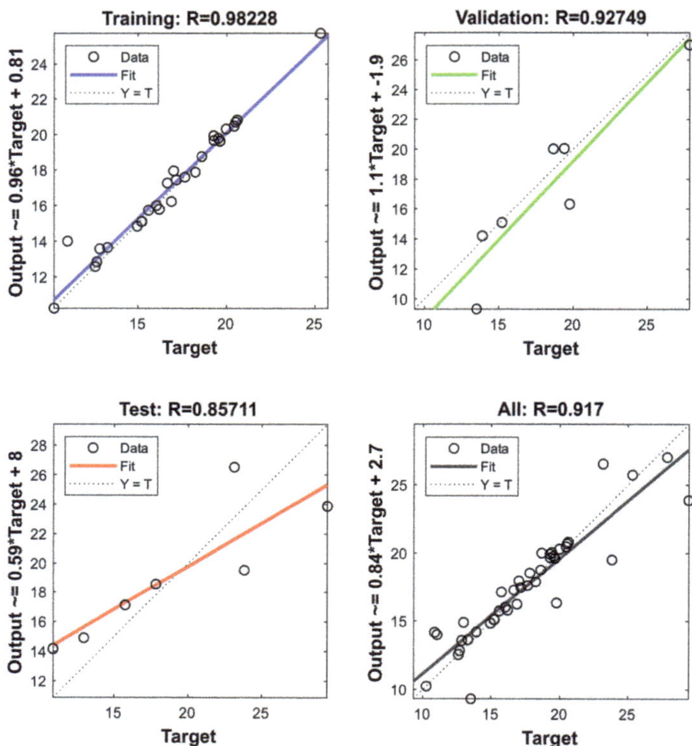

Figure 11. Regression plots for ultimate tensile strength obtained using the ANN model.

Table 8. Comparative evaluation of predictive models.

Exp. Trials	Experimental Results of UTS	ANFIS Model		ANN Model	
		Predicted Results	**% Errors**	**Predicted Results**	**% Errors**
1.	19.983	19.985	−0.0100	20.309	−1.6314
2.	15.585	15.587	−0.0128	15.732	−0.9432
3.	20.443	20.442	0.0049	20.481	−0.1859
4.	12.592	12.593	−0.0079	12.578	0.1112
5.	25.314	25.317	−0.0119	25.735	−1.6631
6.	15.210	15.207	0.0197	15.094	0.7627
7.	19.298	19.292	0.0311	19.672	−1.9380
8.	27.835	27.839	−0.0144	27.017	2.9387
9.	16.879	16.880	−0.0059	16.221	3.8983
10.	14.942	14.941	0.0067	14.835	0.7161
11.	17.006	17.009	−0.0176	17.756	−4.4102
12.	12.960	12.961	−0.0077	13.294	−2.5772
13.	13.283	13.287	−0.0301	13.647	−2.7403
14.	19.758	19.761	−0.0152	18.8662	4.5136
15.	16.654	16.652	0.0120	17.263	−3.6568
16.	23.804	23.812	−0.0336	23.3201	2.0329
17.	10.854	10.860	−0.0553	11.1657	−2.8718

Table 8. *Cont.*

Exp. Trials	Experimental Results of UTS	ANFIS Model		ANN Model	
		Predicted Results	% Errors	Predicted Results	% Errors
18.	17.152	17.150	0.0117	17.443	−1.6966
19.	19.639	19.634	0.0255	19.9011	−1.3346
20.	13.506	13.501	0.0370	13.3181	1.3912
21.	18.604	18.603	0.0054	18.764	−0.8600
22.	29.415	29.414	0.0034	28.8888	1.7889
23.	18.239	18.243	−0.0219	17.876	1.9902
24.	15.211	15.204	0.0460	15.094	0.7692
25.	17.644	17.643	0.0057	17.602	0.2380
26.	15.210	15.203	0.0460	15.094	0.7627
27.	11.027	11.026	0.0091	10.7033	2.9355
28.	12.677	12.676	0.0079	12.851	−1.3726
29.	19.285	19.281	0.0207	19.938	−3.3861
30.	16.198	16.196	0.0123	15.789	2.5250
31.	19.642	19.641	0.0051	19.645	−0.0153
32.	15.220	15.203	0.1117	15.094	0.8279
33.	12.835	12.834	0.0078	12.587	1.9322
34.	10.248	10.246	0.0195	10.236	0.1171
35.	15.200	15.212	−0.0789	15.094	0.6974
36.	19.404	19.405	−0.0052	20.048	−3.3189
37.	18.673	18.672	0.0054	19.1365	−2.4822
38.	15.740	15.739	0.0064	16.111	−2.3571
39.	15.200	15.210	−0.0658	15.094	0.6974
40.	23.129	23.128	0.0043	23.0753	0.2322
41.	17.810	17.811	−0.0056	18.549	−4.1494
42.	20.523	20.522	0.0049	20.698	−0.8527
43.	19.548	19.551	−0.0153	19.785	−1.2124
44.	13.902	13.901	0.0072	14.204	−2.1723
45.	20.608	20.607	0.0049	20.812	−0.9899
46.	16.029	16.026	0.0187	15.995	0.2121
	Average percentage error		0.0163		1.6437

3.4. Comparative Evaluation of the Predictive Models

To compare the ANFIS and ANN, the predictive results and the experimental results of UTS were analyzed by the average percentage error of the responses. The percentages of absolute errors for ANFIS and ANN were computed individually by comparing the predicted values with the test results using Equation (2). The average percentage error of the ANFIS model was 0.0163 %, and that of the ANN model was 1.6437 %, as shown in Table 8. This demonstrates that the ANFIS model is the most accurate or best-predicting model technique.

$$\% \ Absolute \ error = \frac{Actual - predicted}{Actual} \times 100 \tag{2}$$

3.5. Validation of the Models

In order to validate the relative performances of the ANFIS and ANN models, the validation parameters had different values of process parameters for eight new validation specimens. The validation parameters had the raster angle, printing orientation, air gap, raster width, and layer height as shown in Table 9. New specimens were fabricated and tested for tensile strength using the above parameters, and the percentage in deviation was computed using the means of prediction under ANFIS and ANN in MINITAB 2019a. When comparing the actual and predicted outcomes using the ANFIS and ANN approaches, the validation parameter yielded the percentage error as 0.166 and 0.767, respectively. Hence, the ANFIS model's accuracy ranges from 1 to 2%, whereas the ANN model's accuracy ranges from 1 to 5%.

Table 9. Process parameters for validation.

Run Order	Raster Angle (°)	Printing Orientation (°)	Air Gap (mm)	Raster Width (mm)	Layer Height (mm)	Exp. Results of UTS (MPa)	ANFIS Model		ANN Model	
							Predicted Results (MPa)	% Errors	Predicted Results (MPa)	% Errors
1	15	15	0.002	0.4264	0.17	9.690	9.689	0.011	9.558	1.363
2	15	15	0.002	0.4664	0.20	10.500	10.508	−0.072	10.482	0.175
3	15	45	0.004	0.4264	0.17	7.043	7.040	0.045	7.040	0.047
4	15	45	0.004	0.4664	0.20	7.012	7.010	0.034	7.017	−0.063
5	45	15	0.004	0.4264	0.20	7.500	7.581	−1.079	7.173	4.361
6	45	15	0.004	0.4664	0.17	8.931	8.930	0.006	8.932	−0.019
7	45	45	0.002	0.4264	0.20	7.052	7.049	0.048	7.051	0.024
8	45	45	0.002	0.4664	0.17	7.541	7.538	0.040	7.522	0.248

4. Conclusions

This research proposes experimental analysis and the use of adaptive neuro-fuzzy methods and artificial neural networks to forecast the tensile strength for ABS components manufactured using fused deposition modeling. All of the investigations were carried out using a 46 Box–Behnken response surface design to alter the input parameters at different levels. Analysis of variance, main effect plots, 3D, and contour plots were used to investigate the link between input parameters and output results. The experimental outcomes were used to train and evaluate the models that were developed. The MATLAB R2019a fuzzy toolbox and neural toolbox were used to create the neuro-fuzzy system and artificial neural network, respectively. The ability of the models to forecast was tested using percentage errors. The study revealed that layer height, raster angle and layer height combinations, and air gap and layer height combinations were insignificant factors that impacted the ultimate tensile strength of ABS components. Since the p-values of the raster angle, printing orientation, raster width, air gap, and their combinations were less than 0.05, they had a significant impact on the ABS-printed components' tensile strength. The UTS started to decrease with increasing raster angle, printing orientation, and air gap. The UTS also improved as the raster width and layer height started increasing. Enhanced mechanical strength may be achieved by using a lower raster angle, printing orientation, air gap, and larger raster width and layer height. The ANFIS and ANN models can accurately predict tensile strength with average percentage errors of 0.0163 and 1.6437, respectively. The ANFIS model's accuracy ranges from 1 to 2%, whereas the ANN model's accuracy ranges from 1 to 5%. This shows the predicted and experimental data are in good agreement. The arithmetical value indices of the ANFIS model indicated a better predictive performance than that of the ANN model.

Author Contributions: Conceptualization, A.D.T. and H.B.M.; methodology, A.D.T.; software, A.D.T.; validation, A.D.T.; formal analysis, A.D.T. and H.B.M.; investigation, A.D.T. and H.B.M.; resources, A.D.T. and H.G.L.; data curation, A.D.T. and H.G.L.; writing—original draft preparation, A.D.T.; writing—review and editing, A.D.T. and H.G.L.; visualization, H.G.L.; supervision, H.G.L.; project administration, H.B.M.; funding acquisition, H.G.L. All authors have read and agreed to the published version of the manuscript.

Funding: This research received no external funding.

Institutional Review Board Statement: Not applicable.

Informed Consent Statement: Not applicable.

Data Availability Statement: Not applicable.

Conflicts of Interest: The authors declare no conflict of interest.

References

1. Sheoran, A.J.; Kumar, H. Fused Deposition modeling process parameters optimization and effect on mechanical properties and part quality: Review and reflection on present research. *Mater. Today Proc.* **2019**, *10*, 14.
2. Amelia, N.; Ferliadi, F.; Christa, A.; Gunawan, F.E.; Asrol, M. Experimental Investigation on Influence of Infill Density on Tensile Mechanical Properties of Different FDM 3D Printed Materials. *TEM J.* **2021**, *10*, 1195–1201.
3. Wankhede, V.; Jagetiya, D.; Joshi, A.; Chaudhari, R. Experimental investigation of FDM process parameters using Taguchi analysis. *Mater. Today Proc.* **2019**, *27*, 2117–2120. [CrossRef]
4. Scallan, P. Comparative Study of the Sensitivity of PLA, ABS, PEEK, and PETG's Mechanical Properties to FDM Printing Process Parameters. *Crystals* **2021**, *11*, 219–250.
5. Milovanović, A.; Sedmak, A.; Grbović, A.; Golubović, Z.; Mladenović, G.; Čolić, K.; Milošević, M. Comparative analysis of printing parameters effect on mechanical properties of natural PLA and advanced PLA-X material. *Procedia Struct. Integr.* **2020**, *28*, 1963–1968. [CrossRef]
6. Onwubolu, G.C.; Rayegani, F. Characterization and Optimization of Mechanical Properties of ABS Parts Manufactured by the Fused Deposition Modelling Process. *Hindawi* **2014**, *2014*, 1–13. [CrossRef]
7. Mamo, H.B.; Tura, A.D.; Santhosh, A.J.; Ashok, N.; Rao, D.K. Modeling and analysis of flexural strength with fuzzy logic technique for a fused deposition modeling ABS components. *Mater. Today Proc.* **2022**, *57*, 768–774. [CrossRef]
8. Wang, S.; Ma, Y.; Deng, Z.; Zhang, S.; Cai, J. Effects of fused deposition modeling process parameters on tensile, dynamic mechanical properties of 3D printed polylactic acid materials. *Polym. Test.* **2020**, *86*, 106483. [CrossRef]
9. Lluch-Cerezo, J.; Benavente, R.; Meseguer, M.D.; Gutiérrez, S.C. Study of samples geometry to analyze mechanical properties in Fused Deposition Modeling process (FDM). *Procedia Manuf.* **2019**, *41*, 890–897. [CrossRef]
10. Zhou, X.; Hsieh, S.; Ting, C. Modelling and estimation of tensile behaviour of polylactic acid parts manufactured by fused deposition modelling using finite element analysis and knowledge-based library. *Virtual Phys. Prototyp.* **2018**, *13*, 177–190. [CrossRef]
11. Gebisa, A.W.; Lemu, H.G. Influence of 3D printing FDM process parameters on tensile property of ultem 9085. *Procedia Manuf.* **2019**, *30*, 331–338. [CrossRef]
12. Byberg, K.I.; Gebisa, A.W.; Lemu, H.G. Mechanical properties of ULTEM 9085 material processed by fused deposition modeling. *Polym. Test.* **2018**, *72*, 335–347. [CrossRef]
13. Motaparti, K.P.; Taylor, G.; Leu, M.C.; Chandrashekhara, K.; Castle, J.; Matlack, M. Experimental investigation of effects of build parameters on flexural properties in fused deposition modelling parts. *Virtual Phys. Prototyp.* **2017**, *12*, 207–220. [CrossRef]
14. Gebisa, A.W.; Lemu, H.G. Investigating Effects of Fused-Deposition Modeling (FDM) Processing Parameters on Flexural Properties of ULTEM 9085 using Designed Experiment. *Materials* **2018**, *11*, 500. [CrossRef]
15. Christiyan, K.G.J.; Chandrasekhar, U.; Venkateswarlu, K. A study on the influence of process parameters on the Mechanical Properties of 3D printed ABS composite. *IOP Conf. Ser. Mater. Sci. Eng.* **2016**, *114*, 12109. [CrossRef]
16. Hsueh, M.-H.; Lai, C.-J.; Wang, S.-H.; Zeng, Y.-S.; Hsieh, C.-H.; Pan, C.-Y.; Huang, W.-C. Effect of Printing Parameters on the Thermal and Mechanical Properties of 3D-Printed PLA and PETG, Using Fused Deposition Modeling. *Polymers* **2021**, *13*, 1758. [CrossRef]
17. Enemuoh, E.U.; Duginski, S.; Feyen, C.; Menta, V.G. Effect of Process Parameters on Energy Consumption, Physical, and Mechanical Properties of Fused Deposition Modeling. *Polymers* **2021**, *13*, 2406. [CrossRef]
18. Hsueh, M.-H.; Lai, C.-J.; Liu, K.-Y.; Chung, C.-F.; Wang, S.-H.; Pan, C.-Y.; Huang, W.-C.; Hsieh, C.-H.; Zeng, Y.-S. Effects of Printing Temperature and Filling Percentage on the Mechanical Behavior of Fused Deposition Molding Technology Components for 3D Printing. *Polymers* **2021**, *13*, 2910. [CrossRef]
19. Patil, C.; Sonawane, P.D.; Naik, M.; Thakur, D.G. Finite element analysis of flexural test of additively manufactured components fabricated by fused deposition modelling. *In AIP Conf. Proc.* **2020**, *2311*, 070026. [CrossRef]
20. Pazhamannil, R.V.; Govindan, P.S.P. Prediction of the tensile strength of polylactic acid fused deposition models using artificial neural network technique. *Mater. Today Proc.* **2020**, *199*, 9187–9193. [CrossRef]
21. Rayegani, F.; Onwubolu, G.C. Fused deposition modelling (FDM) process parameter prediction and optimization using group method for data handling (GMDH) and differential evolution (DE). *Int. J. Adv. Manuf. Technol.* **2014**, *2*, 509–519. [CrossRef]
22. Srinivasan, R.; Pridhar, T.; Ramprasath, L.S.; Charan, N.S.; Ruban, W. Proceedings Prediction of tensile strength in FDM printed ABS parts using response surface methodology (RSM). *Mater. Today Proc.* **2020**, *3*, 788. [CrossRef]
23. Vishwakarma, S.K.; Pandey, P.; Gupta, N.K. Characterization of ABS Material: A Review. *J. Res. Mech. Eng.* **2017**, *3*, 13–16.
24. Harris, M.; Potgieter, J.; Ray, S.; Archer, R.; Arif, K.M. Acrylonitrile butadiene styrene and polypropylene blend with enhanced thermal and mechanical properties for fused filament fabrication. *Materials* **2019**, *12*, 4167. [CrossRef] [PubMed]
25. Rutkowski, J.V.; Levin, B.C. Acrylonitrile–butadiene–styrene copolymers (ABS): Pyrolysis and combustion products and their toxicity—a review of the literature. *Fire Mater.* **1986**, *10*, 93–105. [CrossRef]
26. Nguyen, H.T.; Crittenden, K.; Weiss, L.; Bardaweel, H. Experimental Modal Analysis and Characterization of Additively Manufactured Polymers. *Polymers* **2022**, *14*, 2071. [CrossRef]

27. Akessa, A.D.; Lemu, H.G.; Gebisa, A.W. Mechanical Property Characterization of Additive Manufactured ABS Material Using Design of Experiment Approach. In Proceedings of the ASME 2017 International Mechanical Engineering Congress and Exposition, Volume 14: Emerging Technologies, Materials: Genetics to Structures, Safety Engineering and Risk Analysis. Tampa, FL, USA, 3–9 November 2017.
28. Croccolo, D.; De Agostinis, M.; Olmi, G. Experimental characterization and analytical modelling of the mechanical behaviour of fused deposition processed parts made of ABS-M30. *Comput. Mater. Sci.* **2013**, *79*, 506–518. [CrossRef]
29. Zadeth, L.A. Fuzzy sets as a basis for a theory of possibility. *Fuzzy Sets Syst.* **1999**, *148*, 9–34. [CrossRef]
30. Sarkar, J.; Prottoy, Z.H.; Bari, M.T.; al Faruque, M.A. Comparison Of Anfis And Ann Modeling For Predicting The Water Absorption Behavior Of Polyurethane Treated Polyester Fabric. *Heliyon* **2021**, *7*, e08000. [CrossRef]
31. Ghorbani, B.; Arulrajah, A.; Narsilio, G.; Horpibulsuk, S.; Leong, M. Resilient moduli of demolition wastes in geothermal pavements: Experimental testing and ANFIS modelling. *Transp. Geotech.* **2021**, *29*, 100592. [CrossRef]
32. Saleh, B.; Maher, I.; Abdelrhman, Y.; Heshmat, M.; Abdelaal, O. Adaptive Neuro-Fuzzy Inference System for Modelling the Effect of Slurry Impacts on PLA Material Processed by FDM. *Polymers* **2021**, *13*, 118. [CrossRef] [PubMed]
33. Nafees, A.; Javed, M.F.; Khan, S.; Nazir, K.; Farooq, F.; Aslam, F.; Musarat, M.A.; Vatin, N.I. Predictive Modeling of Mechanical Properties of Silica Fume-Based Green Concrete Using Artificial Intelligence Approaches: MLPNN, ANFIS, and GEP. *Materials* **2021**, *14*, 7531. [CrossRef] [PubMed]
34. Zhu, H.; Zhu, L.; Sun, Z.; Khan, A. Machine learning based simulation of an anti-cancer drug (busulfan) solubility in supercritical carbon dioxide: ANFIS model and experimental validation. *J. Mol. Liq.* **2021**, *338*, 116731. [CrossRef]
35. Santhosh, A.J.; Tura, A.D.; Jiregna, I.T.; Gemechu, W.F.; Ashok, N.; Ponnusamy, M. Optimization of CNC turning parameters using face centred CCD approach in RSM and ANN-genetic algorithm for AISI 4340 alloy steel. *Results Eng.* **2021**, *11*, 00251. [CrossRef]
36. Tura, A.D.; Mamo, H.B.; Jelila, Y.D.; Lemu, H.G. Experimental investigation and ANN prediction for part quality improvement of fused deposition modeling parts. *IOP Conf. Ser. Mater. Sci. Eng.* **2021**, *1201*, 012031. [CrossRef]
37. Onu, C.E.; Nwabanne, J.T.; Ohale, P.E.; Asadu, C.O. Comparative analysis of RSM, ANN and ANFIS and the mechanistic modeling in eriochrome black-T dye adsorption using modified clay. *S. Afr. J. Chem. Eng.* **2021**, *36*, 24–42. [CrossRef]
38. Deshwal, S.; Kumar, A.; Chhabra, D. Exercising hybrid statistical tools GA-RSM, GA-ANN and GA-ANFIS to optimize FDM process parameters for tensile strength improvement. *CIRP J. Manuf. Sci. Technol.* **2020**, *5*, 11. [CrossRef]

Article

Improvement of the Geometric Accuracy for Microstructures by Projection Stereolithography Additive Manufacturing

Cheng Wen [1], Zhengda Chen [1], Zhuoxi Chen [1], Bin Zhang [1], Zhicheng Cheng [1], Hao Yi [2], Guiyun Jiang [2,*] and Jigang Huang [1,*]

[1] School of Mechanical Engineering, Sichuan University, Chengdu 610065, China; chengwen20010401@163.com (C.W.); zhengdachen0320@163.com (Z.C.); 13666271640@163.com (Z.C.); m14780065195@163.com (B.Z.); czc20020511@163.com (Z.C.)
[2] College of Mechanical and Vehicle Engineering, Chongqing University, Chongqing 400030, China; haoyi@cqu.edu.cn
* Correspondence: gyjiang_1@163.com (G.J.); jigang.huang@scu.edu.cn (J.H.)

Abstract: Projection stereolithography creates 3D structures by projecting patterns onto the surface of a photosensitive material layer by layer. Benefiting from high efficiency and resolution, projection stereolithography 3D printing has been widely used to fabricate microstructures. To improve the geometric accuracy of projection stereolithography 3D printing for microstructures, a compensation method based on structure optimization is proposed according to mathematical analysis and simulation tests. The performance of the proposed compensation method is verified both by the simulation and the 3D printing experiments. The results indicate that the proposed compensation method is able to significantly improve the shape accuracy and reduce the error of the feature size. The proposed compensation method is also proved to improve the dimension accuracy by 21.7%, 16.5% and 19.6% for the circular, square and triangular bosses respectively. While the improvements on the dimension accuracy by 16%, 17.6% and 13.8% for the circular, square and triangular holes are achieved with the proposed compensation method. This work is expected to provide a method to improve the geometric accuracy for 3D printing microstructures by projection stereolithography.

Keywords: projection stereolithography; geometric accuracy; 3D printing

Citation: Wen, C.; Chen, Z.; Chen, Z.; Zhang, B.; Cheng, Z.; Yi, H.; Jiang, G.; Huang, J. Improvement of the Geometric Accuracy for Microstructures by Projection Stereolithography Additive Manufacturing. *Crystals* **2022**, *12*, 819. https://doi.org/10.3390/cryst12060819

Academic Editor: Xiaochun Li

Received: 13 May 2022
Accepted: 8 June 2022
Published: 9 June 2022

Publisher's Note: MDPI stays neutral with regard to jurisdictional claims in published maps and institutional affiliations.

1. Introduction

By accumulating materials layer by layer, additive manufacturing (also known as 3D printing) has the ability to create complex 3D structures with high efficiency [1–3]. Projection stereolithography, as an important 3D printing approach, has attracted great attention and has been applied to various research fields, including metamaterial, biomedicine, microfluidics, actuators and electronics [4–7]. In 2005, Sun et al. [8] reported a high-resolution projection stereolithography process by using a digital micromirror device as a dynamic mask. In their study, the performance of projection stereolithography on fabricating 3D microstructures was illustrated and the 3D printing process was mathematically characterized. In the following years, much progress has been made to push the projection stereolithography to become a mainstream 3D printing technology for fabricating microfeatures. Chen et al. [9] presented a bottom-up projection system for the projection stereolithography process and a two-way linear motion approach has been developed for the quick spreading of liquid resin into uniform thin layers. The fabrication speed of a few seconds per layer has been achieved by their proposed system. Tumbleston et al. [10–12] proposed the digital light synthesis (DLS) (previously known as continuous liquid interface production or CLIP) process for projection stereolithography, which dramatically improves the printing speed and the surface quality of the printed model. Benefiting from the DLS technique, devices can be 3D printed in minutes with high quality. Shao et al. [13] fabricated a customized optical lens in about two minutes via a DLS-based projection stereolithography 3D printing process while the surface roughness was 13.7 nm. Recently, a volumetric projection stereolithography 3D printing method has been proposed [14–17], which prints entire complex objects through one complete revolution, circumventing the need for layering. This method was claimed to be particularly useful for high-viscosity photopolymers and multi-material fabrication. Meanwhile, studies have also focused on the system and material of projection stereolithography 3D printing. For example, Huang et al. [18] built a bottom-up projection stereolithography system with a robot arm and realized conform 3D printing through freeform transformation of layers. Chen et al. [19] presented an electrically assisted projection stereolithography system and Shao et al. [20,21] showed a magnetic material suitable for highly accurate 3D printing.

Previous works have made great progress on the process, system and material of projection stereolithography, while the geometric accuracy lacks sufficient study. Using the grayscale control method, Chen et al. [22] successfully improved the surface accuracy of a printed lens and reached a roughness of 6 nm. Zhou et al. [23] developed a geometric calibration system that can calibrate the position, shape, size, and orientation of a pixel and an energy calibration system that can calibrate the light intensity of a pixel. Based on the two systems, the shape accuracy and the uniformity of the power intensity distribution were improved. Researchers have also tried to improve the 3D printing accuracy of projection stereolithography by optimizing the processes [24,25], structures [26,27] and the photosensitive materials [28,29]. The geometric accuracy on the section shape and the size of the projection stereolithography 3D printing plays an important role for the microstructures and the functions of the 3D printed models for various applications, especially in bioengineering [30], optics [31] and microfluidics [32]. For example, the section shape of a 3D printed micro part has a remarkable effect on fog collection performance [33]. Also, accuracy in shape and size is critical to the function of the 3D printed microneedles [34,35].

However, few studies have investigated the 3D printed accuracy of the section shape and size directly. In this paper, the geometric error of the section shape and the dimension was mathematically analyzed and the projection power intensity distribution was studied with simulation tests. Then, a compensation method based on the structure optimization was proposed according to the simulation results. Also, the performance of the proposed compensation method was investigated by the simulation with the representative projection patterns. Finally, by the 3D printing experiments, the proposed compensation method was proved to be effective for improving the geometric accuracy of the projection stereolithography 3D printing. The results in this study are expected to be widely used to fabricate 3D structures for high precision applications.

2. Analysis and Method

2.1. Geometric Error Analysis

To analyze the geometric error, a projection stereolithography 3D printing system (Figure 1a) was built with a DLP projector that possesses a resolution of 3840 × 2160 pixels and a 405 nm light source (Fuzhou Gyinda Photoelectric Technology Co., Ltd., Fuzhou, China). The pixel resolution of the system was scale to 5 × 5 μm^2 using an UV lens (Universe Kogaku America, Inc., New York, NY, USA). The image of each layer is projected onto the surface of the photosensitive resin and the polymerization process is triggered by the UV light. Since the pattern UV light is composed of pixels, the power intensity distribution of the pixel directly affects the geometric accuracy of the cured layer. Equation (1) describes the power intensity distribution of a single pixel [36], which conforms to the Gaussian distribution in Figure 1b.

$$E(x) = E_0 \exp\left(-\frac{x^2}{w_0^2}\right) \tag{1}$$

where $E(x)$ is the power intensity at given position x, E_0 is the peak intensity, and w_0 is the Gaussian radius. Generally, the size of the single pixel (L) in projection stereolithography 3D printing system is slightly greater than that of the Gaussian radius. Thus, the light source of a pixel has a distinct influence on the power intensity distribution of adjacent pixels. Meanwhile, the rectangular pixel is not conformal to the shape of the power intensity distribution, which can be seen from the qualitative simulation result in Figure 1b. Both the crosstalk between pixels and the inconsistent shapes between the pixel and its power intensity distribution could cause the geometric errors.

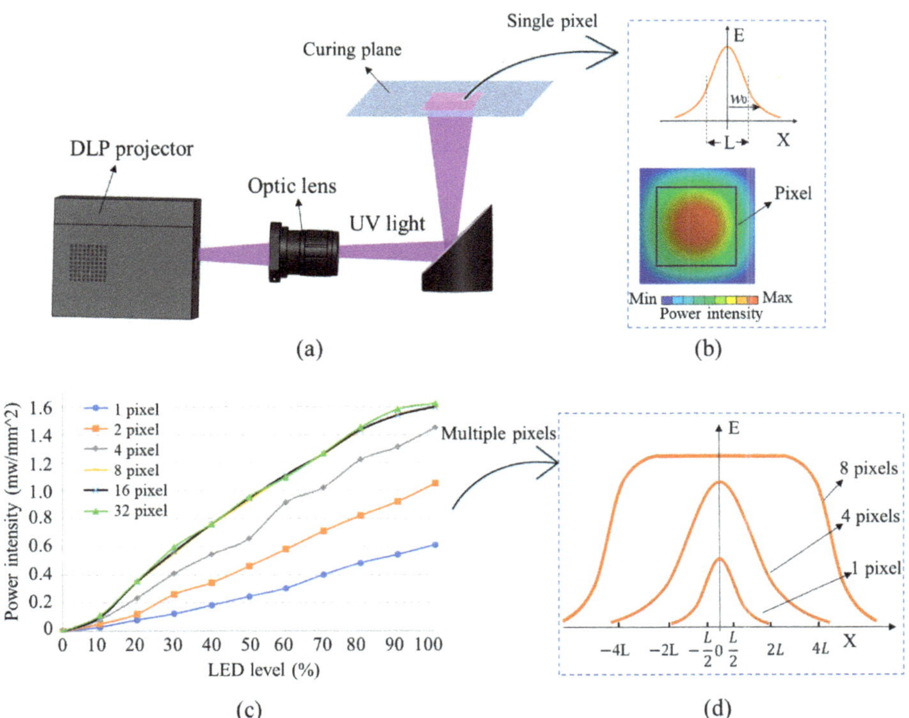

Figure 1. The process analysis. (**a**) The built projection stereolithography 3D printing system, (**b**) the power intensity distribution for single pixel, (**c**) the measurement result of the power intensity and (**d**) the power intensity distribution for multiple pixels.

By projecting a square pattern with different dimensions (ranging from 1 pixel to 32 pixels) onto the focus plane, the power intensities of the single pixel and multiple pixels are measured. From the results shown in Figure 1c, the power intensity rises with the increase of the number of pixels while the power intensity becomes saturated when the length of the square region reaches 8 pixels for the built system. Numerically, the intensity distribution of a region containing multiple pixels is obtained from the summation of the Gaussian distribution of each individual pixel. Thus, the evolution of the intensity profile with increasing pixels can be illustrated in Figure 1d. The "flat-top" shape of the intensity profile is obtained when the total size of pixels is much larger than the Gaussian radius. While the power intensity for the "flat-top" represented with E_{max} can be known by measuring the overall intensity of the projected large square pattern (see Figure 1c) using a dimension much greater than the typical Gaussian radius. Sun et al. [8] gave the general expression of the power intensity at the surface of resin by the Equations (2) and (3).

$$E(x,y) = E_{max}\{f[\sqrt{2}(L/4+x)/w_0] + f[\sqrt{2}(L/4-x)/w_0]\} \times \{f[\sqrt{2}(L/4+y)/w_0] + f[\sqrt{2}(L/4-y)/w_0]\} \tag{2}$$

$$f(x)= \int_0^x exp(-t^2)dt \qquad (3)$$

where $E(x,y)$ is the power intensity at given position (x,y) and $f(x)$ is the error function. As with the case of a single pixel, Figure 1d shows that the power intensity distribution area is always larger than the actual projection area. Moreover, Equation (2) also proves that the effect on the power intensity distribution of the adjacent pixel is nonuniform, depending on the projection pattern. Therefore, due to the unexpected power intensity of the pixels around the projection pattern, the geometric error will be produced between the cured part and the designed model.

2.2. Method

To improve the geometric accuracy of projection stereolithography 3D printing, a compensation method based on structure optimization is proposed in this paper. Three representative section features (circle, square and triangle) are included to verify the effectiveness of the proposed compensation method. The designed structure was optimized according to the simulation result of the power intensity distribution. By 3D printing experiments, the geometric errors between the designed structure and its corresponding 3D printed model were demonstrated. Also, the optimized structure was shown to be available to reduce the geometric errors by characterizing with scanning electron microscope (SEM). The photosensitive resin used in this work is composed of polyethylene glycol diacrylate (PEGDA, CAS: 26570-48-9) with a concentration of 99.45 wt%, ethyl (2,4,6-trimethylbenzoyl) phenylphosphinate (TPO-L, CAS: 84434-11-7) by 0.5 wt% as the photoinitiator, and a photoabsorber of Sudan I (CAS: 842-07-9) by 0.05 wt%. By 3D printing and measuring the thickness of suspend beams (Figure 1a), the curing depth curve of the PEGDA resin can be obtained, which is shown in Figure 2.

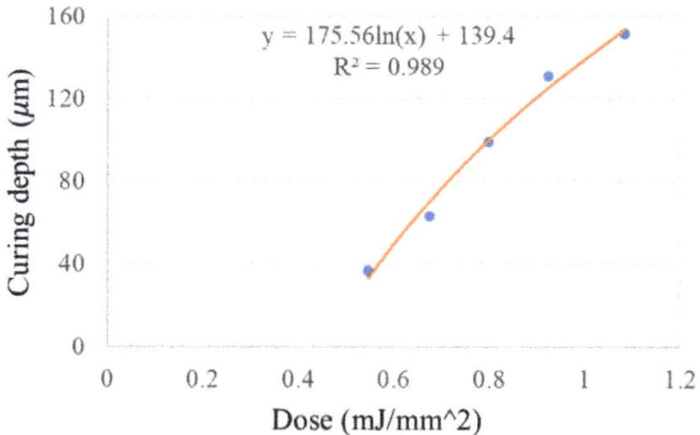

Figure 2. Curing depth of photosensitive resin as a function of UV energy flux.

3. Results and Discussion

3.1. Numerical Simulation

The simulation experiments were performed to investigate the power intensity distribution with Ansys 2020 software. To simplify the simulation model, a flat plane was divided into pixel cells while each cell was set as a light receiver with a size of 5×5 μm^2. The power intensity distribution on the surface of each cell is defined by Equation (1), in which the Gaussian radius is $w_0 = 6$ μm. The power intensity distribution on the surface of the plane is defined by Equation (2). In the simulation study, we assumed that the LED level is 30%, thus the maximum power intensity of a single pixel was 0.124 mw/mm^2 and the saturated value was 0.564 mw/mm^2 according to Figure 3c. Also, we assumed that the threshold power for photopolymerization was 0.37 mJ/mm^2 while the exposure time was 1 s. Then, the threshold power intensity could be calculated, which was 0.37 mw/mm^2. The power intensity distributions of the circular, square and regular triangular patterns were respectively calculated by the simulation process. As shown in Figure 3a, the power intensity distribution of the circle pattern conforms to the projection contour while the area of the threshold for the photopolymerization is remarkably larger than the area of the designed circle pattern, which leads to the dimensional error between the designed structure and its 3D printed model. The dimensional error for the circle pattern can be described by the Equation (4).

$$e_1 = K - R \tag{4}$$

where e_1 represents the dimension error for the circle pattern, K is the distance between the center of the circle and the threshold boundary, and R is the radius for the circle pattern.

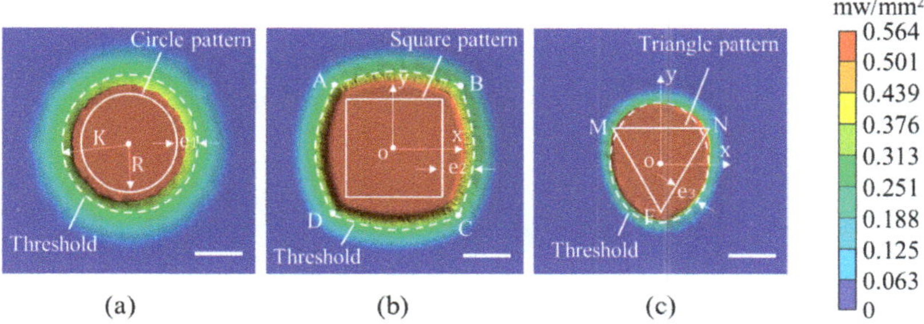

Figure 3. The simulation results for the power intensity distribution. (**a**) The circle pattern with a diameter of 100 μm, (**b**) the square pattern with a length of 100 μm and (**c**) the regular triangle pattern with a length of 100 μm. Scale bar: 50 μm.

For the square pattern in Figure 3b, the power intensity on the surface of the resin shows a barrel distribution with a saturated region that exceeds the projection area, indicating the positive geometric error for this case. The outline of the threshold power intensity composed of the curves AB, BC, CD and DA, can be approximately described by the conical section. The curve AB, which can be illustrated by the Equation (5), is selected for the following analysis.

$$x^2 = \lambda(y - \delta) \tag{5}$$

where δ is the distance from the interaction point between the threshold curve and the coordinate axis to the origin point O. λ is the parameter that determines the curvature, which can be calculated by giving the coordinates of two asymmetrical points on the curve. Based on Equation (5), the geometric error between the threshold outline curve and the projection square pattern can be calculated through Equation (6).

$$e_2 = y - L/2 \tag{6}$$

where e_2 represent the geometric error for the square pattern along the $y+$ direction. L is the length of the square pattern. Due to the symmetry of the projection pattern and the power intensity distribution, the errors for the other three sides of the square pattern are the same as that for the AB side.

From Figure 3c, the power intensity for the triangle pattern presents an elliptic distribution on the surface of the resin and the area with the saturated power intensity is larger than that of the projection pattern. Thus, there will be contour and dimension errors for the 3D printed triangle structure. The geometric error for the MN side can be described as Equation (7).

$$e_3 = \sqrt{b^2 - b^2 x^2 / a^2} - \tau \tag{7}$$

where e_3 is the geometric error for the triangle pattern, τ is the distance from the line MN to the centroid point O, a is the minor axis of the ellipse while b is the major axis. Also, since the symmetric pattern and the power intensity distribution, the geometric errors for the other two sides (MF and NF) can be obtained by the same analysis process.

To better understand the effect of the projection pattern on the power intensity distribution, the projection patterns for 3D printing the hole features with different shapes are simulated. Due to the cumulative effect of the power intensity between pixels, there are unexpected pixels with high power intensity inside the hole features, which can be known from the simulation results in Figure 4, leading to the over-cured 3D printing process. Thus, negative dimension errors can be seen from the simulation results compared to the designed hole features. The power intensity distributions for the circular, square and triangular holes are respectively consistent to those of circle, square and triangle projection patterns. Thus, the geometric errors for the hole features could also be calculated according to the Equations (4), (6) and (7).

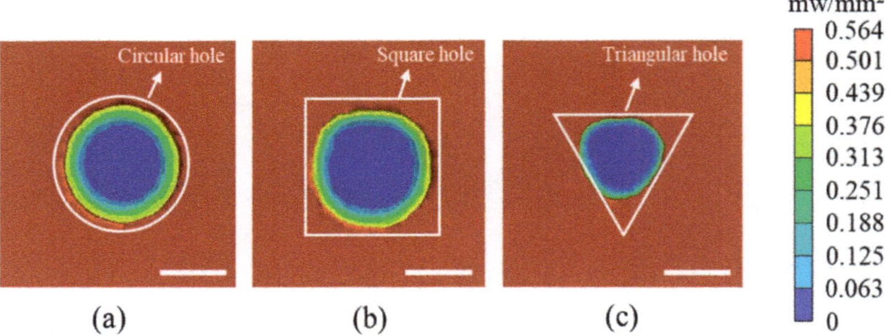

Figure 4. The simulation results for the power intensity distribution. (**a**) The projection pattern for the circular hole with a diameter of 100 μm, (**b**) the pattern for the square hole with a length of 100 μm and (**c**) the pattern for the regular triangle hole with a length of 100 μm. Scale bar: 50 μm.

The simulation results show that the actual power intensity distribution is not exactly consistent with the projection pattern because of the crosstalk between pixels, which brings the geometric errors to the projection stereolithography 3D printing. To improve the geometric accuracy, a compensation strategy based on structure optimization is proposed. For the designed structures with circular, square and triangle sections, the negative compensation strategy for structure optimization is presented due to the positive geometric errors from the numerical simulation tests. For example, the compensation strategy for the square pattern can be illustrated by the Equation (8).

$$L' = L - 2e_2 \tag{8}$$

where L' is the optimized length of the square pattern. While the geometric error is revealed by Equation (5) with the conical section, the optimized structure for the square pattern, which is demonstrated in Figure 5b, can be determined by combining Equations (5), (6) and (8). Likewise, the compensation strategy for the circle pattern is known with the equation of $R' = R - e_1$ (R' is the optimized radius) and Equation (4). Figure 5a shows the optimized result for the circle pattern. The compensation strategy for the triangle pattern is obtained by the equation of $\tau' = \tau - e_1$ (τ' is the optimized distance) and Equation (7), and the optimized structure is shown in Figure 5c.

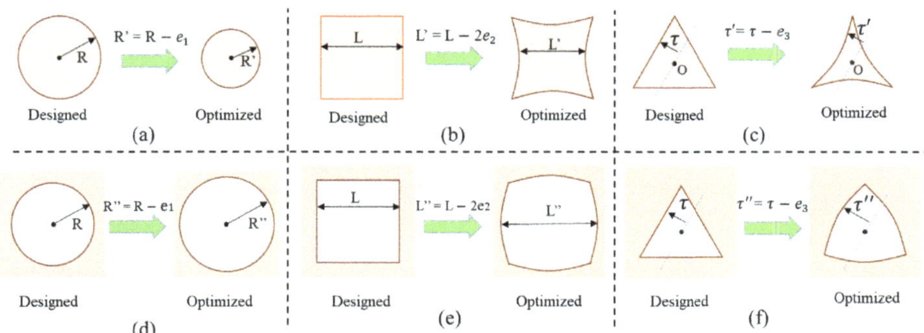

Figure 5. The compensation strategy based on the structure optimization. (**a**) The circle pattern, (**b**) the square pattern, (**c**) the triangle pattern, (**d**) the pattern for the circular hole, (**e**) the pattern for the square hole and (**f**) the pattern for the triangular hole.

With the same compensation strategy, the designed structures for the hole features are optimized according to the equations of $R'' = R - e_1$, $L'' = L - 2e_2$ and $\tau'' = \tau - e_1$. R'', L'' and τ'' are the optimized dimensions for the circular, square and triangular hole features respectively. Figure 5d–f indicate the optimized structures.

To validate the effectiveness of the proposed compensation strategy, the power intensity distributions of the optimized structures were investigated with the simulation tests. The result for the circle pattern shown in Figure 6a indicates that the power intensity distribution of the optimized structure conforms to the designed pattern more precisely compared with the designed structure in Figure 3a. Meanwhile, the power intensity distributions of the optimized square and triangle patterns indicate high 3D printing geometric accuracy, which can be known from Figure 3b,c. Also, comparing with the results in Figure 4, the results of Figure 6d,e show that a higher geometric precision for the circular, square and triangular holes can be reached by the optimized patterns. Thus, the simulation results verify that the proposed compensation strategy is able to dramatically improve the geometric accuracy of the projection stereolithography 3D printing.

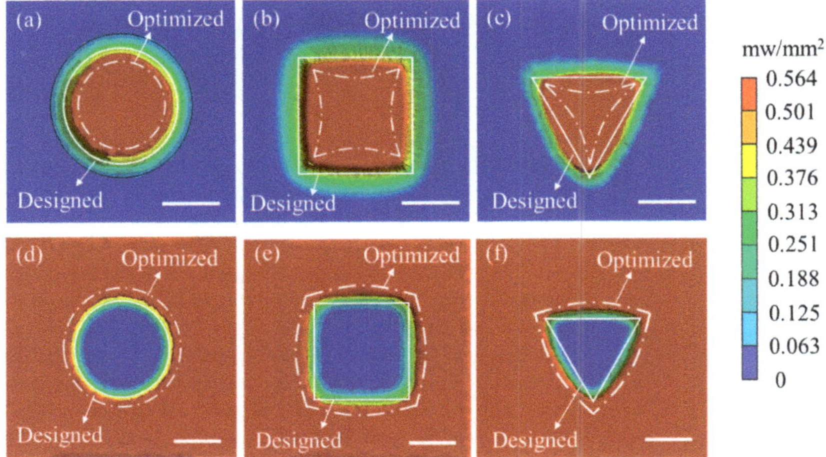

Figure 6. The simulation results for the power intensity distribution. (**a**) The optimized circle pattern, (**b**) the optimized square pattern, (**c**) the optimized triangle pattern, (**d**) the optimized pattern for the circular hole, (**e**) the optimized pattern for the square hole and (**f**) the optimized pattern for the regular triangle hole. Scale bar: 50 μm.

3.2. Experiments of 3D Printed Parts

To further verify the performance of the proposed compensation method, 3D printing experiments were implemented based on the built system. The thickness of 20 μm, the exposure time of 1 s and the power intensity by 0.5 mw/mm^2 were adopted as the key printing parameters for the experiments in this section. The dimensions for the designed and optimized patterns were the same as those in the simulation tests. The 3D printing results for the circle, square and triangle patterns are listed in Figure 7. The 3D printing result of the designed circle pattern shows a satisfactory shape accuracy with a diameter of 128 μm, which is larger than the designed dimension. From Figure 7b,c, significant geometric errors on the contour shape and the dimension can be found for the square and the triangle patterns. However, by optimizing the structures with the proposed compensation method, a high 3D printing geometric accuracy can be achieved, which is supported by the results in Figure 7d–f. The diameter of the 3D printed structure for the optimized circle pattern is 107 μm, which is close to the designed dimension (100 μm), indicating a high dimension accuracy. By projecting the optimized structure for the square pattern, the contour of the 3D printed model is much closer to square comparing to the result in Figure 7b. The dimension accuracy of the optimized square pattern has been improved, which can be seen from the Figure 7e wherein the diameter of the 3D printed model is 103 μm. Also, the 3D printed contour of the triangle pattern has been corrected to the triangular shape with a high precision dimension by optimizing the projection pattern.

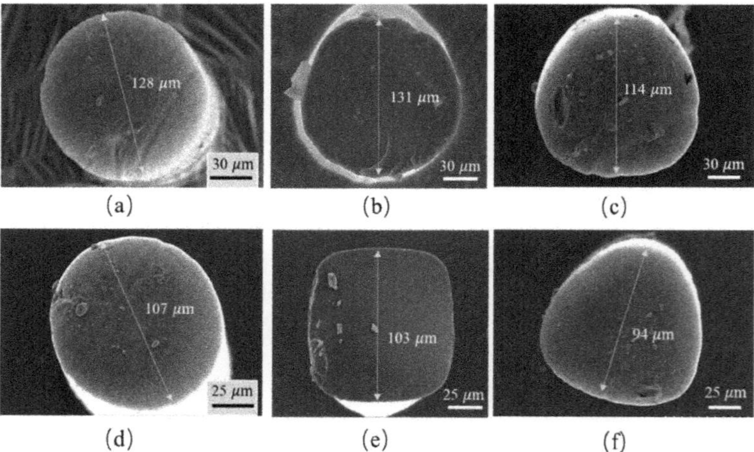

(a)　(b)　(c)

(d)　(e)　(f)

Figure 7. The results of 3D printing experiments. (**a**) the designed circle pattern, (**b**) the designed square pattern, (**c**) the designed triangle pattern, (**d**) the optimized circle pattern, (**e**) the optimized square pattern, (**f**) the optimized triangle pattern.

The comparison between the original designed and the optimized structures on the dimension accuracy was executed. Batches of models for the circle, square and triangle bosses were 3D printed with the original designed and the optimized structures respectively. The dimensions (refer to Figure 7) for six randomly selected samples of each pattern were measured. Figure 8a shows the measured diameter of each sample 3D printed with the circle pattern while the standard diameter is designed as 100 μm. The results illustrate that the samples printed by the original designed structure led to positive dimension errors and the diameters of samples with optimized structures were close to the designed standard dimension. For the square and triangle patterns (Figure 8b,c), the dimensions of the samples printed with the original designed structures were much larger than the designed standard dimensions (100 μm for the length of square and 86.6 μm for the altitude of the triangle pattern), while the printed dimensions were decreased by optimizing the structures. Furthermore, the mean dimension error for the circle pattern with the original designed structure is 23.3% and the mean error is reduced to 1.6% by using the optimized circle structure. For the square pattern, the mean error of the samples printed with the original structure is 21.3% while that with the optimized structure is 4.8%. Similarly, the mean dimension error for the triangle samples printed with the original structure is 22.7%. Using the optimized structure, 3.1% of the mean error is achieved. Thus, the dimension accuracy of projection stereolithography 3D printing could be improved by 21.7%, 16.5% and 19.6% for the circle, square and triangle structures respectively with the proposed compensation method.

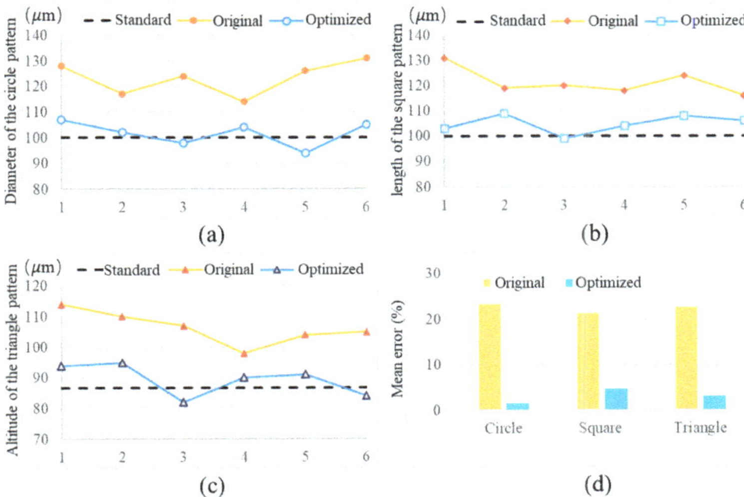

Figure 8. Comparison between the original designed and the optimized structures on dimension accuracy. (**a**) The dimensions of the circle pattern, (**b**) the dimensions for the square pattern, (**c**) the dimensions for the triangle pattern and (**d**) the mean error of each pattern.

For the projection patterns of the hole features, the 3D printing results are shown in Figure 9. By optimizing the pattern for the circular hole according to Figure 5, the 3D printed hole reaches a diameter that meets the design dimension with high precision. For the square hole, the original design pattern leads to the distortion with the small hole size. While the optimized pattern shows the hole feature with the accuracy square shape and the designed size, which can be seen in Figure 9e. Furthermore, the original design pattern for the triangular hole results in an elliptic feature and the smaller size comparing to the design dimension, which is consistent with the simulation result. From the result shown in Figure 9f, it is clear that the 3D printed structure with the optimized pattern possesses an accurate triangular shape with the correct dimension.

The comparison results between the original designed and the optimized structures on the dimension accuracy are demonstrated in Figure 10. Batches of models for the circular, square and triangular holes were 3D printed with the original designed and the optimized structures respectively (refer to Figure 9). The diameter of the designed standard circle is 100 μm, which is the same as the length of the designed square and the length of the designed regular triangle. The negative dimension errors caused by the original designed structures can be detected from the samples for the circular, square and triangular holes, which are shown in Figure 10a–c. However, the samples fabricated with the optimized structure showed the accurate dimensions for each hole feature. Also, the mean dimension errors of samples with different hole features are illustrated in Figure 10d. Specifically, the mean error of the circular hole samples printed with the original designed structure is 16.7% and 0.7% for that of the samples with the optimized circular hole structure. The mean errors before and after the optimization are 18.8% and 1.2% for the samples of the square hole. For the triangular hole, the samples with the original designed structure led to the mean error by 17.1% and this was reduced to 3.3% by using the optimized structure. The results indicate that the proposed compensation method could reduce the dimension error by 16%, 17.6% and 13.8% for 3D printed circular, square and triangular holes respectively.

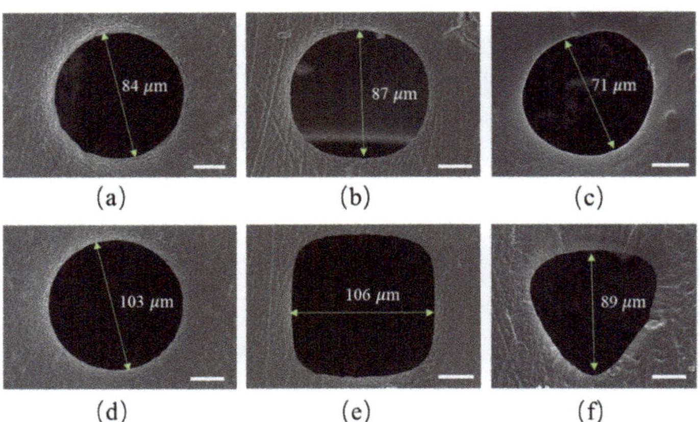

Figure 9. The 3D printing results of the hole features. (**a**) the designed pattern for the circular hole, (**b**) the designed pattern for the square hole, (**c**) the designed pattern for the triangular hole, (**d**) the optimized pattern for the circular hole, (**e**) the optimized pattern for the square hole, (**f**) the optimized pattern for the triangular hole. scale bar: 25 μm.

To further verify the performance of the proposed compensation method, 3D printing experiments with large 3D printing dimension scale were implemented. As shown in Figure 11a,d, two different structures with various features were designed. With the proposed compensation method, the designed structures were regenerated by the transformation of the shape and dimension, which can be seen in Figure 11b,e. The optimized structures were 3D printed (Figure 11c,f) and the models show clear features that are greatly consistent with the designed structures.

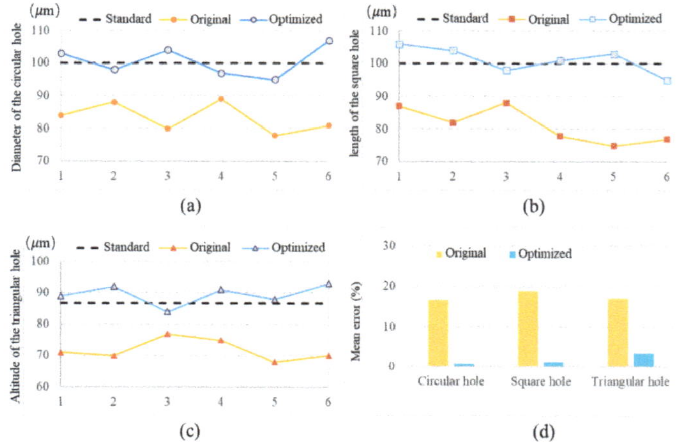

Figure 10. Comparison between the original designed and the optimized structures on dimension accuracy. (**a**) The dimensions for the circular hole, (**b**) the dimensions for the square hole, (**c**) the dimensions for the triangular hole and (**d**) the mean error of each pattern.

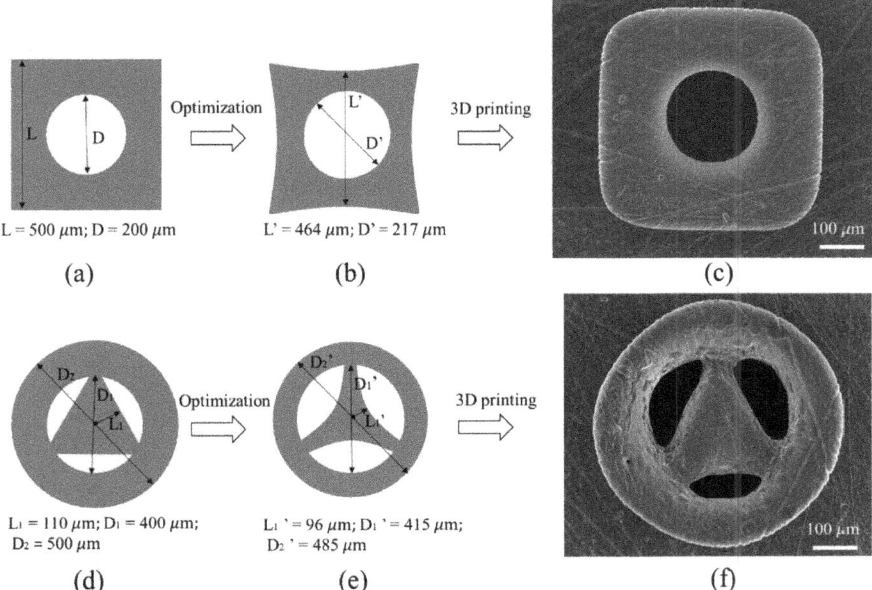

Figure 11. The 3D printing experiments. (**a**,**d**) Designed structures, (**b**,**e**) optimized structures, (**c**,**f**) 3D printed models.

The 3D printing experiments showed the same results as the simulation tests. While all of the results prove that the proposed compensation method could significantly improve the geometric accuracy of the projection stereolithography 3D printing.

4. Conclusions

In this work, the geometric errors of projection stereolithography were mathematically analyzed. Then the simulation tests were conducted to investigate the geometric errors of the projection stereolithography 3D printing with the representative patterns. A compensation method based on the structure optimization was proposed according to the simulation results. By positively or negatively compensating the projection patterns, the geometric errors relative to the design structure could be remarkably decreased. The simulation results and the 3D printing results indicate that the proposed compensation method is able to significantly improve the shape accuracy. The proposed compensation method was also shown to improve the dimension accuracy by 21.7%, 16.5% and 19.6% for the circular, square and triangular bosses respectively. While improvements to the dimension accuracy by 16%, 17.6% and 13.8% for the circular, square and triangular holes could be achieved with the proposed compensation method. The performance of the proposed compensation method was also verified at a large 3D printing dimension scale. This work is expected to provide a method to improve the geometric accuracy for 3D printing microstructures by the projection stereolithography.

Author Contributions: C.W. and Z.C. (Zhengda Chen): Methodology, validation, investigation, writing. Z.C. (Zhuoxi Chen), B.Z. and Z.C. (Zhicheng Cheng): Validation, writing. H.Y., G.J. and J.H.: Conceptualization, investigation, writing, supervision. All authors have read and agreed to the published version of the manuscript.

Funding: This work was supported by the Fundamental Research Funds for the Central Universities (20822041F4049, Sichuan University).

Institutional Review Board Statement: Not applicable.

Informed Consent Statement: Not applicable.

Conflicts of Interest: The authors declare no conflict of interest.

References

1. Huang, J.; Qin, Q.; Wen, C.; Chen, Z.; Huang, K.; Fang, X.; Wang, J. A dynamic slicing algorithm for conformal additive manufacturing. *Addit. Manuf.* **2022**, *51*, 102622. [CrossRef]
2. Kim, S.; Handler, J.J.; Cho, Y.T.; Barbastathis, G.; Fang, N.X. Scalable 3D printing of aperiodic cellular structures by rotational stacking of integral image formation. *Sci. Adv.* **2021**, *7*, eabh1200. [CrossRef] [PubMed]
3. Li, H.; Li, Z.; Li, N.; Zhu, X.; Zhang, Y.F.; Sun, L.; Wang, R.; Zhang, J.; Yang, Z.; Yi, H. 3D Printed High Performance Silver Mesh for Transparent Glass Heaters through Liquid Sacrificial Substrate Electric-Field-Driven Jet. *Small* **2022**, *18*, 2107811. [CrossRef]
4. Huang, J.; Qin, Q.; Wang, J. A review of stereolithography: Processes and systems. *Processes* **2020**, *8*, 1138. [CrossRef]
5. Ge, Q.; Li, Z.; Wang, Z.; Kowsari, K.; Zhang, W.; He, X.; Zhou, J.; Fang, N.X. Projection micro stereolithography based 3D printing and its applications. *Int. J. Extrem. Manuf.* **2020**, *2*, 022004. [CrossRef]
6. Zhu, W.; Ma, X.; Gou, M.; Mei, D.; Zhang, K.; Chen, S. 3D printing of functional biomaterials for tissue engineering. *Curr. Opin. Biotechnol.* **2016**, *40*, 103–112. [CrossRef] [PubMed]
7. Li, Z.; Li, H.; Zhu, X.; Peng, Z.; Zhang, G.; Yang, J.; Wang, F.; Zhang, Y.F.; Sun, L.; Wang, R. Directly Printed Embedded Metal Mesh for Flexible Transparent Electrode via Liquid Substrate Electric-Field-Driven Jet. *Adv. Sci.* **2022**, *9*, 2105331. [CrossRef]
8. Sun, C.; Fang, N.; Wu, D.; Zhang, X. Projection micro-stereolithography using digital micro-mirror dynamic mask. *Sens. Actuators A Phys.* **2005**, *121*, 113–120. [CrossRef]
9. Pan, Y.; Zhou, C.; Chen, Y. A fast mask projection stereolithography process for fabricating digital models in minutes. *J. Manuf. Sci. Eng.* **2012**, *134*, 051011. [CrossRef]
10. Tumbleston, J.R.; Shirvanyants, D.; Ermoshkin, N.; Janusziewicz, R.; Johnson, A.R.; Kelly, D.; Chen, K.; Pinschmidt, R.; Rolland, J.P.; Ermoshkin, A. Continuous liquid interface production of 3D objects. *Science* **2015**, *347*, 1349–1352. [CrossRef]
11. Janusziewicz, R.; Tumbleston, J.R.; Quintanilla, A.L.; Mecham, S.J.; DeSimone, J.M. Layerless fabrication with continuous liquid interface production. *Proc. Natl. Acad. Sci. USA* **2016**, *113*, 11703–11708. [CrossRef] [PubMed]
12. Hossain, M.; Liao, Z. An additively manufactured silicone polymer: Thermo-viscoelastic experimental study and computational modelling. *Addit. Manuf.* **2020**, *35*, 101395. [CrossRef]
13. Shao, G.; Hai, R.; Sun, C. 3D printing customized optical lens in minutes. *Adv. Opt. Mater.* **2020**, *8*, 1901646. [CrossRef]
14. Kelly, B.E.; Bhattacharya, I.; Heidari, H.; Shusteff, M.; Spadaccini, C.M.; Taylor, H.K. Volumetric additive manufacturing via tomographic reconstruction. *Science* **2019**, *363*, 1075–1079. [CrossRef] [PubMed]
15. Loterie, D.; Delrot, P.; Moser, C. High-resolution tomographic volumetric additive manufacturing. *Nat. Commun.* **2020**, *11*, 852. [CrossRef]
16. Shusteff, M.; Browar, A.E.; Kelly, B.E.; Henriksson, J.; Weisgraber, T.H.; Panas, R.M.; Fang, N.X.; Spadaccini, C.M. One-step volumetric additive manufacturing of complex polymer structures. *Sci. Adv.* **2017**, *3*, eaao5496. [CrossRef]
17. Bhattacharya, I.; Toombs, J.; Taylor, H. High fidelity volumetric additive manufacturing. *Addit. Manuf.* **2021**, *47*, 102299. [CrossRef]
18. Huang, J.; Ware, H.O.T.; Hai, R.; Shao, G.; Sun, C. Conformal geometry and multimaterial additive manufacturing through freeform transformation of building layers. *Adv. Mater.* **2021**, *33*, 2005672. [CrossRef]
19. Yang, Y.; Li, X.; Chu, M.; Sun, H.; Jin, J.; Yu, K.; Wang, Q.; Zhou, Q.; Chen, Y. Electrically assisted 3D printing of nacre-inspired structures with self-sensing capability. *Sci. Adv.* **2019**, *5*, eaau9490. [CrossRef]
20. Shao, G.; Ware, H.O.T.; Li, L.; Sun, C. Rapid 3D printing magnetically active microstructures with high solid loading. *Adv. Eng. Mater.* **2020**, *22*, 1900911. [CrossRef]
21. Shao, G.; Ware, H.O.T.; Huang, J.; Hai, R.; Li, L.; Sun, C. 3D printed magnetically-actuating micro-gripper operates in air and water. *Addit. Manuf.* **2021**, *38*, 101834. [CrossRef]
22. Chen, X.; Liu, W.; Dong, B.; Lee, J.; Ware, H.O.T.; Zhang, H.F.; Sun, C. High-speed 3D printing of millimeter-size customized aspheric imaging lenses with sub 7 nm surface roughness. *Adv. Mater.* **2018**, *30*, 1705683. [CrossRef] [PubMed]
23. Zhou, C.; Chen, Y. Calibrating large-area mask projection stereolithography for its accuracy and resolution improvements. In Proceedings of the 2009 International Solid Freeform Fabrication Symposium, Austin, TX, USA, 3–5 August 2009.
24. Qin, Q.; Huang, J.; Yao, J.; Gao, W. Design and optimization of projection stereolithography additive manufacturing system with multi-pass scanning. *Rapid Prototyp. J.* **2021**, *27*, 636–642. [CrossRef]
25. Li, X.; Xie, B.; Jin, J.; Chai, Y.; Chen, Y. 3D printing temporary crown and bridge by temperature controlled mask image projection stereolithography. *Procedia Manuf.* **2018**, *26*, 1023–1033. [CrossRef]
26. Lian, Q.; Wu, X.; Li, D.; He, X.; Meng, J.; Liu, X.; Jin, Z. Accurate printing of a zirconia molar crown bridge using three-part auxiliary supports and ceramic mask projection stereolithography. *Ceram. Int.* **2019**, *45*, 18814–18822. [CrossRef]
27. Dar, U.A.; Mian, H.H.; Abid, M.; Topa, A.; Sheikh, M.Z.; Bilal, M. Experimental and numerical investigation of compressive behavior of lattice structures manufactured through projection micro stereolithography. *Mater. Today Commun.* **2020**, *25*, 101563. [CrossRef]
28. Revilla-León, M.; Mostafavi, D.; Methani, M.M.; Zandinejad, A. Manufacturing accuracy and volumetric changes of stereolithography additively manufactured zirconia with different porosities. *J. Prosthet. Dent.* **2021**, *in press*. [CrossRef]

29. Wang, Z.; Chen, L.; Chen, Y.; Liu, P.; Duan, H.; Cheng, P. 3D printed ultrastretchable, hyper-antifreezing conductive hydrogel for sensitive motion and electrophysiological signal monitoring. *Research* **2020**, *2020*, 1426078. [CrossRef] [PubMed]
30. Wang, Y.; Ahmed, A.; Azam, A.; Bing, D.; Shan, Z.; Zhang, Z.; Tariq, M.K.; Sultana, J.; Mushtaq, R.T.; Mehboob, A. Applications of additive manufacturing (AM) in sustainable energy generation and battle against COVID-19 pandemic: The knowledge evolution of 3D printing. *J. Manuf. Syst.* **2021**, *60*, 709–733. [CrossRef]
31. Hong, Z.; Ye, P.; Loy, D.A.; Liang, R. Three-dimensional printing of glass micro-optics. *Optica* **2021**, *8*, 904–910. [CrossRef]
32. Mohamed, M.G.; Kumar, H.; Wang, Z.; Martin, N.; Mills, B.; Kim, K. Rapid and inexpensive fabrication of multi-depth microfluidic device using high-resolution LCD stereolithographic 3D printing. *J. Manuf. Mater. Processing* **2019**, *3*, 26. [CrossRef]
33. Liu, L.; Liu, S.; Schelp, M.; Chen, X. Rapid 3D printing of bioinspired hybrid structures for high-efficiency fog collection and water transportation. *ACS Appl. Mater. Interfaces* **2021**, *13*, 29122–29129. [CrossRef]
34. Krieger, K.J.; Bertollo, N.; Dangol, M.; Sheridan, J.T.; Lowery, M.M.; O'Cearbhaill, E.D. Simple and customizable method for fabrication of high-aspect ratio microneedle molds using low-cost 3D printing. *Microsyst. Nanoeng.* **2019**, *5*, 42. [CrossRef] [PubMed]
35. Tareq, M.S.; Rahman, T.; Hossain, M.; Dorrington, P. Additive manufacturing and the COVID-19 challenges: An in-depth study. *J. Manuf. Syst.* **2021**, *60*, 787–798. [CrossRef] [PubMed]
36. Kumar, G.; Sayanagi, K. Measurement of optical transfer function by its moments. *J. Opt. Soc. Am.* **1968**, *58*, 1369–1374. [CrossRef]

Article

Fabrication and Formability of Continuous Carbon Fiber Reinforced Resin Matrix Composites Using Additive Manufacturing

Lining Yang *, Donghao Zheng, Guojie Jin and Guang Yang *

School of Mechanical Engineering, Hebei University of Science and Technology, Shijiazhuang 050018, China; hh1185994756@163.com (D.Z.); jinguojie98@163.com (G.J.)
* Correspondence: yang_li_ning@126.com (L.Y.); y_guang@126.com (G.Y.); Tel.: +86-180-1000-5590 (L.Y.)

Abstract: In the current process for additive manufacturing of continuous carbon fiber reinforced resin matrix composites, the fiber and resin matrix are fed into the molten chamber, and then impregnated and compounded in the original position, and finally extruded and deposited on the substrate. It is difficult to control the ratio of fiber and resin, and to achieve good interface fusion, which results in an unsatisfactory enhancement effect. Therefore, an additive manufacturing process based on continuous carbon fiber reinforced polylactic acid composite prepreg filament was explored in this study. The effects of various process parameters on the formability of composites were studied through systematic process experiments. The results showed that the process parameters of additive manufacturing have a systematic influence on the forming quality, accuracy and efficiency, and on the mechanical properties of CFRP. Through the experimental optimization of various process parameters, a continuous and stable forming process was achieved when the nozzle aperture was 0.8 mm, the nozzle printing temperature was 240 °C, the substrate temperature was 60 °C, the wire feeding speed was 5 mm/s, the nozzle moving speed was 5 mm/s, the path bonding rate was 40%, and the printing layer thickness was 0.7 mm. Based on the optimized process parameters, direct additive manufacturing of a lightweight and high-strength composite cellular load-bearing structure could be realized. Its volume fraction of carbon fiber was approximately 7.7%, and the tensile strength was up to 224.3 MPa.

Keywords: additive manufacturing; continuous carbon fiber reinforced; resin matrix composites; fused deposition modeling (FDM); formability

Citation: Yang, L.; Zheng, D.; Jin, G.; Yang, G. Fabrication and Formability of Continuous Carbon Fiber Reinforced Resin Matrix Composites Using Additive Manufacturing. *Crystals* **2022**, *12*, 649. https://doi.org/10.3390/cryst12050649

Academic Editors: Hao Yi, Huajun Cao, Menglin Liu, Le Jia and Zhao Zhang

Received: 14 March 2022
Accepted: 26 April 2022
Published: 2 May 2022

Publisher's Note: MDPI stays neutral with regard to jurisdictional claims in published maps and institutional affiliations.

1. Introduction

Recently, emerging additive manufacturing technology [1–3] has provided the possibility of small-batch and customized manufacturing of composite components with complex structures and special functional requirements. For example, Valerio et al. [4,5] investigated the application of an additive manufactured hybrid metal composite shock absorber panel to a military seat ejection system, and studied the effects of core microstructure on the energy absorbing capabilities of sandwich panels intended for additive manufacturing. Fused deposition molding, a widely used additive manufacturing technology, has excellent potential for rapidly manufacturing fiber reinforced resin matrix composite components, and has become a research hotspot. The most common materials used in this type of forming process are amorphous thermoplastics, with acrylonitrile–butadiene–styrene (ABS) and polylactic acid (PLA) being the most common [6,7]. Recently, a few studies have reported on the use of short- or long- fiber reinforced thermal plastics as the feedstock of the FDM process. Tekinalp et al. [8], Zhong et al. [9], and Ning et al. [10], investigated short fiber reinforced acrylonitrile–butadiene–styrene (ABS) composites as a feedstock for 3D printing in terms of their processibility, microstructure, and mechanical performance. Gray et al. [11,12] developed polypropylene (PP) strands reinforced with thermotropic liquid crystalline polymer (TLCP) fibers for an FDM process, and investigated the effects of

FDM processing conditions on short TLCP fiber reinforced parts. Luo et al. [13] investigated the impregnation and interlayer bonding behaviors of 3D-printed continuous carbon fiber reinforced poly-ether-ketone composites.

Continuous carbon fiber reinforced resin matrix composites (CCFRRMCs) have the advantages of high specific strength, high specific stiffness, low thermal expansion coefficient, excellent fatigue performance, and corrosion resistance, and are critical substrates for the main load-bearing structures and high-precision substructures of satellites and spacecraft. CCFRRMCs have been also used in the manufacturing process for satellite load-bearing cylinders, rocket motor cases, and other components [14,15]. At present, the traditional forming processes of CCFRRMC components mainly include laying forming [16], molding forming [17,18], and weaving forming [19,20]. It not only has complicated processing procedures, a long processing cycle, and high manufacturing costs, but it also cannot achieve flexible manufacturing of parts with complex structures. Therefore, it has become the bottleneck for composite component processing in the aerospace field.

Based on the fusion deposition molding method, most research institutions are currently feeding continuous carbon fiber dry filaments and resin matrix filaments into the heating block fusion chamber, which are impregnated and compounded in the original position, and then extruded from the end of the nozzle and deposited layer by layer. Li et al. [21] investigated the rapid prototyping of continuous carbon fiber reinforced poly-lactic acid composites for 3D printing, and performed comparison experiments of printed samples with or without a preprocessed carbon fiber bundle; the properties were also measured. Tian et al. [22] researched the interface and performance of 3D printed continuous carbon fiber reinforced PLA composites. Wang et al. [23,24] improved the performance of continuous carbon fiber reinforced plastic composites by optimizing the printing strategy, and they also investigated the fiber–matrix impregnation behavior during the additive manufacturing process. Liang et al. [25] adopted a bionic structure design to further improve the impact resistance of additive manufacturing composites. However, due to the poor rigidity of carbon fiber dry filaments, it is difficult to precisely control the ratio of fiber to resin during the composite process. In addition, the impregnation time for fiber and resin in the melting chamber is short, and a favorable interface combination cannot be acquired, making it difficult to provide the desired reinforcement when using continuous carbon fibers within composite parts.

In order to further improve the mechanical properties of fused deposition molding CCFRRMCs, we employed pre-prepared continuous carbon fiber/PLA composite filaments to systematically investigate the effects of printing speeds, nozzle height, path bonding rate, printing temperature, printing layer thickness, and other parameters on the forming quality of single channel, single-layer, and solid composites. As a result, direct additive manufacturing of honeycomb CCFRRMCs was realized. Furthermore, the research results reported in this paper reveal the coupling mechanism for multiple process parameters in the additive manufacturing process of CCFRRMCs, and lay the foundation for the optimization of each process parameter. This study achieved structure–function integrated design for lightweight composite parts with complex structures and high strengths for use in the aerospace field.

2. Materials and Methods

2.1. Materials

In the present research, the 1 mm diameter continuous carbon fiber reinforced resin matrix composite filaments were all prepared using self-developed filament-making equipment. Polyacrylonitrile T300 (PAN) 1K carbon fiber bundles from TORAY Corp. in Japan, with a monofilament diameter of 6~8 μm, were used as the raw material for the continuous carbon fiber. The raw material for the thermoplastic resin was polylactic acid (PLA) pellets, type 4032D, from NATUREWORKS Corp. in the USA, with a melting temperature of 175 °C.

For the preparation of the composite filaments, first the surface of the carbon fiber was treated with a sizing agent. A bundle of 1K treated carbon fiber was then fed into the filament-making equipment under a certain tension. In this equipment, the carbon fiber was spread evenly onto the surface of a set of impregnated rollers. At the same time, the molten PLA in the impregnation cavity was allowed to penetrate into the filament under the continuous high pressure. Finally, the impregnated carbon fibers were pulled out from a certain diameter outlet and cured for a certain time to form the final composite filaments.

The prepared composite filaments and their cross-sectional microscopic morphology are shown in Figure 1. It can be seen that PLA rings wrap the continuous carbon fiber filament bundles, and the PLA forms an effective homogeneous impregnation between the carbon fiber monofilaments. The tensioning state of the continuous carbon fibers is difficult to control during the preparation of the composite filaments and therefore, the stability of the neutral distribution of continuous carbon fiber in the composite wire was imperfect. However, this problem can be compensated for during the secondary melting and extrusion of the composite filaments.

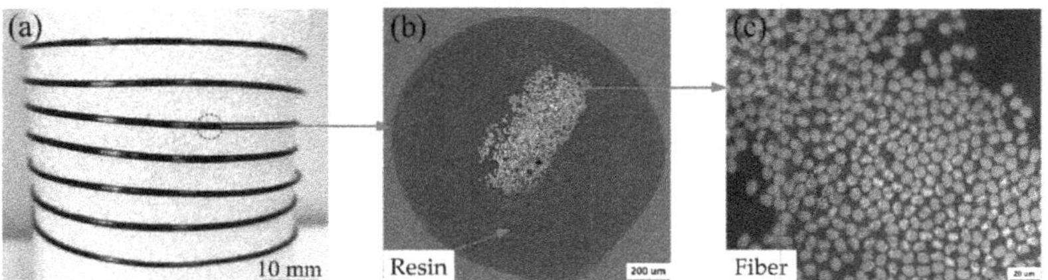

Figure 1. (**a**) Continuous carbon fiber reinforced resin matrix composite wire; (**b**) composite wire microstructure (cross section); (**c**) distribution of fibers in the resin.

2.2. Methods

2.2.1. Additive Manufacturing Process for Continuous Carbon Fiber Reinforced Resin Matrix Composites

The process for additive manufacturing of CCFRRMCs used in this study is shown in Figure 2a. The prepreg composite filament was driven by the wire feeding wheel into the melt extrusion nozzle at the bottom, and was melted by the heating block at the bottom of the nozzle in real time and with high efficiency. The molten composite material was extruded via the nozzle end under the driving force of the solid wire at the upper end. The nozzle moved according to the section profile and filling trajectory of the preprepared parts, so that the composite materials were selectively stacked on the substrate, layer by layer. Finally, complete CCFRRMC parts with a specific complex shape were obtained. Figure 2b shows the extrusion nozzle used in the experimental process.

2.2.2. Single Channel Composite Forming Experiment

Additive manufacturing is a systematic forming process in which materials are formed from point to line, from line to surface, and from surface to body. The process of forming lines from points is the basis of additive manufacturing technology. Therefore, this study first used continuous carbon fiber/polylactic acid composite wire to conduct a single channel forming experiment to study the relationship between wire feeding speed, nozzle moving speed, the influence of nozzle height to substrate height on forming quality and on the single channel deposition path width of the single channel composite.

Based on reading many references and carrying out preliminary process experiments, the process parameters in the single channel forming experiment were determined as follows [21–25]: nozzle aperture 0.8 mm, nozzle printing temperature 220 °C, and substrate temperature 60 °C; and the range of the study parameters was as follows: wire feeding speed 3~10 mm/s, moving speed of the nozzle 3~10 mm/s, and the height of the nozzle from the substrate 0.3~1 mm.

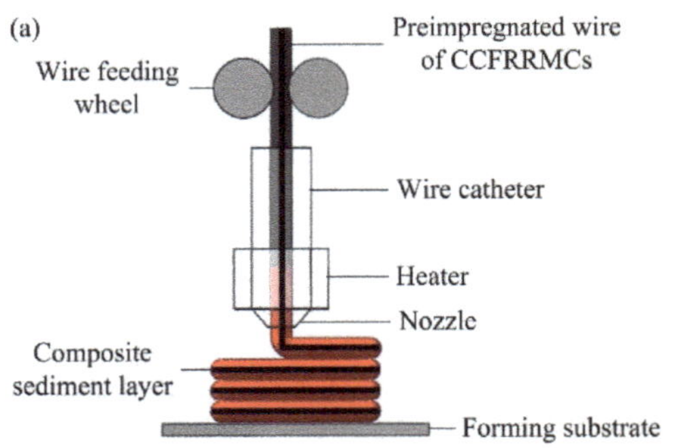

Figure 2. (**a**) Additive manufacturing process for continuous carbon fiber reinforced resin matrix composites; (**b**) the composite wire melting extrusion nozzle.

2.2.3. Forming Experiment of Monolayer Composite

Path bonding rate affects the surface quality, fiber volume fraction, and mechanical properties of composite material. Therefore, based on the single channel forming experiment, we conducted an experiment on the forming process of the single-layer composite material to investigate the path bonding rate parameters. The surface quality of the monolayer composite was studied under the conditions of wire feeding speed of 5 mm/s, moving speed of the nozzle of 5 mm/s, and height of the nozzle from the substrate of 0.7 mm.

Figure 3 illustrates the overlap zone formed between two adjacent deposition paths. In the Figure, w is the single path deposition path width, l is the width of the lap zone, and the path bonding rate δ can be expressed as:

$$\delta = \frac{l}{w} \tag{1}$$

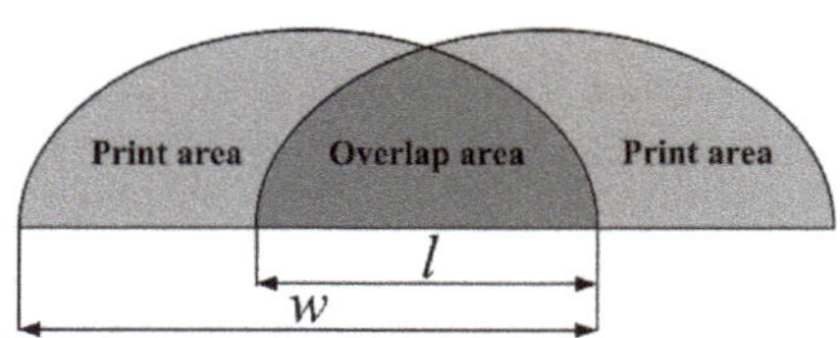

Figure 3. Schematic diagram of the overlap region formed between two adjacent trajectories.

The range of the path overlap rate studied in the experiment was 10–60%.

2.2.4. Composite Material Solid Forming Experiment

The nozzle printing temperature and printing layer thickness significantly affect the forming accuracy, forming efficiency, and mechanical properties of composites. Therefore, based on the optimized path bonding rate, experiments on solid forming of composite materials were carried out for these two process parameters. In accordance with standard GB/T 100.44-2006 (Determination of tensile properties of plastics—Part 4: Isotropic and orthotropic fiber reinforced composite material), three layers of composite tensile test specimens were prepared under different nozzle print temperatures and layer thickness parameters. The tensile properties of the specimens were tested using an electronic universal testing machine (UTM6503, SANSI ZHONGHENG Technology Corp, Shenzhen, China), and the effects of printing temperature and printing layer thickness on the tensile properties of solid specimens were studied. At the same time, the interlaminar bonding, the carbon fiber and polylactic acid composite situation, and the distribution of carbon fiber at the fracture point of the tensile specimen were observed using a scanning electron microscope (S-4800, HITACHI Corp., Hitachi, Japan), to further explore the effect of printing temperature and printing layer thickness on the tensile properties of solid specimens.

The range of printing temperatures studied in the experiment was 200~260 °C, and the range of printing layer thicknesses was 0.6~0.9 mm.

2.2.5. Forming Experiment for Composite Cellular Load-Bearing Structure

Finally, we verified the practicality and stability of the additive manufacturing process based on continuous carbon fiber reinforced resin matrix composite prepreg filaments. Based on the optimized parameters, nozzle aperture was 0.8 mm, nozzle printing temperature was 240 °C, substrate temperature was 60 °C, wire feeding speed was 5 mm/s, nozzle moving speed was 5 mm/s, path bonding rate was 40% and printing layer thickness was 0.7 mm, with the aim to realize the formation of typical honeycomb load-bearing structures used in the aerospace field.

3. Experimental Results and Analysis
3.1. Experimental Results and Analysis of Single Channel Composite Forming

Figure 4 shows the theoretical distribution state of continuous carbon fiber in the molten chamber at the end of the nozzle, and the corresponding images of the actual single channel forming when the wire feeding speed was set at 5 mm/s and different moving speeds of the nozzle were adopted.

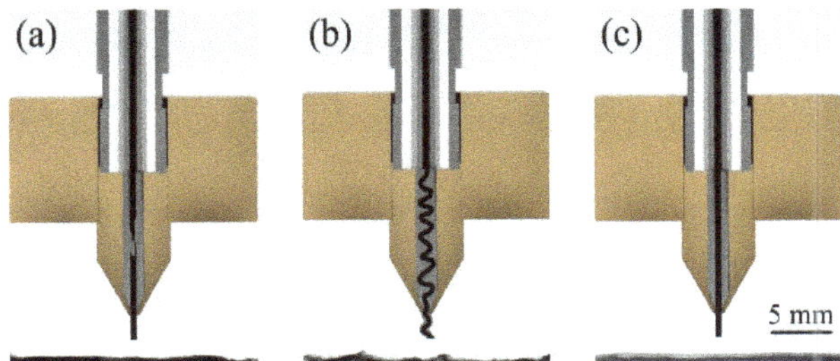

Figure 4. (**a**) The theoretical distribution state of continuous carbon fiber in nozzle melting chamber and the corresponding image of actual single channel when the the moving speed of the nozzle was greater than the wire feeding speed; (**b**) the moving speed of the nozzle was smaller than the wire feeding speed; (**c**) the moving speed of the nozzle was equal to the wire feeding speed. With equal values for the wire feed speed and nozzle moving speed, it is often necessary to increase the speed to improve the forming efficiency, but a faster wire feed and nozzle moving speed will also impact the single-pass forming quality.

As shown in Figure 4a, when the moving speed of the nozzle was 7 mm/s higher than the set wire feeding speed, the length of carbon fiber provided by the wire feeding pair wheel per unit time was less than the length of carbon fiber required by single path filling. At this moment, the drag force of the carbon fiber in the molten cavity caused by the composite material which has been deposited and solidified at the end of the nozzle was also increasing. Finally, there were breaks in the continuous carbon fiber in the nozzle melting chamber, which affected the single channel forming quality. In the opposite case, Figure 4b shows that when the nozzle moving speed was set at 3 mm/s lower than the wire feed speed, the carbon fiber length provided by the wire was longer than the single channel path. Such a case caused the over-accumulation problem of continuous carbon fiber in the nozzle melting chamber, which is easy to cause the nozzle blockage and affect the single channel forming quality. Figure 4c shows the situation when the nozzle moving speed was equal to the set speed, 5 mm/s. Per unit time, the carbon fiber length provided by the wire was equal to the length of the single channel path. Now, homogeneous and continuous carbon fiber was distributed in the nozzle melt chamber and deposited in the middle of the single channel path. Single channel forming quality was also better.

In summary, in order to get a good single-pass forming quality, the wire feed speed needed to be set at the same value as the nozzle moving speed.

Figure 5 shows images of a single path forming when the moving speed of the nozzle is unequal to the wire feed rate. It can be seen that when the moving velocity of the wire feeder and the nozzle is relatively small at 5 mm/s, as shown in Figure 5a, the edge of the single channel deposition path is relatively uniform. When the moving velocity of the wire feed and nozzle is relatively large at 10 mm/s, as shown in Figure 5b, the edge uniformity of the single channel deposition path is poor. The reasons were as follows: on the one hand, at a higher wire feeding speed, the volume of molten polylactic acid extruded from the nozzle will increase per unit time, and the extrusion stability will also decrease, resulting in an uneven deposition path edge. On the other hand, the stability of the height between the nozzle and the substrate decreases due to the fast-moving speed of the nozzle, which further aggravates the inhomogeneity of the edge of the single channel deposition path.

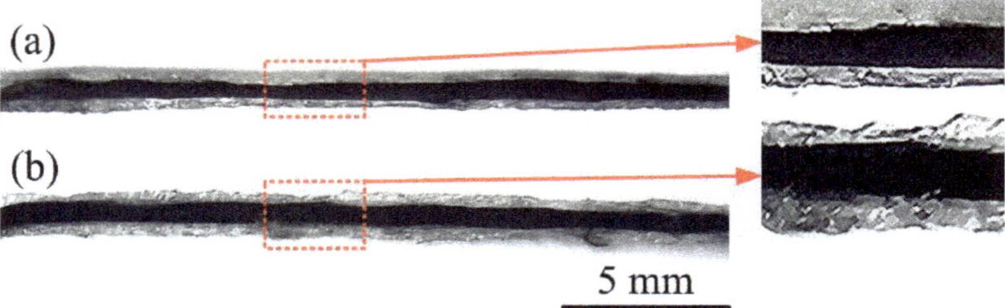

Figure 5. (**a**) Photograph of single channel forming path when the nozzle moving speed and wire feeding speed were both 5 mm/s; (**b**) the nozzle moving speed and wire feeding speed were both 10 mm/s. In the single channel composite forming process, the height of the nozzle from the substrate also had a significant impact on the single channel deposition path width and forming quality. Therefore, we set the wire feed speed and nozzle moving speed at 5 mm/s and tested different nozzle height conditions in single-pass forming experiments. Figure 6 shows the resulting experimental data and forming paths.

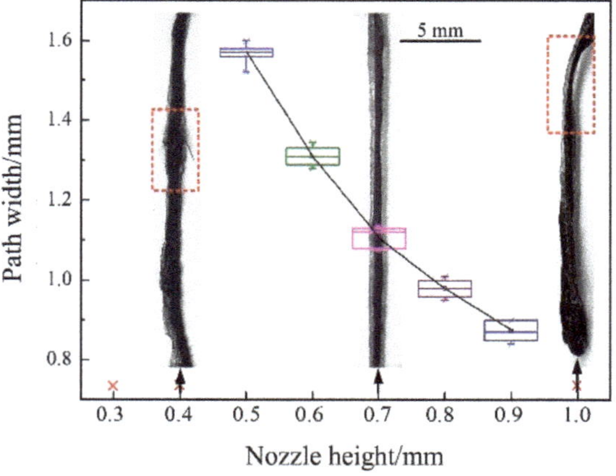

Figure 6. Single channel deposition path width and its resulting image for different nozzle heights.

When the height of the nozzle was small (0.3~0.4 mm), the deposition path was uneven, resulting in poor uniformity of the path and carbon fiber that is easy to scrape off. When the nozzle height was extended (1 mm), the extruded composite could not be effectively compacted, so the deposition path could not adhere firmly to the substrate, thus affecting the subsequent forming process. When the height of the nozzle was in an appropriate range (0.5~0.9 mm), a uniform deposition path was obtained; the width of the deposition path decreased with an increase in the of the height of the nozzle.

3.2. Experimental Results and Analysis of Single Layer Composite Forming

Figure 7 shows the surface morphologies of monolayer composites formed with different path overlap rates.

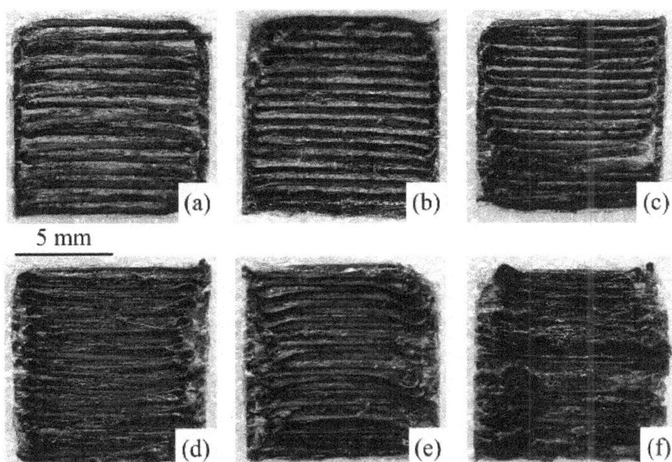

Figure 7. Surface morphology of monolayer composites with different path overlap rates. (**a**) 10%; (**b**) 20%; (**c**) 30%; (**d**) 40%; (**e**) 50%; (**f**) 60%.It is obvious that when the path bonding rate was 10%, 20% and 30%, as shown in (**a–c**), an overlap between the two deposited single paths could not be perfectly achieved and there were obvious gaps between the paths; when the path bonding rate was 40%, as shown in (**d**), the gaps between the deposited paths were filled and the surface quality of the monolayer composite was also good; when the path bonding rate was 50% and 60%, as shown in (**d,e**), an excessive amount of composite material accumulated and formed bumps in the overlap area, and the movement of the nozzle formed scratches on the bumps, resulting in poor surface quality of the monolayer composite. Under the experimental conditions of this work, the optimal path lap ratio was 40%.

3.3. Experimental Results and Analysis of Solid Forming of Composite Materials

Figure 8 shows the forming path, preparation, and tensile test process used for the composite tensile test samples. The length and width of the tensile test samples were 85 and 10 mm, respectively. In order to prevent damage to the clamping part of the composite material caused by the testing machine fixture during the tensile process, and to ensure that the carbon fiber inside the entity can bear the tensile force uniformly, both ends of the prepared tensile specimen were coated with strong foundry glue, and the cured foundry glue was polished with sandpaper to make the two ends of the tensile specimen relatively flat. The gauge length of the sample was 25 mm. Finally, the tensile properties were tested. As shown in Figure 8e, an electronic universal testing machine effectively stretched the treated tensile sample entity, and the position where the specimen was broken was within the gauge length. Under the tensile force, the composite specimen sustained fracture and pull-out of the fibers. No continuous carbon fibers were found to be pulled out as complete clusters. The results show that the continuous carbon fiber in the composite sample was fully impregnated with PLA, and a good composite was formed between these two materials.

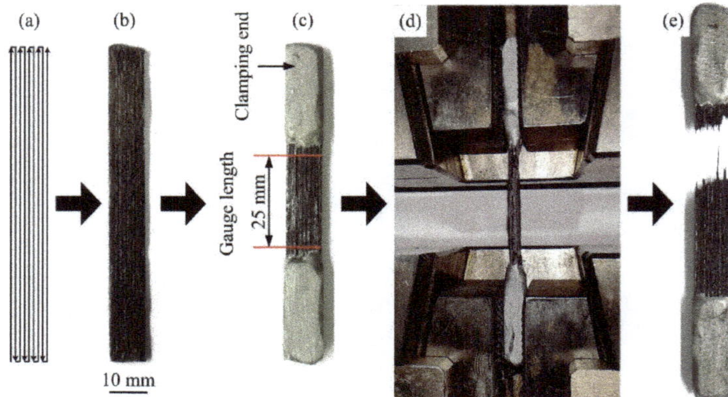

Figure 8. Composite tensile specimen entity and its tensile process. (**a**) S-shaped forming path along the long side; (**b**) the original tensile sample; (**c**) both ends of the prepared tensile specimen were coated with strong foundry glue; (**d**) sample clamping and stretching; (**e**) the specimen after being pulled.

Figure 9 shows the results for the tensile properties of the three solid layers prepared at different printing temperatures with a set layer thickness of 0.7 mm, and photographs of the specimens after fracture. Figure 10 shows the microscopic morphology of the tensile specimen fractures at different magnifications; these were manufactured with print temperatures of either 210 °C or 240 °C.

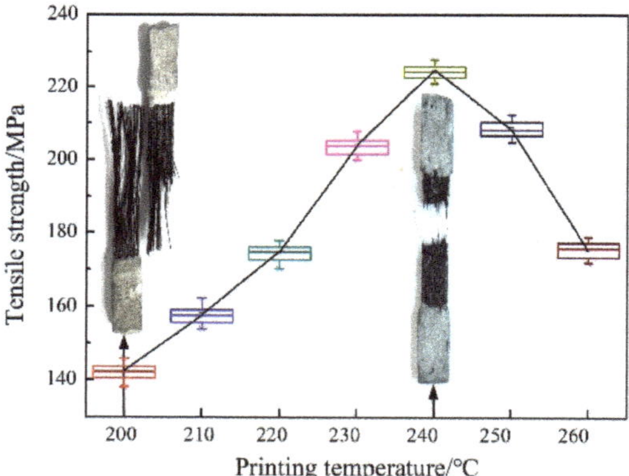

Figure 9. The test results of the tensile properties of the three-layer entities prepared at different printing temperatures and the photos of the specimens after pulling.

It can be seen that the tensile strength of the composite material is 2.2~3.5 times that of the single polylactic acid material (65 MPa). When the printing temperature was low (200~210 °C), due to the high viscosity and low fluidity of the re-melted PLA, it was impossible to achieve the secondary impregnation compound using continuous carbon fibers. The lower printing temperature also made it impossible to achieve sufficient diffusion of the molecular chains and effective interfacial fusion between the left and right deposition monolayers and the upper and lower deposition monolayers. Therefore, when the specimen was pulled, many continuous carbon fibers were pulled out directly from the composite specimen in complete clusters instead of fracturing. As shown in Figure 9, the interlayer interface fusion effect was poor after the specimen snapped, and as shown in Figure 10a, this eventually led to a lower tensile strength value of 142.3~157.6 MPa. As shown on the left of Figure 9, the interlayer interface of the specimen after pulling was poorly fused, and as shown in Figure 10a, this eventually led to a lower tensile strength value of 142.3~157.6 MPa. When the temperature of the PLA solution was increased from 200 °C to 240 °C, its non-Newtonian index increased from 0.1922 to 0.4262. Thus, we observed that, when the printing temperature was increased to 210~240 °C, the fluidity of PLA was enhanced, which promoted the impregnation of the PLA into the continuous carbon fibers, and improved the effective fusion between the deposited single channels and interlayers. Therefore, the tensile strength of the specimen also increased. When the printing temperature was 240 °C, the composite specimen sustained fracture and pull-out of the fibers under the tensile force. No continuous carbon fibers were found to be pulled out in complete clusters, as shown on the right side of Figure 9. After the specimen was snapped, we observed a better interfacial fusion between the specimen layers. As shown in Figure 10c, the resulting specimens had a maximum tensile strength value of 224.3 MPa, an increase of approximately 57.6% compared to the specimens printed at 200 °C.

Figure 10. (**a**,**b**) The micromorphology of tensile test specimen at different magnifications—210 °C; (**c**,**d**) the micromorphology of tensile test specimen at different magnifications—240 °C.

When the printing temperature was between 240 and 260 °C, the tensile strength of the specimen decreased with an increase in printing temperature. At the printing temperature of 260 °C, the tensile strength value of the specimen decreased to 175.8 MPa, which was approximately 21.6% lower than that of the specimen printed at 240 °C. This is because when the temperature of the matrix material PLA approached its thermal decomposition temperature of 260 °C, decomposition of the macromolecular chains started to occur, which resulted in an overall decline in the mechanical properties of the composite specimen.

As can be seen in Figure 10a,c, many fibers were detached from the resin matrix. This is because in the printing process, due to the extrusion of the nozzle, the fibers of the composite material were more fully dispersed. At the same time, the distribution direction of some fibers was at a certain angle to the printing path. In the process of tensile testing, these fibers were not only subjected to axial tension, but also subjected to radial shear force, so it was easy for them to be detached from the matrix, and to fracture.

As can be seen in Figure 10b,d, the clusters of continuous carbon fibers were fully impregnated by PLA to form an excellent composite inside the composite specimens prepared at different printing temperatures. This was mainly attributed to the preparation of the prepreg filaments of the CCFRRMCs in the early stages.

Figure 11 shows the results for the tensile properties of the three-layer solids prepared at different printing layer thicknesses with the printing temperature set at 240 °C, and the microscopic morphology of the carbon fiber distribution pattern at the fracture point of the specimen after pulling.

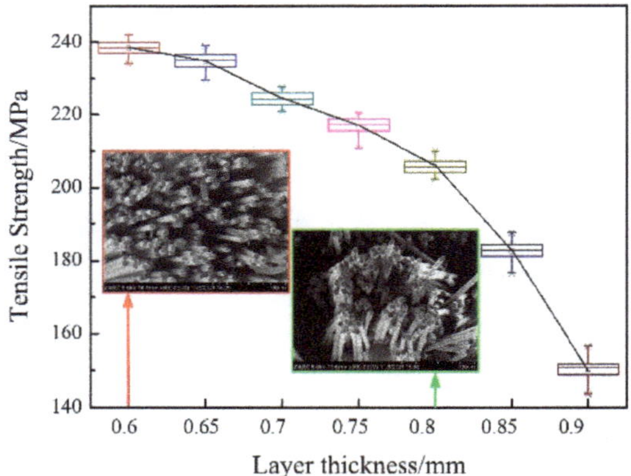

Figure 11. The test results for the tensile properties of the three-layer entities prepared at different printing layer thicknesses and the microstructures of carbon fiber distribution at the fracture point of the sample. The red arrow points to the microstructure of the sample formed when the layer thickness is 0.6 mm. The green arrow points to the microstructure of the sample formed when the layer thickness is 0.8 mm.

Figure 12 shows the calculated theoretical values for fiber volume ratio of the three-layer entities prepared at different printing layer thicknesses.

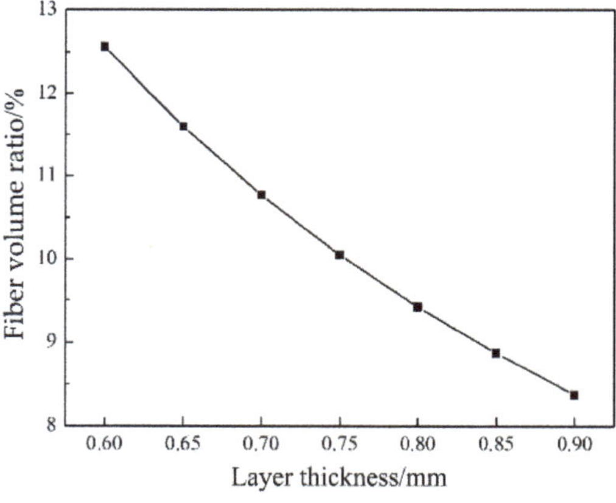

Figure 12. The fiber volume ratios of the three-layer entities prepared at different printing layer thicknesses.

It can be seen that the tensile strength of the composite tensile test specimen decreased with an increase in printed layer thickness. When the printed layer thickness was increased from 0.6 mm to 0.9 mm, the tensile strength of the specimen decreased from 238.4 MPa to 150.8 MPa, which is a decrease of approximately 36.7%. When the printed layer thickness was smaller (0.6 mm), on the one hand, the fiber volume ratio in the sample was relatively large; on the other hand, the nozzle had a better compaction effect on the extruded composite, which not only facilitated the effective fusion between the deposition channels and the interlayer interface, but also promoted better dispersion and spreading of the clustered carbon fibers in the deposition channels. As shown in Figure 11, when the layer thickness was 0.6 mm, the micromorphology of carbon fiber distribution at the fracture point of sample was relatively ordered, thus obviously improving the tensile strength of the composite specimens. At the same time, the effective and uniform dispersion of fibers in the matrix was conducive to the more effective and uniform dispersion of the force to each fiber when the sample was subjected to radial load, so as to realize an increase in the shear strength of the sample. In contrast, when the print layer was thicker (0.8 mm), on the one hand, the fiber volume ratio in the sample decreased, while on the other hand, the nozzle could not compact the extruded composite, and the carbon fiber clusters in the specimen were not further dispersed and spread, so the mechanical properties of the composite specimen were poor. Under the experimental conditions of this work, the optimal layer thickness was 0.7 mm.

3.4. Experimental Results of Forming of Composite Cellular Load-Bearing Structural Parts

The honeycomb structure is the best topological structure covering two-dimensional planes; high in strength, light in weight, with good sound and heat insulation properties, and other excellent performance. Therefore, the current space shuttles, artificial satellites, and spacecraft adopt large numbers of honeycomb structures, and the shells of satellites are almost all made with honeycomb structures.

Continuous carbon fiber reinforced resin matrix composite has the advantages of light weight and high strength. Under the optimized parameter system, direct additive manufacturing of a composite honeycomb load-bearing structure was realized, as shown in Figure 13.

Figure 13. Additive manufacturing of continuous carbon fiber reinforced resin matrix composite cellular load-bearing structure.

The boundary dimension of the structure was 77 mm × 67 mm × 17.5 mm, the wall thickness was 2 mm, the density was 1.28×10^3 kg/m^3, and the volume fraction of carbon fiber was approximately 7.7%.

4. Conclusions

An experimental study on a systematic additive manufacturing process was conducted using continuous carbon fiber reinforced PLA composite prepreg filaments, and we explored the effects of each process parameter on the composite forming properties. We can finally come to a number of noteworthy conclusions:

In order to obtain an excellent single channel forming quality, the wire feed speed needed to be set at the same value as the nozzle moving speed, and the speed should not be too large, to avoid unevenness of the deposition path edges. The optimal wire feeding speed and nozzle moving speed were 5 mm/s under the experimental conditions.

An appropriate setting of the path bonding rate could effectively fill gaps between adjacent deposited monolayers and avoid excessive accumulation of composites in the overlap region, thus obtaining monolayer composites with good surface quality. The optimal path bonding rate value was 40% under the experimental conditions.

The nozzle printing temperature had a significant effect on the macroscopic mechanical properties of the composites. An optimized printing temperature could promote the impregnation of the composite of molten matrix material and continuous carbon fiber, and the effective fusion between deposition channels and interlayer interfaces. Changing this parameter influenced the fracture mode of the composites under tensile forces, ultimately reflecting an improvement in mechanical properties. Under the experimental conditions, the optimal print temperature was 240 °C.

Nozzle height and print layer thickness had evident influence on forming accuracy, and on the mechanical properties of the composite materials. Nozzle height and print layer thickness were critical parameters for the forming process, and when set effectively, enabled avoidance of the phenomenon of poor path uniformity and of the carbon fiber being scratched, either due to abrasion by adjacent deposition in the single channel, or between the nozzle and the deposition layer. Furthermore, proper nozzle height and print layer thickness could realize adequate compaction of the nozzle to the composite material, facilitate the effective fusion between deposition channels and interlayer interfaces, and promote the dispersion and spread of clustered carbon fiber in the deposition channel. Finally, the mechanical properties of the sample were greatly improved. Under experimental conditions, the optimal nozzle height and print layer thickness was 0.7 mm.

Based on the optimized process parameters, direct additive manufacturing of honeycomb load-bearing structural parts with lightweight and high-strength composites could be achieved. The volume fraction of carbon fiber was approximately 7.7%, and the tensile strength reached 224.3 MPa.

Author Contributions: Writing—original draft preparation, L.Y.; formal analysis and data curation, D.Z.; writing—review and editing, G.J.; supervision and project administration, G.Y. All authors have read and agreed to the published version of the manuscript.

Funding: This research was funded by the Key Research and Development Program of Hebei Province, No. (21351002D); Hebei University of Science and Technology School of UAV R & D Special Project, No. (2019WRJ11).

Institutional Review Board Statement: Not applicable.

Informed Consent Statement: Not applicable.

Data Availability Statement: Not applicable.

Conflicts of Interest: The authors declare no conflict of interest.

References

1. Lu, B.H.; Li, D.C. Development of the additive manufacturing (3D printing) technology. *J. Mach. Build. Autom.* **2013**, *42*, 1–4.
2. Wei, Q.S.; Shi, Y.S. *The Theory and Application of Additive Manufacturing*; Science Press: Beijing, China, 2017; pp. 1–7.
3. Yao, J.F.; Zhang, J.; Que, J.L. *The Theory and Application of 3D Printing*; Science Press: Beijing, China, 2017; pp. 15–37.
4. Acanfora, V.; Corvino, C.; Saputo, S.; Sellitto, A.; Riccio, A. Application of an additive manufactured hybrid metal/composite shock absorber pan-el to a military seat ejection system. *J. Appl. Sci.* **2021**, *11*, 6473. [CrossRef]

5. Acanfora, V.; Castaldo, R.; Riccio, A. On the effects of core microstructure on energy absorbing capabilities of sandwich panels intended for additive manufacturing. *Materials* **2022**, *15*, 1291. [CrossRef] [PubMed]
6. Turner, B.N.; Strong, R.; Gold, S.A. A review of melt extrusion additive manufacturing processes: 1. Process design and modeling. *Rapid Prototyp. J.* **2014**, *20*, 192–204. [CrossRef]
7. Turner, B.N.; Gold, S.A. A review of melt extrusion additive manufacturing processes: II. Materials, dimensional accuracy, and surface roughness. *Rapid Prototyp. J.* **2015**, *21*, 250–261. [CrossRef]
8. Tekinalp, H.L.; Kunc, V.; Velez-Garcia, G.M.; Duty, C.E.; Love, L.J.; Naskar, A.K.; Blue, C.A.; Ozcan, S. Highly oriented carbon fiber-polymer composites via additive manufacturing. *Compos. Sci. Technol.* **2014**, *105*, 144–150. [CrossRef]
9. Zhong, W.; Li, F.; Zhang, Z.; Song, L.; Li, Z. Short fiber reinforced composites for fused deposition modeling. *Mater. Sci. Eng. A* **2001**, *301*, 125–130. [CrossRef]
10. Ning, F.; Cong, W.; Qiu, J.; Wei, J.; Wang, S. Additive manufacturing of carbon fiber reinforced thermoplastic composites using fused deposition modeling. *Compos. Part B Eng.* **2015**, *80*, 369–378. [CrossRef]
11. Gray, R.W., IV; Baird, D.G.; Bohn, J.H. Thermoplastic composites reinforced with long fiber thermotropic liquid crystalline polymers for fused deposition modeling. *Polym. Compos.* **1998**, *19*, 383–394. [CrossRef]
12. Gray, R.W., IV; Baird, D.G.; Bohn, J.H. Effects of processing conditions on short TLCP fiber reinforced FDM parts. *Rapid Prototyp. J.* **1998**, *4*, 14–25. [CrossRef]
13. Luo, M.; Tian, X.; Shang, J.; Zhu, W.; Li, D.; Qin, Y. Impregnation and interlayer bonding behaviors of 3D-printing continuous carbon-fiber-reinforced poly-ether-ketone composites. *Compos. Part A Appl. Sci. Manuf.* **2019**, *121*, 130–138. [CrossRef]
14. Yu, T.; Zhang, Z.; Song, S.; Bai, Y.; Wu, D. Tensile and flexural behaviors of additively manufactured continuous carbon fiber-reinforced polymer composites. *Compos. Struct.* **2019**, *225*, 111–147. [CrossRef]
15. Wang, J.; Ge, X.; Liu, Y.; Qi, Z.; Li, L.; Sun, S.; Yang, Y. A review on theoretical modelling for shearing viscosities of continuous fibre-reinforced polymer composites. *Rheol. Acta* **2019**, *58*, 321–331. [CrossRef]
16. He, Y.F.; Jiao, W.C.; Yang, F.; Liu, W.B.; Wang, R.G. The development of polymer composites forming process. *Chin. J. Fiber Compos.* **2011**, *28*, 7–13.
17. Vaidya, U.K.; Chawla, K.K. Processing of fiber reinforced thermoplastic processing. *Int. Mater. Rev.* **2008**, *53*, 185–218. [CrossRef]
18. Ma, Q.; Chen, Z.; Zheng, W.; Hu, H.F.; Xiao, J.Y. Resin transfer molding: A novel shaping process for composite materials. *Mater. Sci. Eng.* **2000**, *18*, 92–97.
19. Kang, H.; Shan, Z.; Zang, Y.; Liu, F. Effect of yarn distortion on the mechanical properties of fiber-bar composites rein-forced by three-dimensional weaving. *Appl. Compos. Mater.* **2016**, *23*, 119–138. [CrossRef]
20. Shan, Z.; Chen, S.; Zhang, Q.; Qiao, J.; Wu, X.; Zhan, L. Three-dimensional Woven forming technology and equipment. *J. Compos. Mater.* **2016**, *50*, 1587–1594. [CrossRef]
21. Li, N.Y.; Li, Y.G.; Liu, S.T. Rapid prototyping of continuous carbon fiber reinforced polylactic acid composites by 3D printing. *J. Mater. Processing Technol.* **2016**, *238*, 218–225. [CrossRef]
22. Tian, X.; Liu, T.; Yang, C.; Wang, Q.; Li, D. Interface and performance of 3D printed continuous carbon fiber reinforced PLA composites. *Compos. Part A Appl. Sci. Manuf.* **2016**, *88*, 198–205. [CrossRef]
23. Wang, F.; Zheng, J.; Wang, G.; Jiang, D.; Ning, F. A novel printing strategy in additive manufacturing of continuous carbon fiber reinforced plastic composites. *Manuf. Lett.* **2021**, *27*, 72–77. [CrossRef]
24. Wang, F.; Wang, G.; Ning, F.; Zhang, Z. Fiber-matrix impregnation behavior during additive manufacturing of continuous carbon fiber reinforced polylactic acid composites. *Addit. Manuf.* **2021**, *37*, 101661. [CrossRef]
25. Liang, Y.; Liu, C.; Zhao, Q.; Lin, Z.; Han, Z.; Ren, L. Bionic design and 3D printing of continuous carbon fiber-reinforced polylactic acid composite with barbicel structure of eagle-owl feather. *Materials* **2021**, *14*, 3618. [CrossRef]

Article

Research on the Preparation Process and Performance of a Wear-Resistant and Corrosion-Resistant Coating

Xianbao Wang [1] and **Mingdi Wang** [2,*]

1   School of Mechanical and Electric Engineering, Soochow University, Suzhou 215000, China; 18972666977@163.com

2   Director of Laser Intelligent Manufacturing Joint R&D Center, Soochow University, Suzhou 215000, China

*   Correspondence: wangmingdi@suda.edu.cn

Abstract: In order to study the wear resistance and corrosion resistance of a composite material with a Fe316L substrate and Co-Cr-WC coating, Co-Cr alloy coatings with different mass fractions of WC (hard tungsten carbide) were prepared on a Fe316L substrate by laser cladding technology. The phase composition, microstructure and element distribution were analyzed by X-ray diffraction (XRD), scanning electron microscopy (SEM) and energy dispersive spectroscopy (EDS). The hardness of the samples was tested by a Vickers microhardness tester, the friction coefficient and wear amount of the samples were tested by a friction and wear tester, and the corrosion resistance of the samples was tested by an electrochemical corrosion workstation. The results showed that the macroscopic appearance of the coating surface was good without obvious cracks, and the microstructures were mostly equiaxed crystals, cellular crystals and dendrites. With the addition of WC, the structures near the particles became more refined and extended from the surface of the WC particles. When the WC content was 40%, defects such as fine cracks appeared in the coating. The average microhardness of the 30%WC-Co-Cr coating was 732.6 HV, which was 2.29 times that of the Fe316L matrix; the friction coefficient was 0.16, and the wear amount was 14.64×10^{-6} mm^3 N^{-1} m^{-1}, which were 42.1% and 44.47% of the matrix, respectively; the self-corrosion voltage of the cladding layer was 120 mV, and the self-corrosion current was 7.263×10^{-4} A/cm^2, which were 30.3% and 7.62% of the substrate, respectively. The experimental results showed that the laser cladding Co-Cr-WC composite cladding layer could significantly improve the wear resistance and corrosion resistance of the Fe316L matrix under the optimal laser process parameters.

Keywords: laser cladding; composite alloy powder; wear resistance; corrosion resistance

Citation: Wang, X.; Wang, M. Research on the Preparation Process and Performance of a Wear-Resistant and Corrosion-Resistant Coating. *Crystals* **2022**, *12*, 591. https://doi.org/10.3390/cryst12050591

Academic Editors: Hao Yi and José L. García

Received: 24 March 2022
Accepted: 20 April 2022
Published: 22 April 2022

Publisher's Note: MDPI stays neutral with regard to jurisdictional claims in published maps and institutional affiliations.

1. Introduction

Laser cladding technology utilizes a composite alloy powder material with a high-performance ratio melted on the surface of the substrate by the high energy of the laser to improve the hardness, wear resistance, corrosion resistance and other properties of the surface of the parts [1]. The powder feeding method of laser cladding technology is divided into preset powder and synchronous powder feeding laser cladding. Preset powder cladding is when the powder is pre-arranged on the surface of the workpiece before the laser action forms the molten pool, and synchronous powder cladding is when the powder is simultaneously fed into the molten pool during the laser action process [2,3]. Both methods can prepare composite materials on the surface of the substrate. The alloy cladding layer, among the traditional surface engineering techniques of thermal spraying and electroplating, is considered to be the most efficient method for preparing coatings. Both of these methods are surface treatment processes used to improve the wear resistance, corrosion resistance, high-temperature resistance and heat insulation properties of the surface of equipment or components, and are widely used in the production of hydraulic supports, rolls, valves and other industrial products [4,5]. Due to the non-environmentally friendly nature and low reliability of traditional surface engineering techniques, coating processes that can replace these traditional techniques have received extensive attention. Under the action of the laser beam, the alloy powder or ceramic powder and the surface of the substrate are rapidly heated and melted, and the powder alloy in the molten state is cooled to form a surface coating with a very low dilution rate and a metallurgical aggregate with the substrate material. The thermal influence on the surface of the part is very low [6]. During the solidification process of the cladding layer from the bottom to the top, non-equilibrium solidification structures such as saturated solid solution and crystallites will be formed. These crystals can greatly improve the comprehensive mechanical properties of the coating, so as to significantly improve the wear resistance, corrosion resistance, heat resistance, antioxidant and electrical properties of the substrate surface [7,8].

In the current research on laser cladding powder materials, self-fluxing alloy powders are mostly used in practical engineering applications [9]. Self-fluxing alloy powders generally include iron-based, nickel-based, and cobalt-based powders, to which elements such as silicon and boron are added to make the alloy powders have better self-fluxing properties. The Co-Cr-based alloy powder has good wear resistance and corrosion resistance [10]. In the preparation process of the cladding layer, Co and Cr react to form a solid solution, and other elements in the Co-based alloy react to form carbon and boron compounds dispersed in the Co-Cr solid solution, resulting in the comprehensive performance being greatly improved, so it is widely used in practical production. The surface hardness and wear resistance of the composite alloy cladding layer prepared by VP Biryukov using Co-Cr-based powder are significantly higher than those of the matrix babbitt steel [11]. The research results show that the laser cladding technology can not only be used to repair the surface of the workpiece, but also can be used to repair the inner wall of parts such as pump equipment, motor shafts, crank journals, and other parts such as sliding friction pairs and cladding, to increase their reliability and life. Li Yongquan, Jining, et al. studied the phase structure and friction and wear properties of Ni alloy coatings prepared by laser cladding on the surface of 45 steel [12]. They found that the coating and the substrate were well bonded, the two were metallurgically bonded, and the coating phase was mainly Ni_3Cr_2, NiTi, SiC, TiC and γ-Ni, etc. Under the same conditions, the wear quality of the coating was about 1/8 of the wear quality of the substrate, and the wear resistance of the coating was greatly improved compared with that of the substrate [13]. Wang Kaiming et al. prepared WC-Ni matrix composite cladding layers on the surface of Q235 steel substrate and studied the effect of different WC contents on the microstructure and wear resistance of the coating. They concluded that when the WC mass fraction is 20%, the cladding the layer dilution rate is the smallest [14]. With the increase of WC content, the grain refinement strengthening effect of the cladding layer is enhanced, and its wear resistance is improved. The phases of the coating containing WC particles are mainly γ-Ni, M_7C_3, $M_{23}C_6$, CrB, WC and W_2C [15].

Based on the current knowledge in the literature, it is found that laser cladding technology has been gradually developed to strengthen the comprehensive performance of parts, but the related process research only discusses the influence of a single alloy material on the properties of the cladding layer [16]. Therefore, it is designed to mix Co-Cr alloy powder and hard phase WC with stronger hardness and corrosion resistance to form a composite powder material in order to enhance the hardness, wear-resistance and corrosion resistance of the Co-Cr-based cladding layer, respectively. Different mass fractions of WC hard phases were added to enhance the comprehensive mechanical properties of the composite powders. The WC hard phase has a high melting point, high hardness, and excellent wear resistance and corrosion resistance. It was used to strengthen the surface properties of the parts to study the microstructure, microscopic appearance, microhardness, resistance of laser cladding layers, abrasiveness and electrochemical corrosion, thereby improving the comprehensive surface properties of the cladding layer.

2. Powder Materials and Experimental Equipment

2.1. Powder Material

The Fe316L stainless steel plate was selected as the base material. The substrates were treated with sandpaper and acetone solution, respectively, then ultrasonically cleaned and dried with ethanol. The powder material used in the experiment is Co-Cr-based spherical powder with a particle size of 48–106 μm. The powder is a self-fluxing alloy powder, which has good self-fluxing properties because of B and Si elements, and has excellent characteristics such as deoxidation and slag formation, and prevention of oxidation of the cladding layer. The Co-Cr-based powder has excellent wear resistance, corrosion resistance and toughness, and its physical properties are still stable in high-temperature environments [17]. The chemical composition of the powder is shown in Table 1.

Table 1. Chemical composition of Co-Cr based powder (Quality Score %).

Element	C	W	Ni	Fe	Cr	Si	Co
Unit	2.2	12.7	0.2	0.3	29.8	1.3	margin
MIN	2.2	11.5	0	0	29.0	1.0	margin
MAX	2.5	13.5	2.0	3	32.0	1.5	margin

The prepared Co-Cr-based powder and the weighed WC ceramic hard phase with mass fractions of 10%, 20%, 30%, and 40% were put into a ball mill for mixing, and the ball milling speed was 45 r·min^{-1} for 1 h. After mixing evenly, it was put in a drying cabinet for 1 h to ensure the drying of the mixed powder and marked as samples N2, N3, N4, and N5, respectively; the Co-Cr-based powder was marked as sample N1.

2.2. Laboratory Equipment

The laser cladding equipment used in this experiment was mainly comprised of a three-axis motion platform and its wide-spot laser cladding head, a fiber laser, a double-barrel powder feeder, a water cooler, and an inert gas delivery system [18]. Under the action of the control system, the laser cladding head can make an optimized motion trajectory according to the shape of the workpiece under the movement of the three-axis platform. The principle of the laser cladding processing is shown in Figure 1. At the same time, the control system opened the powder feeder and the inert gas conveying system to ensure that the powder ejected from the powder feeding port could accurately enter the molten pool generated by the laser beam, and melt and solidify on the surface of the substrate. The laser cladding head was fixed on a three-axis mobile platform, and the cladding head could make a corresponding movement trajectory according to the strengthening position required by the part through teaching programming or manual drawing [19].

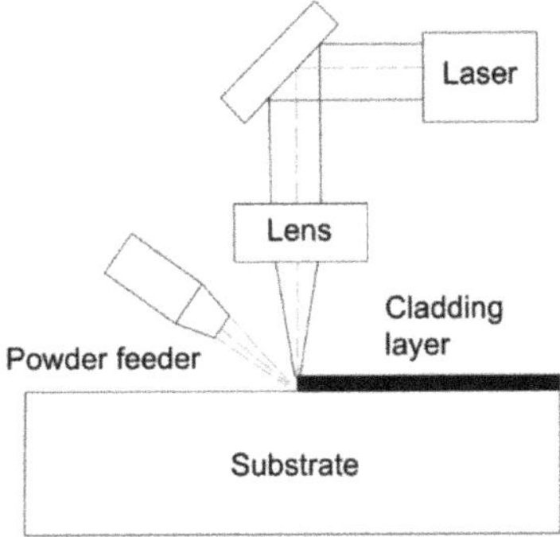

Figure 1. Laser cladding processing principle.

2.3. Characterization Methods

The sample was prepared by wire cutting, with a size of 15 mm × 15 mm × 2 mm, a tube pressure of 45 KV, a tube flow of 45 mA, and a diffraction angle range of 10° to 90° [20]. The selected target was Cu, and the scanning speed was 20°/min. The microscopic appearance and composition of the cladding layer were analyzed by a ZEISS EVO18 scanning electron microscope and its own energy dispersive spectrometer. The microhardness of the cladding layer was tested by an MH-5 digital Vickers microhardness tester with a load of 500 g and a loading time of 10 s. The friction properties of the cladding layer were tested by the ball-disk friction and wear testing machine [21]. The normal load was 9.8 N, the friction time was 45 min, the friction radius was 2 mm, and the friction disk speed was 1440 r/min. The abrasive material was Si_3N_4 high-hardness ceramic balls. The CHI electrochemical corrosion workstation was used to conduct the corrosion resistance test of the sample. The electrolytic cell included 3.5% NaCl solution, graphite auxiliary electrode and 3.5 mol/L AgCl electrode. The effective test area of the sample was 1 cm^2, and the range of the scanning motor selected for the polarization curve test was −0.75~−0.25 v (vs. SCE), and the scan rate was 2 mV/s. Although a rate of 2 mV/s was adopted in this stage of the experimentations, it is remarked that the potential scan rate has no substantial provided distortions in the polarization curves obtained [22–25]. Besides, no deleterious effect was verified when polarization parameters were obtained (e.g., corrosion current densities and potentials). However, it is worth noting that potential scan rate has an important role in minimizing the effects of distortion in Tafel slopes and corrosion current density analyses, as previously reported [22–25].

3. Results and Analysis

3.1. XRD Phase Analysis of Cladding Layer

Figure 2 shows the XRD pattern of the Co-Cr-based alloy cladding layer. By analyzing the diffraction peaks of the pattern, it can be seen that the phase composition of the Co-Cr alloy cladding layer includes γ-Co, $M_{23}C_6$, Cr_7C_3, $FeNi_3$, Co_3W, in which M mainly includes Ni, Cr, Fe elements, which is basically the same as the element composition in the alloy powder, but also contains $FeNi_3$, etc. This is because, during the laser cladding process, Fe316L in the molten pool penetrated into the molten pool through convection. In the cladding layer, some elements of Fe and Mn appear. Among them, γ-Co can not only enhance the hardness of the cladding layer, but also have wrapping and supporting effects on carbides such as $M_{23}C_6$ and Cr_7C_3. The crystal structures of the alloys are body-centered (bcc) and face-centered (fcc) cubic structures, respectively [26]. The analyte phase shows that the sample is a cobalt-chromium-iron-based cladding layer reinforced by various carbide hard phases. These main phases are determined. The comprehensive mechanical properties of the cladding layer are guaranteed, and the high hardness and high wear resistance of the cladding layer are guaranteed. In addition, due to the high content of Cr in the Co-Cr-based alloy powder, the diffraction peaks of carbides such as $M_{23}C_6$ and Cr_7C_3 are very high, indicating that their content in the cladding layer is large, and these carbides can ensure that the cladding layer has a higher hardness and wear resistance. When different mass fractions of WC hard phases were added to the Co-Cr alloy powder, carbide phases with higher hardness such as CCo_2W_4, W_2C and WC appeared in the cladding layer. Because there are no elements such as C and Fe in the Co-Cr powder, the dissolved and diffused C elements of the WC particles form austenite by solid solution with elements such as Cr, Fe, and Co, and part of the austenite is rapidly cooled to form $M_{23}C_6$. When the mass fraction is 40%, a large amount of C makes the diffraction peaks of the Cr_7C_3 and $M_{23}C_6$ phases the most obvious. With the increase of WC content, more and more kinds of carbides are formed, and the diffraction peaks of the cladding layer become more obvious, indicating that the addition of WC can promote the production of carbides in the cladding layer, and the amount of these carbides can be increased. To sum up, it can be seen from the phase analysis that the hardness of the composite alloy cladding layer with the addition of the WC ceramic hard phase has been greatly improved, while the hardness of the Co-Cr-40%WC cladding layer has reached its highest value, and its toughness also can be improved [27].

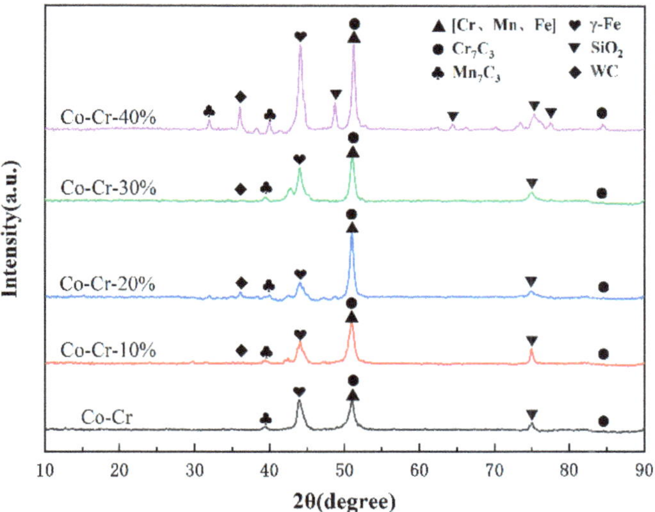

Figure 2. XRD pattern of Co-Cr composite cladding layer.

3.2. Microstructure Analysis of Cladding Layer

Figure 3 shows the metallographic microstructure of the junction, middle and top of the Co-Cr-based alloy powder cladding coating respectively. From Figure 3b, it can be seen that the cladding layer and the matrix are all bonded in a uniform bonding line, there are no obvious defects such as pores and cracks on the surface of the cladding layer, indicating that the laser cladding process parameters are suitable, and the Co-Cr alloy cladding layer has good adhesion to the Fe316 stainless steel substrate. Because when the heat input is too high due to the inappropriate laser power, scanning speed and powder feeding speed, the morphology of the bonding zone between the cladding layer and the substrate will be wavy due to the high temperature; when the heat input is too low, the powder feeding the powder sprayed by the nozzle into the spot cannot be completely melted, resulting in defects such as cracks and pores in the cladding layer. During the laser cladding process, the cladding powder flow in the cladding state cools first on the surface of the substrate, the temperature difference between the bottom and the top of the cladding layer is extremely large, and the temperature difference and the solidification speed form a large contrast [28]. The crystal growth rate is much faster than the nucleation rate. It can be seen from the bottom of the cladding layer in Figure 3g that the crystals extend to the interior of the cladding layer in the form of planar crystal growth at the junction with the substrate. With the progress of the cladding process, the matrix itself absorbs the huge amount of heat generated by the laser energy, and the temperature rises; the molten metal powder gradually solidifies, the temperature drops, the temperature difference gradually decreases, and the degree of microstructure grain refinement is much greater than that of the molten metal [29]. The part of the cladding layer close to the substrate is shown in Figure 3j; as shown in Figure 3h, it is the crystal state of the top of the Co-Cr alloy cladding layer. During the solidification process of this area, the molten state of the powder is still surrounded by high-temperature crystals that have just been formed, so the temperature difference between the two is not large, and its solidification reaches its maximum value. The crystals are rapidly cooled and formed before they grow. From the figure, we can see that there are many cellular and equiaxed crystals that extend in all directions and have smaller crystal sizes. This is also the reason why laser cladding can refine the microstructure of the cladding layer.

When the WC hard phase with different mass fractions was added, the growth state of dendrites changed [30]. When the heat input was too high due to inappropriate laser power, scanning speed and powder feeding speed, the high temperature between the cladding layer and the substrate caused the morphology of the bonding zone to be wavy; when the heat input is too low, the alloy powder cannot be completely melted, and metallurgical bonding occurs with the matrix during the solidification process, causing defects such as cracks and pores in different areas of the cladding layer. During the laser cladding process, the cladding powder flow in the cladding state cools first on the surface of the substrate, and the temperature difference between the bottom and the top of the cladding layer is extremely large. The cold theory shows that the temperature difference and the solidification speed have a large contrast, and the crystal growth speed in the cladding layer is much greater than the nucleation speed. As shown in Figure 3h, the central area of the Co-Cr-20%WC cladding layer has many WC particles distributed on the surface. The unmelted WC particles in the molten alloy powder have the lowest temperature and dendrites. Due to the hindered growth direction of the unmelted WC particles, the growth direction is no longer from the matrix to the top, but from the surface of the WC particles to the surrounding area, and many small, neatly arranged stacked cellular crystals appear, and the original needle-like dendrites and the columnar crystals become more refined, the volume of the interdendritic network becomes larger, and the overall structure becomes denser, as shown in the microstructure around the WC particles in Figure 3i [31]. When the WC mass fraction is high, as shown in Figure 3l, the content of unmelted WC particles is also higher, the grains of the cross-section of the cladding layer are more refined, and most of them are cellular crystals with smaller volumes. The structure is neatly stacked, dense and without obvious defects, so the comprehensive physical properties are also higher.

As shown in Figure 4, the SEM images of the cross-sectional joint area, the middle and the top area of the cladding layers of the samples N1, N3, and N4, respectively, are shown. From the micro-SEM image of the whole section of the cladding layer in the figure, it can be seen that the sizes of dendrites in different parts of the cladding layer of the sample are significantly different. It is because during the solidification process of the molten pool, the temperature gradient of different positions has different solidification rates, and the temperature of the molten pool decreases from the bonding area to the top of the cladding layer, as shown in Figure 4d,e,g. As shown in Figure 4h, the crystal state of the middle and top regions of the cladding layer is compared, so the dendrite matrix in different positions has different nucleation and growth rates, and the temperature in the top region changes most rapidly, so the grain size is smaller than that in the middle region [32]. Furthermore, the microstructures at the top of the samples are all cellular crystals, which are smaller in volume and denser than the needle-like dendrites that exist in large quantities in the middle and binding areas; as shown in Figure 4d, the sample (0% WC) microstructure is mainly composed of needle-like dendrites and columnar crystals and contains a small number of cellular crystals. When the WC hard phase with different mass fractions is gradually added, under the optimal process parameters, the morphology of WC particles in the cladding layer is mainly divided into two types, one is the slightly dissolved-diffusion type, in which the WC particles are relatively complete and the surface morphology is relatively clear, but the edges and corners are more dissolved and the original shape is basically maintained. The reason is that the heat generated by the laser beam is not enough to dissolve the WC particles at the bottom due to high melting point elements such as Cr and W. For the dissolution-diffusion type, the edges and corners of WC have basically been dissolved, resulting in more protruding needle-like structures. Dissolution and diffusion occur on the surface [33]. With the increase of WC content, the structure of the cladding layer becomes denser, as shown in Figure 4f, the micrograph of the cladding layer with 30%WC addition and the dendrites near the unmelted WC particles. The growth was hindered, and a large number of needle-like dendrites and columnar crystals without a WC cladding layer were gradually refined into cellular crystals, and the cellular crystals were smaller than the original grains, indicating that the unmelted WC particles were melting the microstructure of the cladding, playing a role in grain refinement. In addition, at different positions of the cladding layer, dendrites and interdendritic networks will form different phases due to the inclusion of W, C and other elements, such as carbides of different elements, resulting in different sizes of organizational structures. Therefore, it can be observed that the structure of each part of the Co-Cr-30%WC composite cladding layer is the smallest and densest, and the comprehensive physical properties are optimized [34].

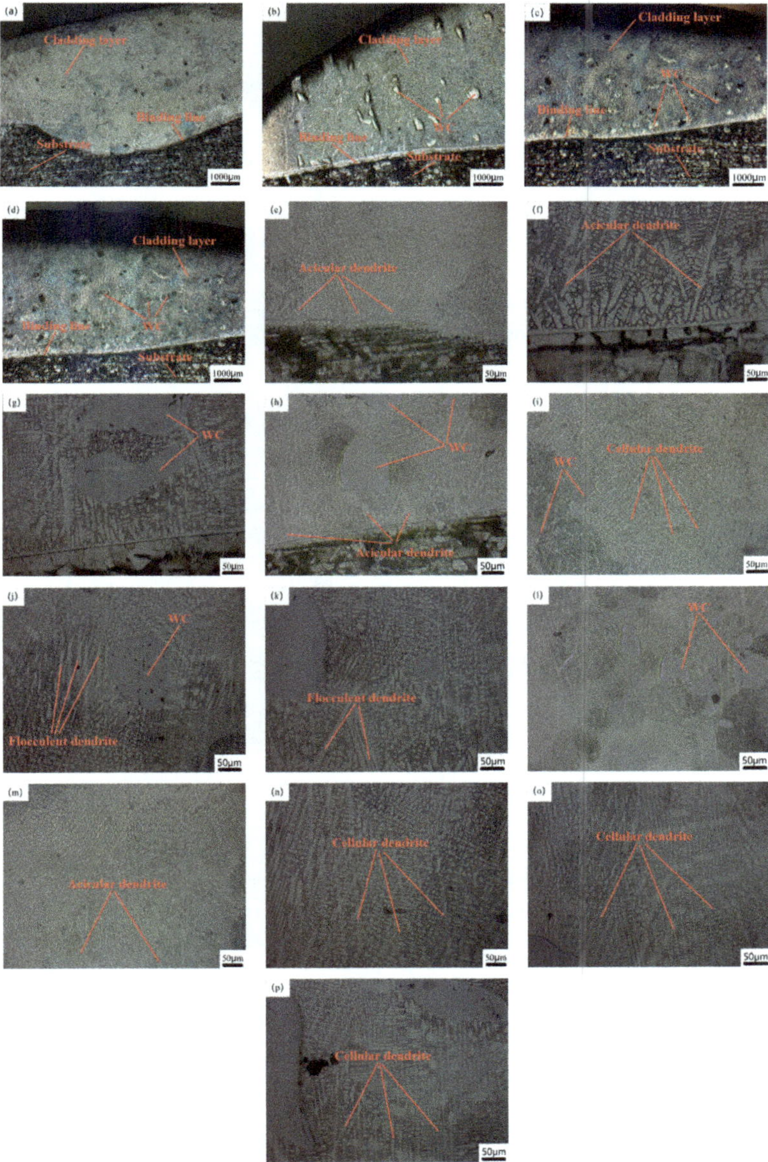

Figure 3. Metallographic diagram of Co-Cr-WC composite cladding layer. (**a,e,i,m**) Co-Cr; (**b,f,j,n**) Co-Cr-10%WC; (**c,g,k,o**) Co-Cr-30%WC; (**d,h,l,p**) Co-Cr-40%WC.

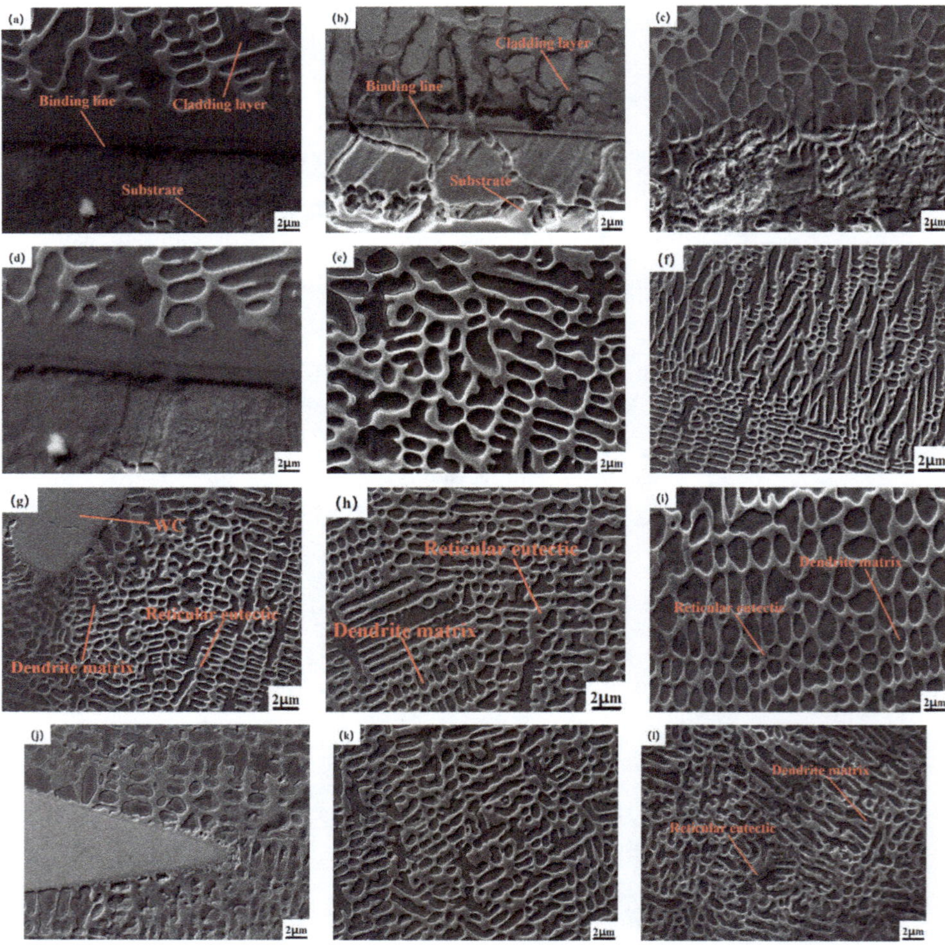

Figure 4. SEM image of Co-Cr-WC composite cladding layer. (**a**,**e**,**i**) Co-Cr; (**b**,**f**,**j**) Co-Cr-10%WC; (**c**,**g**,**k**) Co-Cr-30%WC; (**d**,**h**,**l**) Co-Cr-40%WC.

Figure 5 shows the micro-dissolution-diffusion WC particle diagram of the Co-Cr-30%WC cladding layer. When the composite cladding layer contains WC hard phases with different mass fractions, the microstructure of the WC-based cladding layer needs to be measured on its surface elements to analyze the changes in the elements of the cladding layer and the elements in the powder, as well as various elements in the dendrite and network structure. Figure 6a shows the element surface analysis of the Co-Cr-WC composite cladding layer, and Table 2 shows the element point analysis of Figure 5, and contains elements with smaller mass fractions such as Ni and Si. The interdendritic network eutectic structure contains a large amount of Cr elements, mainly in the form of Cr_7C_3. At the same time, the surface microstructure of the cladding layer is a columnar and flaky interdendritic structure, which is denser. These hard phases can greatly improve the melting point, hardness, wear and corrosion resistance of the coating. It can be found in the XRD pattern that the diffraction peaks are mainly in the γ-Co solid solution; this is because, in the process of solidification and forming of the alloy powder in the molten state, elements such as Ni and Cr combine with Fe elements to form an iron-containing solid solution, thereby making the alloy cladding. The microstructure of the layer is denser, and its hardness, wear resistance and corrosion resistance are also greatly improved in terms of macroscopic properties. Co-Cr alloy powder originally contains a small amount of the Si element. In the process of laser cladding, the Si element is mainly used for deoxidation during the solidification of the alloy in the molten state and is evenly distributed in the intradendritic and interdendritic network's eutectic structure. When adding different mass fractions of WC hard phase, it can be seen in the SEM image that the WC particles in the cladding layer mainly include micro-dissolution-diffusion type and dissolution-diffusion type. The point-scanning EDS analysis at the four points of D all detected W and C elements, and the diffraction peak of the W element appeared on the diffraction pattern. This is because, in addition to the unmelted WC particles, the elements of the other melted WC diffused into the dendrites of the intradendritic and interdendritic cladding layers. The molten W element is mainly distributed in the crystal, and the W element mainly exists in the crystal as a Fe-W-C-Cr carbide, which gives the cladding layer a higher hardness and wear resistance. In Figure 5b, it can be seen that there are no cracks and melting pits on the surface of the unmelted WC particles, and the crystal growth is hindered by the particles, making the original Co and Cr elements more uniform. A large amount of Cr element in the microstructure crystal can improve the toughness of the material surface, and the carbides between grain boundaries such as Cr_7C_3, $M_{23}C_6$, Co_3W, CCo_2W_4, W_2C, and WC can improve the hardness and wear resistance of the phase, so the WC is hard. The increase of the phase content can comprehensively improve the comprehensive mechanical properties of the cladding layer [35].

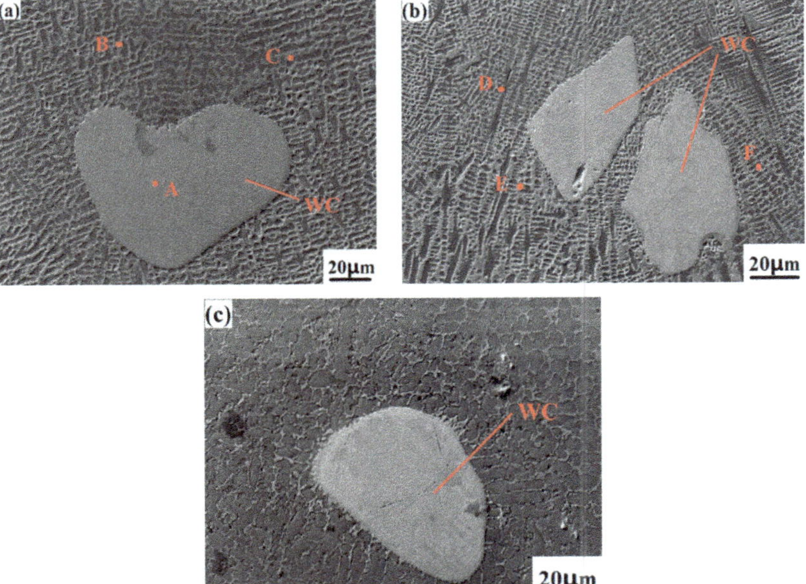

Figure 5. Microstructure image around WC particles in Co-Cr-30%WC cladding layer. (**a**) Slightly dissolving-diffusing WC particles; (**b**) Slightly dissolving-diffusing WC particles; (**c**) Slightly dissolving-diffusing WC particles.

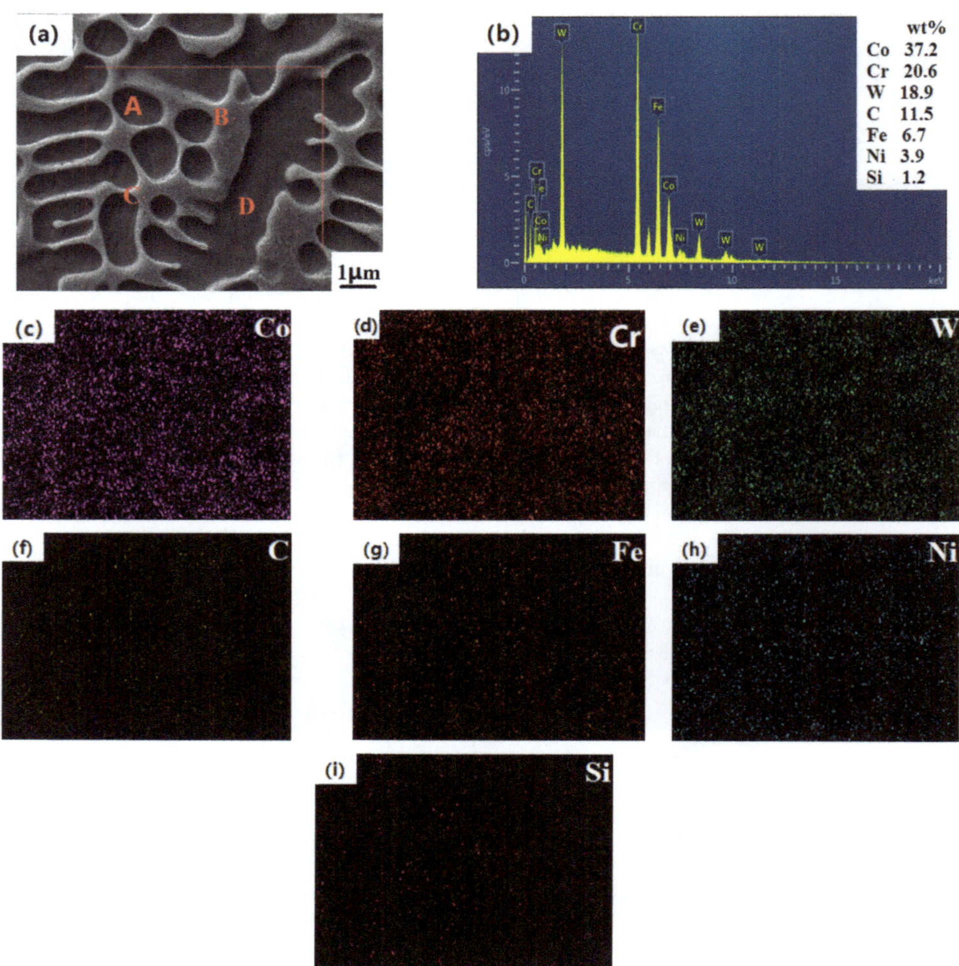

Figure 6. EDS surface scanning analysis of the middle area of the Co-Cr-30%WC cladding section. (**a**) Surface scanning area; (**b**) EDS surface total spectrum; (**c**) Co element distribution; (**d**) Cr element distribution; (**e**) W element distribution; (**f**) C element distribution; (**g**) Fe element distribution; (**h**) Ni element distribution; (**i**) Si element distribution.

Table 2. Analysis of EDS points in the section of Co-Cr-30%WC cladding layer (wt.%).

Point	Co	Cr	W	Fe	Ni	C	Si
A	-	-	86.5	-	-	13.5	-
B	26.3	23.6	35.1	5.2	4.4	4.2	1.2
C	32.9	19.6	29.9	6.3	3.6	6.1	1.6
D	23.9	25.2	36.4	4.7	3.4	4.9	1.5
E	28	18.6	34.5	6.1	4.8	5.5	2.5
F	22.5	23.4	33.7	5.7	5.3	6.5	2.9

3.3. Microhardness of Cladding Layer

Figure 7 shows the microhardness curves of different heights of the N1-N5 cladding layer section of the sample. The microhardness curve of the cladding layer is divided into three stages, which are the cladding layer, the heat-affected zone and the Fe316L matrix.

N1, N2, and N3 in the cladding layer and the heat-affected zone are first low and then high. The microhardness is 712.8HV0.5, 748.9HV0.5, and 749.8HV0.5, respectively. Among them, the microhardness of the surface of the cladding layer is the highest in the sample N5 with a mass fraction of WC of 40%, the average microhardness is 775.6HV0.5, and the average microhardness of the Fe316L matrix is 320.6HV0.5. The average microhardness of the cladding layers of 0%, 10%, 20%, and 30% are 552.6HV0.5, 649.8HV0.5, 694.3HV0.5, and 732.6HV0.5, respectively. The average measured microhardness of N1, N2, N3, N4 and N5 samples was 1.7, 2.0, 2.16, 2.28 and 2.42 times higher than that of the Fe316L matrix, respectively. With the addition of WC, the elements W and C generated by the dissolution-diffusion of WC particles interact with elements such as Co, Cr, and Fe to produce new hard phases such as Cr_7C_3, $M_{23}C_6$, Co_3W, CCo_2W_4, W_2C, and WC. The strength of the cladding layer is far greater than that of the pure Co-Cr-based cladding layer; in addition, the undissolved WC particles in the cladding layer can hinder the growth of columnar crystals and dendrites, which can refine the grains and make them denser and more uniform, which can be seen from the fact that with the increase of WC mass fraction, the microhardness curve of the cladding layer is smoother in the three parts of the cladding layer area, the heat-affected zone and the matrix material, and the fluctuation range is reduced [36]. It shows that the addition of WC makes the overall structure of the cladding layer more uniform and dense, and forms a good metallurgical bond with the matrix. In summary, the addition of WC ceramic hard phase can improve the microhardness of the cladding layer, make the microstructure more uniform, and make a smooth transition in the bonding area between the matrix and the cladding layer, reducing the hardness difference in each area.

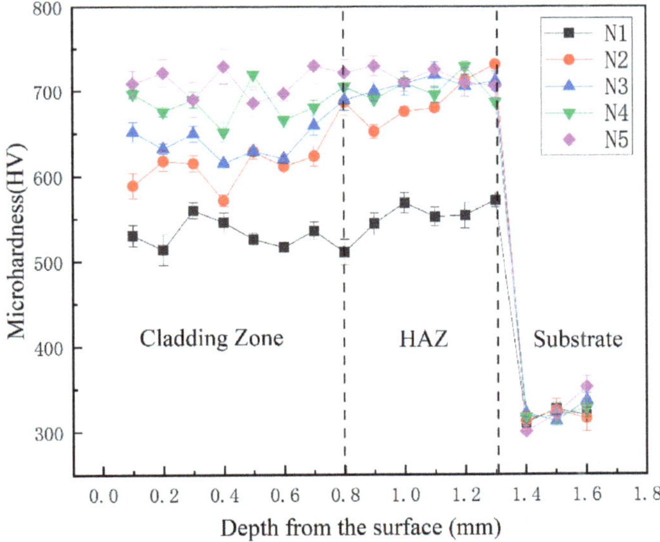

Figure 7. Microhardness of Co-Cr-WC composite cladding layer.

3.4. Friction and Wear Properties of Cladding

Figure 8a shows the friction coefficient of the Fe316L matrix and the Co-Cr-WC composite alloy cladding layer with different mass fractions of WC under the same test conditions. The friction coefficient curve of the sample is divided into the starting stage and the stable stage. At the beginning of the test, the grinding ball begins to contact the surface of the sample. The curve changes sharply up and down, so when the curve enters the stable stage, it will go through the stage of rising or falling in vain. When friction pits appear on the surface of the cladding layer, the curve tends to be stable. Under the condition of

25 N load and 45 min friction time, the average friction coefficient of Fe316L is 0.38, and the average friction coefficient of Co-Cr-WC composite cladding layer is 0.25, 0.23, 0.18, 0.16, and 0.15, respectively. It can be seen from the figure that the Fe316L matrix friction coefficient curve is at the top of the whole graph, and the friction coefficient is the largest; the friction coefficient curve of the Co-Cr-WC composite alloy cladding layer with different mass fractions increases with the increase of the WC mass fraction [37]. The coefficient decreases gradually, and the friction coefficient of the Co-Cr-40%WC cladding layer is the smallest, but the friction coefficient of the alloy cladding layer of the five samples has little difference. Therefore, with the continuous increase of the WC mass fraction, the microstructure of the Co-Cr-WC composite alloy cladding layer is more refined due to the addition of WC to the dendrite structure, and the W and C elements diffuse into the Co and Cr structure, and the reaction generates fused chromium carbides Cr_7C_3, $M_{23}C_6$, Co_3W, and CCo_2W_4 forms hard carbides that are widely and evenly distributed in the cladding layer, so the hardness and wear resistance of the cladding layer are high, meaning its microhardness is increasing. The average friction coefficient is also decreasing.

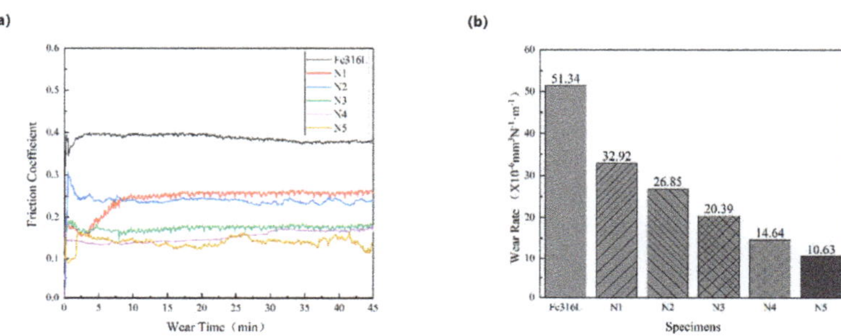

Figure 8. Friction and wear coefficient diagram and wear amount diagram of different cladding layers. (**a**) Friction and wear coefficient diagram of different cladding layers; (**b**) Comparison diagram of wear amount of different cladding layers.

Figure 8b shows the comparison of wear rates of Fe316L matrix and Co-Cr-WC composite alloy cladding layers with different mass fractions of WC under the same test conditions, and the wear rates are 51.34, 26.85, 19.63, 16.21, 11.02 and 5.21 $\times$ 10^{-6} mm^3 N^{-1} m^{-1}, respectively. The wear rate of the N6-N10 cladding layers of the sample is 52.3%, 38.24%, 31.57%, 21.46% and 10.15% of the base Fe316L, respectively. When the powder contains the WC cemented carbide phase, due to the presence of unmelted WC particles in the cladding layer, the growth of dendrites is hindered during the cooling process, and the original dendrites become small-grained cellular crystals; the elements W and C diffuse into the Co and Cr structures and react to form fused chromium carbides Cr_7C_3, CCo_2W_4, W_2C, and WC, thereby improving the hardness and wear resistance of the cladding layer. In addition, the Cr, Si, W and C in the Co-Cr-WC powder are weakly bonded between the graphite-like layers, and dislocation slip easily occurs along the basal plane direction. When the surface is loaded, the dislocation shift energy greatly increases, which increases the wear resistance of the cladding layer.

Figure 9 shows the surface wear morphologies of the Fe316L matrix and samples N1-N5 under the same conditions. The surface wear of the base Fe316L is relatively serious, the depressions are large, and a large amount of tissue falls off, and the pits left by the falling off of wear debris can be observed on the surface of the base, showing the tearing appearance of plastic deformation, as shown in Figure 9a. Figure 9b shows the wear profile of the Co-Cr alloy cladding layer. Compared with the wear profile of the substrate, there is less material falling off the surface, and the size and depth of the pits are also smaller. The hardness of the coating is greater than that of the substrate, and the degree of softening of the substrate by heat during the grinding process is low, so the surface material falls off less, and the abrasive particles formed by the fallen material will also be relatively weak. When the cladding layer contains WC hard phases with different mass fractions, with the increase of WC content and the pit depth of the wear morphology of the N2, N3, N4, and N5 cladding layers decreases, and the surface material of the cladding layer falls off. It is also decreasing; an important reason for this is that as the proportion of WC increases, the hardness of the cladding layer is also increasing, and the wear resistance is also stronger. In addition, the surface of the Co-Cr alloy cladding layer in the interdendritic network structure phases is mainly chrome compounds such as $M_{23}C_6$ and Cr_7C_3. During the wear process, these chrome compounds act as lubricants between the grinding ball and the cladding layer, so that the friction coefficient of the friction pair is small and the friction is greatly reduced to protect the wear scar furrow from being further damaged by the lateral shear stress. When the WC hard phase is added, the W and C elements generated by the dissolution-diffusion of WC particles interact with Co and Cr to produce WC, Cr_7C_3 and other high hardness materials, and further enhance the wear resistance of the cladding layer. Figure 9a shows the overall wear appearance of the Co-Cr-10%WC composite alloy cladding layer. From the figure, it can be observed that the wear pits mainly show concave furrow-like wear marks. This is because the wear debris that fell off during the wear process is thrown out of the furrow by the grinding ball, and the residual wear debris in the furrow flows along the sliding track of the grinding ball, and acts as the abrasive particle between the friction pair between the grinding ball and the substrate. Under pressure, the surface of the substrate is extruded and sheared, so that the surface material undergoes directional plastic deformation, so the metal surface forms a pit shape with low middle and high sides. When the WC mass fraction increased to 40%, as shown in Figure 9d, the number of micro-dissolved-diffused WC particles on the surface of the cladding layer reached the maximum, and the wear pattern of the Co-Cr-WC composite cladding layer was the depth and size of the furrows which were significantly smaller than those of the pure Co-Cr-based alloy cladding layer without WC, indicating that the addition of WC hard phase enhances the hardness and strength of the cladding layer, and the wear is greatly reduced. To sum up, the dissolution-diffusion and slight dissolution-diffusion type WC hard particles can strengthen the Co-Cr cladding layer and improve its hardness and wear resistance.

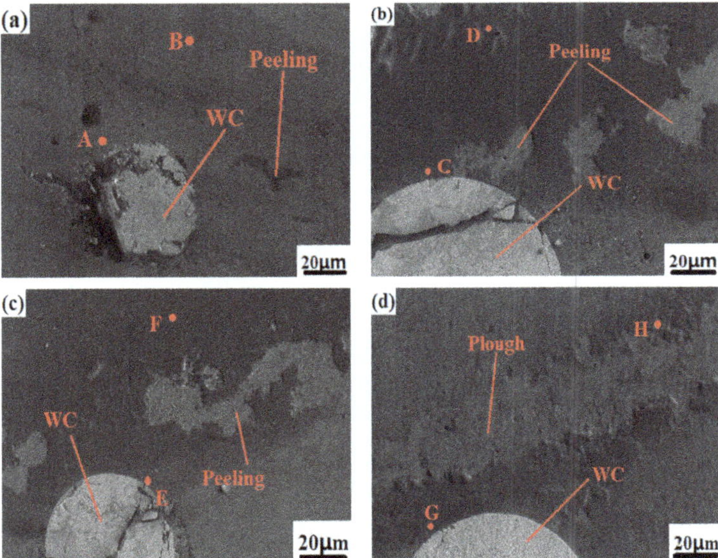

Figure 9. SEM images of surface wear of samples. (**a**) Co-Cr; (**b**) Co-Cr-10%WC; (**c**) Co-Cr-30%WC; (**d**) Co-Cr-40%WC.

Figure 10 shows the scanning diffraction pattern of the EDS surface after Co-Cr-30%WC wear, and Table 3 show the EDS element point analysis of the five points. A, B, C, D, E, and F in the table show that after 45 min of friction and wear, it is the same as the Co-Cr-WC composite alloy cladding layer with WC added. Oxygen elements were detected in the furrows of the wear scars, indicating that the friction and wear of the three samples all had oxidation wear. At the initial starting stage, when the cladding layer was in contact with the counter-grinding ball, the micro-morphology of the cladding layer was always the same. When the grinding ball was pressed and rubbed against the cladding layer, the contact part of the friction pair was slightly deformed, so the cladding layer material in the contact part fell off into granules, resulting in the melting of the surface material and the cladding layer being lost, and the wear mechanism is mainly adhesive wear. So the friction and wear of the cladding layer is actually a process of oxidative wear and mechanical wear. Due to the protective effect of the oxide film, the degree of oxidative wear on the sample is far less than that of adhesive wear. So in general, the oxide film can prevent the metal surface from sticking and play the role of protecting the friction pair. After adding different mass fractions of WC, the wear mechanism and EDS element detection content are similar to those of the Co-Cr-WC composite alloy cladding layer. Increasingly, the incompletely dissolved WC particles are evenly distributed on the surface of the cladding layer. During the wear test, the tissue material of the cladding layer is continuously peeled off, so that the WC particles with the highest hardness directly rub against the grinding ball, resulting in a large degree of wear. Therefore, after the mass fraction of the WC hard phase reaches a certain level, the oxidative wear and adhesive wear are greatly reduced to achieve the purpose of improving the wear resistance of the cladding layer. However, the Cr element existing in the Co-Cr cladding layer has the effect of lubrication and wear resistance. The chromium compounds generated by the interaction with elements such as W, C, and Fe are evenly distributed in the dendrite and interdendritic network structure. In the process, the friction coefficient of the cladding layer is reduced, and the degree of adhesive wear is further weakened to achieve the purpose of improving the wear resistance of the composite cladding layer.

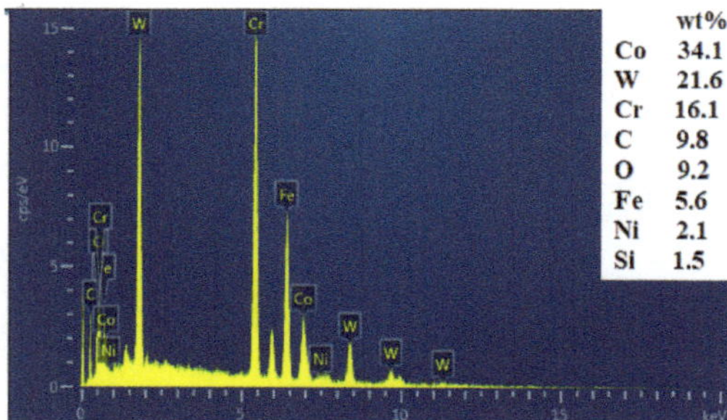

Figure 10. Scanning diffraction pattern of EDS surface after Co-Cr-30%WC wear.

Table 3. EDS point analysis of the worn surface of Co-Cr-30%WC cladding layer.

Point	Co	Cr	W	O	Fe	Ni	C	Si
A	33.8	23.9	15.6	8.6	5.8	5.2	5.6	1.5
B	31	26.7	16.9	7.9	4.5	6.8	4.4	1.8
C	24.5	21.9	24.9	9.5	6.7	3.5	6.5	2.5
D	19.1	27.5	26.4	10.1	4.8	5.7	4.5	1.9
E	22.2	20.1	28.6	6.9	6.8	4.7	7.5	3.2
F	23.3	18.1	30.4	7.0	5.4	5.6	7.8	2.4
G	24.8	15.9	29.6	8.6	3.5	5.3	10.5	1.8
H	14.4	21.3	28.9	9.1	8.1	4.9	11.8	1.5

Figure 11 shows the surface profile of the substrate and N1-N5 after friction and wear. Figure 11a,b shows the worn surface of Fe316L of the substrate. It can be seen from the figure that the contour of the substrate after wear is relatively rough, with a large number of particles and uneven places, and the wear scars are deep, indicating that its hardness is relatively high. In the process of wear, the tissue is worn and shed more, and the plastic deformation of the furrow is serious, so the surface morphology is irregular. Figure 11c,d shows the wear surface of the Co-Cr alloy cladding layer. Since its hardness is higher than that of the substrate, the degree of plastic deformation caused by extrusion during friction and wear is small, so the wear surface of the Co-Cr alloy cladding layer is higher. The marks are relatively shallow and the edges of the pits are well-defined. When the WC hard phase with different mass fractions was added, the surface roughness of the friction dent of the Co-Cr-based alloy cladding layer decreased significantly. The lower the degree of plastic deformation of the cladding layer, the stronger its ability to resist wear damage, and the surface integrity of the cladding layer becomes better under the same wear conditions. When there is no WC in the cladding layer, the wear type of the cladding layer is mainly for adhesive wear and oxidative wear, but when there are slightly dissolved-diffusion WC particles in the cladding layer due to the high hardness of the WC particles, they can directly contact the counter-grinding balls, thereby reducing the wear degree of the bottom and making the pits and furrows; the edge appears smooth. On the other hand, when the degree of wear decreases, less wear debris is generated during the friction process, and the degree of wear to the pits is gradually reduced, so with the increase of WC mass fractions N6, N7, and N8, the worn surface roughness of N9 and N10 cladding layers will also gradually decrease.

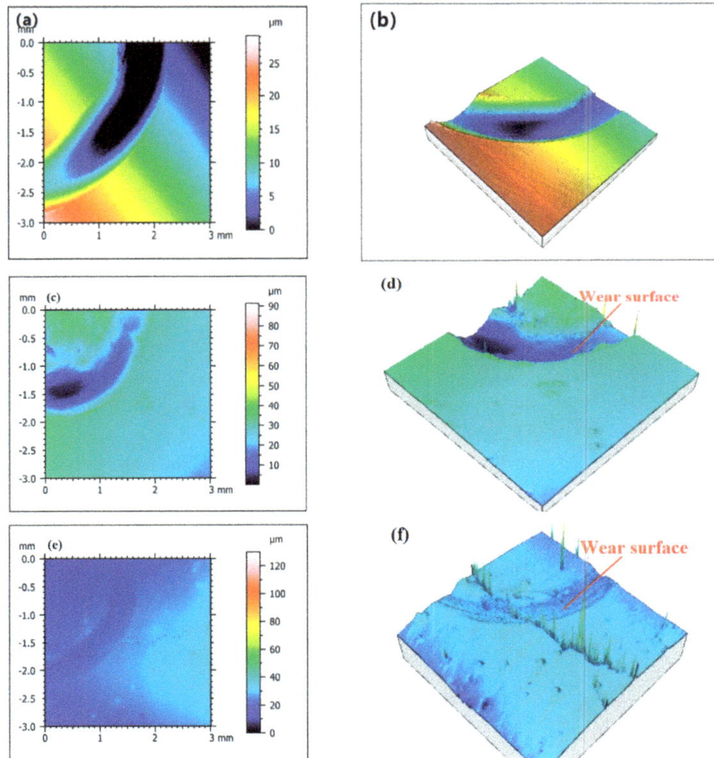

Figure 11. *Cont.*

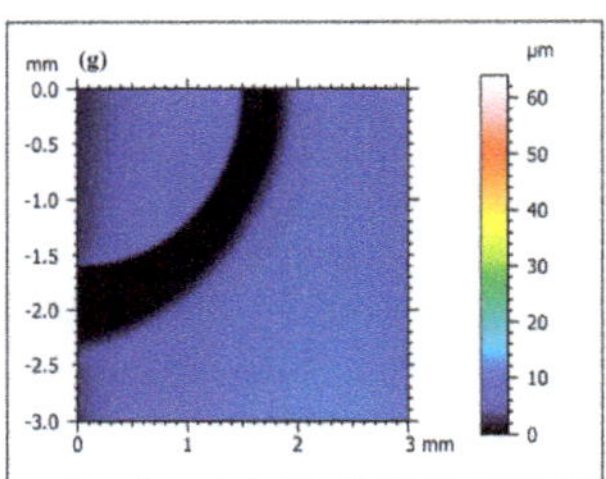 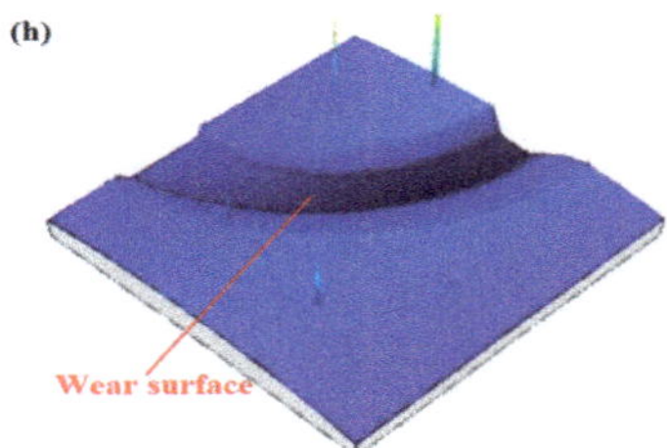

Figure 11. Profiles of worn surfaces of different cladding layers. (**a,b**) Fe316L; (**c,d**)Co-Cr; (**e,f**) Co-Cr-20%WC; (**g,h**) Co-Cr-40%WC.

3.5. Corrosion Resistance of Cladding

Figure 12 shows the potentiodynamic polarization curves of the Fe316L matrix and samples N1–N5 in a 3.5%NaCl solution. It can be seen from the figure that the corrosion current density of Fe316L increases significantly on the right side of the curve as the potential rises, indicating that the Fe element rapidly dissolves and iron ions are generated, and the dissolution rate of the material surface is greater than the formation rate of the passivation film, so the surface cannot form an effective passivation film. Corrosion electric density and self-corrosion potential are important parameters for evaluating the corrosion resistance of materials. The smaller the corrosion current density, the greater the self-corrosion potential, and the better the corrosion resistance of the material. Compared with Fe316L, the self-corrosion potential of the Co-Cr alloy cladding layer of sample N1 is significantly positive, the corrosion current density is 3.146×10^{-3} A/cm^2, and the corrosion resistance is greatly improved. This is because the Cr content in the Co-Cr alloy powder reaches 29.8%. When the Cr content on the surface of the sample exceeds 12%, a dense passivation film can be formed on the surface of the sample during the test, which prevents the NaCl solution from diffusing into the interior and reduces the concentration of the surface of the sample and the cross-section of the solution improves the corrosion resistance of the cladding layer. Furthermore, under the optimal laser cladding parameters, the cladding layer has a dense structure and no obvious defects such as pores and cracks, so the corrosion resistance of the Co-Cr alloy cladding layer is much greater than that of the Fe316L matrix.

When different mass fractions of the WC hard phase were added, the self-corrosion voltages of samples N2–N5 gradually shifted positively, which were 280 mV, 243 mV, 120 mV, and 71 mV, respectively. As shown in Table 4, the self-corrosion currents are 1.592×10^{-3} A/cm^2, 9.832×10^{-4} A/cm^2, 7.263×10^{-4} A/cm^2, and 4.263×10^{-4} A/cm^2, respectively. The self-corrosion potential moves positively, the corrosion current density is significantly reduced, and the corrosion resistance is greatly improved. In the anode section of the curve, there is a passivation area in the process of increasing the corrosion current density, and the curve tends to be stable, indicating that a passivation film is formed on the surface of the cladding layer, which weakens the corrosion of the electrolyte and enhances the corrosion resistance of the material. When the mass fraction of the WC hard phase is 40%, the corrosion current density is the smallest, because the potential of the WC itself is much higher than that of the alloy cladding layer. It acts as a passivation film. Moreover, the microstructure of the cladding layer after adding WC is more uniform and dense, and the unmelted WC particles can well hinder the extension of the grains and play the role of refining the grains, thereby improving the corrosion resistance of the cladding layer.

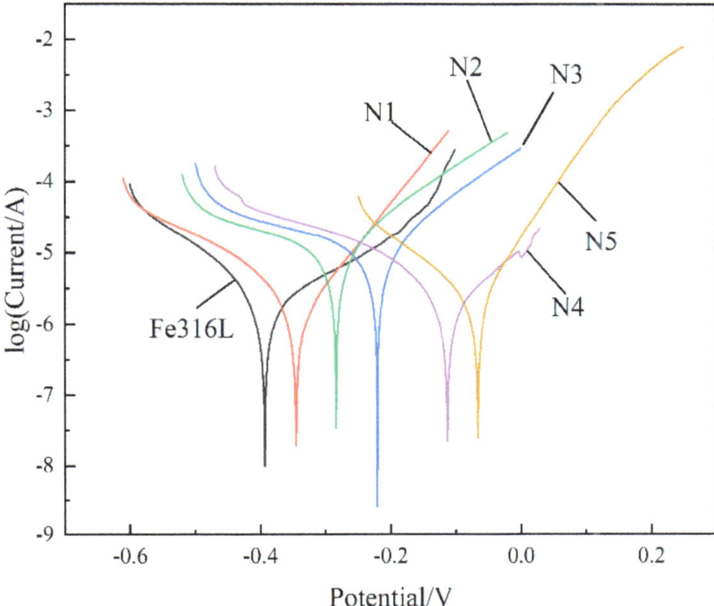

Figure 12. Potentiodynamic polarization curves of different cladding layers.

Table 4. Corrosion current density comparison of different cladding layers.

Sample	E_{corr}/mV	J_{corr}/(um·cm^{-2})
Fe316L	396 mV	9.534×10^{-3} A/cm^2
Co-Cr	351 mV	3.146×10^{-3} A/cm^2
Co-Cr-10%WC	280 mV	1.592×10^{-3} A/cm^2
Co-Cr-20%WC	243 mV	9.832×10^{-4} A/cm^2
Co-Cr-30%WC	120 mV	7.263×10^{-4} A/cm^2
Co-Cr-40%WC	71 mV	4.263×10^{-4} A/cm^2

Figure 13 shows the appearance of the Co-Cr-WC composite alloy cladding layer with different mass fractions of WC after electrochemical corrosion, and Table 5 shows the corrosion of the Co-Cr-WC composite alloy cladding layer EDS point scan analysis. After the sample is electrochemically tested, it needs to be soaked in distilled water, rinsed with alcohol, and rinsed with acetone in order to remove the oxide film on the surface of the sample and the surface attachments after corrosion. It can be seen from the figure that the five samples have different degrees of corrosion. With the increase of the WC mass fraction, the corrosion resistance of the composite cladding layer is getting stronger and stronger. Among them, the Co-Cr-40%WC composite cladding layer has the strongest corrosion resistance, and the corrosion surface basically has no defects such as material peeling and corrosion pits. Different from the Ni-WC composite cladding layer, the Co-Cr-WC composite cladding layer contains more than 12% Cr, and colorless and transparent chromium-containing oxide will spontaneously form on the surface of the cladding layer. This layer of film can isolate the cladding layer from the medium and reduce the corrosion rate of the cladding layer. Therefore, it can be seen that the Co-Cr-WC composite cladding layer has stronger corrosion resistance than the Ni-WC composite cladding layer. Thanks to the Cr and microstructure of the Co-Cr-WC cladding layer, the compactness of the cladding layer can enhance the corrosion resistance of the cladding layer, and its passivation range is also the largest. There are dendrites and columnar crystals extending from the particle surface around the WC itself. Under the corrosion of the electrolyte, these crystal grains are dissolved to different degrees, but the WC particles themselves do not appear slightly dissolved, indicating that the WC hard phase has an anti-corrosion effect. In addition, due to the addition of WC, the overall microstructure of the cladding layer is transformed into a phase with higher strength such as carbide and chrome, and the grains are more uniform and dense, so that it is difficult for the electrolyte to enter the interior of the sample and improve the corrosion resistance of the cladding layer. As shown in Figure 13c, the surface of the Co-Cr-40%WC composite cladding layer has almost no traces of corrosion after electrochemical corrosion, and no obvious corrosion pits are found on the surface of the cladding layer. Under the SEM scanning electron microscope, WC particles and their surrounding grains are basically complete, and the interdendritic network structure is still evenly distributed on the surface of the cladding layer, which can well prevent the electrolyte from entering the cladding layer and prevent the interior of the cladding layer's corrosion from being continued by the electrolyte; as such, this layer has the highest corrosion resistance.

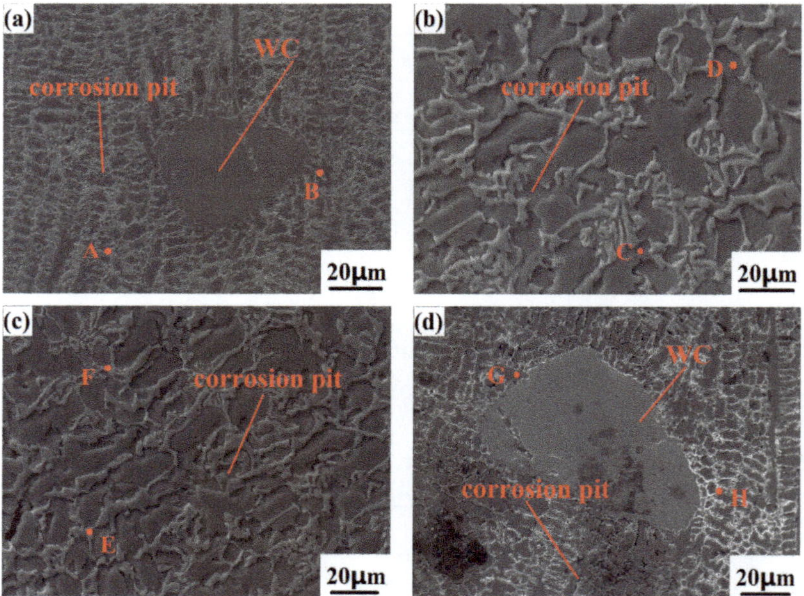

Figure 13. SEM surface corrosion morphologies of Co-Cr-WC composite alloy cladding layers with different mass fractions of WC. (**a**) Co-Cr-10%WC; (**b**) Co-Cr-20%WC; (**c**) Co-Cr-30%WC; (**d**) Co-Cr-40% WC.

Table 5. EDS point scanning analysis of Co-Cr-WC composite alloy cladding layer after corrosion.

Point	Co	Cr	W	Fe	Ni	C	Si
A	43.8	25.9	14.8	5.5	3.6	5.2	1.2
B	47.8	24.5	17.5	3.2	2.9	3.2	0.9
C	30.8	22.9	26.5	7.2	4.1	6.3	2.2
D	28.4	28.1	27.8	5.1	3.5	5.1	2.0
E	21	25.6	33.6	6.1	5.6	6.5	1.6
F	18.9	19.9	36.5	8.2	7.2	6.9	2.4
G	19.9	25.7	35.2	4.6	5.3	7.8	1.5
H	23.2	21.5	32.5	6.7	4.9	9.6	1.6

4. Conclusions

(1) The Co-Cr alloy cladding layer was prepared on the surface of Fe316L by laser cladding technology, and WC hard phases with different mass fractions were added to the Co-Cr alloy powder. Under the optimal laser cladding parameters, the main phases of the Co-Cr-based alloy cladding layer are γ-Co, $M_{23}C_6$, Cr_7C_3, $FeNi_3$, Co_3W, and the microstructure and crystallinity of the Ni-based and Co-Cr-based alloy cladding layers. The grains mainly include dendrites, cell-like crystals, strip-like crystals and interdendritic network structures. When the WC ceramic hard phase is added, the Co_3W, CCo_2W_4, W_2C, WC and other phases with higher hardnesses appear in the Co-Cr-WC composite cladding layer. The growth of dendrites near the unmelted WC particles is hindered, and the original large number of needle-like dendrites and columnar crystals are gradually refined into cellular crystals. The WC particles play a role in grain refinement in the microstructure of the cladding layer.

(2) The microhardness of the cladding layer increases with the increase of WC content. The maximum microhardness of the 30%WC-Co-Cr alloy cladding layer is 732.6HV0.5, which is 2.28 times that of the matrix; the minimum friction coefficient is 0.16, which is 42.11% of the matrix, and the wear amount is 14.64×10^{-6} mm^3 N^{-1} m^{-1}, which is 44.47% of the matrix. The profile of the cladding layer after friction and wear is relatively smooth, the furrow is relatively regular, and there is no obvious deformation; the wear mechanism is mainly adhesive wear.

(3) In the electrochemical corrosion test, the corrosion resistance of the Co-Cr-30%WC alloy cladding layer was better, the self-corrosion voltage was 276 mv positive relative to the substrate, and the self-corrosion current density was 7.263×10^{-4} A/cm^2, which was the substrate of 7.62%. The Cr element and carbides on the surface of the cladding layer play a role similar to the passivation film in the electrolyte, and the microstructure was more uniform and dense, thereby improving the corrosion resistance of the cladding layer.

Author Contributions: Conceptualization, M.W. and X.W.; methodology, M.W.; software, X.W.; validation, X.W.; resources, M.W.; data curation, X.W.; writing—original draft preparation, X.W.; writing—review and editing, X.W.; project administration, M.W.; funding acquisition, M.W. All authors have read and agreed to the published version of the manuscript.

Funding: This research was funded by Suzhou key core technology research and development projects(Grant No. SGC2021010), Suzhou Gusu Technology Entrepreneurship Angel Program(Grant No. CYTS2020094).

Institutional Review Board Statement: Not applicable.

Informed Consent Statement: Not applicable.

Data Availability Statement: Not applicable.

Conflicts of Interest: The authors declare no conflict of interest.

References

1. Lu, X.; Yang, Y.; Qi, K.; Hu, G.; Li, J. Research on the properties of Fe-Ni-C Invar alloy cladding layer by laser cladding of 42CrMo steel substrate. *Hotworkingprocess* **2022**, *18*, 103–107+111. [CrossRef]

2. Wang, W.; Sun, W.; Zhou, H.; Xing, X.; Yang, K.; Yu, J. Study on crack formation mechanism and matching method of laser cladding nickel-based alloy coating. *Hot Work. Process* **2022**, *215*, 83–88. [CrossRef]

3. Yang, Y.; Li, Y.; Liang, Z.; Bai, P.; Nie, J.; Liu, S.; Chen, B.; Wei, S.; Guan, Q.; Cai, J. Continuous hot corrosion behaviour of an FeCrAlSi coating prepared by laser cladding. *Surf. Coat. Technol.* **2021**, *421*, 127424. [CrossRef]

4. Thawari, N.; Gullipalli, C.; Katiyar Jitendra, K.; Gupta, T.V.K. Effect of multi-layer laser cladding of Stellite 6 and Inconel 718 materials on clad geometry, microstructure evolution and mechanical properties. *Mater. Today Commun.* **2021**, *28*, 102604. [CrossRef]

5. Wang, H.-P.; Mo, J.-L.; Mu, S.; Zhang, M.-Q.; Duan, W.-J.; Li, J.-B.; Zhou, Z.-R. Effects of interfacial trapezoidal grooves on the mechanical properties of coatings by laser cladding. *Surf. Coat. Technol.* **2021**, *421*, 127425. [CrossRef]

6. Zhao, H.; Sun, X.; Du, C.; Wei, L.; Li, Y.; Hong, J. Research progress of laser cladding materials on TC4 alloy surface. *J. Shenyang Univ. Sci. Technol.* **2022**, *41*, 31–37.

7. Wang, T.; Wang, C.; Zhu, L.; Wang, H.; Zhang, Y. Effects of different basal plane angles on the wear resistance of laser clad Ni25/WC coatings. *Surf. Technol.* **2022**, 1–11.

8. Wang, X.; Liu, S.S.; Zhao, G.L.; Zhang, M. Fabrication of in-situ TiN ceramic particle reinforced high entropy alloy composite coatings by laser cladding processing. *J. Tribol.* **2021**, *144*, 031402. [CrossRef]

9. Chen, L.; Zhao, Y.; Guan, C.; Yu, T. Effects of CeO$_2$ addition on microstructure and properties of ceramics reinforced Fe-based coatings by laser cladding. *Int. J. Adv. Manuf. Technol.* **2021**, *155*, 2581–2593. [CrossRef]

10. Bai, Q.; Ouyang, C.; Zhao, C.; Han, B.; Liu, Y. Microstructure and Wear Resistance of Laser Cladding of Fe-Based Alloy Coatings in Different Areas of Cladding Layer. *Materials* **2021**, *14*, 2839. [CrossRef]

11. Zhang, Q.; Xu, P.; Zha, G.; Ouyang, Z.; He, D. Numerical simulations of temperature and stress field of Fe-Mn-Si-Cr-Ni shape memory alloy coating synthesized by laser cladding. *Optik* **2021**, *242*, 167079. [CrossRef]

12. Wu, H.; Shu, L. Influence of laser power on the microstructure and properties of MoS_2+FeCrNiSi composite coating. *J. Mater. Heat Treat.* **2021**, *42*, 134–141. [CrossRef]

13. Meng, X.; Shen, Y.; Zhang, S.; Xu, H.; Wang, Y.; Liu, X. Microstructure and Tribological Properties of Co-Ti_3SiC_2 Composite Coatings by Laser Cladding of Ti6Al4V Alloy. *Met. Heat Treat.* **2021**, *46*, 199–203. [CrossRef]

14. Wang, H.; Sun, Y.; Qiao, Y.; Du, X. Effect of Ni-coated WC reinforced particles on microstructure and mechanical properties of laser cladding Fe-Co duplex coating. *Opt. Laser Technol.* **2021**, *142*, 107209. [CrossRef]

15. Hu, D.; Liu, Y.; Chen, H.; Liu, J.; Wang, M.; Deng, L. Microstructure and properties of Ta-reinforced NiCuBSi + WC composite coating deposited on 5Cr5MoSiV1 steel substrate by laser cladding. *Opt. Laser Technol.* **2021**, *142*, 107210. [CrossRef]

16. Lu, H.; Li, W.; Qin, E.; Liu, C.; Liu, S.; Zhang, W.; Wu, S. The Gradient Microstructure and High-Temperature Wear Behavior of the CoCrMoSi Coating by Laser Cladding. *J. Therm. Spray Technol.* **2021**, *30*, 968–976. [CrossRef]

17. Wang, T.; Wang, N.; Zhu, L.; Lei, J. Influence of laser scanning speed on microstructure and friction and wear properties of IN718 coating. *Hot Processing Technol.* **2022**, *15*, 79–84. [CrossRef]

18. Bao, Y.; Guo, L.; Zhong, C.; Song, Q.; Yang, K.; Jiang, Y.; Wang, Z. Effects of WC on the cavitation erosion resistance of FeCoCrNiB0.2 high entropy alloy coating prepared by laser cladding. *Mater. Today Commun.* **2021**, *26*, 102154. [CrossRef]

19. Zhang, L.; Zhao, Z.; Bai, P.; Du, W.; Liao, H.; Li, Y.; Liang, M.; Han, B.; Huo, P. Microstructure and Properties of In situ Synthesized TiC/Graphene/Ti6Al4V Composite Coating by Laser Cladding. *Trans. Indian Inst. Met.* **2021**, *74*, 891–899. [CrossRef]

20. Li, Y.; Shi, Y. Microhardness, wear resistance, and corrosion resistance of AlxCrFeCoNiCu high-entropy alloy coatings on aluminum by laser cladding. *Opt. Laser Technol.* **2021**, *134*, 106632. [CrossRef]

21. Zhang, P.-X.; Hong, Y.; Sun, Y.-H. Microstructure, microhardness and corrosion resistance of laser cladding Al 2 O 3 @Ni composite coating on 304 stainless steel. *J. Mater. Sci.* **2021**, *56*, 8209–8224. [CrossRef]

22. Osório, W.R.; Freitas, E.S.; Garcia, A. EIS and potentiodynamic polarization studies on immiscible monotectic Al–In alloys. *Electrochim. Acta* **2013**, *102*, 436–445. [CrossRef]

23. Osório, W.R.; Peixoto, L.C.; Moutinho, D.J.; Gomes, L.G.; Ferreira, I.L.; Garcia, A. Corrosion resistance of directionally solidified Al–6Cu–1Si and Al–8Cu–3Si alloys castings. *Mater. Des.* **2011**, *32*, 3832–3837. [CrossRef]

24. Zhang, X.L.; Jiang, Z.H.; Yao, Z.P.; Song, Y.; Wu, Z.D. Effects of scan rate on the potentiodynamic polarization curve obtained to determine the Tafel slopes and corrosion current density. *Corros. Sci.* **2009**, *51*, 581–587. [CrossRef]

25. McCafferty, E. Validation of corrosion rates measured by the Tafel extrapolation method. *Corros. Sci.* **2005**, *47*, 3202–3215. [CrossRef]

26. Zhao, P.; Li, J.; Lei, R.; Yuan, B.; Xia, M.; Li, X.; Zhang, Y. Investigation into Microstructure, Wear Resistance in Air and NaCl Solution of AlCrCoNiFeCTax High-Entropy Alloy Coatings Fabricated by Laser Cladding. *Coatings* **2021**, *11*, 358. [CrossRef]

27. Yang, N.; Li, L. Effects of laser power and powder content on hardness of (V_2O_5+B_2O_3+C) nickel-based coatings. *Hot Work. Process* **2021**, *50*, 103–105. [CrossRef]

28. Wang, Y.; Gong, S.; Tang, M.; Zhang, Y. Experimental study on laser cladding of Fe-based C-Ti-W composite coatings. *J. Heilongjiang Univ. Sci. Technol.* **2021**, *31*, 719–724.

29. Li, X.; Zhang, Q.; Xie, H.; Feng, Y.; Yang, X. Microstructure and Properties of Laser Cladding Gradient High-Entropy Alloy Coatings. *J. Cent. South Univ. (Nat. Sci. Ed.)* **2021**, *52*, 3835–3842.

30. Yao, F.; Fang, L.; Li, J.; Jin, J. Influence of laser power on temperature field and stress field of laser cladding Ni-based coating. *Chin. J. Plast. Eng.* **2021**, *28*, 87–94.

31. Wu, T.; Wu, M.; Chen, Y.; Hua, Q.; Shi, S.; Li, J.; Mao, P.; Chen, X. Laser Cladding AlCoCrFeNiW_x High Entropy Alloy Coating on H13 Steel and Its Properties. *Met. Heat Treat.* **2021**, *46*, 241–244. [CrossRef]

32. Zhang, H.; Zhang, J.; Zhu, L.; Wang, T. Effect of WC content on the microstructure and properties of laser cladding TC4 coatings. *Hot Processing Technol.* **2022**, *146*, 83–87+93. [CrossRef]

33. Zhang, Z.; Hu, J.; Cheng, Z.; Liu, S. Comparison of two kinds of high-strength steel powders by laser cladding on the surface of 30CrMnSiNi2A steel. *Hot Work. Process* **2016**, *13*, 6. [CrossRef]

34. Shi, S.; Chen, J.; Shu, L.; Liu, Z. Process optimization and performance research of Fe-Cr-Ni powder laser cladding. *Appl. Laser* **2021**, *41*, 955–960. [CrossRef]

35. Hu, J.; Sui, X.; Zhang, L.; Zhao, W.; Zhang, W. Influence of TiN content on the microstructure and properties of Ti_(0.8)CoCrFeNiAl_(0.5) high-entropy alloy coatings. *Aviat. Manuf. Technol.* **2021**, *64*, 71–79. [CrossRef]

36. Wang, H.-Z.; Cheng, Y.-H.; Song, W.; Yang, J.-Y.; Liang, X.B. Research on the influence of laser scanning speed on Fe-based amorphous coating organization and performance. *Intermetallics* **2021**, *136*, 107266. [CrossRef]

37. Wang, D.; Chu, J.; Qu, G.; Zhou, X. Preparation and Characterization of Laser Clad NiCrBSi Coating Reinforced with WC. *J. Phys. Conf. Ser.* **2021**, *1965*, 012110. [CrossRef]

Article

Comparison and Determination of Optimal Machine Learning Model for Predicting Generation of Coal Fly Ash

Chongchong Qi [1,2,*], Mengting Wu [1], Xiang Lu [2], Qinli Zhang [1] and Qiusong Chen [1]

[1] School of Resources and Safety Engineering, Central South University, Changsha 410083, China; qinli.zhang@csu.edu.cn (Q.Z.); qiusong.chen@csu.edu.cn (Q.C.)
[2] State Key Laboratory of Coal Resources and Safe Mining, China University of Mining and Technology, Xuzhou 221116, China; xiang.lu@cumt.edu.cn
* Correspondence: chongchong.qi@csu.edu.cn

Abstract: The rapid development of industry keeps increasing the demand for energy. Coal, as the main energy source, has a huge level of consumption, resulting in the continuous generation of its combustion byproduct coal fly ash (CFA). The accumulated CFA will occupy a large amount of land, but also cause serious environmental pollution and personal injury, which makes the resource utilization of CFA gradually to be attached importance. However, given the variability of the amount of CFA generation, predicting it in advance is the basis to ensure effective disposal and rational utilization. In this study, CFA generation was taken as the target variable, three machine learning (ML) algorithms were used to construct the model, and four evaluation indices were used to evaluate its performance. The results showed that the DNN model with the $R = 0.89$, $R^2 = 0.77$ on the testing set performed better than the traditional multiple linear regression equation and other ML algorithms, and the feasibility of DNN as the optimal model framework was demonstrated. Applying this model framework to the engineering field enables managers to identify the next step of the disposal method in advance, so as to rationally allocate ways of recycling and utilization to maximize the use and sales benefits of CFA while minimizing its disposal costs. In addition, sensitivity analysis further explains ML's internal decisions and verifies that coal consumption is more important than installed capacity, which provides a certain reference for ensuring the rational utilization of CFA.

Keywords: CFA; generation; machine learning; multiple linear regression; sensitivity analysis; utilization

Citation: Qi, C.; Wu, M.; Lu, X.; Zhang, Q.; Chen, Q. Comparison and Determination of Optimal Machine Learning Model for Predicting Generation of Coal Fly Ash. *Crystals* **2022**, *12*, 556. https://doi.org/10.3390/cryst12040556

Academic Editors: Hao Yi, Huajun Cao, Menglin Liu, Le Jia and Francesco Montalenti

Received: 11 March 2022
Accepted: 12 April 2022
Published: 15 April 2022

Publisher's Note: MDPI stays neutral with regard to jurisdictional claims in published maps and institutional affiliations.

1. Introduction

The acceleration of industrial processes has led to the rapid development of the energy industry as a large player. As the main energy supplier, power stations based on coal and lignite provide massive amounts of energy, and coal consumption has soared [1]. Although electricity demand and coal emissions experienced a small decline in 2020, as the COVID-19 outbreak depressed energy demand, the economic stimulus package and the rollout of vaccines promoted the economic rebound, leading to a 9% increase in coal-fired power generation in 2021, its highest level ever [2]. In the first half of 2021, coal market consumption showed an 11% year-on-year growth. Coal consumption in the European Union is expected to increase by 4% by the end of the year [3], and coal may remain a mainstay of international energy in the short term.

The huge consumption of coal makes the generation of coal fly ash (CFA), as a byproduct of coal combustion [4], continue to increase [5], particularly in India. Over a 10-year span (2009–2010 to 2018–2019), CFA in the power sector increased by nearly 76% and is now producing around 217 million tones [6]. A large amount of deposition of CFA takes up land resources and directly pollutes the soil [7], in addition to causing serious impacts on water and air due to inappropriate treatment. On the one hand, a large amount of rainfall makes CFA landfills a potentially dangerous place, producing toxic leachate that seeps into groundwater and pollutes water resources [8]; on the other hand, toxic elements may be discharged into the air with the flue gas produced, endangering air quality. Li et al. have pointed out that solid Hg waste produced by fly ash from coal burning is the main source of Hg in the environment [9]. Moreover, human health is also at risk from long-term exposure to CFA diffusing into the air or from drinking contaminated groundwater [10]. As a result, academia and industry are paying increasing attention to the resource utilization of CFA.

In recent years, CFA has gradually been effectively utilized in various fields, among which the construction industry is the most widely used. CFA is used as the supplementary cementitious material to partially replace cement in concrete or to prepare geopolymers [11] and is also used as coarse and fine aggregate in asphalt pavement [12]. Due to its potential for soil improvement and heavy metal adsorption [13], CFA has a good development prospect in the agricultural field. Moreover, CFA also has a large presence in the manufacturing of ceramic glass [14], metal matrix composites, and metal coatings.

However, the amount of generation of CFA is variable, so predicting the generation of CFA in advance is the basis for ensuring its effective disposal and rational utilization. Some scholars used neural networks in MATLAB and linear regression statistical analysis in IBM SPSS to predict the generation of CFA in power plants in five or ten years [15]. However, the description of this method is too simple and general, and the accuracy of the prediction has not been verified. Others predicted the average annual output of hazardous wastes by multiplying the amount of industrial hazardous wastes generated in the base year by the average annual growth rate index [16], which is too complicated and time-consuming, and the accuracy cannot be guaranteed either.

In view of the limitations of the above prediction methods, this paper used advanced machine learning (ML) algorithm [17] to predict the generation of CFA by constructing three different regression models, of which installed capacity and coal consumption were input variables, and the generation of CFA was output variable. The established model framework can be applied to the engineering site after thorough evaluation and comparison, which can quickly and accurately predict the amount of CFA generation, thus saving time for further planning of CFA disposal, and is the basis for reasonable recycling of CFA.

2. Dataset

Data are the basis for the development of machine learning algorithms, and any ML algorithms need data to evaluate their effects. How to collect a comprehensive and appropriate dataset and analyze it was key to this research.

2.1. Data Collection

In this study, domestic and foreign databases and related academic websites were searched, a large number of studies in the literature and academic reports related to CFA were consulted, the relevant data were sorted out and recorded, and finally, the dataset used in this paper was obtained through screening. This dataset was extracted from a report documenting CFA generation and utilization in coal-fired power plants across India in 2019–2020 and contained data from 183 power plants across 17 states (outliers with a coal consumption value of 0 were removed) [18], as shown in Figure 1. Chhattisgarh was the most sampled state, with 27 power plants, accounting for 14.8%, but only one power plant was sampled in Assam. From a holistic perspective, the distribution of sampling sites was relatively uniform.

2.2. Data Analysis

The ultimate purpose of the algorithm is to fit the distribution of the data and predict the trend of change. Different datasets have different feature distributions, so the statistical distribution and correlation analysis of data serve as sources of reference for establishing the optimal algorithm model.

In this paper, the dataset with a distribution characteristic presented in the form of a bubble chart included two features: installed capacity and coal consumption, and the target variable was the generation of CFA. As shown in Figure 2, the size of bubbles represents the CFA generation. Data points were mainly distributed in the lower-left corner, meaning that when the installed capacity was between 0 and 2000 MW, and the coal consumption varied from 0 to 5 MT, the amount of generation of CFA was small, less than 2.95 MT. With the increase in installed capacity and coal consumption, the CFA generation also increased. The maximum generation of 8.85 MT was realized when the installed capacity was 4760 MW, and the coal consumption was about 25 MT.

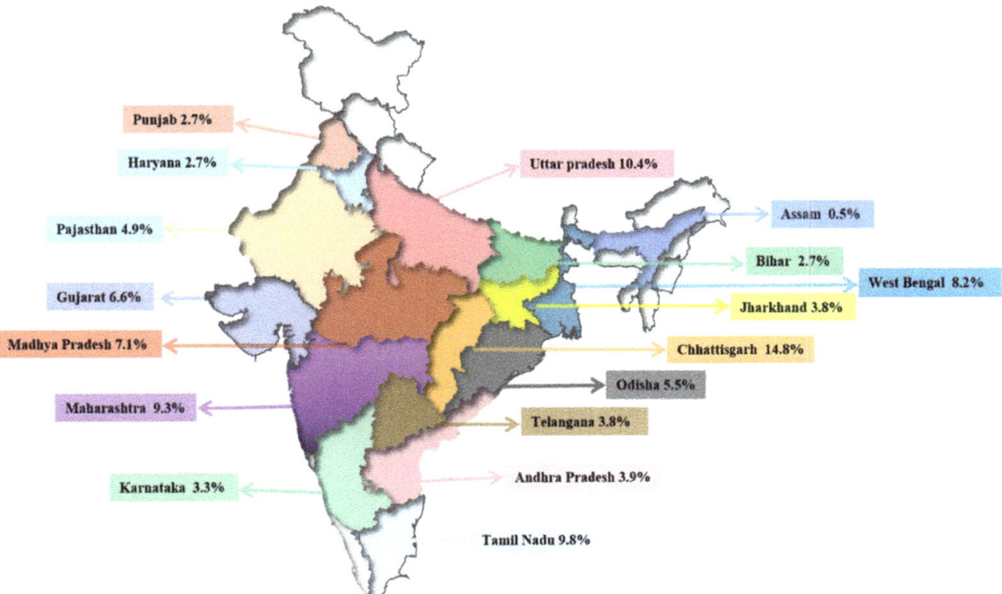

Figure 1. Distribution of data sampling in India.

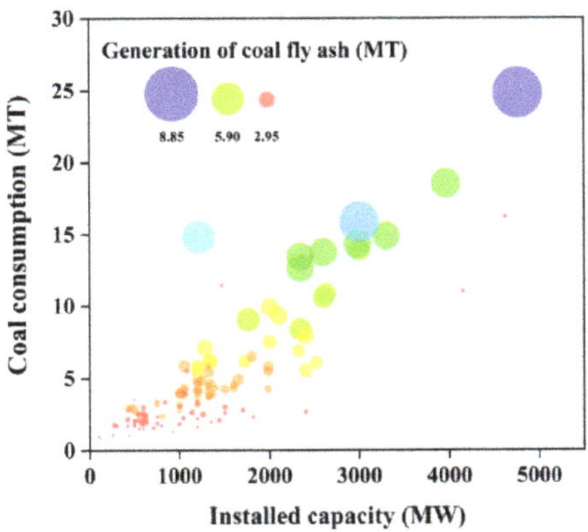

Figure 2. Bubble chart of data distribution. Note that the size of the bubble represents the generation of CFA. Purple bubbles represent CFA maximum generation and red represents minimum generation.

Correlations between features or between features and target variables were measured by Pearson correlation coefficients (R), which were between -1 and 1, as shown in Figure 3. The correlation degree between coal consumption and generation of CFA was the highest, and R was 0.9. Meanwhile, R between installed capacity and target variable was 0.73, indicating a strong correlation between installed capacity and CFA generation.

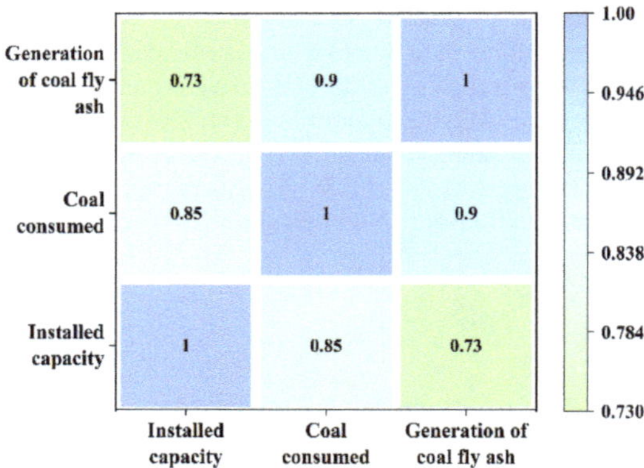

Figure 3. Correlation heat map.

3. Methodology

To achieve rapid and accurate prediction of CFA generation, Python 3.8 programming language was used in this paper, and three machine learning algorithms were selected to construct the model framework using the scikit-learn library [19]. Four evaluation indices were used to measure the performance of the model [20]. Finally, the optimal model was determined according to the evaluation results and compared with traditional methods to verify the feasibility and superiority of the prediction framework. The specific methodology is shown in Figure 4.

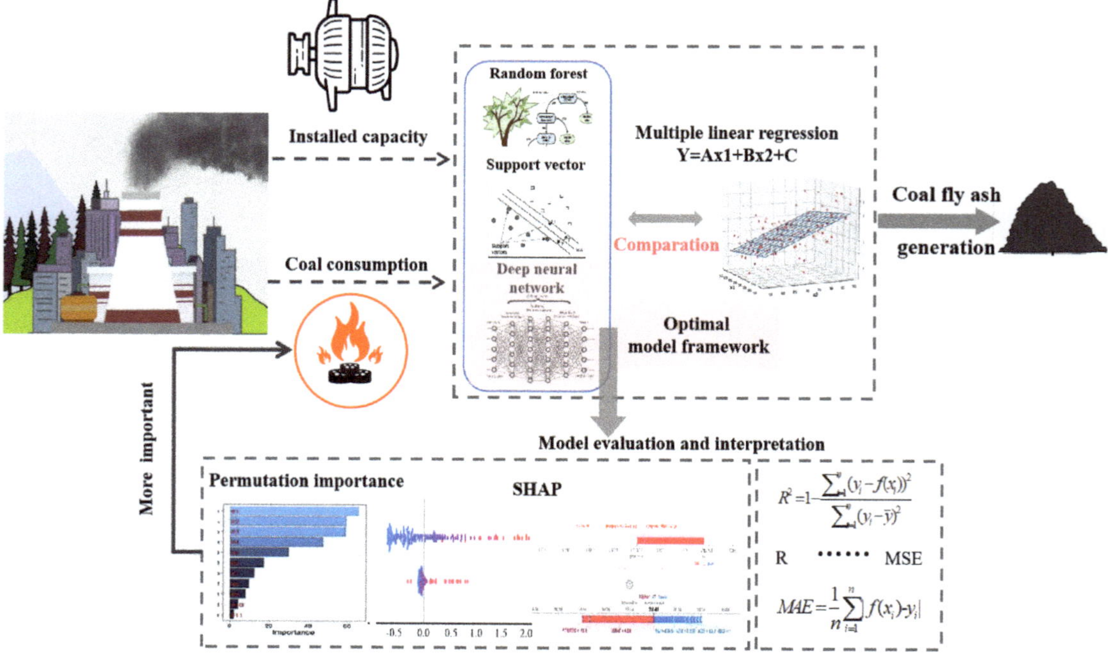

$$R^2 = 1 - \frac{\sum_{i=1}^{n}(y_i - f(x_i))^2}{\sum_{i=1}^{n}(y_i - \bar{y})^2}$$

$$R \quad \cdots\cdots \quad MSE$$

$$MAE = \frac{1}{n}\sum_{i=1}^{n}|f(x_i) - y_i|$$

Figure 4. Complete diagram of methodology.

3.1. Modeling Methods

Machine learning algorithms used in this study had a general modeling process. Firstly, the original data were preprocessed, including the removal of outliers or normalization (this part is explained in detail in Section 3.2). Then, coal consumption and installed capacity after treatment were taken as input variables, and CFA generation was taken as the output variable. Then, training, evaluation, and prediction were carried out by random forest, support vector machine, and neural network. The specific principles and steps of the three algorithms are as follows:

3.1.1. Random Forest

Random forest (RF) is an integration algorithm that combines the outputs of multiple decision trees into one result to deal with classification and regression problems [21]. It has the characteristics of ease of use and flexibility [22]. The construction of RF includes the following four main steps:

1. Random sampling and training decision tree: The original data population with sample size N is randomly sampled N times, and each time, the samples need to be put back [23]. N samples formed at last are used to train a decision tree;
2. Randomly selected attributes as node-splitting attributes: When the nodes of the decision tree are split, m attributes ($m << M$) should be randomly selected from the M attributes of each sample, and then some strategies (such as information gain) should be adopted to select one attribute as the final split attribute of the node;
3. Step 2 is repeated until the tree cannot be split, noting that no pruning occurs during the entire decision tree formation process;
4. A large number of decision trees are established according to steps 1~3 to form an RF.

3.1.2. Support Vector Regression

Support vector regression (SVR) is an important branch of support vector machine (SVM) [24]. SVR has only one type of sample point in the end. The optimal hyperplane it seeks is to minimize the total deviation of all sample points from the hyperplane.

Different from traditional regression methods, SVR indicates that, as long as the deviation degree of $f(x) = \omega^T \Phi(x) + b$ and y is not too large, the prediction can be considered correct without calculating the loss. SVR can obtain a regression model in the form of $f(x) = \omega^T \Phi(x) + b$ by inputting the training sample set $X = (x_i, y_i)_{i = 1 \sim N, y_i \in R}$ [25], where $\Phi(x)$ is the vector mapped to X, $\omega = (\omega_1, \omega_2, \ldots \omega_n)$ is the normal vector, and b is the intercept. Then, an interval band with a distance of ε is created on both sides of the linear function (tolerance deviation) [26]. The loss is not calculated when all samples fall into the interval band but is calculated only when the absolute value of the gap between $f(x)$, and y is greater than ε. Finally, the optimized model is obtained by minimizing the total loss and maximizing the interval.

3.1.3. Deep Neural Network

A deep neural network (DNN) is an extension based on perceptron. The internal structure of DNN has only one input layer and one output layer, but there are multiple hidden layers in the middle [27]. Each layer of the neural network has several neurons. The neurons between layers are connected to each other but are not within a layer, and the neurons in the next layer are connected to all the neurons in the previous layer [28,29].

Generally speaking, the steps of constructing a DNN structure include the following three points: (1) network construction, (2) assignment parameters, and (3) iterative calculation. The main principles of iterative calculation include forward-propagation (FP) and back-propagation (BP) algorithms [30].

The FP algorithm uses several weighted coefficient matrices W and bias vector B to carry out a series of linear operations and activation operations with input vector X. Starting from the input layer, the output of the previous layer is used to calculate the output of the next layer, and then one layer after another is calculated until it reaches the output layer, and the predicted value Y is obtained. In comparison, the BP algorithm uses the gradient descent method to iteratively optimize the loss function to obtain the minimum value [31]. Additionally, it then seeks the appropriate linear coefficient matrix W and bias vector B

corresponding to the hidden layer and output layer, so that the output calculated by all the input of training samples is equal or close to the sample label as far as possible [32].

In short, FP is the recognition process of the predicted value Y, while BP is the reverse adjustment of parameters W and B according to the difference between the target y and the predicted Y. After repeated forward- and back-propagation training, the neural network model with high accuracy is finally formed.

3.2. Dataset Preprocessing and Splitting

The dimensionality and its unit of evaluation index (feature) affect the result of data analysis. To eliminate the influence of dimension between indicators, standardizing or normalizing data to achieve comparability between data indicators are generally adopted [33].

The variance ratio between features and target variables of the dataset in this paper was 200:4:2. There are several orders of magnitude differences between the variances, which leads to features with large variances dominating the algorithm, resulting in poor modeling performance [34]. Therefore, the "processing" module in the sklearn was used to standardize data (sklearn.preprocessing.scale) whose outliers has been removed.

The preprocessed dataset was divided into training and testing sets. Among them, the training set was used to train the model, whereas the testing set was used to verify the final effect of the model. In view of the impact of the division ratio of the dataset on model performance, the size of the testing set in this paper varied from 10% to 45% with an interval of 5%, and R was used as the evaluation index to determine the optimal division ratio [22].

3.3. Model Evaluation

After the model was constructed, it was necessary to evaluate its effect and then select the optimal model by comparison. In this paper, four common indicators—namely, R, R squared (R^2), mean-squared error (MSE), and mean absolute error (MAE)—were used to evaluate the model. The calculation formulas are as follows:

$$R = \frac{\sum\limits_{i=1}^{n}(f(x_i) - \overline{f(x_i)})(y_i - \overline{y})}{\sqrt{\sum\limits_{i=1}^{n}(f(x_i) - \overline{f(x_i)})^2}\sqrt{\sum\limits_{i=1}^{n}(y_i - \overline{y})^2}} \tag{1}$$

$$R^2 = 1 - \frac{\sum_{i=1}^{n}(y_i - f(x_i))^2}{\sum_{i=1}^{n}(y_i - \overline{y})^2} \tag{2}$$

$$MSE = \frac{1}{n}\sum_{i=1}^{n}(f(x_i) - y_i)^2 \tag{3}$$

$$MAE = \frac{1}{n}\sum_{i=1}^{n}|f(x_i) - y_i| \tag{4}$$

where n represented the number of samples, y_i was the real observed value, $\overline{y}$ represented the average of the real value, and $f(x_i)$ was the predicted value, with a mean value of $\overline{f(x_i)}$.

As introduced in Section 2.2, R is used to reflect the degree of linear correlation between two variables; in addition, R^2 is used to judge the degree of fit between the prediction model and the real data [35]. The best value of R^2 is 1 and can be negative. MSE calculates the mean of the sum of squares of sample point errors corresponding to the fitting data and original data, and the smaller the value is, the better the fitting effect is [36]. MAE is used to evaluate how close the predicted results are to the real dataset, and the smaller the MAE, the better the model [37].

4. Result and Discussion

4.1. Determination of Dataset Division Ratio

To avoid the randomness of the evaluation results, in this paper, we evaluated each division ratio of the dataset 50 times repeatedly and took the mean value of the correlation coefficient R as the final performance of the model under a specific partition. As shown in Figure 5, the RF model was taken as an example. For the training set, the influence of the division ratio on the modeling performance was small, and the R fluctuated, by a small margin, around 0.98. Focusing on the testing set, when it accounted for 10% of the dataset, R was 0.84; when the size of the testing set was 15%, the performance of the model reached the highest, satisfying $R = 0.87$. After that, R generally decreased, with a further increase in division ratio up to 45%. In summary, the RF model performed best when the size of the testing set was 15%, and the analysis results of SVR and DNN were consistent with it. Therefore, the training set:testing set = 0.85:0.15 ratio was determined as the optimal division ratio.

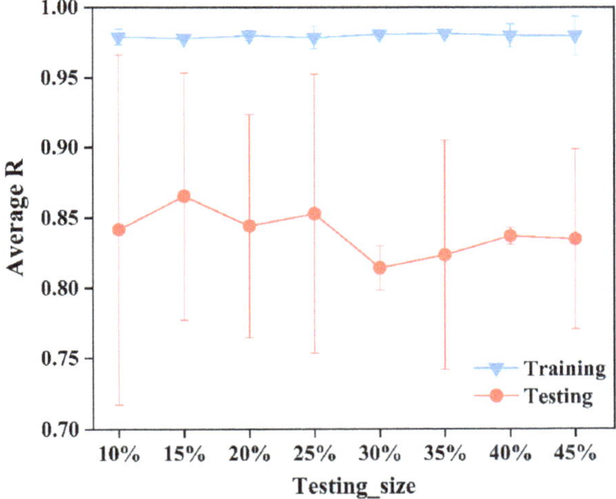

Figure 5. Optimal division ratio of dataset on RF model.

4.2. Parameters of the Model

In this study, RF and SVR, as traditional machine learning regression algorithms, were trained with corresponding default parameters in the ensemble module of sklearn, as shown in Table 1. DNN is a deep learning model in which performance is greatly affected by network structure and parameters [38]. Based on the trial-and-error method and suggestions in references [39,40], the neural network layer, learning rate, activation function, and epoch were constantly changed during the model training process, and 10% of the data were separated for performance verification. As shown in Figure 6, when the loss on the validation set tends to fluctuate stably with the increase in steps, the DNN model that included one input layer, five hidden layers, and one output layer was finally determined. The number of neurons in each layer was 2→8→32→64→16→8→1. To speed up convergence based on the gradient descent method and prevent overfitting, two "Batch normalization" layers and one "dropout" layer were also included. The specific network structure and parameters are shown in Figure 7 and Table 2.

Table 1. Default hyperparameters for RF and SVR models.

RF		SVR	
Parameters	**Default Value**	**Parameters**	**Default Value**
n_ estimators	100	kernel	'rbf'
min_ samples_ split	2	degree	3
min_ samples_ leaf	1	gamma	scale
max_ features	'auto'	C	1
max_ depth	None	epsilon	0.1

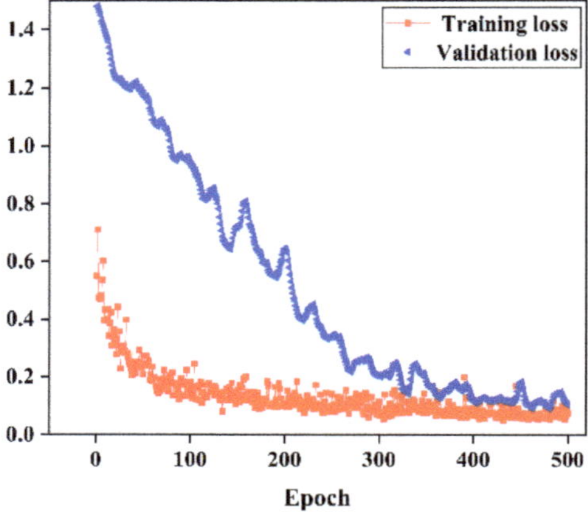

Figure 6. Loss of training set and testing set.

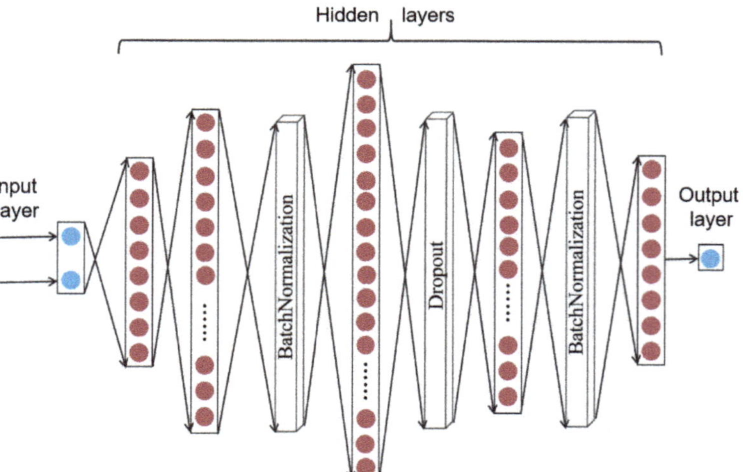

Figure 7. Network structure of optimal DNN model.

Table 2. Specific parameters of DNN structure.

Parameters	Option or Value	Implication
Activation function	Relu	The output is no longer a linear combination of the inputs and can approximate any function
Optimizer	Adam	A hybrid of momentum gradient descent and RMSprop.
Learning rate	0.0005	The weight of neural network input is adjusted.
Batch size	128	Number of samples is used for training.
Epoch	500	One epoch is equal to training with all the samples in the training set.

4.3. Comparative Analysis of Model Performance

To obtain a reliable model, fivefold-cross-validation was adopted for RF and SVR models (cross_val_predict), while DNN used the parameter "validation_split" to perform simple cross-validation. Moreover, as the result of a simple random partition is accidental, it cannot represent the actual performance of the model. Therefore, modeling and evaluation for three ML models were repeated 50 times on the training and testing sets, respectively, and the evaluation indexes were averaged as the final performance of the model.

As shown in Figure 8, the linear fitting functions between the prediction results of RF model on the training and testing sets, and the real values were $y = 0.878x + 0.134$ and $y = 0.861x + 0.094$, while those of the DNN model were $y = 0.790 + 0.375$ and $y = 0.837x + 0.316$. All data points were relatively concentrated on the two curves, and the p values were 4.68E-30, 1.30E-27, 8.35E-73, and 0.00000000000109, respectively, which were less than the significance level of 0.05. In addition, R and R^2 were relatively high, indicating the good performance of the models. Moreover, the linear regression between actual and SVR-estimated generation of CFA was $y = 0.451x + 0.614$ and $y = 0.861x + 0.094$, respectively, on the training and testing sets. Compared with RF and DNN models, the SVR model had relatively discrete data distribution on the training set, and its performance was slightly worse.

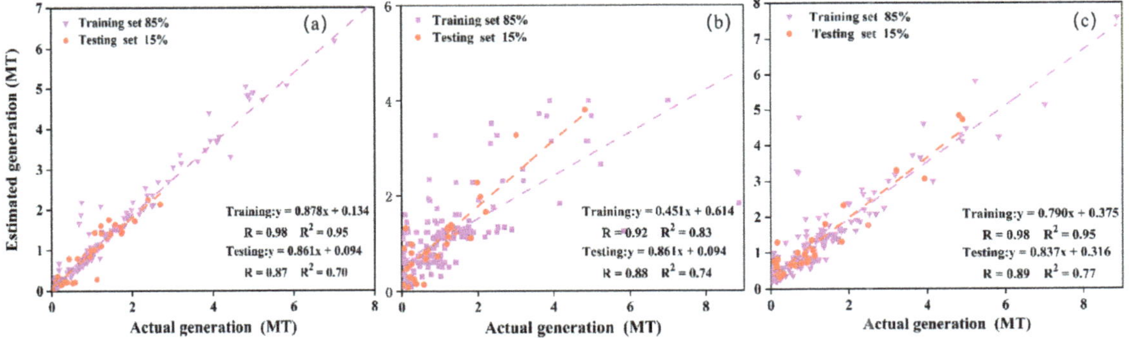

Figure 8. Scatter distribution fitting of generation of CFA under three modeling frameworks: (**a**) RF model, (**b**) SVR model, and (**c**) DNN model.

As can be seen from Figure 9, the difference between the actual and estimated generation of CFA was small in the three models, and the data were mostly concentrated around 0, indicating the good prediction performance of ML models. The probability of data points in RF and DNN models appearing in the small interval $[-0.1, 0.1]$ was close to 0.9, while that of SVR was only 0.45. In addition, for the testing set, the data points on the DNN model were more concentrated in the areas with smaller differences, which implied that the DNN model had higher prediction accuracy.

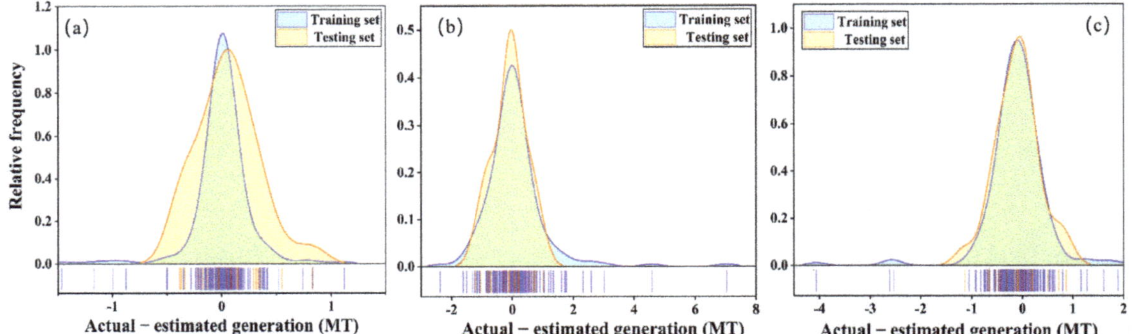

Figure 9. The relative frequency of the difference between actual and estimated CFA generation: (**a**) RF model, (**b**) SVR model, and (**c**) DNN model.

Figure 10 shows a comparison of the performance of the three models more intuitively with four evaluation indices. For the training set in Figure 10a, R and R^2 values of RF and DNN models were the same, which were 0.98 and 0.95, respectively, and slightly higher than those of SVR models, which were 0.92 and 0.83. Meanwhile, the MSE and MAE values of the RF model were the smallest of the three models. On the contrary, the RF model had the lowest R and R^2 values on the testing set of Figure 10b, which were 0.87 and 0.7. However, R and R^2 values of the DNN model were the highest, which were 0.89 and 0.77, and MSE and MAE were relatively low. In general, The DNN model was the optimal model framework suitable for the CFA dataset in this study.

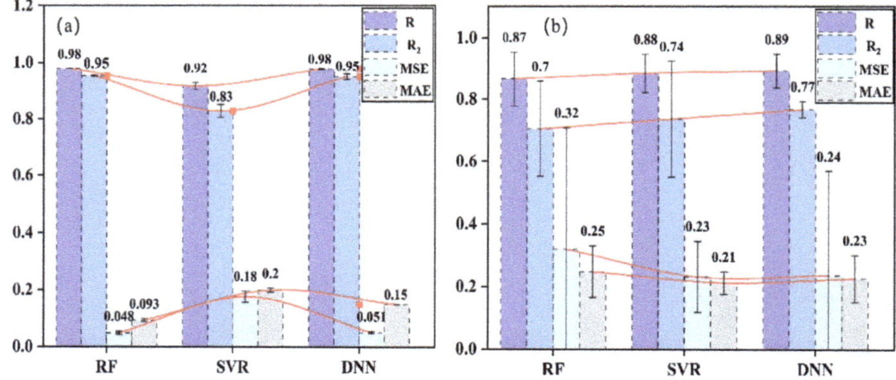

Figure 10. Comparison of four evaluation indicators under three modeling frameworks: (**a**) training set, (**b**) testing set.

4.4. Comparison with Multiple Linear Regression

Multiple linear regression is a conventional data analysis method that uses multiple independent variables to predict or estimate dependent variables [41]. In this method, the dataset was repeatedly divided 50 times according to the same ratio of 0.85:0.15, and the multiple linear regression equation $Y = Ax_1 + Bx_2 + C$ was established. The average results of statistical analysis are shown in Table 3. After 50 evaluations, the mean R^2 and R of the multiple regression training set were 0.82 and 0.90, which were lower than the results of the three ML models using fivefold cross-validation. Then, the data of the testing set were put into the equation for verification, and the mean values of R and p-value were 0.86 and 0.0000643805, respectively, indicating a significant correlation between the results. However, the mean value of R^2 was 0.76, which was higher than the RF and SVR models but lower than that of the DNN models, which once again proved that the DNN model was more suitable for the dataset. The specific results are in the attachment.

Table 3. Statistical parameter analysis of multiple linear regression.

	$Y = A \times 1 + B \times 2 + C$					
	Regression Coefficient	**95% LCL**	**95% UCL**	**SE**	**T**	***p*-Value**
Installed capacity (A)	−0.221283602	−4.6257086	−2.89796	2.5330302	−2.4167858	0.05265861
Coal consumption (B)	0.365218825	0.310336	0.404628	0.0228592	15.6097245	3.49436E-29
Constant (C)	0.173085366	0.1951878	0.318641	0.0759174	2.188839	0.041769567

4.5. Feature Analysis

In this section, the analysis of the sensitivities of two features that affect the generation of CFA is presented using the permutation importance provided by sklearn and eli5, and "TreeExplainer" and "KernelExplainer" in the Shapley Additive Interpretation (SHAP) library.

4.5.1. Permutation Importance

The evaluation of the sensitivity of the feature depends on the degree of degradation of the model performance score after the feature is randomly rearranged [42]. As shown in Figure 11, after the values of coal consumption were randomly shuffled, the decrease in *MSE* of RF, SVR, and DNN algorithms were 1.73, 1.07, and 1.89, respectively, which were generally higher than those in the case of installed capacity randomly disturbed. This proved that the three models reached a consensus on the view that coal consumption had a greater impact on the generation of CFA.

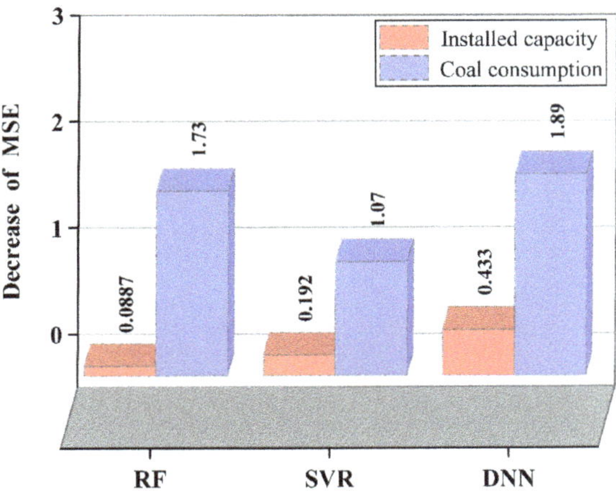

Figure 11. Permutation importance under three machine learning algorithms.

4.5.2. SHAP

SHAP is a model agnostic interpretation method that can be used for both global and individual applications. SHAP can judge which feature is more important, as well as reflect the positive and negative influences of features on the target variable [43]. The model generates a predictive value for each sample, and the SHAP value is the contribution value assigned to each feature in the sample [44].

To better understand the overall pattern, Figure 12a shows the results of calculated SHAP values for each feature of each sample. Among them, features were arranged from top to bottom in order of importance on the y axis [45], which indicated that coal consumption had a greater influence on the model, consistent with permutation importance.

In addition, the color represented the feature value (red was high, blue was low [46]). It can be seen that, under the three algorithm models, higher coal consumption increased the predicted generation of CFA. However, for installed capacity, the results were different. In RF and SVR models, larger installed capacity increased the predicted generation of CFA, but in DNN, the result was completely opposite. As shown in Figure 12b, the first sample for which preprocessed feature values were 0.8059 and 1.363 was used as an example to explain the generation details of a single prediction. In the figure, the red bar represents the range in which a feature played a positive role in the prediction of the model [47]; the base value was the mean value of the target variables of all samples, and $f(x)$ was the final predicted value for this sample, which satisfies $f(x)$ = base value + $\sum$SHAP value. The analysis showed that the prediction results of the three algorithm models were slightly different for the same sample, which may be affected by the algorithm principle and disrupted data. However, coal consumption and installed capacity both played positive, driving roles in the prediction of the model, but coal consumption had a greater impact.

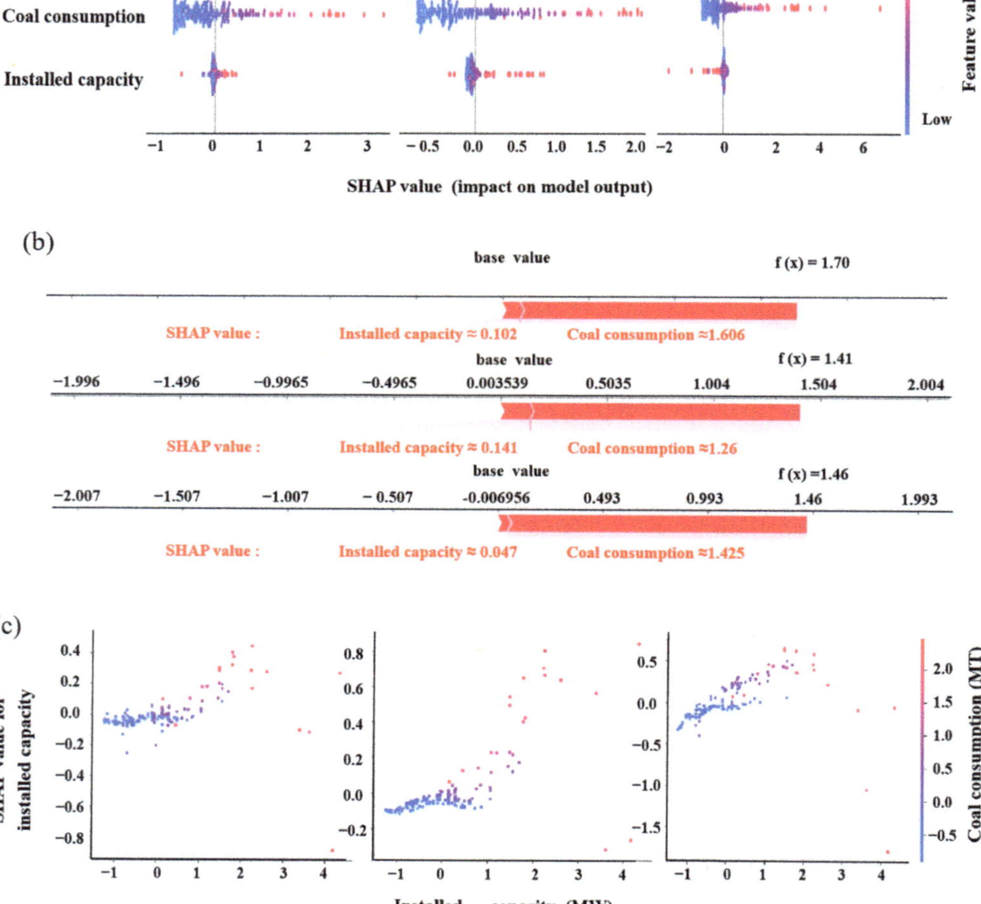

Figure 12. Features interpretation from three perspectives using SHAP: (**a**) global explanation, (**b**) local explanation, and (**c**) feature interaction. It is worth noting that the RF, SVR, and DNN models are from top to bottom or left to right in the three subgraphs.

In addition to explaining the model globally and locally, Figure 12c revealed hidden relationships among features through quick, precise interactions. The analysis showed that the interaction between coal consumption and installed capacity had positively correlated influences on CFA generation prediction. Specifically, when both coal consumption and installed capacity were high, the installed capacity had a great influence on the generation of CFA, except for some outliers. On the contrary, when the coal consumption and installed capacity were relatively small, the installed capacity contributed little to the variation in the model output and even hindered the prediction.

As indicated above, the effect of installed capacity on the CFA generation was not always positive, compared with that of coal consumed. In real life, the installed capacity is the designed capacity for one specific powder station. The actual capacity is influenced by many external factors, such as coal production, the market, policies, etc. Therefore, the correlation between installed capacity and CFA generation was not as close as the correlation between coal consumed and CFA generation. Moreover, it is possible that a power station with a large installed capacity produced a relatively small amount of power, and thus CFA, due to the influence of the above-mentioned factors. ML models based on datasets with such special cases might indicate the negative influence of installed capacity for some data samples.

5. Significance and Outlook

High energy consumption leads to increased generation of solid wastes such as CFA, posing a potential threat to the environment and human health. Meanwhile, more CFA byproducts are gradually being recycled and utilized to achieve sustainability [48]. However, the uncertainty of CFA generation poses difficulties to the rational planning and design of its disposal and utilization. The optimal model framework constructed in this study can quickly and accurately predict the generation of CFA only by inputting coal consumption and installed capacity, which is feasible and efficient. Applying this model framework to the engineering field enables managers to identify the next step of the disposal method in advance, so as to rationally allocate ways of recycling and utilization to maximize the use and sales benefits of CFA while minimizing its disposal costs. However, due to the small size of the dataset and few input variables, the results of this model framework lack further validation, and its general application needs to be improved. Subsequent studies can expand the search scope and consider various factors affecting CFA generation.

6. Conclusions

DNN was determined as the optimal ML model through comparative evaluation, which can accurately predict the generation of CFA. In addition, the sensitivity analysis of the features also provided a certain point of reference for ensuring the rational utilization of CFA. The specific conclusions are as follows:

(1)   Among the three model algorithms, the DNN model had the best performance. R and R^2 on the training set were 0.98 and 0.95, whereas these on the testing set were 0.89 and 0.77, respectively;

(2)   The R^2 of the traditional multiple linear regression equation on the testing set was 0.76, higher than those of RF and SVR models, but lower than that of the DNN model;

(3)   Permutation importance and SHAP both indicated that coal consumption had a greater positive effect on the generation of CFA. As influenced by other factors, the influence of installed capacity on CFA generation was as significant as coal consumed and could be negative for some special data samples.

Author Contributions: Conceptualization, C.Q.; methodology, M.W. and C.Q.; validation, M.W.; investigation, M.W. and C.Q.; data curation, M.W. and C.Q.; writing—original draft preparation, M.W.; writing—review and editing, All authors; visualization, All authors; supervision, C.Q. All authors have read and agreed to the published version of the manuscript.

Funding: This study was funded by the Research Fund of The State Key Laboratory of Coal Resources and Safe Mining, CUMT (No. SKLCRSM21KF004).

Institutional Review Board Statement: Not applicable.

Informed Consent Statement: Not applicable.

Data Availability Statement: Not applicable.

Acknowledgments: The authors gratefully acknowledge the CUMT for financial support. C. Q. acknowledgements the unwavering support from Y.S. Liu and the help from Andy Fourie (The University of Western Australia).

Conflicts of Interest: The authors declare no conflict of interest.

References

1. Naqvi, S.Z.; Ramkumar, J.; Kar, K.K. Coal-based fly ash. In *Handbook of Fly Ash*; Kar, K.K., Ed.; Butterworth-Heinemann: Oxford, UK, 2022; pp. 3–33. [CrossRef]
2. IEA. *Electricity Market Report*; IEA: Paris, France, 2022; Available online: https://www.iea.org/reports/electricity-market-report-january-2022 (accessed on 25 February 2022).
3. IEA. *Global Energy Review*; IEA: Paris, France, 2021; Available online: https://www.iea.org/reports/global-energy-review-2021 (accessed on 25 February 2022).
4. Wang, X.Y. Evaluation of the hydration heat and strength progress of cement-fly ash binary composite. *J. Ceram. Process. Res.* **2020**, *21*, 622–631. [CrossRef]
5. Mathapati, M.; Amate, K.; Durga Prasad, C.; Jayavardhana, M.L.; Hemanth Raju, T. A review on fly ash utilization. *Materials* **2022**, *50*, 1535–1540. [CrossRef]
6. Arora, S. *An Ashen Legacy: India's Thermal Power Ash Mismanagement*; Centre for Science and Environment: New Delhi, India, 2020.
7. Blaha, U.; Sapkota, B.; Appel, E.; Stanjek, H.; Rösler, W. Micro-scale grain-size analysis and magnetic properties of coal-fired power plant fly ash and its relevance for environmental magnetic pollution studies. *Atmos. Environ.* **2008**, *42*, 8359–8370. [CrossRef]
8. Chowdhury, A.; Naz, A.; Chowdhury, A. Waste to resource: Applicability of fly ash as landfill geoliner to control ground water pollution. *Materials* **2021**, *20*, 897. [CrossRef]
9. Chen, Q.; Chen, L.; Li, J.; Guo, Y.; Wang, Y.; Wei, W.; Liu, C.; Wu, J.; Tou, F.; Wang, X.; et al. Increasing mercury risk of fly ash generated from coal-fired power plants in China. *J. Hazard. Mater.* **2022**, *429*, 128296. [CrossRef]
10. Jiang, A.; Zhao, J. Experimental Study of Desulfurized Fly Ash Used for Cement Admixture. In *Proceedings of Civil Engineering in China–Current Practice and Research Report*; Hindawi: Hebei, China, 2010; pp. 1038–1042.
11. Ragipani, R.; Escobar, E.; Prentice, D.; Bustillos, S.; Simonetti, D.; Sant, G.; Wang, B. Selective sulfur removal from semi-dry flue gas desulfurization coal fly ash for concrete and carbon dioxide capture applications. *Waste Manag.* **2021**, *121*, 117–126. [CrossRef]
12. Shanmugan, S.; Deepak, V.; Nagaraj, J.; Jangir, D.; Jegan, S.V.; Palani, S. Enhancing the use of coal-fly ash in coarse aggregates concrete. *Mater. Today Proc.* **2020**, *30*, 174–182. [CrossRef]
13. Kotelnikova, A.D.; Rogova, O.B.; Karpukhina, E.A.; Solopov, A.B.; Levin, I.S.; Levkina, V.V.; Proskurnin, M.A.; Volkov, D.S. Assessment of the structure, composition, and agrochemical properties of fly ash and ash-and-slug waste from coal-fired power plants for their possible use as soil ameliorants. *J. Clean. Prod.* **2022**, *333*, 130088. [CrossRef]
14. Zhu, M.; Ji, R.; Li, Z.; Wang, H.; Liu, L.; Zhang, Z. Preparation of glass ceramic foams for thermal insulation applications from coal fly ash and waste glass. *Constr. Build. Mater.* **2016**, *112*, 398–405. [CrossRef]
15. Zahari, N.M.; Mohamad, D.; Arenandan, V.; Beddu, S.; Nadhirah, A. Study on prediction fly ash generation using statistical method. In Proceedings of the 3rd International Sciences, Technology and Engineering Conference (ISTEC), Penang, Malaysia, 17–18 April 2018.
16. Widyarsana, I.; Tambunan, S.A.; Mulyadi, A.A. Identification of Fly Ash and Bottom Ash (FABA) Hazardous Waste Genera-tion From the Industrial Sector and Its Reduc-tion Management in Indonesia. *Res. Sq.* **2022**. [CrossRef]
17. Cakir, M.; Guvenc, M.A.; Mistikoglu, S. The experimental application of popular machine learning algorithms on predictive maintenance and the design of IIoT based condition monitoring system. *Comput. Ind. Eng.* **2021**, *151*, 106948. [CrossRef]
18. Prakash, M. *Report on Fly Ash Generation at Coal/Lignite Based Thermal Power Stations and its Utilization in The Country for The Year 2019–2020*; Central Electricity Authority Government of India Ministry of Power: New Delhi, India, 2020.
19. Pedregosa, F.; Varoquaux, G.; Gramfort, A.; Michel, V.; Thirion, B.; Grisel, O.; Blondel, M.; Prettenhofer, P.; Weiss, R.; Dubourg, V. Scikit-learn: Machine learning in Python. *J. Mach. Learn. Res.* **2011**, *12*, 2825–2830.
20. Meiyazhagan, J.; Sudharsan, S.; Venkatesan, A.; Senthilvelan, M. Prediction of occurrence of extreme events using machine learning. *Eur. Phys. J. Plus* **2022**, *137*, 16. [CrossRef]
21. Li, H.; Lin, J.; Lei, X.; Wei, T. Compressive strength prediction of basalt fiber reinforced concrete via random forest algorithm. *Mater. Today Commun.* **2022**, *30*, 103117. [CrossRef]
22. Qi, C.; Wu, M.; Zheng, J.; Chen, Q.; Chai, L. Rapid identification of reactivity for the efficient recycling of coal fly ash: Hybrid machine learning modeling and interpretation. *J. Clean. Prod.* **2022**, *343*, 130958. [CrossRef]

23. Pi, J.; Jiang, D.; Liu, Q. Random Forest Algorithm for Power System Load Situation Awareness Technology. In *Application of Intelligent Systems in Multi-modal Information Analytics*; Sugumaran, V., Xu, Z., Zhou, H., Eds.; Springer International Publishing: Cham, Switzerland, 2021; pp. 925–929.

24. Wang, L. *Support Vector Machines: Theory and Applications*. In *Proceedings of Machine Learning and Its Applications*; Advanced Lectures; Springer: Berlin/Heidelberg, Germany, 2001.

25. Thalib, R.; Bakar, M.A.; Ibrahim, N.F. Application of support vector regression in krylov solvers. *Ann. Emerg. Technol. Comput.* **2021**, *123*, 178–186. [CrossRef]

26. Xia, Z.; Mao, K.; Wei, S.; Wang, X.; Fang, Y.; Yang, S. Application of genetic algorithm-support vector regression model to predict damping of cantilever beam with particle damper. *J. Low Freq. Noise Vib. Act. Control* **2017**, *36*, 138–147. [CrossRef]

27. Phapatanaburi, K.; Wang, L.; Oo, Z.; Li, W.; Nakagawa, S.; Iwahashi, M. Noise robust voice activity detection using joint phase and magnitude based feature enhancement. *J. Ambient. Intell. Humaniz. Comput.* **2017**, *8*, 845–859. [CrossRef]

28. Feng, C. Robustness Verification Boosting for Deep Neural Networks. In Proceedings of the 6th International Conference on Information Science and Control Engineering (ICISCE), Shanghai, China, 20–22 December 2019; pp. 531–535.

29. Liu, L.; Chen, J.; Xu, L. Realization and application research of BP neural network based on MATLAB. In Proceedings of the International Seminar on Future Biomedical Information Engineering, Wuhan, China, 18 December 2008; pp. 130–133.

30. Silaban, H.; Zarlis, M. Sawaluddin Analysis of Accuracy and Epoch on Back-propagation BFGS Quasi-Newton. In Proceedings of the International Conference on Information and Communication Technology (ICONICT), Singapore, 27–29 December 2017.

31. Han, T.; Lu, Y.; Zhu, S.-C.; Wu, Y.N. Alternating Back-Propagation for Generator Network. In Proceedings of the Thirty-First Aaai Conference on Artificial Intelligence, San Francisco, CA, USA, 4–9 February 2017; pp. 1976–1984.

32. Yan, Z. Research and Application on BP Neural Network Algorithm. In Proceedings of the 2015 International Industrial Informatics and Computer Engineering Conference, Shaanxi, China, 10–11 January 2015; pp. 1444–1447.

33. Upadhyay, A.; Singh, M.; Yadav, V.K. Improvised number identification using SVM and random forest classifiers. *J. Inf. Optim. Sci.* **2020**, *41*, 387–394. [CrossRef]

34. Mo, C.; Cui, H.; Cheng, X.; Yao, H. Cross-Scale Registration Method Based on Fractal Dimension Characterization. *Acta Opt. Sin.* **2018**, *38*, 1215001. [CrossRef]

35. Mittlböck, M.; Heinzl, H. A note on R2 measures for Poisson and logistic regression models when both models are applicable. *J. Clin. Epidemiol.* **2001**, *54*, 99–103. [CrossRef]

36. Patnana, A.K.; Vanga, N.R.V.; Chandrabhatla, S.K.; Vabbalareddy, R. Dental Age Estimation Using Percentile Curves and Regression Analysis Methods–A Test of Accuracy and Reliability. *J. Clin. Diagn. Res.* **2018**, *12*, ZC1–ZC4. [CrossRef]

37. Qi, J.; Du, J.; Siniscalchi, S.M.; Ma, X.; Lee, C.-H. On Mean Absolute Error for Deep Neural Network Based Vector-to-Vector Regression. *IEEE Signal Process. Lett.* **2020**, *27*, 1485–1489. [CrossRef]

38. Shinozaki, T.; Watanabe, S. Structure Discovery of Deep Neural Network Based on Evolutionary Algorithms. In Proceedings of the IEEE International Conference on Acoustics, Speech, And Signal Processing (ICASSP), Queensland, Australia, 19–24 April 2015; pp. 4979–4983.

39. Panchagnula, K.K.; Jasti, N.V.K.; Panchagnula, J.S. Prediction of drilling induced delamination and circularity deviation in GFRP nanocomposites using deep neural network. *Materials* **2022**, in press. [CrossRef]

40. Beniaguev, D.; Segev, I.; London, M. Single cortical neurons as deep artificial neural networks. *Neuron* **2021**, *109*, 2727–2739.e2723. [CrossRef] [PubMed]

41. Wang, G.; Wu, J.; Yin, S.; Yu, L.; Wang, J. Comparison between BP Neural Network and Multiple Linear Regression Method. *Inf. Comput. Appl.* **2010**, *6377*, 365–370.

42. Afanador, N.L.; Tran, T.N.; Buydens, L.M.C. Use of the bootstrap and permutation methods for a more robust variable importance in the projection metric for partial least squares regression. *Anal. Chim. Acta* **2013**, *768*, 49–56. [CrossRef]

43. Rodriguez-Perez, R.; Bajorath, J. Interpretation of machine learning models using shapley values: Application to compound potency and multi-target activity predictions. *J. Comput.-Aided Mol. Des.* **2020**, *34*, 1013–1026. [CrossRef]

44. Rodríguez-Pérez, R.; Bajorath, J. Interpretation of Compound Activity Predictions from Complex Machine Learning Models Using Local Approximations and Shapley Values. *J. Med. Chem.* **2020**, *63*, 8761–8777. [CrossRef]

45. Peng, J.; Zou, K.; Zhou, M.; Teng, Y.; Zhu, X.; Zhang, F.; Xu, J. An Explainable Artificial Intelligence Framework for the Deterioration Risk Prediction of Hepatitis Patients. *J. Med. Syst.* **2021**, *45*, 61. [CrossRef]

46. Wang, F.; Wang, Y.; Zhang, K.; Hu, M.; Weng, Q.; Zhang, H. Spatial heterogeneity modeling of water quality based on random forest regression and model interpretation. *Environ. Res.* **2021**, *202*, 111660. [CrossRef] [PubMed]

47. Chen, Y.; Cheng, A.; Zhang, C.; Chen, S.; Ren, Z. Rapid mechanical evaluation of the engine hood based on machine learning. *J. Braz. Soc. Mech. Sci. Eng.* **2021**, *43*, 345. [CrossRef]

48. Qi, C.; Xu, X.; Chen, Q. Hydration reactivity difference between dicalcium silicate and tricalcium silicate revealed from structural and Bader charge analysis. *Int. J. Miner. Metall. Mater.* **2022**, *29*, 335–344. [CrossRef]

Article

Compression Behavior of SLM-Prepared 316L Shwartz Diamond Structures under Dynamic Loading

Qingyan Ma [1], Zhonghua Li [1,*] and Jiaxin Li [2]

[1] School of Mechanical Engineering, North University of China, Taiyuan 030051, China; maqingyan@nuc.edu.cn
[2] School of Materials Science and Engineering, North University of China, Taiyuan 030051, China; jiaxin_li_nuc@163.com
* Correspondence: lizhonghua6868@163.com

Abstract: In this paper, the compression behavior of a triply minimal periodic surface (Shwartz Diamond) fabricated by selective laser melting (SLM) under different loading rates was studied. A quasi-static strain rate of 2.22×10^{-3}/s was tested using a universal testing machine, and a strain rate of 650/s was tested by Hopkinson pressure bar (SHPB). The results showed that the yield stress of all structures increased under dynamic load, and the DIF of sheet structure was higher than that of the skeleton structure, among which the DIF of GSHD was the largest and most sensitive to strain rate. However, the normalized SEA of USHD was the highest.

Keywords: diamond structure; compression; strain rate; specific energy absorption

Citation: Ma, Q.; Li, Z.; Li, J. Compression Behavior of SLM-Prepared 316L Shwartz Diamond Structures under Dynamic Loading. *Crystals* **2022**, *12*, 447. https://doi.org/10.3390/cryst12040447

Academic Editors: Hao Yi, Huajun Cao, Menglin Liu, Le Jia and Shujun Zhang

Received: 25 February 2022
Accepted: 21 March 2022
Published: 23 March 2022

Publisher's Note: MDPI stays neutral with regard to jurisdictional claims in published maps and institutional affiliations.

1. Introduction

With the increasing maturity of additive manufacturing technology, lattice structures with a triply periodic minimal surface (TPMS) can be fabricated to further their applications in lightweight [1], filtration [2], energy absorption [3], and Bio-implants [4]. Researchers usually use selective laser melting (SLM) and electron beam melting (EBM) to prepare lattice structures, including truss structures [5–8] and TPMS structures [9–12] due to their high fabrication accuracy. TPMS structures have a clean mathematical representation and a very convenient gradient design process. Their advantages are having a continuous topology that can avoid stress concentration and have good mechanical properties, and having a surface with zero mean curvature that is similar to natural structures (e.g., porous bone) and thus excellent bionic properties. The commonly used TPMS structures such as Primitive, Diamond and Gyroid structures have attracted researchers' interest. Researchers have found that the gradient structure exhibited superior energy absorption capacity, larger strain before densification, and significantly different deformation behavior compared to the uniform structure. Yang et al. [13] investigated the effect of loading direction on the compressive properties of gradient Gyroid structures and developed a mathematical model to predict the mechanical properties of gradient structures. The gradient skeletal Gyroid structure had a greater initial densification strain, and the energy absorption was greater than that of the uniform skeletal structure. Additionally, the deformation behavior of the gradient structure changed with a changing loading direction. Liu et al. [9] investigated the deformation behavior of gradient cell-size Diamond and Gyroid, and the results showed that 45 shear damage occurred first to the larger cell. Zhang et al. [14] investigated energy absorption characteristics and established energy absorption diagrams of three 316L sheet TPMS structures under compressive loading using experiments and finite element simulations. The results showed that Diamond structures exhibited better mechanical properties and energy-absorption capabilities. Therefore, it was necessary to study the mechanical behavior and deformation mechanism of gradient sheet with a Diamond structure.

The dynamic compression behavior of porous materials was very important when they were used for energy absorption, intrusion resistance and blast protection [15]. The dynamic compression behavior of foams has been thoroughly investigated by researchers. Deshpande et al. [16] investigated the high strain rate compression behavior of two foams (Alulight and Duocel) and showed that the platform stress and densification strain were independent of the strain rate but were sensitive to the relative density. However, the results of ALPORAS aluminum foam showed a high strain-rate sensitivity for platform stresses. Duarte et al. [17] reduced the high dispersion of mechanical properties caused by structural defects by adding a skin to the aluminum foam surface and found that the foam density and pore size distribution have a significant effect on the mechanical properties of the foam cubes. Mukai et al. [18] showed that the yield stress of ALPORAS aluminum foam exhibited significant strain rate sensitivity and the per-unit volume energy absorption was 50% higher at high strain rates than at quasi-static rates. Ozdemir et al. [19] showed that the lattice structure was able to diffuse the impact load and reduce the peak impact stress in time. Novak et al. [10] investigated four common 316L uniform sheet TPMS structures and found that the platform stress of Diamond structure was the largest under quasi-static and dynamic loading. However, strain rate hardening caused an increase in platform stress and specific energy absorption. Yin et al. [20] investigated the crushing behavior of four 316L sheet TPMS structures under dynamic loading using experiments and finite element simulations, and the results showed that the specific energy absorption, average crushing force, and peak crushing force of all structures increased with the increase in shell thickness. TPMS structures had a continuous surface, which was accompanied by energy absorption during surface deformation and therefore had superior energy absorption performance.

Steel is one of the most widely used metallic materials due to its superior mechanical properties, such as corrosion resistance, compression hardening and high strain rate sensitivity. Park et al.'s results [21,22] showed that ellipsoidal pores inside the foam led to anisotropy of stainless steel foam, the yield strength of foam steel was sensitive to strain rate at higher strain rates, and the corrugation of cell walls in foam steel changed the dependence on the density. A study by Zong et al. [23] showed that the impact toughness of SLM-prepared 316L stainless steel cross-section was 62.69% higher than that of the longitudinal section due to lower tough-brittle transition temperature by grain refinement perpendicular to the building direction. A study by Cao et al. [24] showed that the SLM-prepared rhombic dodecahedron 316L lattice structure had a higher compressive strength, platform stress and specific energy. Additionally, the mechanical properties had a certain dependence on the strain rate. The porous structure might be subjected to impact load or explosion load, and the mechanical behavior and deformation behavior under a dynamic load were more complex. However, the dynamic compression characteristics of TPMS structures manufactured by additive manufacturing were less studied. Additionally, the reason for stress enhancement and deformation mechanism of TPMS structures under dynamic load was not clear, especially for skeleton-based TPMS and gradient TPMS structures. Therefore, it was necessary to study the properties of skeleton-based TPMS and gradient TPMS structures under dynamic loading.

Therefore, this paper discusses the mechanical properties of SLM-manufactured 316L gradient sheet Diamond (GSHD), gradient skeletal Diamond (GSKD), uniform sheet Diamond (USHD), and uniform skeletal Diamond (USKD) structures under nominal strain rates ranging from 2.22×10^{-3}/s to 650/s. Additionally, dynamic performance was tested using the split Hopkinson pressure bar (SHPB).

2. Materials and Methods

2.1. TPMS Lattice Structure Design

In this work, we used MATLAB 2019a, which has strong programming and plotting capabilities, to generate Diamond-equivalent surface cells described by mathematical expressions as shown in Equation (1). The STL format of the solid model can be generated in Materialise magic 22.0 by complementing, thickening, intersecting, and aggregating the

Diamond equivalent surface. We designed a total of four structures, namely uniform sheet diamond (USHD) structure, uniform skeletal diamond (USKD) structure, gradient sheet diamond (GSHD) structure, and gradient skeletal diamond (GSKD) structure, as shown in Figure 1. USHD and USKD structures were composed of 5 mm × 5 mm × 5 mm unit cells arrayed three times in x-, y-, and z-directions to obtain the total structure size. The relative density of GSHD and GSKD unit cells varied uniformly in the z-direction, and GSHD and GSKD structures with relative density ranging from 30% at the top to 10% at the bottom were designed. GSHD is obtained by subtracting USKD with a relative density of 10% from GSKD with a relative density of 40–20%, and USHD is obtained by subtracting USKD with relative density of 40% from USKD with relative density of 20%. In order to keep the relative densities of the four structures consistent, the relative densities of uniform sheet diamond (USHD) and uniform skeletal diamond (USKD) were both 20%.

$$D(x, y, z) = \sin \frac{2\pi x}{L_x} \sin \frac{2\pi y}{L_y} \sin \frac{2\pi z}{L_z} + \sin \frac{2\pi x}{L_x} \cos \frac{2\pi z}{L_y} \cos \frac{2\pi z}{L_z} + \cos \frac{2\pi x}{L_x} \sin \frac{2\pi y}{L_y} + \cos \frac{2\pi z}{L_z} + \cos \frac{2\pi x}{L_x} \cos \frac{2\pi y}{L_y} \sin \frac{2\pi z}{L_z} = C \tag{1}$$

where C is the value of the equation controlling the position of the equivalence surface, and L_x, L_y, and L_z are the unit length values in the x-, y-, and z-directions, respectively.

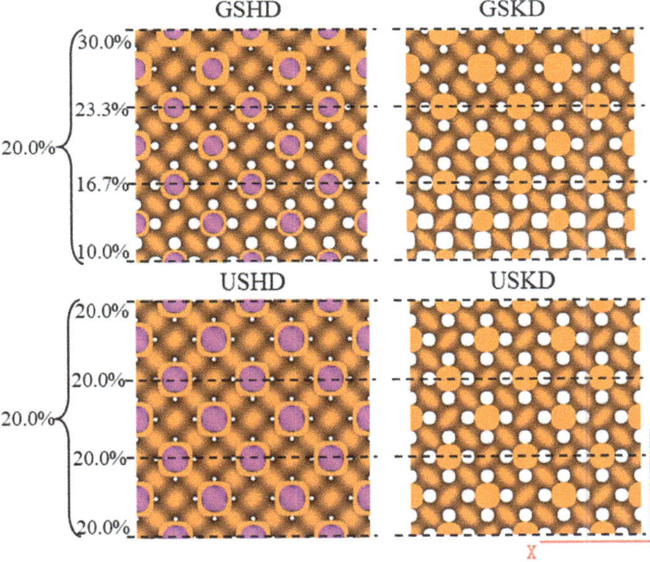

Figure 1. Front view of the four designed Diamond structures.

2.2. Diamond Structure Prepared by SLM

Diamond lattice structure samples were prepared from 316L powder of AVIMETAL AM. The chemical composition of 316L powder is shown in Table 1. As analyzed by Mastersizzer (S3500, Microtrac, Osaka, Japan), its particle size distribution ranged from 20.91 μm (*D*10) to 56.62 μm (*D*90), with an average particle size of 33.71 μm, as shown in Figure 2. Samples of 316L of Diamond porous structures were fabricated by EP-M150 (E-PLUS 3D, Beijing, China) in a building chamber of Φ150 × 150 mm². The SLM system was equipped with a maximum power of 500 W water-cooled fiber laser with a beam diameter of 70 μm. The specimens were manufactured on 304 substrates. The SLM process parameters were as follows: laser power 185 W, layer thickness 30 μm, hatch forming 110, hatch type rotation 10°–352°–19°–344°–31°–326°–43°–315°, and scanning speed 620 mm/s. The SLM process was carried out in high-purity argon gas to avoid oxygen contamination

($O_2 \leq 100$ ppm). These parameters were given by the equipment manufacturer to obtain the highest density and low concentration of surface defects (e.g., spherification). Subsequently, the fabricated samples, shown in Figure 3a, were removed from the substrate by electric discharge machining (EDM) followed by ultrasonic cleaning with pure ethanol to remove the powder from the sample pores and impurities left on the sample surface after EDM cutting. A SU5000 (Hitachi, Tokyo, Japan) SEM was used to characterize the microstructure of the 316L Diamond structure. Figure 3b shows the front view of a strut node, in which incomplete melting of powder and significant step effect were evident. This was mainly caused by the rapid transition of the Diamond structure at the node. The microstructure of GSKD structure contained honeycombed grains with a size of about 1 μm, as shown in Figure 3c. In general, the smaller the grain size, the higher the strength of the material. Therefore, the strength of the Diamond structure prepared this time was relatively high.

Table 1. Chemical composition of 316L.

Alloying Element	Fe	Cr	Ni	Mo	Mn	Si	P	S	C	O
Wt (%)	bal	17.16	12.2	2.71	1.45	0.56	0.01	0.008	0.022	≤0.07

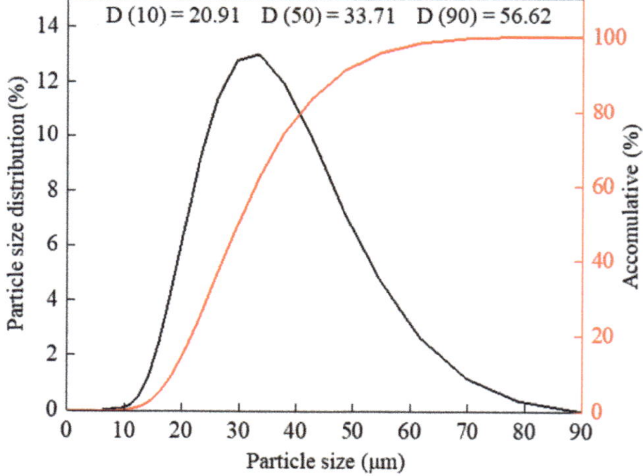

Figure 2. 316L powders particle size distribution.

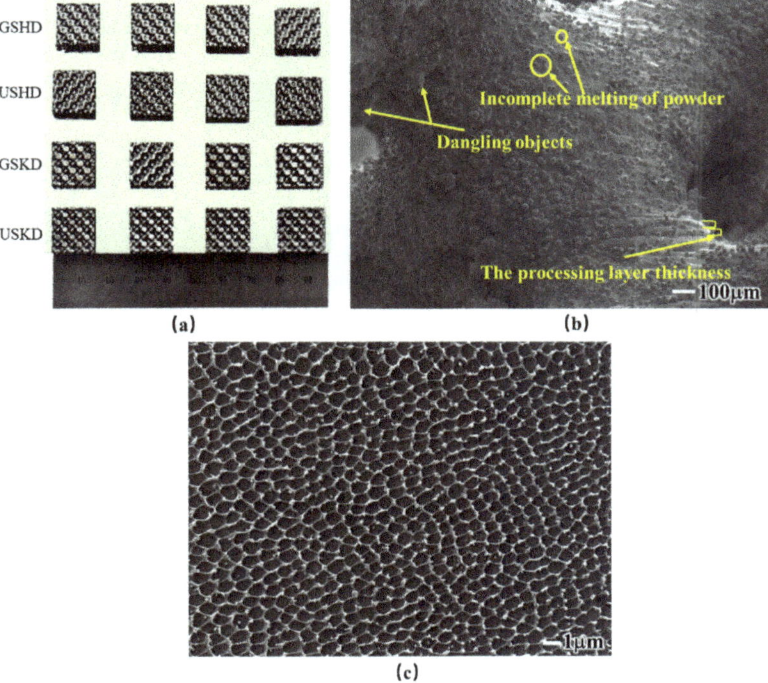

Figure 3. (a) Samples of Diamond lattice structures manufactured by SLM; SEM images of SLM-manufactured GSKD structure: (b) the microscopic structure of the upper surface; (c) the front view of a strut node.

2.3. Mechanical Properties Experiments

Uniaxial compression tests were carried out on CMT5101 Electronic Universal Testing Machine (MTS, Eden Prairie, USA) at room temperature with a speed of 2 mm/min. Additionally, the loading direction was opposite to the construction direction. The dynamic compression test was performed using a split Hopkinson compression bar as shown in Figure 4. The lengths of the bullet, incident rod, transmission rod, and absorption rod used in this test were 500 mm, 2500 mm, 2500 mm, and 1000 mm, respectively, all with a diameter of 37 mm. The two strain gauges used to collect and output signals were located at 1250 mm from the test-incident–rod interface and the projection-rod–sample interface, respectively, as shown in Figure 4. Additionally, the construction direction of the sample was opposite to the impact direction. The strain gauges collected the incident stress wave and the transmission wave signal. The signals were processed and amplified by the dynamic strain test device and then recorded and displayed on the computer. The mechanical properties of the impacted dotted structure were measured at strain rate of 650/s. The relevant mechanical derivation was as follows.

$$\varepsilon(t) = \frac{2C_0}{l_0} \int_0^t \varepsilon_r(t)dt \tag{2}$$

$$\dot{\varepsilon}(t) = \frac{2C_0}{l_0} \varepsilon_r(t) \tag{3}$$

$$\sigma(t) = -\frac{AE}{A_0} \varepsilon_t(t) \tag{4}$$

where $\varepsilon_r(t)$ and $\varepsilon_t(t)$ are the obtained reflected and transmitted waves, respectively. E, A, and C_0 are the modulus of elasticity, cross-sectional area and wave velocity (5050 m/s) of the steel rod, respectively. A_0 and l_0 are the initial surface area and length of the specimen, respectively.

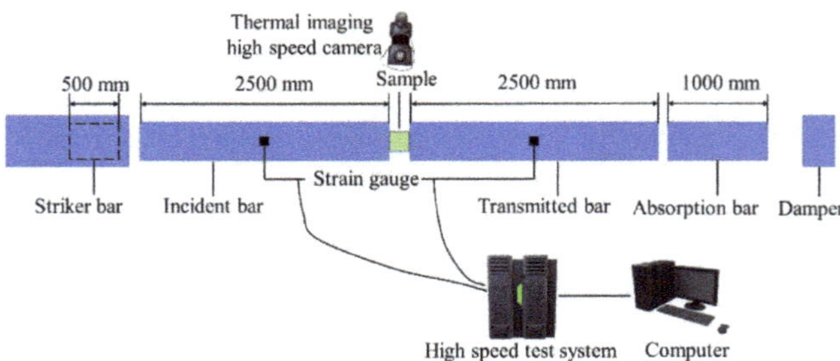

Figure 4. Schematic diagram of the separated Hopkinson compression bar.

According to the standard IOS 13314:2011, the yield strength was the stress corresponding to 0.2% plastic deformation in the quasi-static experiments, and the dynamic experiments all show significant yielding with the initial peak stress as the yield strength. The energy absorption capacity of the structure was assessed using the specific energy absorption (SEA) [20,25,26], which is defined as follows.

$$\mathrm{SEA} = \frac{W}{\rho} \tag{5}$$

$$W = \int_0^\varepsilon \sigma(\varepsilon)d\varepsilon \tag{6}$$

where W is the energy absorbed per unit volume, ρ is the density of the 316L stainless steel, σ is the stress, and ε is the strain.

3. Results and Discussion

In order to reduce the effect of signal fluctuations, the stress–strain curves at nominal strain rates of 650/s were obtained by smoothing the obtained incident wave with 170 points, as shown in Figure 5. Unfortunately, the densification of the lattice structures was not seen due to the low dynamic loading speed.

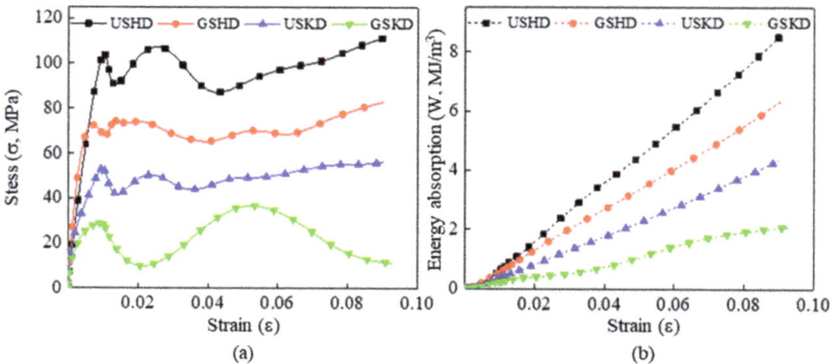

Figure 5. (**a**) Stress–strain and (**b**) strain–energy absorption curves under a strain rate of 650/s.

It can be seen from Figures 5 and 6 that, at nominal strain rates of 650/s, all four structures exhibited significant yielding and stress drop, while under quasi-static loading, only the GSKD structure experienced a stress drop. This might be due to the strain rate effect of the structure. The strain-hardening properties of the 316L substrate led to an overall rise in stress after a stress drop. In the work of Li et al. [25], the Gyroid structure also showed a stress drop in the SHPB experiment. Saremian et al. [27] suggest that the reflection of stress waves within the cell structure caused slight fluctuations after the stress drop. The energy absorption–strain curve under dynamic load was closer to a straight line, excluding a GSKD structure. However, the energy absorption–strain curve under quasi-static load was closer to a parabola. This might be mainly caused by strain rate effect.

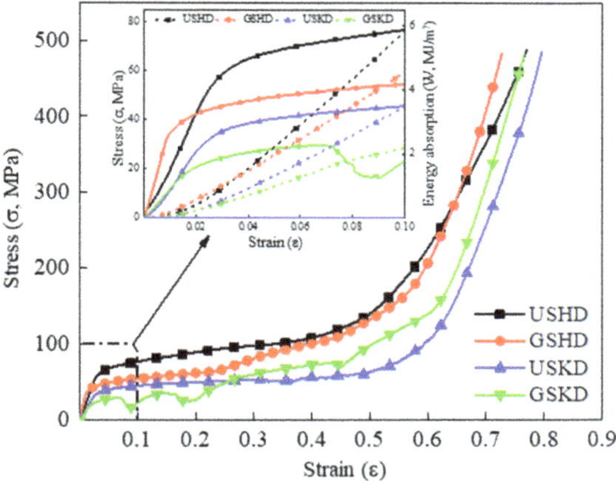

Figure 6. Stress–strain and energy absorption-strain curves under a strain rate of $2.22 \times 10^{-3}/\text{s}$.

Wang [28] defined a dimensionless parameter-stress drop factor λ to quantify the degree of stress drop, i.e., $\lambda = \Delta\sigma/\sigma_d$, where $\Delta\sigma$ is the interpolation of the initial stress peak and valley and σ_d is the initial stress peak. The corresponding stress drop factors of the dotted structure under different velocity loading can be obtained in Table 2, and it can be clearly seen that the GSHD structure had the smallest stress drop factor of 0.06 and the GSKD structure had the largest of 0.65. This was mainly due to the fact that the GSKD had the lowest relative density, where the deformation occurred first and was more difficult to contact with the adjacent parts' site.

Table 2. Stress drop factor and dynamic increase factor of the lattice structures.

Structure	GSKD	GSHD	USKD	USHD
Dynamic increase factor (DIF)	0.38	0.98	0.60	0.84
Stress drop factor (λ)	0.65	0.06	0.22	0.13

There were four main reasons recognized by researchers for strain rate sensitivity [29,30]: the sensitivity of the matrix material itself to strain rate, the local deformation of the structure, inertial effects, the action of gases within the closed-cell foam, and the magnitude of the strain rate. Dynamic increase factor (DIF, the ratio of the difference between dynamic and static collapse strength to static collapse strength) was one of the important parameters to measure whether the material was sensitive to strain rate. The Dynamic increase factor of the four structures are given in Table 2, and it can be clearly seen that the GSHD structure had the largest value of stress enhancement factor (0.98) and the highest sensitivity to strain rate. Similarly, the GSKD structure has the lowest sensitivity to strain rate. This was mainly due to the different material distribution characteristics of the four structures and unique stress-distribution law. A study by Yin et al. [20] showed that the shell thickness had a significant effect on the performance of TPMS structures under dynamic loading. In addition, the air in our added protective casing contributed to the strain rate sensitivity of the structure.

Previous studies have shown that TPMS structures exhibit excellent performance in terms of energy absorption. The results of Novak et al. [10] showed that sheet Diamond structures of the same relative density exhibit the highest energy absorption due to their extremely high plateau stress. In our quasi-static experiments, the SEA of the USHD structure was 5.42 J/g. However, the initial densification strain was not reached due to the short loading time of the SHPB experiment. We analyzed the energy absorption of the four structures at $\varepsilon = 0.03$ and $\varepsilon = 0.08$, as shown in Figure 7. Interestingly, the SEA of USHD was lower than that of GSHD at quasi-static loading at $\varepsilon = 0.03$; however, the opposite result was exhibited under dynamic loading. This was mainly due to the stress fluctuations under dynamic loading. As shown in Figure 8, all SEAs are normalized by corresponding quasi-static SEAs. The normalized SEA of the USHD structure was the largest, and that of GSKD was the smallest. This was mainly due to the higher stresses in the USHD structure.

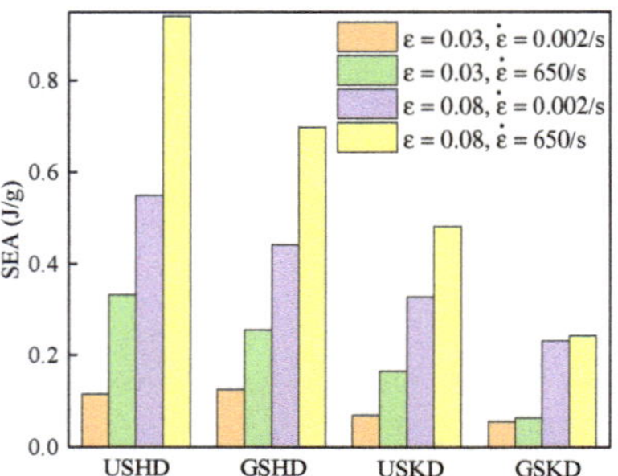

Figure 7. Specific energy absorption of the four structures at $\varepsilon = 0.03$ and $\varepsilon = 0.08$.

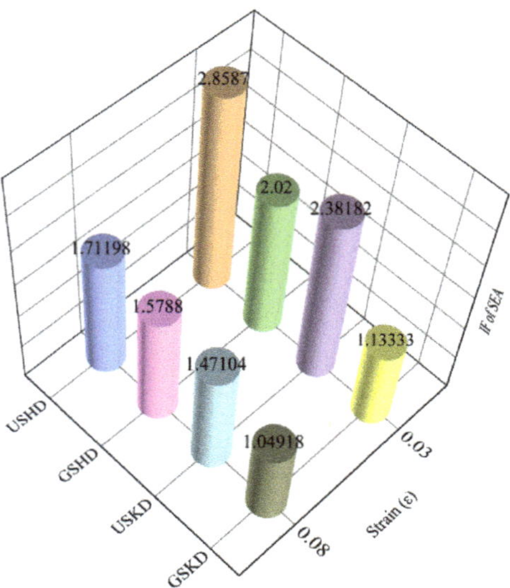

Figure 8. SEA after normalization of four structures.

4. Conclusions

In this paper, four Diamond lattice structures with different material distributions were prepared using a selective laser melting technique. The mechanical behavior of gradient structure under quasi-static and dynamic loads was studied and compared with that of uniform structure. The following conclusions were made.

1. When the strain rate was 650/s, all structures showed a stress drop, and the stress drop factor of GSKD structure was the largest, mainly because the relative density of GSKD at the beginning of deformation was the lowest and it took a long time to contact the adjacent parts.
2. At nominal strain rates of 650/s, the DIF of sheet structures was higher than that of the skeleton structure, and the DIF of GSHD structure was the largest, reaching 0.98, which was the most sensitive to strain rate. Additionally, the USHD structure showed the best energy absorption capability.
3. After the corresponding static SEA normalization, SEA showed different rules at $\varepsilon = 0.03$ and $\varepsilon = 0.08$, which was mainly caused by the different structural stiffness.

Author Contributions: Conceptualization, Q.M. and Z.L.; methodology, J.L.; formal analysis, J.L.; investigation, Q.M. and Z.L.; writing—original draft preparation, Q.M. and J.L.; writing—review and editing, Q.M. and Z.L. All authors have read and agreed to the published version of the manuscript.

Funding: This research was funded by the National Natural Science Foundation of China (Grant No. 51905497, No. 52075502) and the Support Program for Young Academic Leaders of North University of China (No. QX201902).

Institutional Review Board Statement: Not applicable.

Informed Consent Statement: Not applicable.

Data Availability Statement: Not applicable.

Conflicts of Interest: The authors declare no conflict of interest.

References

1. Li, D.; Liao, W.; Dai, N.; Dong, G.; Tang, Y. Optimal design and modeling of gyroid-based functionally graded cellular structures for additive manufacturing. *Comput.-Aided Des.* **2018**, *104*, 87–99. [CrossRef]
2. Wei, Q.; Li, H.; Liu, G.; He, Y.; Wang, Y.; Tan, Y.E.; Wang, D.; Peng, X.; Yang, G.; Tsubaki, N. Metal 3D printing technology for functional integration of catalytic system. *Nat. Commun.* **2020**, *11*, 4098. [CrossRef] [PubMed]
3. Andrew, J.J.; Schneider, J.; Ubaid, J.; Velmurugan, R.; Gupta, N.K.; Kumar, S. Energy absorption characteristics of additively manufactured plate-lattices under low- velocity impact loading. *Int. J. Impact Eng.* **2021**, *149*, 103768. [CrossRef]
4. Alzahrani, M.; Choi, S.K.; Rosen, D.W. Design of truss-like cellular structures using relative density mapping method. *Mater. Des.* **2015**, *85*, 349–360. [CrossRef]
5. Li, P. Constitutive and failure behaviour in selective laser melted stainless steel for microlattice structures. *Mater. Sci. Eng.* **2015**, *622*, 114–120. [CrossRef]
6. Cheng, X.Y.; Li, S.J.; Murr, L.E.; Zhang, Z.B.; Hao, Y.L.; Yang, R.; Medina, F.; Wicker, R.B. Compression deformation behavior of Ti-6Al-4V alloy with cellular structures fabricated by electron beam melting. *J. Mech. Behav. Biomed. Mater.* **2012**, *16*, 153–162. [CrossRef]
7. Isaenkova, M.G.; Yudin, A.V.; Rubanov, A.E.; Osintsev, A.V.; Degadnikova, L.A. Deformation behavior modelling of lattice structures manufactured by a selective laser melting of 316L steel powder. *J. Mater. Res. Technol.* **2020**, *9*, 15177–15184. [CrossRef]
8. Smith, M.; Guan, Z.; Cantwell, W.J. Finite element modelling of the compressive response of lattice structures manufactured using the selective laser melting technique. *Int. J. Mech. Sci.* **2013**, *67*, 28–41. [CrossRef]
9. Liu, F.; Mao, Z.; Zhang, P.; Zhang, D.Z.; Jiang, J.; Ma, Z. Functionally graded porous scaffolds in multiple patterns: New design method, physical and mechanical properties. *Mater. Des.* **2018**, *160*, 849–860. [CrossRef]
10. Novak, N.; Al-Ketan, O.; Krstulović-Opara, L.; Rowshan, R.; Abu Al-Rub, R.K.; Vesenjak, M.; Ren, Z. Quasi-static and dynamic compressive behaviour of sheet TPMS cellular structures. *Compos. Struct.* **2021**, *266*, 113801. [CrossRef]
11. Santos, J.; Pires, T.; Gouveia, B.P.; Castro, A.P.G.; Fernandes, P.R. On the permeability of TPMS scaffolds. *J. Mech. Behav. Biomed. Mater.* **2020**, *110*, 103932. [CrossRef] [PubMed]
12. Yan, C.; Hao, L.; Hussein, A.; Bubb, S.L.; Young, P.; Raymont, D. Evaluation of light-weight AlSi10Mg periodic cellular lattice structures fabricated via direct metal laser sintering. *J. Mater. Process. Technol.* **2014**, *214*, 856–864. [CrossRef]

13. Yang, L.; Mertens, R.; Ferrucci, M.; Yan, C.; Shi, Y.; Yang, S. Continuous graded Gyroid cellular structures fabricated by selective laser melting: Design, manufacturing and mechanical properties. *Mater. Des.* **2019**, *162*, 394–404. [CrossRef]

14. Zhang, L.; Feih, S.; Daynes, S.; Chang, S.; Wang, M.Y.; Wei, J.; Lu, W.F. Energy absorption characteristics of metallic triply periodic minimal surface sheet structures under compressive loading. *Addit. Manuf.* **2018**, *23*, 505–515. [CrossRef]

15. Merrett, R.P.; Langdon, G.S.; Theobald, M.D. The blast and impact loading of aluminium foam. *Mater. Des.* **2013**, *44*, 311–319. [CrossRef]

16. Deshpande, V.S.; Fleck, N.A. High strain rate compressive behaviour of aluminium alloy foams. *Int. J. Impact Eng.* **2000**, *24*, 277–298. [CrossRef]

17. Duarte, I.; Vesenjak, M.; Krstulović-Opara, L. Variation of quasi-static and dynamic compressive properties in a single aluminium foam block. *Mater. Sci. Eng.* **2014**, *616*, 171–182. [CrossRef]

18. Mukai, T.; Kanahashi, H.; Miyoshi, T.; Mabuchi, M.; Nieh, T.G.; Higashi, K. Experimental study of energy absorption in a closed-celled aluminum foam under dynamic loading. *Scr. Mater.* **1999**, *40*, 921–927. [CrossRef]

19. Ozdemir, Z.; Hernandez-Nava, E.; Tyas, A.; Warren, J.A.; Fay, S.D.; Goodall, R.; Todd, I.; Askes, H. Energy absorption in lattice structures in dynamics: Experiments. *Int. J. Impact Eng.* **2016**, *89*, 49–61. [CrossRef]

20. Yin, H.; Liu, Z.; Dai, J.; Wen, G.; Zhang, C. Crushing behavior and optimization of sheet-based 3D periodic cellular structures. *Compos. Part B Eng.* **2020**, *182*, 107565. [CrossRef]

21. Park, C.; Nutt, S.R. Strain rate sensitivity and defects in steel foam. *Mater. Sci. Eng.* **2002**, *323*, 358–366. [CrossRef]

22. Park, C.; Nutt, S.R. Anisotropy and strain localization in steel foam. *Mater. Sci. Eng.* **2001**, *299*, 68–74. [CrossRef]

23. Zong, X.; Liu, W.; Yang, Y.; Zhang, S.; Chen, Z. Anisotropy in microstructure and impact toughness of 316L austenitic stainless steel produced by melting. *Rare Metal Mater. Eng.* **2020**, *49*, 4031–4040.

24. Cao, X.; Xiao, D.; Li, Y.; Wen, W.; Zhao, T.; Chen, Z.; Jiang, Y.; Fang, D. Dynamic compressive behavior of a modified additively manufactured rhombic dodecahedron 316L stainless steel lattice structure. *Thin-Walled Struct.* **2020**, *148*, 106586. [CrossRef]

25. Li, X.; Xiao, L.; Song, W. Compressive behavior of selective laser melting printed Gyroid structures under dynamic loading. *Addit. Manuf.* **2021**, *46*, 102054. [CrossRef]

26. Bai, L.; Zhang, J.; Xiong, Y.; Chen, X.; Sun, Y.; Gong, C.; Pu, H.; Wu, X.; Luo, J. Influence of unit cell pose on the mechanical properties of Ti6Al4V lattice structures manufactured by selective laser melting. *Addit. Manuf.* **2020**, *34*, 101222. [CrossRef]

27. Saremian, R.; Badrossamay, M.; Foroozmehr, E.; Kadkhodaei, M.; Forooghi, F. Experimental and numerical investigation on lattice structures fabricated by selective laser melting process under quasi-static and dynamic loadings. *Int. J. Adv. Manuf. Technol.* **2021**, *112*, 2815–2836. [CrossRef]

28. Wang, P.F. Study on Dynamic Mechanical Response and Temperature Dependence of Porous Metals. Ph.D. Thesis, University of Science and Technology of China, Hefei, China, 2012.

29. Zhao, H.; Elnasri, I.; Abdennadher, S. An experimental study on the behaviour under impact loading of metallic cellular materials. *Int. J. Mech. Sci.* **2005**, *47*, 757–774. [CrossRef]

30. Atturan, U.A.; Nandam, S.H.; Murty, B.S.; Sankaran, S. Deformation behaviour of in-situ TiB2 reinforced A357 aluminium alloy composite foams under compressive and impact loading. *Mater. Sci. Eng.* **2017**, *684*, 178–185. [CrossRef]

Article

Effect of Nano-Si$_3$N$_4$ Reinforcement on the Microstructure and Mechanical Properties of Laser-Powder-Bed-Fusioned AlSi10Mg Composites

Zhongliang Lu [1,2], Yu Han [1], Yunpeng Gao [1], Fusheng Cao [1], Haitian Zhang [1], Kai Miao [1,*], Xin Deng [3,*] and Dichen Li [1]

[1] State Key Laboratory for Manufacturing Systems Engineering, Xi'an Jiaotong University, Xi'an 710049, China; zllu@xjtu.edu.cn (Z.L.); yu_hann@163.com (Y.H.); gao194@163.com (Y.G.); caofs0122@163.com (F.C.); zht4869@163.com (H.Z.); dcli@mail.xjtu.edu.cn (D.L.)
[2] Guangdong Xi'an Jiaotong University Academy, Foshan 528399, China
[3] School of Electromechanical Engineering, Guangdong University of Technology, Guangzhou 510006, China
* Correspondence: kaimiao@xjtu.edu.cn (K.M.); dengxin@gdut.edu.cn (X.D.)

Abstract: Laser powder bed fusion (LPBF) technology is of great significance to the rapid manufacturing of high-performance metal parts. To improve the mechanical behavior of an LPBFed AlSi10Mg alloy, the influence of nano-Si$_3$N$_4$ reinforcement on densification behavior, microstructure, and tensile property of AlSi10Mg was studied. The experimental results show that 97% relative density of the 3 vol.% nano-Si$_3$N$_4$/AlSi10Mg composite was achieved via optimization of the LPBF process. With the increase in the nano-Si$_3$N$_4$ content, the tensile strength and the yield strength of the composite steadily increase as per the Orowan strengthening mechanism while the elongation decreases. In addition, nano-Si$_3$N$_4$ reinforcement reduces the width of the coarse cell structure region and the thermal influence region of the AlSi10Mg matrix. After annealing, the tensile strength of the nano-Si$_3$N$_4$/AlSi10Mg composite decreases and the elongation increases significantly.

Keywords: laser powder bed fusion (LPBF); nano-Si$_3$N$_4$; AlSi10Mg; mechanical properties; heat treatment

Citation: Lu, Z.; Han, Y.; Gao, Y.; Cao, F.; Zhang, H.; Miao, K.; Deng, X.; Li, D. Effect of Nano-Si$_3$N$_4$ Reinforcement on the Microstructure and Mechanical Properties of Laser-Powder-Bed-Fusioned AlSi10Mg Composites. *Crystals* **2022**, *12*, 366. https://doi.org/10.3390/cryst12030366

Academic Editors: Hao Yi, Huajun Cao, Menglin Liu and Le Jia

Received: 25 January 2022
Accepted: 2 March 2022
Published: 9 March 2022

Publisher's Note: MDPI stays neutral with regard to jurisdictional claims in published maps and institutional affiliations.

1. Introduction

Additive manufacturing (AM) technology is revolutionizing the way that traditional products are manufactured, via its unique layer-by-layer superposition fabrication manner based on a 3D model [1] that can help create complex-shaped products with good physical and mechanical properties [2]. Laser powder bed fusion (LPBF) is one of the advanced AM technologies that provide outstanding advantages in combining shape forming and sintering/melting/casting in a single step. At present, it has wide application prospects in aerospace, biomedicine, automobile engineering, and other fields [3–6]. However, the bottleneck in LPBF is the relatively limited number of metals suitable for this technology, such as some aluminum, titanium, and nickel alloys and stainless steel [7]. Therefore, expanding the material range is becoming more and more important for the development of LPBF.

Due to their high specific strength, corrosion resistance, and preparation convenience, Al–Si casting alloys are now the main aluminum alloy for LPBF and have been widely investigated for applications in aerospace and automotive fields. Among all the Al–Si casting alloys, AlSi10Mg attracts the most attention [8,9]. AlSi10Mg has been proven to be a typical material suitable for LPBF due to its good weldability, admirable hardenability, high thermal conductivity, and favorable corrosion resistance. The reasons for these excellent properties are that the near-eutectic composition of Al and Si reduces the solidification range and a small amount of Mg makes AlSi10Mg suitable for heat treatment to harden and strengthen the matrix [10]. Some scholars have studied the LPBF process of AlSi10Mg, often focusing on the laser parameters, porosity, microstructure, mechanical properties, and heat treatment [11–15]. For example, Cauwenbergh [16] studied the threefold inter-relationship between the process (LPBF and heat treatment), the microstructure at different scales (macro-, meso-, micro-, and nano-scale), and the macroscopic material properties of AlSi10Mg. Zhou [17] investigated the effects of curvatures on the geometrical performance, defects, microstructure, and mechanical properties of as-built AlSi10Mg parts with curved surfaces.

With the development of the automobile industry and other fields, the mechanical performance requirements for aluminum alloy are getting more stringent, which makes the development of ceramic-particle-reinforced aluminum alloy matrix composites extremely urgent. Powder metallurgy was once the essential technology to make particle-enhanced metal matrix composites (MMCs), but LPBF has become an important alternative way to produce MMCs at present [18,19]. Recently, high-density ceramic-particle-reinforced aluminum composites have been successfully fabricated by LPBF. Li et al. [7] prepared an 11.6 wt.% nano-TiB2-decorated AlSi10Mg composite (NTD–Al) with a crack-free microstructure, high tensile strength, excellent plasticity, and high microhardness. Gu et al. [20] prepared nano-TiC/AlSi10Mg composites that are suitable for SLM. Wang et al. [21] also prepared 3 wt.% TiC/AlSi10Mg composites by LPBF. Their experimental results showed that nano-TiC particles can significantly improve tensile properties but increase the porosity of nano-TiC/AlSi10Mg composites. Jiang et al. [22] prepared 1 wt.% CNT AlSi10Mg via SLM and found a positive relationship between laser scanning speed and composite mechanical properties. Gao et al. [23] prepared nano-TiN-enhanced AlSi10Mg. Their tests showed that the evenly distributed particles brought about excellent microhardness and wear properties to the composite. Chang et al. [24] used micro-SiC to produce the AlSi10Mg composite by SLM, with improved wear resistance. With the help of good crystal atomic matching at the Al–LaB6 interface, Tan et al. [25] achieved high-plasticity LaB6/AlSi10Mg composites. The above research manifests that suitable ceramic particles can make metal-based composites achieve better mechanical properties through LPBF. However, among these state-of-art LPBF studies, the nano-Si_3N_4-particle-reinforced Al matrix composite has rarely been investigated and little attention has been paid to the effect of heat treatment on the performance of nano-scale-particle-enhanced metal-based composites.

In this work, nano-silicon nitride, due to its high strength, low thermal expansion coefficient, superior corrosion resistance, and good thermal stability [26,27], was designed as an enhancement phase for AlSi10Mg. This study was focused on the LPBF process parameters, microstructure, and mechanical properties of nano-Si_3N_4-enhanced AlSi10Mg composites. Moreover, the effect of heat treatment on the mechanical properties of composites was explored.

2. Materials and Methods

Gas-atomized AlSi10Mg powder with a median particle size of 37.43 μm, as shown in Figure 1a,c, provided by AVIC Maite Powder Metallurgy Technology (Gu'an) Co., Ltd., (Langfang, China), and nano-Si_3N_4 powder with a median particle size of 50 nm, provided by Beijing Guoxin Hongyu Curtain Wall Material Co., Ltd., (Beijing, China), were used as the feedstock powders in this study. Table 1 shows the composition of AlSi10Mg powder. The contents of Si_3N_4 powders in nano-Si_3N_4/AlSi10Mg composites were set as 1, 3, and 5 vol.%, respectively. The Si_3N_4 and AlSi10Mg powders were mixed by ball milling, using a ball-to-powder weight ratio of 1:1, a rotation speed of 200 rpm, and a mixing time of 10 h. Figure 2 shows the SEM of mixed powders. AlSi10Mg powder maintained its original spherical shape after ball milling, while nano-Si_3N_4 particles were uniformly distributed on the surface of AlSi10Mg particles.

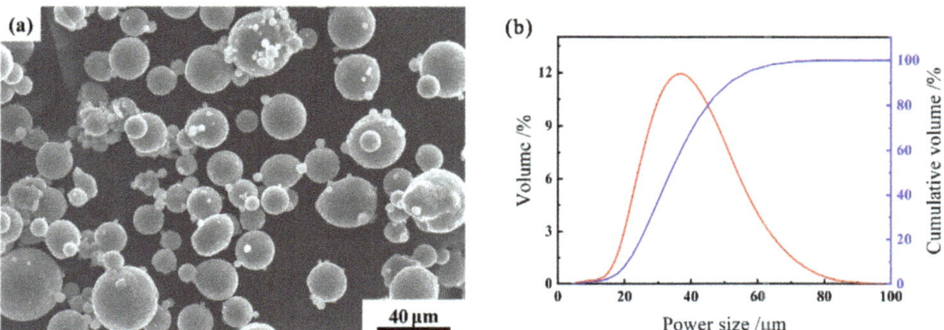

Figure 1. The morphology of AlSi10Mg (**a**) and the particle size distribution of AlSi10Mg powder (**b**).

Table 1. Chemical composition of AlSi10Mg powders (in wt.%).

Element	Al	Si	Mg	Fe	Cu	Mn	Pb	Zn	Ni	Be
Content (wt.%)	Bal.	9.5	0.39	<0.1	<0.1	<0.45	<0.1	<0.1	<0.1	<0.1

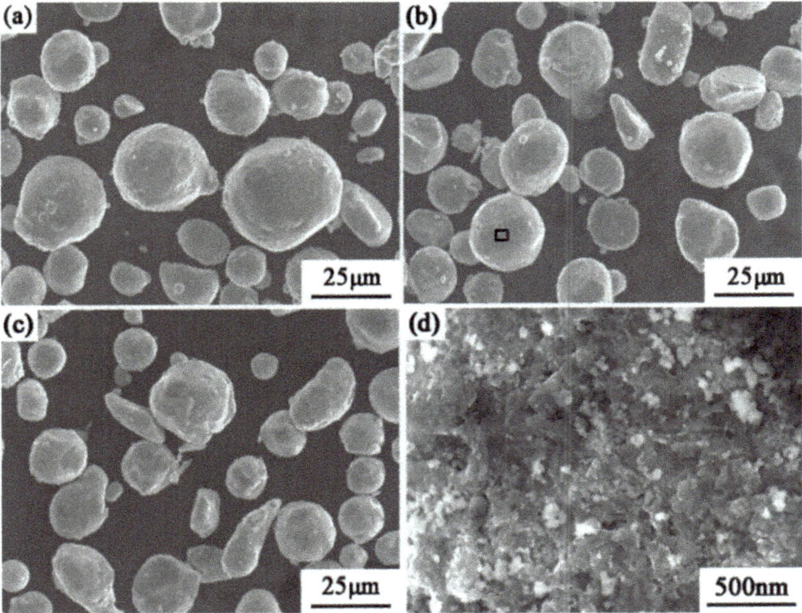

Figure 2. SEM images for nano-Si_3N_4/AlSi10Mg mixed powders with different Si_3N_4 contents: (**a**) 1 vol.%, (**b**) 3 vol.%, and (**c**) 5 vol.% Si_3N_4, and (**d**) enlarged view of (**b**).

Pure AlSi10Mg and nano-Si_3N_4/AlSi10Mg composites were produced by LPBF via SLM equipment (SLM-100C, Guangdong Hanbang 3d Tech Co., Ltd., Zhongshan, China). The substrate was sandblasted and preheated to 150 °C before LPBF. During the LPBF process, argon was continuously filled in the chamber to ensure that the oxygen content was less than 50 ppm. Various combinations of laser powers (180–300 W), scanning speeds (300–800 mm/s), hatching spaces (0.03–0.07 mm), and layer thicknesses (0.03 mm) were explored to obtain the optimal LPBF parameters in order to fabricate samples with the highest possible relative density. The chess scanning strategy was used to fabricate block samples (6 mm × 6 mm × 5 mm) and tensile bars.

The relative densities of LPBF specimens were measured using the Archimedes methods (in accordance with ASTM B962), and 4 samples were measured at a time. The microstructure was observed and analyzed by a scanning electron microscope (SEM, Gemini500). In metallographic preparation, the polished surface was etched with Keller's reagent (95% deionized water + 1.5% hydrochloric acid + 2.5% nitric acid + 1% HF) for 10–30 s. LPBF specimens were characterized using X-ray diffraction (XRD, Bruker, D8 ADVANCE A2) with a scanning range of 20–80°. Heat treatment at 180, 230, and 280 °C annealing temperatures was applied to the LPBFed specimens, respectively. The samples were heated in the furnace for 1 h and then kept in the corresponding annealing temperature for 6 h, followed by air cooling. The micro-Vickers hardness was tested using a microhardness tester (Taiming, HXD-1000TMC/LCD) with a load of 200 g and a dwell time of 20 s. The tensile properties were evaluated using a universal testing machine (Letry, PLD-5) with a displacement speed of 0.5 mm/min.

3. Results and Discussion

3.1. Densification Behavior

Figure 3 shows the relative density of nano-Si_3N_4/AlSi10Mg composites as the function of laser power (at laser scanning speed 500 mm/s and hatching space 0.05 mm), laser scanning speed (at laser power 200 W and hatching space 0.05 mm), and hatching space (at laser power 200 W and laser scanning speed 500 mm/s). It can be seen that as the content of Si_3N_4 nanoparticles increases, the relative density of the samples decreases and the optimum process parameter window shrinks greatly. Due to the strong agglomeration tendency caused by their high specific surface area, the nanoparticles may lead to an increase in the composite porosity. With the increase in the nanoparticle content, the fluidity of the powders becomes worse, which leads to more gas trapping between nanoparticles.

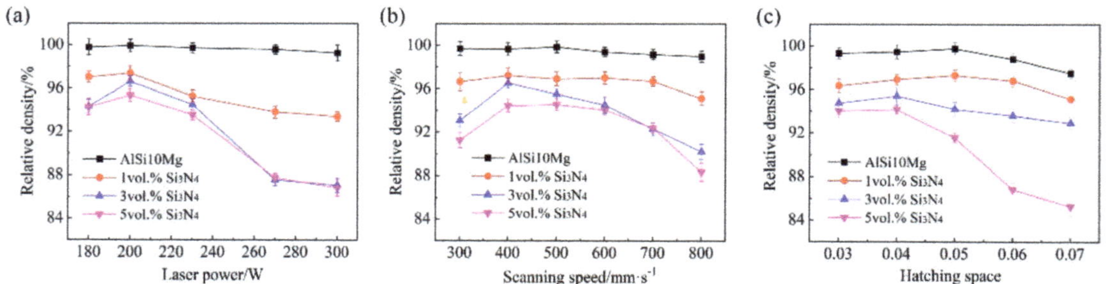

Figure 3. Influence of process parameters on the sample's relative density: (**a**) laser power, (**b**) scanning speed, and (**c**) hatching space.

For all the nano-Si_3N_4/AlSi10Mg composites with different nano-Si_3N_4 contents, laser power, laser scanning speed, as well as hatching space have optimum ranges corresponding to the peak relative density. Although the energy input during LPBF can be monotonically controlled by laser power, laser scanning speed, hatching space, and layer thickness, the relative density of the nano-Si_3N_4/AlSi10Mg composites does not simply increase or decrease with these LPBF parameters. If the laser power is too low, porosity and incomplete melting will lead to a low relative density. If the laser power is too high, keyhole, spheroidization, crack, and other defects can also result in a low relative density. Other LPBF parameters have a similar effect on relative density. Based on the LPBF parameter evaluation in this study, the optimum LPBF parameters of each nano-Si_3N_4/AlSi10Mg composite are shown in Table 2.

Table 2. The optimum LPBF parameters for the maximum relative densities of AlSi10Mg and nano-Si$_3$N$_4$/AlSi10Mg composites.

Samples	Laser Power/W	Scanning Speed/(mm·s^{-1})	Hatching Space/mm
AlSi10Mg	200	500	0.05
1 vol.% Si$_3$N$_4$	200	400	0.05
3 vol.% Si$_3$N$_4$	200	400	0.04
5 vol.% Si$_3$N$_4$	200	500	0.04

3.2. Microstructure

As shown in Figure 4, the microstructure of LPBFed nano-Si$_3$N$_4$/AlSi10Mg composites is mainly divided into three regions: (1) the inner part of the molten pool, which is a fine-cell-structure area with an average cell size of about 1 μm, (2) the boundary of the molten pool, which is a coarse-cell zone, and (3) the heat-affected zone. With the increase in the Si$_3$N$_4$ content, the cell structure becomes smaller. At the same time, the widths of its coarse-cell zone and heat-affected zone also decrease. Figure 4e shows that complete, undecomposed, Si$_3$N$_4$ particles still exist in the composite sample. From a theoretical point of view, the addition of Si$_3$N$_4$ particles affects the energy fluctuation and heat transfer in the molten pool and it also changes the heat transfer in the solidified region to a certain extent. Thus, it reduces the widths of the coarse-cell zone and the heat-affected zone. Furthermore, as Si$_3$N$_4$ particles are dispersed in the molten pool, the solid–liquid interface encounters and engulfs them during solidification at the boundary of the molten pool (see Figure 4e). Si$_3$N$_4$ particles hinder the movement of solid–liquid interface and eventually refine the cell structure. With the increase in the Si$_3$N$_4$ content, the distance between particles further decreases and the influence of refinement also increases.

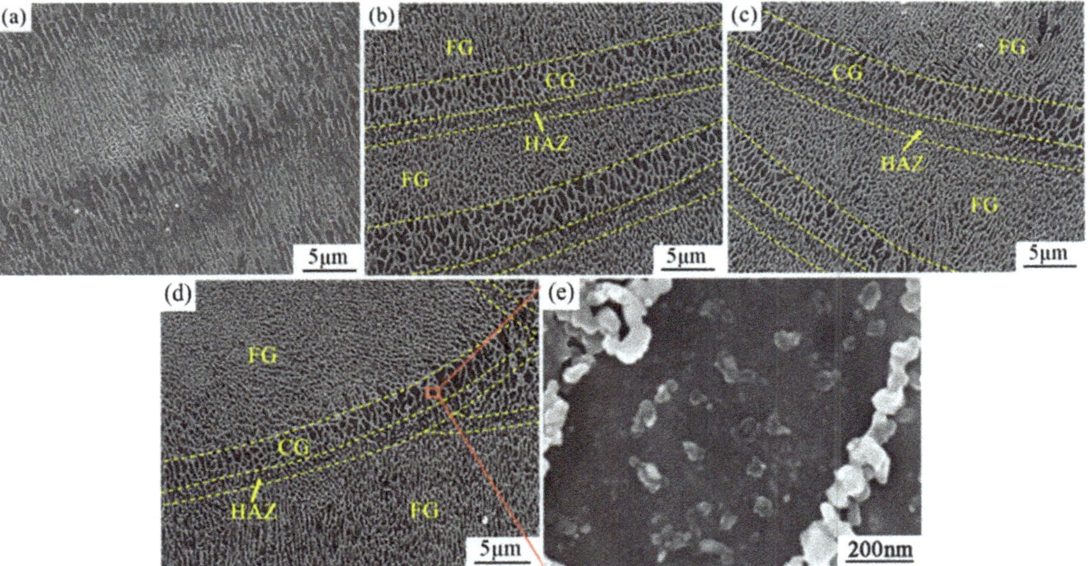

Figure 4. SEM images of nano-Si$_3$N$_4$/AlSi10Mg composites with different Si$_3$N$_4$ contents: (**a**) Pure AlSi10Mg, (**b**) 1 vol.% Si$_3$N$_4$, (**c**) 3 vol.% Si$_3$N$_4$, (**d**) 5 vol.% Si$_3$N$_4$, and (**e**) microstructure inside a single cell.

3.3. Mechanical Properties

The tensile stress–strain curves of AlSi10Mg and nano-Si$_3$N$_4$/AlSi10Mg composites are shown in Figure 5, and the tensile test results are shown in Figure 6. The alloys tested in

Figures 5 and 6 were prepared using the optimized process parameters in Table 2. For example, a 1 vol.% Si_3N_4 sample was manufactured under the conditions of laser power = 200 W, scanning speed = 400 mm·s^{-1}, and hatching speed = 0.05 mm. As compared with pure AlSi10Mg, 1 vol.% Si_3N_4 does not result in noticeable change in elastic modulus and elongation. When the Si_3N_4 content is increased to 3 vol.%, the strength and elastic modulus increase slightly, while the elongation decreases slightly. Compared with the test results of the nano-Si_3N_4/AlSi10Mg composite with 3 vol.% Si_3N_4, the strength and elastic modulus of the 5 vol.% Si_3N_4 composites increased steadily, but the relative elongation decreased. In Figure 7, we used the same batch of powder as the previous experiment. This figure shows the tensile fractograph of the composite samples. The fractographs of the composites with different Si_3N_4 contents are similar to that of AlSi10Mg. In addition, there are a few small tearing edges and cleavage steps, which are typical features of ductile and brittle fracture, respectively. Figure 8 shows the XRD patterns of AlSi10Mg and nano-Si_3N_4/AlSi10Mg samples. Because the Si_3N_4 content is low, the diffraction peak of the Si_3N_4 phase is not detected in XRD patterns. The peak intensity and angle of Al and Si phase diffraction peaks are not influenced by Si_3N_4 particles.

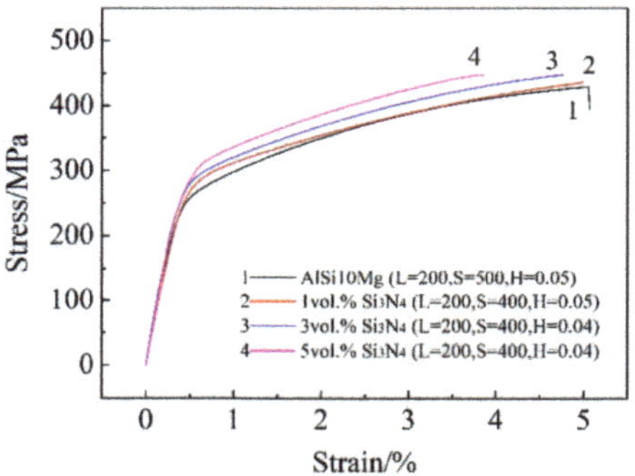

Figure 5. Tensile stress–strain curves of pure AlSi10Mg and nano-Si_3N_4/AlSi10Mg composites (L represents the laser power/W, S is the scanning speed/(mm·s^{-1}), and H is the hatching space/mm).

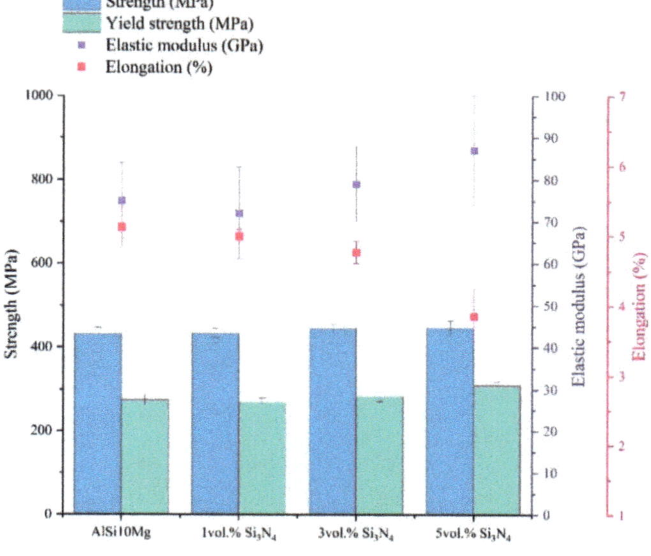

Figure 6. Mechanical properties of AlSi10Mg and nano-Si_3N_4/AlSi10Mg composites.

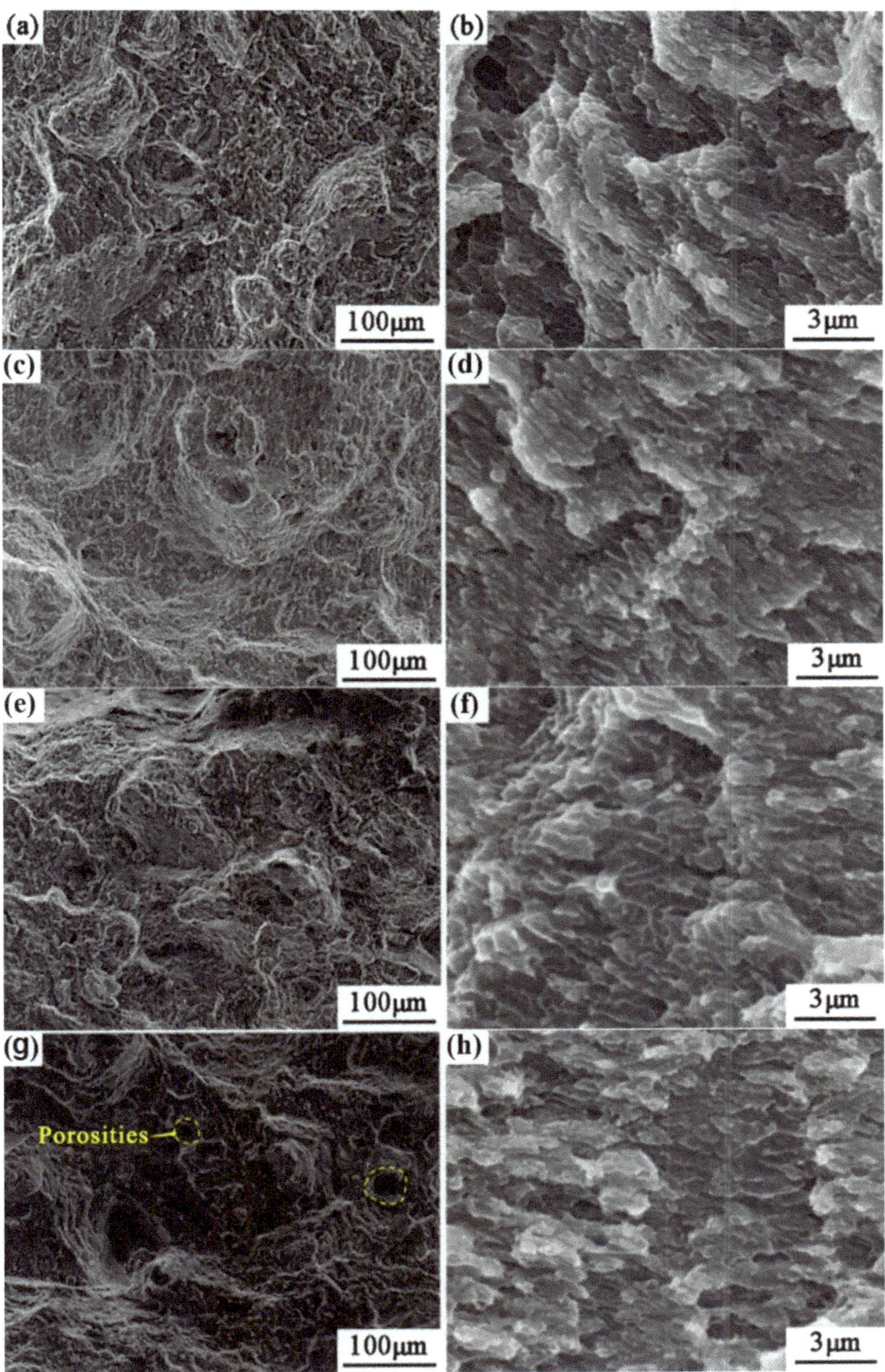

Figure 7. Tensile fractograph of pure AlSi10Mg and nano-Si$_3$N$_4$/AlSi10Mg composites with different Si$_3$N$_4$ contents, (**a**,**b**) pure AlSi10Mg, (**c**,**d**) 1 vol.% Si$_3$N$_4$, (**e**,**f**) 3 vol.% Si$_3$N$_4$, and (**g**,**h**) 5 vol.% Si$_3$N$_4$.

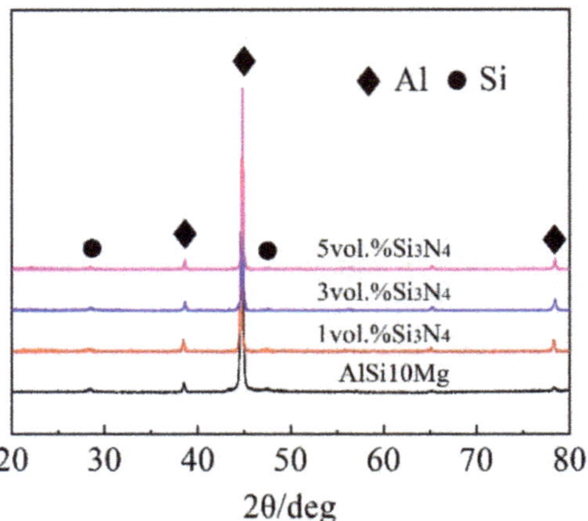

Figure 8. XRD diffraction patterns of AlSi10Mg and nano-Si_3N_4/AlSi10Mg composites.

As for the composite's strengthening mechanism, the second phase strengthening and the fine grain strengthening are included. On the one hand, the hard/brittle Si_3N_4 phase uniformly dispersed in the matrix and blocked dislocation movement during deformation, increasing strength and reducing elongation. That is to say, the second-phase strengthening was the main strengthening factor. Moreover, with the increase in the Si_3N_4 content, the more obvious the increase in strength, the more obvious the decrease in elongation. On the other hand, as shown in Figure 4, the grain size decreased to a certain extent after the addition of Si_3N_4 and fine grain strengthening could also improve the strength of the material. In addition, the porosities of different alloys were tested (Table 3), which indicated that the high Si_3N_4 content corresponds to the high porosity. The defects caused the alloy mechanical properties to deteriorate. Considering all the above factors, the strength of the alloy did not increase significantly. Therefore, the nano-Si_3N_4/AlSi10Mg composite with a 5 vol.% Si_3N_4 content shows the highest tensile strength and yield strength.

Table 3. The relative density of Si_3N_4/AlSi10Mg samples.

Samples	1 vol.% Si_3N_4	3 vol.% Si_3N_4	5 vol.% Si_3N_4
Relative density	97.94 ± 0.31	97.68 ± 0.22	96.36 ± 0.65

3.4. Heating Treatment

The 3 vol.% nano-Si_3N_4/AlSi10Mg composite was used for heat treatment tests because it had a high relative density (>97%) and better mechanical properties compared with the other nano-Si_3N_4/AlSi10Mg composites. Figure 9 demonstrates the XRD patterns of 3 vol.% nano-Si_3N_4/AlSi10Mg composite samples upon heat treatment under different annealing temperatures. The microstructure of a 3 vol.% nano-Si_3N_4/AlSi10Mg composite after annealing is shown in Figure 10. Annealing at 180 or 230 °C does not change the microstructure noticeably, while annealing at 280 °C seriously decomposes the eutectic structure and spheroidizes the Si phase because of the significant diffusion of the Si phase. When the annealing temperature is low, the diffusion ability of Si is not strong enough to change the eutectic structure. In fact, the meshed eutectic structure distributed along the cell boundary is not uniform and the eutectic Si phase at these locations is more likely to diffuse at a higher annealing temperature, resulting in the decomposition of the eutectic structure. In addition, Si in the solid solution begins to precipitate during annealing, contributing to the Si phase aggregation and granulation.

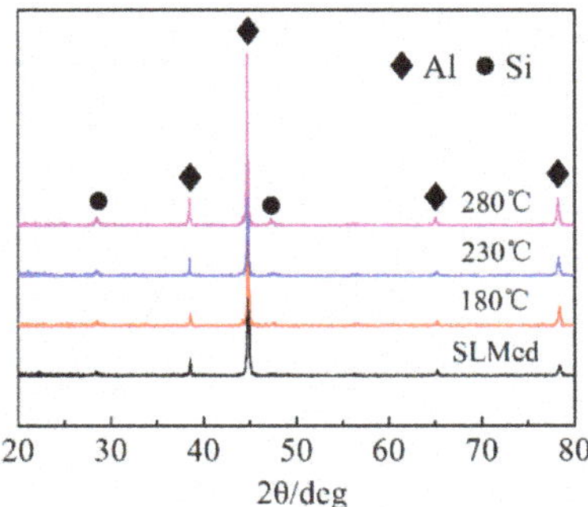

Figure 9. XRD patterns of nano-Si$_3$N$_4$/ AlSi10Mg composites after annealing.

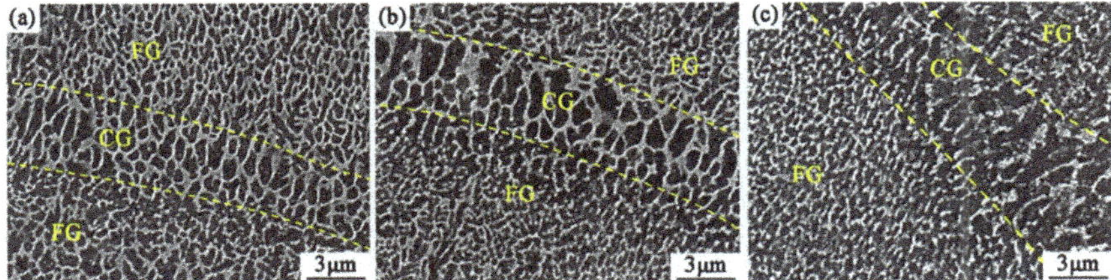

Figure 10. The microstructure of a 3 vol.% nano-Si$_3$N$_4$/AlSi10Mg composite after annealing at (**a**) 180 °C, (**b**) 230 °C, and (**c**) 280 °C.

Figure 11 shows the tensile test results of an LPBFed 3 vol.% nano-Si$_3$N$_4$/AlSi10Mg composite at different annealing temperatures. To ensure the reliability of the data, twelve 3 vol.% Si$_3$N$_4$/AlSi10Mg samples were divided into four groups. One group was taken as the control group, and the rest were heat-treated at different temperatures. As the relevant changes in the mechanical properties of the sample could be clearly seen from the original experimental data, Figure 11 was drawn to show these changes: with the increase in the annealing temperature, the strength of the samples decreases while the elongation increases. The tensile fractograph is shown in Figure 12. We can see that, as the annealing temperature increases, the size and depth of the dimple increase, indicating the ductility of the composite increases. At the same time, the microhardness after annealing decreases slightly. The microhardness values of the composite after annealing at 180, 230, and 280 °C are 130 ± 1 HV, 113 ± 2 HV, and 108 ± 2 HV, respectively.

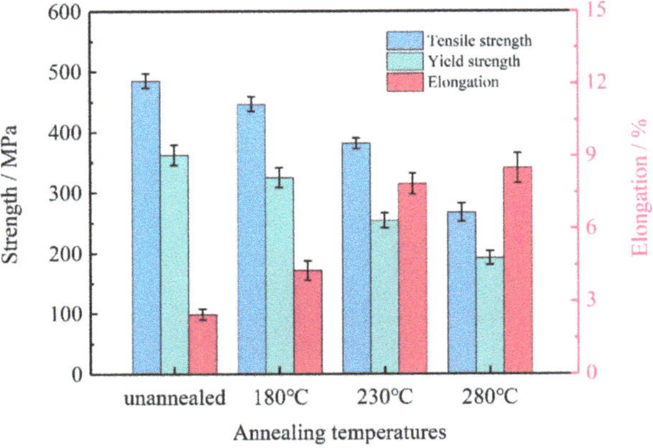

Figure 11. Effect of annealing temperature on tensile properties of a 3 vol.% nano-Si$_3$N$_4$/AlSi10Mg composite.

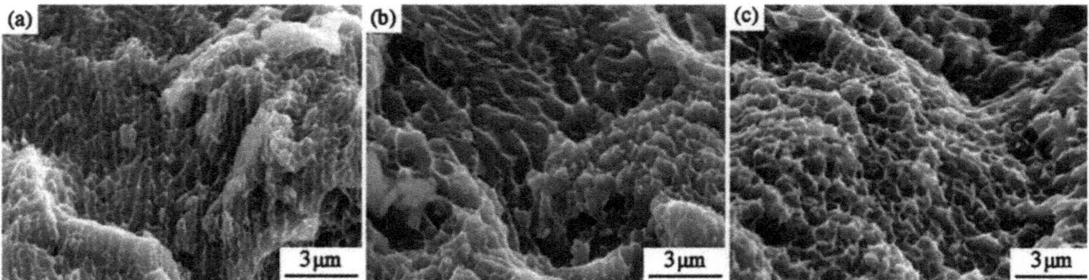

Figure 12. The tensile fractograph of a 3 vol.% nano-Si$_3$N$_4$/AlSi10Mg composite after annealing at (**a**) 180 °C, (**b**) 230 °C, and (**c**) 280 °C.

4. Conclusions

In this study, laser powder bed fusion (LPBF) was used to fabricate nano-Si$_3$N$_4$/AlSi10Mg composites with different Si$_3$N$_4$ contents. Firstly, the LPBF process parameters were optimized to achieve a higher density. Then, the effects of nano-Si$_3$N$_4$ content and heat treatment on the microstructure and mechanical properties of LPBFed composites were investigated. The main conclusions can be summarized as follows:

(1) The relative density of nano-Si$_3$N$_4$/AlSi10Mg composites decreases with the increase in the Si$_3$N$_4$ content. LPBF parameters have been optimized to achieve a high relative density (above 97%) for 1–3 vol.% nano-Si$_3$N$_4$/AlSi10Mg composites.

(2) Nano-Si$_3$N$_4$ particles noticeably refine the microstructure of nano-Si$_3$N$_4$/AlSi10Mg composites and improve the tensile strength as per the Orowan strengthening mechanism. The tensile strength, yield strength, and elongation of 5 vol.% nano-Si$_3$N$_4$/AlSi10Mg reach 448 ± 10 MPa, 311 ± 12 MPa, and 3.85 ± 0.16%, respectively.

(3) After annealing, the strength of the 3 vol.% nano-Si$_3$N$_4$/AlSi10Mg composite decreases while the elongation greatly increases. Furthermore, annealing decreases the microhardness of the 3 vol.% nano-Si$_3$N$_4$/AlSi10Mg composite and a higher annealing temperature results in lower microhardness.

Author Contributions: Conceptualization, Z.L. and Y.H.; methodology, Z.L. and Y.H.; software, F.C. and H.Z.; validation, Z.L., Y.H. and Y.G.; formal analysis, Y.G.; investigation, F.C.; resources, Z.L., X.D. and K.M.; data curation, H.Z.; writing—original draft preparation, Z.L., Y.H. and Y.G.; writing—review and editing, Y.H., F.C. and H.Z.; supervision, D.L.; project administration, Z.L. All authors have read and agreed to the published version of the manuscript.

Funding: This research was funded by the Science and Technology Projects of Guangdong Province, China (Grant No. 2017B090911006) and China Postdoctoral Science Foundation (Grant No. 2019M663684).

Institutional Review Board Statement: Not applicable.

Informed Consent Statement: Not applicable.

Data Availability Statement: Not applicable.

Conflicts of Interest: The authors declare no conflict of interest. Zhongliang Lu is a professor at Xi'an Jiaotong University. His research focus is additive manufacturing of metals and ceramics.

References

1. Thijs, L.; Kempen, K.; Kruth, J.-P.; Van Humbeeck, J. Fine-structured aluminium products with controllable texture by selective laser melting of pre-alloyed AlSi10Mg powders. *Acta Mater.* **2013**, *61*, 1809–1819. [CrossRef]
2. Schanz, J.; Hofele, M.; Ruck, S.; Schubert, T.; Hitzler, L.; Schneider, G.; Merkel, M.; Riegel, H. Metallurgical investigations of laser remelted additively manufactured AlSi10Mg parts: Metallurgische Untersuchungen von laserstrahlumgeschmolzenen additiv hergestellten AlSi10Mg Bauteilen. *Mater. Und Werkst.* **2017**, *48*, 463–476. [CrossRef]
3. Trevisan, F.; Calignano, F.; Lorusso, M.; Pakkanen, J.; Aversa, A.; Ambrosio, E.P.; Lombardi, M.; Fino, P.; Manfredi, D. On the selective laser melting (SLM) of the AlSi10Mg alloy: Process, microstructure, and mechanical properties. *Materials* **2017**, *10*, 76. [CrossRef] [PubMed]

4.   Herzog, D.; Seyda, V.; Wycisk, E.; Emmelmann, C. Additive manufacturing of metals. *Acta Mater.* **2016**, *117*, 371–392. [CrossRef]

5.   Emelogu, A.; Marufuzzaman, M.; Thompson, S.M.; Shamsaei, N.; Bian, L. Additive manufacturing of biomedical implants: A feasibility assessment via supply-chain cost analysis. *Addit. Manuf.* **2016**, *11*, 97–113. [CrossRef]

6.   Riza, S.H.; Masood, S.H.; Rashid, R.; Chandra, S. Selective laser sintering in biomedical manufacturing. *Met. Biomater. Process. Med. Device Manuf.* **2020**, *1*, 193–233.

7.   Li, X.P.; Ji, G.; Chen, Z.; Addad, A.; Wu, Y.; Wang, H.; Vleugels, J.; Van Humbeeck, J.; Kruth, J.-P. Selective laser melting of nano-TiB2 decorated AlSi10Mg alloy with high fracture strength and ductility. *Acta Mater.* **2017**, *129*, 183–193. [CrossRef]

8.   Zhou, L.; Mehta, A.; Schulz, E.; McWilliams, B.; Cho, K.; Sohn, Y. Microstructure, precipitates and hardness of selectively laser melted AlSi10Mg alloy before and after heat treatment. *Mater. Charact.* **2018**, *143*, 5–17. [CrossRef]

9.   Qi, Y.; Hu, Z.; Zhang, H.; Nie, X.; Zhang, C.; Zhu, H. High strength Al–Li alloy development for laser powders bed fusion. *Addit. Manuf.* **2021**, *47*, 102249. [CrossRef]

10. Ashkenazi, D.; Inberg, A.; Shacham-Diamand, Y.; Stern, A. Gold, Silver, and Electrum Electroless Plating on Additively Manufactured Laser Powder-Bed Fusion AlSi10Mg Parts: A Review. *Coatings* **2021**, *11*, 422–450. [CrossRef]

11. Schneller, W.; Leitner, M.; Springer, S.; Grün, F.; Taschauer, M. Effect of HIP treatment on microstructure and fatigue strength of selectively laser melted AlSi10Mg. *J. Manuf. Mater. Process.* **2019**, *3*, 16. [CrossRef]

12. Tradowsky, U.; White, J.; Ward, R.; Read, N.; Reimers, W.; Attallah, M. Selective laser melting of AlSi10Mg: Influence of post-processing on the microstructural and tensile properties development. *Mater. Des.* **2016**, *105*, 212–222. [CrossRef]

13. Wu, J.; Wang, X.; Wang, W.; Attallah, M.; Loretto, M. Microstructure and strength of selectively laser melted AlSi10Mg. *Acta Mater.* **2016**, *117*, 311–320. [CrossRef]

14. Liu, B.; Li, B.-Q.; Li, Z. Selective laser remelting of an additive layer manufacturing process on AlSi10Mg. *Results Phys.* **2019**, *12*, 982–988. [CrossRef]

15. Wang, L.-Z.; Wang, S.; Wu, J.-J. Experimental investigation on densification behavior and surface roughness of AlSi10Mg powders produced by selective laser melting. *Opt. Laser Technol.* **2017**, *96*, 88–96. [CrossRef]

16. Van Cauwenbergh, P.; Samaee, V.; Thijs, L.; Nejezchlebová, J.; Sedlák, P.; Iveković, A.; Schryvers, D.; Van Hooreweder, B.; Vanmeensel, K. Unravelling the multi-scale structure–property relationship of laser powder bed fusion processed and heat-treated AlSi10Mg. *Sci. Rep.* **2021**, *11*, 6423–6438. [CrossRef] [PubMed]

17. Zhou, Y.; Ning, F.; Zhang, P.; Sharma, A. Geometrical, microstructural, and mechanical properties of curved-surface AlSi10Mg parts fabricated by powder bed fusion additive manufacturing. *Mater. Des.* **2021**, *198*, 109360–109373. [CrossRef]

18. Dai, D.; Gu, D. Influence of thermodynamics within molten pool on migration and distribution state of reinforcement during selective laser melting of AlN/AlSi10Mg composites. *Int. J. Mach. Tools Manuf.* **2016**, *100*, 14–24. [CrossRef]

19. Dadbakhsh, S.; Mertens, R.; Hao, L.; Van Humbeeck, J.; Kruth, J.P. Selective laser melting to manufacture "in situ" metal matrix composites: A review. *Adv. Eng. Mater.* **2019**, *21*, 1801244. [CrossRef]

20. Gu, D.; Wang, H.; Dai, D.; Yuan, P.; Meiners, W.; Poprawe, R. Rapid fabrication of Al-based bulk-form nanocomposites with novel reinforcement and enhanced performance by selective laser melting. *Scr. Mater.* **2015**, *96*, 25–28. [CrossRef]

21. Wang, Y.; Shi, J. Effect of hot isostatic pressing on nanoparticles reinforced AlSi10Mg produced by selective laser melting. *Mater. Sci. Eng. A* **2020**, *788*, 139570. [CrossRef]

22. Jiang, L.; Liu, T.; Zhang, C.; Zhang, K.; Li, M.; Ma, T.; Liao, W. Preparation and mechanical properties of CNTs-AlSi10Mg composite fabricated via selective laser melting. *Mater. Sci. Eng. A* **2018**, *734*, 171–177. [CrossRef]

23. Gao, C.; Wang, Z.; Xiao, Z.; You, D.; Wong, K.; Akbarzadeh, A. Selective laser melting of TiN nanoparticle-reinforced AlSi10Mg composite: Microstructural, interfacial, and mechanical properties. *J. Mater. Process. Technol.* **2020**, *281*, 116618. [CrossRef]

24. Chang, F.; Gu, D.; Dai, D.; Yuan, P. Selective laser melting of in-situ Al4SiC4+ SiC hybrid reinforced Al matrix composites: Influence of starting SiC particle size. *Surf. Coat. Technol.* **2015**, *272*, 15–24. [CrossRef]

25. Tan, Q.; Zhang, J.; Mo, N.; Fan, Z.; Yin, Y.; Bermingham, M.; Liu, Y.; Huang, H.; Zhang, M.-X. A novel method to 3D-print fine-grained AlSi10Mg alloy with isotropic properties via inoculation with LaB6 nanoparticles. *Addit. Manuf.* **2020**, *32*, 101034. [CrossRef]

26. Mattli, M.R.; Matli, P.R.; Shakoor, A.; Amer Mohamed, A.M. Structural and mechanical properties of amorphous Si3N4 nanoparticles reinforced Al matrix composites prepared by microwave sintering. *Ceramics* **2019**, *2*, 126–134. [CrossRef]

27. Zhang, C.; Yao, D.; Yin, J.; Zuo, K.; Xia, Y.; Liang, H.; Zeng, Y.-P. Effects of β-Si3N4 whiskers addition on mechanical properties and tribological behaviors of Al matrix composites. *Wear* **2019**, *430*, 145–156. [CrossRef]

Article

Experimental Analysis of Wax Micro-Droplet 3D Printing Based on a High-Voltage Electric Field-Driven Jet Deposition Technology

Yanpu Chao [1,*], Hao Yi [2,3,*], Fulai Cao [1], Yaohui Li [1], Hui Cen [1] and Shuai Lu [1]

[1] Department of Mechanical Engineering, Xuchang University, Xuchang 461002, China; 13271247588@163.com (F.C.); lyh@xcu.edu.cn (Y.L.); hui.cen@foxmail.com (H.C.); zdlu996@126.com (S.L.)
[2] State Key Laboratory of High Performance Complex Manufacturing, Central South University, Changsha 410083, China
[3] State Key Laboratory of Mechanical Transmission, Chongqing University, Chongqing 400044, China
* Correspondence: chaoyanpu@163.com (Y.C.); haoyi@cqu.edu.cn (H.Y.)

Abstract: High-voltage electric field-driven jet deposition technology is a novel high resolution micro scale 3D printing method. In this paper, a novel micro 3D printing method is proposed to fabricate wax micro-structures. The mechanism of the Taylor cone generation and droplet eject deposition was analyzed, and a high-voltage electric field-driven jet printing experimental system was developed based on the principle of forming. The effects of process parameters, such as pulse voltages, gas pressures, pulse width, pulse frequency, and movement velocity, on wax printing were investigated. The experimental results show that the increasing of pulse width and duration of pulse high voltage increased at the same pulse frequency, resulting in the micro-droplet diameter being increased. The deposited droplet underwent a process of spreading, shrinking, and solidifying. The local remelting and bonding were acquired between the contact surfaces of the adjacent deposited droplets. According to the experiment results, a horizontal line and a vertical micro-column were fabricated by adjusting the process parameters; their size deviation was controlled within 2%. This research shows that it is feasible to fabricate the micro-scale wax structure using high-voltage electric field-driven jet deposition technology.

Keywords: micro-droplet; high-voltage electric field-driven jet; Taylor cone; micro-scale wax structure

Citation: Chao, Y.; Yi, H.; Cao, F.; Li, Y.; Cen, H.; Lu, S. Experimental Analysis of Wax Micro-Droplet 3D Printing Based on a High-Voltage Electric Field-Driven Jet Deposition Technology. *Crystals* **2022**, *12*, 277. https://doi.org/10.3390/cryst12020277

Academic Editor: Indrajit Charit

Received: 28 January 2022
Accepted: 15 February 2022
Published: 17 February 2022

Publisher's Note: MDPI stays neutral with regard to jurisdictional claims in published maps and institutional affiliations.

1. Introduction

Micro/nano 3D printing technology is a new micro/nano manufacturing method based on the principle of additive manufacturing. It can directly print and form functional products with a micro/nano characteristic structure [1]. Compared with traditional LIGA (Lithographie, Galvanoformung and Abformung) technology [2], nano-lithography [3], micro/nano embossing [4], high-speed micro-milling, micro-electrical discharge machining (EDM) [5], and other micro/nano manufacturing technologies, micro/nano 3D printing technology has the advantages of being a simple manufacturing process, being low in cost, and of allowing for high utilization of materials, wide material application, and direct formation.

At present, the micro/nano 3D printing technologies developed mainly include: direct laser writing based on two-photon polymerization [6], micro-stereophotolithography, electrohydrodynamic jet (E-jet) printing [7], laser-induced forward transfer (LIFT) [8], focused electron beam (FEB) induced deposition [9], electrochemical deposition [10], focused-ion-beam direct writing (FIBDW) [11], and microplasma deposition [12–14]. The printing scales include micron scales, sub-micron scales, and nano scales. Micro/nano 3D printing technologies have been applied to aerospace, micro-electro-mechanical systems, biomedicine, intelligent materials, solar cells, flexible electronics, microfluidic chips, nano-catalysis, and other fields and industries, demonstrating favorable prospects for industrial application [15–18].

However, the above existing micro/nano 3D printing technologies are unable to print multiple materials simultaneously, especially as it is difficult to achieve macro/micro/nano scale 3D printing and printing of multi-material complex 3D structures. Compared with other 3D printing processes, the freeform fabrication with micro-droplet on-demand sprayed deposition has shown outstanding advantages and potential in multi-material and multi-scale structure manufacturing.

However, the traditional inkjet printing [19] process is limited by the material, the viscosity of material ink being low, and the resolution of the printed graph being generally more than 50 microns. The metal droplet on-demand jet printing process [20] can realize the ejecting and deposition forming of a variety of metal materials, but the size of the droplet is larger than the diameter of the nozzle, which does not allow for high printing accuracy and resolution [21].

Electrohydrodynamic jet (E-jet) printing [22] is a micro-droplet spray forming technology based on electro fluid dynamics (EHD), which was proposed and developed by Park and Rogers et al. in 2007 [23]. It has the advantages of good material compatibility, low cost, simple system structure, and high resolution. At present, global researchers have carried out in-depth research on electrohydrodynamic jet printing and direct writing and have made considerable breakthroughs in many aspects [24]. Huang et al. [25–27] applied the electrohydrodynamic direct writing technology to print high-resolution nanofibers for the production of stretchable flexible electronic devices on a specific material substrate. Zhang et al. [28–30] fabricated tissue engineering scaffolds by electrohydrodynamic jet printing. Han et al. [31] printed the columnar array structure with high depth ratio by utilizing the phase transformation properties of materials. Jiang et al. [32] developed an all electrospray printing process for perovskite solar cells (PSCs) in ambient air below 150 °C. García-Farrera et al. [33] used electrohydrodynamic deposition technology to print sub-100 nm ceramic piezoelectric films. Taylor et al. [34] reported a graphene oxide low-cost nano flake conductometric gas sensor, which was fabricated by atmospheric pressure electrospray printing.

Although electrohydrodynamic jet printing technology has high printing accuracy and resolution, the printing process needs to ensure that the distance between the nozzle and the substrate is limited to 3 mm. Therefore, the electrohydrodynamic jet printing technology still has many shortcomings and limitations in printing materials, nozzle materials, deposition substrates, forming height, printing stability, and other aspects, especially as it is difficult to realize the integrated printing of macro/microstructures [35].

In order to solve the existing problems of 3D printing in multi-material and multi-scale complex 3D structure manufacturing, Lan et al. proposed and established a novel micro/nano 3D printing technology using high-electric field-driven jet deposition (HEFD) [36]. Its preliminary exploration has been carried out in the application of transparent electrode, transparent electric heating, transparent electromagnetic shielding, vascular stent, flexible electronics, paper-based electronics, and many other fields [37–40]. High-voltage electric field-driven jet deposition technology is a kind of high-resolution micro/nano scale 3D printing, which is based on electrostatic induction and electrohydrodynamic jetting behavior. The molten viscoelastic liquid in the nozzle tip is stretched into a Taylor cone under an electric field force; the high-resolution micro/nano scale 3D printing can thus be achieved, which is an order of magnitude lower than the nozzle diameter. Two working modes: pulsed cone-jet mode and continuous cone-jet mode, have been adopted for implementing multi-scale 3D printing, which are suitable for various solutions and melted materials with ultrahigh viscosity, macro/microstructures, and multi-scale fabrication.

In recent years, the research on the fabrication process of microfluidic devices based on silicon polymers has attracted increasingly more attention from researchers in different fields. The process principle is mainly to fabricate micro-channel structures by using a 3D printing wax mold combined with PDMS micro-transfer technology [41]. The analytical accuracy of the microfluidic device is directly affected by the width and section the shape of microfluidic channel structure. Given that the traditional 3D printing wax mold is fabricated by the inkjet printer, the printing accuracy is limited by the nozzle, the printing resolution is generally close to the nozzle diameter, and the width of the micro-channel after printing is generally more than 200 μm. It is difficult to fabricate micro-channel structures that are tens of microns in width.

In this study, microcrystalline wax is used as experimental material. The working mechanism of high-voltage electric field-driven jet deposition technology was analyzed. The ejection and deposition process of the pulsed cone-jet mode was studied. Based on this, the densified wax line and the large ratio of height to diameter micro-cylinder were printed. This research proposes a new technical approach for a fabricated micro-scale wax part with tens of microns and has important application prospects in the fabrication of micro-channel structures using a 3D printing wax mold combined with PDMS micro-transfer technology.

2. Process Principle

High-voltage electric field-driven jet deposition technology is a novel high-resolution micro/nano scale 3D printing, which is based on electrostatic induction and electrohydrodynamic jetting behavior. Figure 1 shows a schematic diagram of the principle of on-demand high-voltage electric field-driven jet deposition.

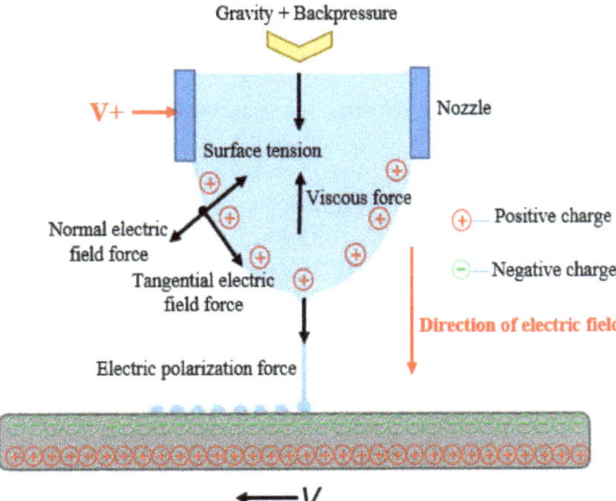

Figure 1. Schematic diagram of the principle of high-voltage electric field-driven jet deposition on demand.

The forming principle is as follows: the wax liquid at the bottom of the nozzle is formed into a meniscus shape under the action of gravity and gas backpressure, the conductive nozzle is connected to the positive pole of the high-voltage pulse power supply, and a high voltage is applied to the conductive nozzle to give it high potential. The wax liquid material inside the nozzle is polarized under the high voltage, and the positive charges are accumulated on meniscus shape liquid surface. When the high voltage nozzle is close to the deposition substrate, the electrostatic induction is generated between the nozzle and the substrate, resulting in the charge being migrated on the surface and inside the substrate. The charge is then redistributed on the substrate. The negative charge is distributed on the upper surface of the substrate, and the positive charge is distributed on the lower surface of the substrate. The interaction is generated between the positive charge on the meniscus shape liquid and the negative charge on the upper surface of the substrate, resulting in the electric field being formed between the meniscus liquid surface and the substrate. The meniscus liquid surface is stretched and deformed gradually into a Taylor cone under the action of electric field force, surface tension, viscosity force, and electric polarization force. As the high voltage increases, the tip charges continually accumulate. Once the electric field force exceeds the surface tension and viscosity force of the liquid, a very fine Taylor cone jet is generated. The diameter size of jet is usually 1~2 orders of magnitude lower than the nozzle diameter. Finally, under the action of liquid surface tension, the micro-jet is broken into individual droplets and deposited on the substrate according to the frequency of a high-voltage pulse.

3. Experiment System

Based on the abovementioned principles, an experimental system was developed herein. The experimental system mainly included a high voltage power supply (DMC-200, Dalian Dingtong Technology China, Dalian, China), an electrode ring, an air pump (BLT-5A/7A, Bolaite, Shanghai, China), an air pressure control unit (IR2000, King Kong Pneumatic Technology, Wenzhou, China), an XY-axis movement platform (G3TH764, KingWonda, Shenzhen, China), a Z-axis movement platform, a deposition substrate, an annular heating furnace (customized), a quartz crucible (customized), a print head (stainless steel precision musashi needle), a nozzle heating block, and a temperature controller (OHR-A300, HongYun instrument, Shunchang, China).

Figure 2 shows a schematic diagram of a high-voltage electric field-driven jet printing experimental system. The positive pole of the high-voltage power supply was connected with the electrode ring. The electrode ring was fixed on the outer surface of the print head. The high voltage power supply can realize two power supply modes: high-voltage pulse and high-voltage DC. The annular heating furnace was fixed on the Z-axis moving platform through a connecting frame, and the quartz crucible was installed inside the annular heating furnace. The printing material was placed inside the quartz crucible, and the temperature of the annular heating furnace could be controlled through the temperature controller to achieve the melting of the printing material. The nozzle heating block was installed on the outer surface of the printing head. The nozzle temperature could be precisely controlled to meet the requirement of the injection temperature of the printing material by the temperature controller. The control accuracy of the temperature is ±0.5 °C. The deposited substrate was installed on the XY-axis moving platform, and the droplet material was precisely deposited on the substrate by coordination controlling the movement of the XY-axis moving platform and controlling the ejection frequency of the printing material. The repeated positioning accuracy of the XY-axis moving platform is ±0.01 mm. After one layer was printed, the printing head moved upward by a thickness layer, and a stable electrostatic induced electric field was generated between the conductive nozzle and the printed layer; the next layer was then printed. The above process was repeated until the manufacturing of the entire structure was completed.

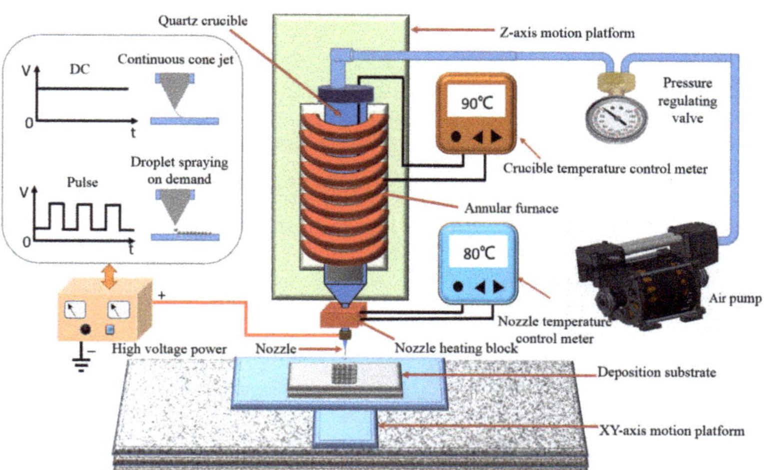

Figure 2. Schematic diagram of a high-voltage electric field-driven jet printing experimental system.

The injection and deposition process of micro-droplets is a high-speed nonlinear dynamic process using high-voltage electric field-driven jet deposition technology. In order to monitor the real dynamic process of Taylor cone ejection and droplet deposition in real time, a high-speed monitoring system was designed and developed, which can guide the experimental process. Figure 3 shows the working principle system. The system is mainly composed of a high-performance computer, high-speed image capture card (DH-CG410, Daheng, Beijing, China), high-speed CCD camera (Sp-5000M-CXP2, Daheng, Beijing, China), integrated microscope (RetroZoom65, Optem, Medina, OH, USA), stroboscope (DA/DB, Monado, St. Louis, MO, USA), synchronous trigger, and so on. The high-speed CCD uses ID's Os4 camera, with a resolution of 1024 × 1024 and a frame rate of up to 3000 fps, making it easy to capture images of high-speed moving objects. The image acquisition of the Silicon Software Company in Germany can transfer the collected images to the computer hard disk through PCI bus and save them for analysis and calculation. A stroboscope is used as the light source. In the experiment, the simultaneous exposure method of stroboscope and CCD camera was adopted to realize the shooting of the droplet ejecting and deposition instant image. The high-speed CCD camera and the stroboscope light source were simultaneously driven by the TTL signal provided by an external trigger to realize synchronous operation.

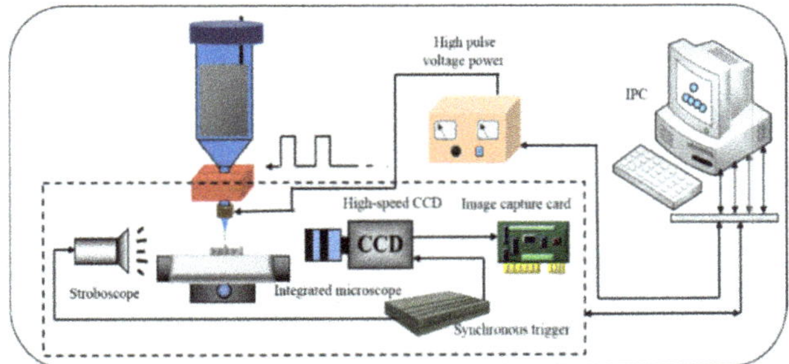

Figure 3. Schematic diagram of a high-speed monitoring system.

4. Experiment Results and Discussion

4.1. Arc Discharge

In order to verify the ability of accurately controlling the high-voltage electric field, an arc discharge experiment of the high-voltage electric field was carried out first in a 3D printing experiment system. A stainless-steel precision Musashi needle was selected for the nozzle, and no printing materials were in the nozzle. The conductive nozzle was connected with the positive pole of the high-voltage pulse power supply, and a high voltage was applied to the conductive nozzle to create a high potential. A copper clad plate was used for the deposition substrate without connecting the ground electrode. The frequency of the high voltage pulse was set as 1 Hz, and the voltage of the high voltage pulse was gradually changed to 1000 V, 1500 V, 2000 V, 2500 V, 3000 V, 3500 V, 4000 V and 5000 V. The arc discharge phenomenon was observed by the high-speed monitoring system.

In the experiment, when the nozzle gradually approached the deposition substrate, the electric field between the nozzle and the substrate was enhanced. Under the action of the discharge energy, the air medium between the poles was ionized and broken down, forming an arc discharge. Figure 4 shows the pictures of arc discharge obtained by a high-speed camera system under different pulse voltages. The measured lengths of discharge arc were 0 mm, 0.165 mm, 0.395 mm, 0.580 mm, 0.860 mm, 1.085 mm, 1.260 mm and 1.755 mm, respectively. The experimental results show that the phenomenon of arc discharge becomes more intense with the increase of pulse voltage. When the voltage was 1000 V, the phenomenon of arc discharge did not occur, indicating that the voltage was too low.

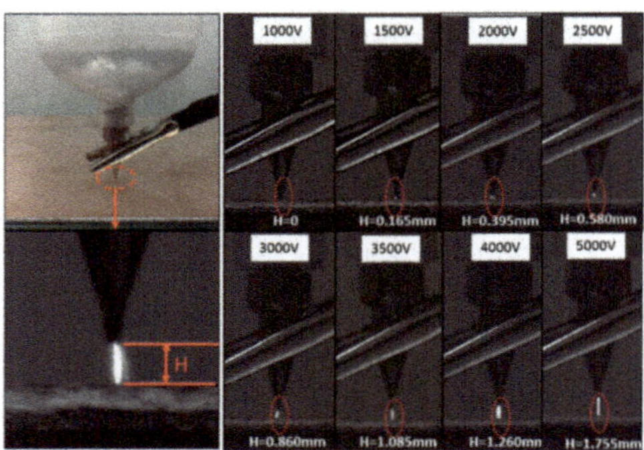

Figure 4. Arc discharge phenomenon under different pulse voltages.

4.2. Liquid Morphology at the Bottom of the Nozzle

In the process of micro-droplet ejecting and depositing driven by a high electric field, it is necessary to ensure that the liquid surface at the bottom of the nozzle is a meniscus shape. On this basis, a Taylor cone can be generated under the action of a high-voltage electric field, and micro-droplet on-demand ejecting can be achieved. Microcrystalline wax 80$^{\#}$ (CAS NO: 63231-60-7) was used as the printing material in the experiment, which can be obtained by solvent fractionation of the still-bottom fraction of petroleum by suitable dewaxing or deoiling. Microcrystalline wax is generally composed of fine needle-like or granular crystals. The main components are cycloalkanes and aromatic hydrocarbons with larger molecular weight and longer carbon chains. Its molecular formula is $C_{11}H_{14}N_2$, and its molecular weight is 174.24226 g/mol. Its melting point is T_m is 80 °C, the glass transition temperature is T_g is 67.5 °C, the dynamic viscosity is 16.5 cP, and the surface tension is about 0.245 N·m^{-1} after being melted into a liquid state at 80 °C. Under the action of gravity, the wax liquid will naturally drop from the nozzle or form large wrapped droplets at the nozzle. Therefore, it is necessary to utilize a technical approach to make the wax liquid form a meniscus shape at the nozzle and maintain a stable meniscus shape. At present, there are two commonly used methods: (1) applying negative pressure to balance the wax liquid's own gravity and (2) coating the nozzle with a hydrophobic film. In this paper, the negative pressure balance gravity method is used to make the wax liquid form meniscus shape at the nozzle and maintain a stable meniscus shape.

Figure 5 shows the morphologies of wax liquid at the bottom of the nozzle under different gas pressures. When the negative pressure is set as 0 Pa, the bottom of the nozzle is completely wrapped by wax liquid, which affects the printing stability and accuracy as shown in Figure 5a. As shown in Figure 5b, when the negative pressure is set as 450 Pa, the larger wrapped droplet at the bottom of the nozzle essentially disappear; a stable meniscus shape liquid is generated. When the negative pressure increases to −750 Pa, the meniscus shape liquid disappears due to excessive negative pressure as shown in Figure 5c. The Taylor cone cannot be generated, and micro-droplet on-demand ejecting cannot be achieved under the action of a high voltage electric field. Therefore, the optimal meniscus shape liquid can be guaranteed under −450 Pa negative pressure, and too high or too low negative pressure is not conducive to the droplet spray forming.

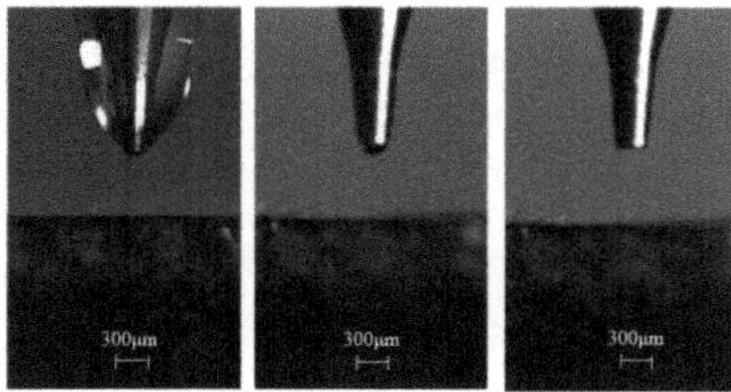

Figure 5. Morphology of wax liquid at the bottom of the nozzle under different gas pressures: (**a**) fully wrapped nozzle; (**b**) meniscus shape; (**c**) no meniscus shape.

4.3. Taylor Cone and Droplet Ejecting

Two working modes, pulsed cone-jet mode and continuous cone-jet mode, are adopted in high voltage electric field driven jets, according to different printing materials and printing characteristic structures. The pulsed cone-jet mode uses a high-voltage pulse as a driving signal, and the continuous cone-jet mode uses a high-voltage DC as a driving signal.

The high-voltage pulse is a digital signal, the pulse frequency, pulse width, and duty cycle can be adjusted in real time, which makes it more robust as there are numerous possibilities. When using a high-voltage pulse, the printing discretized into voxels. For low viscosity materials or fine structures, the pulsed cone-jet printing mode is used to ensure the accuracy of the printed micro-characteristic structures. The high-voltage DC is a continuous analog signal, where the electric field is continuous. The printed material is deposited in a continuous jet stream and will not be separated into droplets. For high-viscosity materials or macroscopic structures, the continuous cone-jet mode is adopted to achieve efficient printing under the premise of meeting the required accuracy.

In this paper, the pulse cone-jet mode was studied. Under the action of high-voltage pulse, the wax liquid material at the bottom of the conductive nozzle went through the following processes: from tensile deformation to Taylor cone to jet retraction and finally to fracture into a single droplet. The diameter of the droplet produced by the Taylor cone jet is much smaller than the diameter of the nozzle, this is achieved using precise control of the pulse width and the pulse frequency.

The experimental process parameters are presented in Table 1. The microcrystalline wax was selected as the printing material, and the stainless-steel precision Musashi needle with inner diameter 200 μm and outer diameter 350 μm was selected for the nozzle. The negative pressure in the crucible was set as 450 Pa, the temperature of the crucible T_c was set as 90 °C, and the temperature of the nozzle T_n was set as 80 °C. The Pulse frequency f was set as 1 Hz, the pulse width W_t was set as 10 ms, and the applied pulse voltage U was set as 1500 V. The experiment process was observed with a high-speed camera. The arc discharge phenomenon was observed by the high-speed monitoring system.

Table 1. Process parameters of Taylor cone and droplet ejecting.

Printing Material:	Microcrystalline Wax
Back pressures of crucible: P (Pa)	−450
Diameter of the nozzle: D (μm)	200
Temperature of the crucible: T_c (°C)	90
Temperature of the nozzle: T_n (°C)	80
Pulse frequency: f (Hz)	1
Pulse width: W_t (ms)	10
Pulse voltage: U (V)	1500

Figure 6 shows the whole process from Taylor cone jet generation to droplet deposition in the high-voltage electric field-driven jet printing experiment photographed by a high-speed camera system. At 0 ms, because the high-voltage pulsed electric field was not applied, the liquid meniscus at the bottom of the nozzle was not particularly obvious, and the height of the liquid meniscus was lower. At 5 ms, the meniscus liquid at the bottom of the nozzle began to be stretched and deformed, the height of the meniscus liquid was increased, and the meniscus liquid appeared as a cone under the action of a high-voltage electric field. At 6–7 ms, the meniscus liquid at the bottom of the nozzle was further stretched into a Taylor cone jet. At 8–9 ms, the Taylor cone jet at the bottom of the nozzle began to retract and fracture into a single droplet deposited on the substrate. At 11 ms, the Taylor cone jet retraction fracture was completed, and the meniscus liquid surface at the bottom of the nozzle was restored to the minimum height. The single droplet electric field driven injecting was completed and prepared for the next injecting. The morphology of the micro-droplet deposited on the substrate was observed; the diameter of the single micro-droplet after deposition and solidification was about 42 μm.

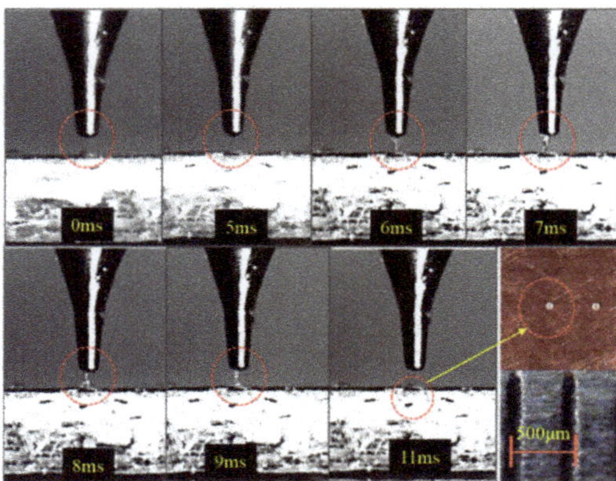

Figure 6. Image of Taylor cone generation and droplet eject deposition process.

4.4. Pulse Width

By adjusting the frequency and pulse width of the high-voltage pulse signal, the desired duration and interval time of the high-voltage pulse can be obtained, and then the droplet size can be controlled. In the experiment, the duration of the high-voltage pulse was changed by adjusting the pulse width when the pulse frequency remained constant, and the effect of the process parameters on the forming size of the droplet was investigated.

The microcrystalline wax was selected as printing material, and the copper clad plate was used as a deposition substrate. The inner diameter of the nozzle was 150 μm, and the outer diameter of the nozzle was 300 μm. The same experimental process parameters, as shown in Table 1, were employed except for the pulse width W_t. Figure 7 shows the experimental results of droplets deposition with different pulse widths. The pulse width was set and adjusted by a precision function generator (AFG2000-SC, TektroniX, Shanghai, China), the parameter error was ±0.1 ms. The pulse widths W_t is set as 10 ms, 20 ms, 30 ms, 50 ms, 100 ms, 300 ms and 500 ms in the experiment. The diameter of deposited droplets under different pulse widths were measured by Image Pro software, with the measurement error set as ±0.01 μm. Each measurement droplet was measured three times and the average values were calculated as shown in Figure 7. The average diameters of the measured droplets were 37.69 μm, 43.98 μm, 48.75 μm, 56.54 μm, 60.58 μm, 69.11 μm and 75.39 μm, respectively. The droplet diameter versus pulse width is an approximate linear relationship in different ranges. Figure 8 shows the linear relationship plot.

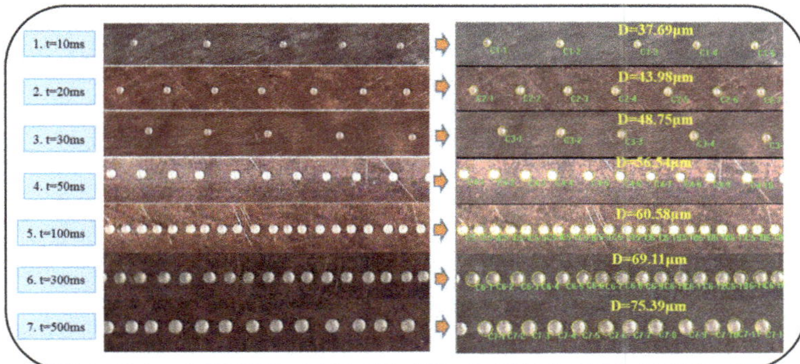

Figure 7. The experimental results of the droplet deposition at different pulse widths.

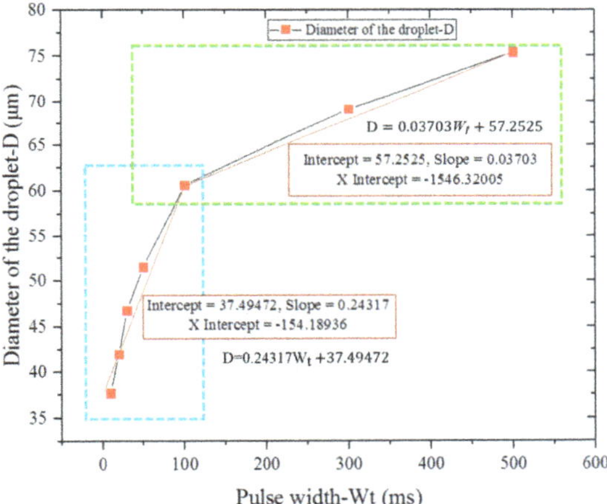

Figure 8. The approximate linear relation of droplet diameter versus pulse width in different ranges.

Theoretically, a complete droplet jetting process needs a high voltage to last for a certain period of time, which mainly includes electric charge accumulation release and deformation fracture time of the Taylor cone jet. If the duration of the high-voltage pulse is too short, the deformation fracture of the Taylor cone jet cannot be completed, and the droplet jet cannot be realized. With the increasing of pulse width W_t, the duration of the pulse high voltage was increased at the same pulse frequency, and the amount of liquid ejected from the Taylor cone was also aggrandized, resulting in the micro-droplet diameter being increased.

4.5. Horizontal Direction Deposition

In the experiment, the movement velocity of the deposition platform V_p was changed to observe its influence on the deposition distance of droplets. The experimental process parameters are shown in Table 2. Different movement velocities of the deposition platform V_P were set (the values are listed in Table 2). Figure 9 shows the images of the droplet printing and deposition process at different velocities in a horizontal direction obtained by a high-speed camera system. The morphology of the deposited lines at each movement velocity was captured using a macro lens (shown in Figure 10).

Table 2. Process parameters of droplet deposited lines.

Printing Material		Printing Substrate		Diameter of the Nozzle: D (µm)		Pulse Frequency: f (Hz)	
Microcrystalline wax		PET (Polyethylene terephthalate)		200		1	
Pulse width: W_t (ms)		Temperature of the crucible: T_c (°C)		Temperature of the nozzle: T_n (°C)		Pulse voltage: U (V)	
10		90		80		1500	
Movement velocity of the deposition platform: V_p (Plus/s)							
1	2	3	4	5	6	7	8
F100	F90	F70	F50	F30	F10	F5	F2

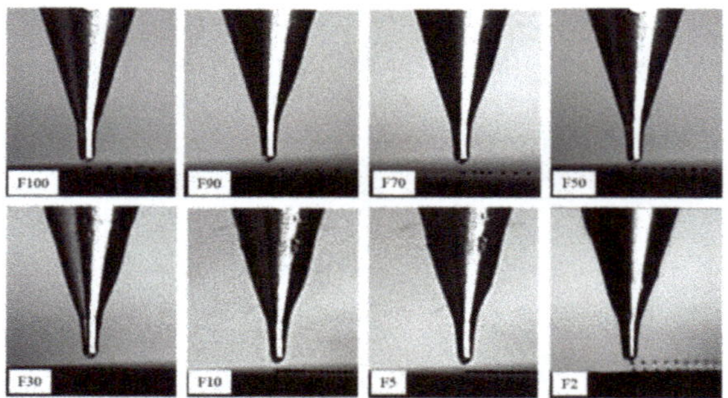

Figure 9. Image of droplets deposition process at different velocities in horizontal direction.

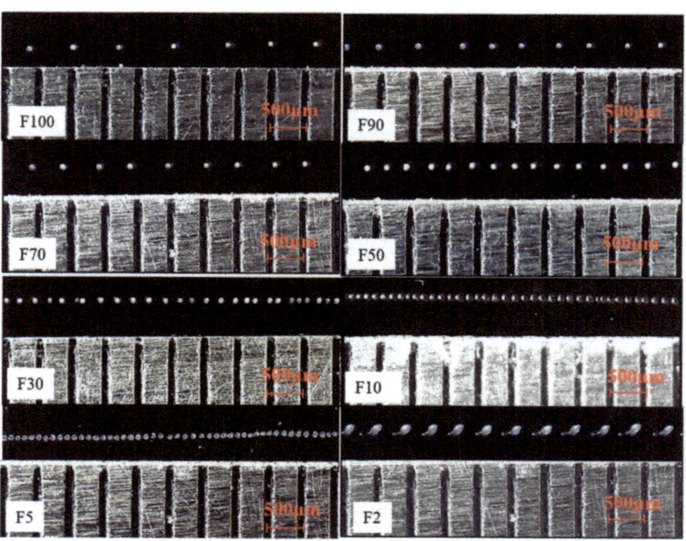

Figure 10. The morphology of the deposited lines at different movement velocities.

The movement velocity of the deposition platform was set and adjusted by a precision motion controller (y, GUC, GooGoLTECH, Shenzhen, China), with a parameter error of ± 1 Plus/s. The movement velocities of the deposition platform V_P were set as F100, F90, F70, F50, F30, F10, F5 and F2 in the experiment. The spacing of the deposited droplets under different movement velocities were measured by Image Pro software, and the measurement error was set as ± 1 µm. Each measurement object was measured three times and the average values are calculated. The average measured spacing of the deposited droplets S were 668 µm, 580 µm, 475 µm, 332 µm, 234 µm, 124 µm, 85 µm and 34 µm, respectively. The droplet spacing versus speed of the deposition platform is an approximate linear relationship; Figure 11 shows the linear relationship plot.

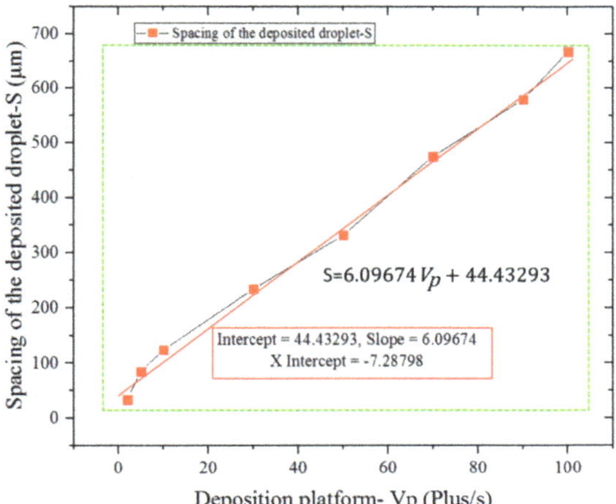

Figure 11. The approximate linear relation of droplet spacing versus speed of the deposition platform.

It can be seen from the experimental results that, with the decreasing of the movement velocities (V_p), the distance between droplets after deposition was also gradually decreased. When the movement velocitiy, V_p, was set as F5, the deposited droplets form a continuous line. However, there were still small gaps between adjacent droplets, the main reason being that the average droplet spacing of 85 µm was larger than the average droplet diameter of 72 µm after deposition. When the movement velocity, V_p, was set as F2, the deposited droplets formed an array of uniformly spaced micro-cylinders, the main reason being that the average droplet spacing of 34 µm was less than the average droplet diameter of 72 µm after deposition. The overlap rate of two adjacent droplets was too high, which led to the deposited droplets growing upwards and being unable to form a continuous line. The diameter of the deposited droplets did not change very much under different movement velocity, V_p, the average diameter of the deposited droplets was determined to be 72 µm using software measurement. These experimental results indicate that the movement velocities (V_p) have almost no effect on the diameter of droplet after deposition.

4.6. Vertical Direction Deposition

Figure 12 shows the image of the droplets vertical deposition process in the vertical direction obtained by a high-speed camera system. In the experiment, the deposition substrate remained stationary, and the Z-axis movement velocity was set as F100. Other experimental process parameters are shown in Table 2. Figure 12a–d show the deposition process of the first, second, third, and fourth droplet, respectively. It can be seen that the jetting process of each wax droplet has undergone four processes: tensile deformation to Taylor cone to jet retraction then to fracture into a single droplet.

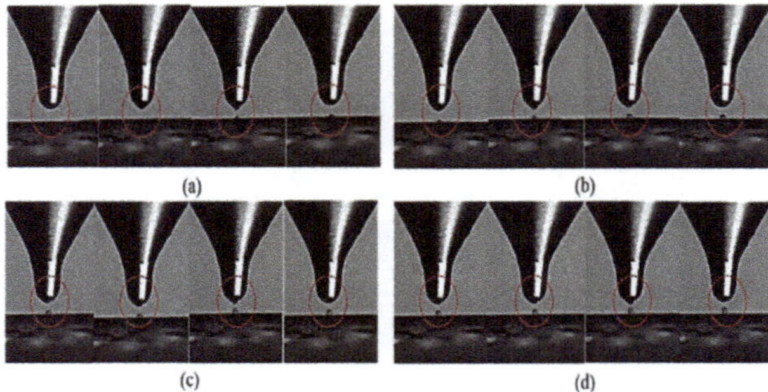

Figure 12. A vertical deposition process of droplets on demand: (**a**) the first droplet deposited; (**b**) the second droplet deposited; (**c**) the third droplet deposited; (**d**) the fourth droplet deposited.

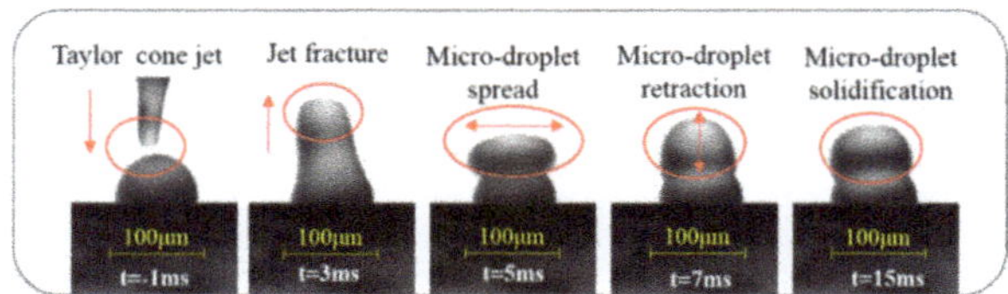

Figure 13. The process of droplet spreading, shrinking, and solidifying in vertical deposition.

4.7. Deposition Line and Column

In order to further verify the feasibility of micro/nano 3D printing based on high-voltage electric field-driven jet deposition, on the basis of the above experimental research and analysis, the horizontal line and vertical column were fabricated using wax micro-droplets.

The microcrystalline wax was selected as printing material, and the copper clad plate was used as a deposition substrate. The inner diameter of nozzle D was 200 μm, the negative pressure in the crucible was set as 450 Pa, the temperature of the crucible T_c was set as 90 °C, and the temperature of the nozzle T_n was set as 80 °C. The pulse frequency f was set as 10 Hz, and the applied pulse voltage U was set as 1600 V.

Figure 14 shows the deposited lines of wax droplets on a copper clad plate, the deposited line with one layer is shown in Figure 14a, and the deposited line with ten layers is shown in Figure 14b. It can be seen that there is no obvious gap between the adjacent deposited droplets, and a good remelting and bonding state of droplets has been obtained. The formed wax lines have a compact structure and good straightness. Four positions on each deposited line were selected as line width measurement points. According to the software measurement analysis, the line widths of the deposited line with one layer were L1 = 87.65 μm, L2 = 88.63 μm, L3 = 84.68 μm and L4 = 86.68 μm in the four positions; the line widths of the deposited line with ten layers were L5 = 85.64 μm, L6 = 87.35 μm, L7 = 86.35 μm and L8 = 85.86 μm in the four positions. According to the analysis and measurement results, the line widths of the deposited lines were essentially the same, and the size deviation was controlled within 2%.

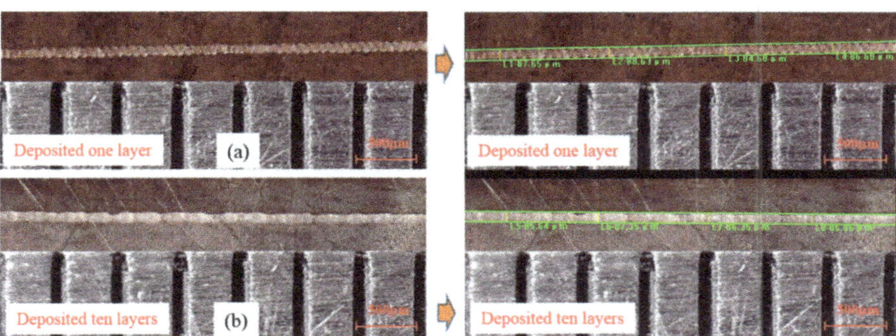

Figure 14. The deposited lines by wax droplets on a copper clad plate: (**a**) deposited one layer; (**b**) deposited ten layers.

Figure 15 shows the deposited micro-cylinder of wax droplets in a vertical direction with a large ratio of height to diameter. It can be seen that the local remelting was acquired between the contact surfaces of deposited droplets, and a relatively strong bonding was formed. The formed wax micro-cylinder with a large ratio of height to diameter has a compact structure without pores, a smooth surface, and good verticality. Four positions (the location of the refusion region of two adjacent droplets) on the deposited micro-cylinder were selected as diameter measurement points. According to the software measurement analysis, the diameter of the deposited micro-cylinders were D1 = 80.34 μm, D2 = 83.65 μm, D3 = 82.86 μm and D4 = 81.32 μm in the four positions. According to the analysis and measurement results, the diameter of the deposited micro-cylinders were essentially the same, and the size deviation was controlled within 2%.

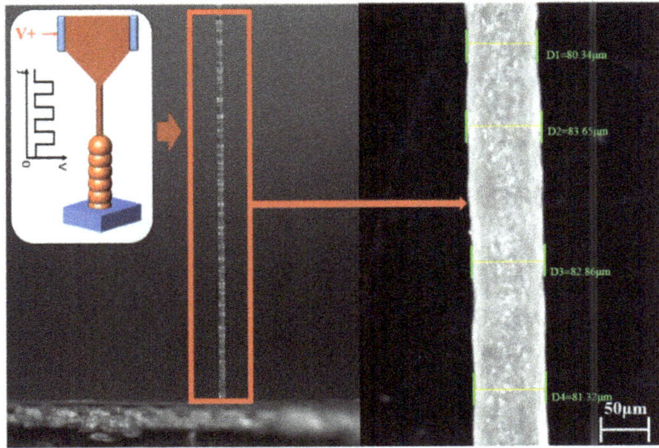

Figure 15. The deposited micro-cylinder of wax droplets with large ratio of height to diameter.

In conclusion, the deposition experiment results of the above line and micro-column verify the feasibility of micro/nano 3D printing based on high-voltage electric field-driven jet deposition. This research proposes a new technical approach for a fabricated micro scale wax structure with tens of microns and has important application prospects for the fabrication of micro-channel structures by using a 3D printing wax mold combined with PDMS micro-transfer technology.

In order to improve the spraying efficiency of micro-droplets driven by a high-voltage-electric field, a kind of parallelization, multiplexed, and array injection device will be developed. It could increase the magnitude of the throughput by two or more orders without sacrificing the quality of the print, greatly extending the capabilities of the technology.

5. Conclusions

The conclusions drawn from this paper are as follows:

(1) A high-voltage electric field-driven jet deposition technology was proposed to fabricate wax micro-structures. The mechanism of the Taylor cone generation and droplet eject deposition were analyzed, and an experimental system was developed.

(2) The droplet on-demand ejecting of a pulse cone-jet mode was studied under the action of a high-voltage pulse, the production of a droplet required the following four processes: tensile deformation to Taylor cone to jet retraction and finally to fracture into a single droplet. The diameter of the droplet was much smaller than the diameter of the nozzle.

(3) A meniscus shape liquid at the bottom of the nozzle is necessary to generate the Taylor cone jet; the optimal meniscus shape wax liquid can be guaranteed under 450 Pa negative pressure. With the increasing pulse width W_t, the duration of the pulse high-voltage plus was increased at the same pulse frequency, resulting in the micro-droplet diameter being increased. Under the combined action of inertia force and gravity, the deposited droplet underwent a process of spreading, shrinking and solidifying. Local remelting and bonding were acquired between the contact surfaces of adjacent deposited droplets.

(4) The horizontal line and vertical micro-column were fabricated using wax micro-droplets, and the size deviation between them was controlled within 2%. The correctness and feasibility of micro 3D printing based on high-voltage electric field-driven jet deposition were verified.

Author Contributions: Y.C. and H.Y. conceived the idea, developed the theory, performed the experiments and parameter optimization, and analyzed the results. Y.C. wrote the paper, and H.Y. provided corrections. H.C. and S.L. helped to evaluate the idea and engaged in discussions regarding the outcome. Y.L. and F.C. assisted with the editing of the paper. All authors have read and agreed to the published version of the manuscript.

Funding: This work was financially supported by the National Natural Science Foundation of China (grant nos. 51305128 and 52005059), Open Research Fund of State Key Laboratory of High-Performance Complex Manufacturing, Central South University (Kfkt2020-10), the key scientific research projects of the colleges and universities of Henan province (grant no.18A4600050), and the Outstanding Young Backbone Teachers projects of Xuchang University.

Data Availability Statement: All the data reported here can be made available on request.

Conflicts of Interest: The authors declare no conflict of interest.

References

1. Liu, P.P.; Guo, Y.W.; Chen, J.Y.; Yang, Y.B. A Low-Cost Electrochemical Metal 3D Printer Based on a Microfluidic System for Printing Mesoscale Objects. *Crystals* **2020**, *10*, 257. [CrossRef]
2. Hamid, H.M.A.; Celik-Butler, Z. A novel MEMS triboelectric energy harvester and sensor with a high vibrational operating frequency and wide bandwidth fabricated using UV-LIGA technique. *Sens. Actuators A Phys.* **2020**, *313*, 112175. [CrossRef]
3. Sigita, G.; Aukse, N.; Edvinas, S.; Mangirdas, M.; Angels, S.; Jolita, O. Vegetable Oil-Based Thiol-Ene/Thiol-Epoxy Resins for Laser DirectWriting 3D Micro-/Nano-Lithography. *Polymers* **2021**, *13*, 872.
4. Xu, J.; Su, Q.; Shan, D.B.; Guo, B. Sustainable micro-manufacturing of superhydrophobic surface on ultrafine-grained pure aluminum substrate combining micro-embossing and surface modification. *J. Clean. Prod.* **2019**, *232*, 705–712. [CrossRef]
5. Kuo, C.L.; Yeh, T.H.; Nien, Y.P.; Chen, Y. Multi-objective optimization of edge quality and surface integrity when wire electrical discharge machining of polycrystalline diamonds in cutting tool manufacture. *J. Manuf. Process.* **2022**, *74*, 520–534. [CrossRef]
6. Adriano, J.G.O.; Tomazio, N.B.; Paula, K.T.; Mendonça, C.R. Two-Photon Polymerization: Functionalized Microstructures, Micro-Resonators, and Bio-Scaffolds. *Polymers* **2021**, *13*, 1994.
7. Farjam, N.; Cho, T.H.; Dasgupta, N.P.; Barton, K. Subtractive patterning: High-resolution electrohydrodynamic jet printing with solvents. *Appl. Phys. Lett.* **2020**, *117*, 133702. [CrossRef]

8. Liang, P.; Shang, L.D.; Wang, Y.T.; Booth, M.J.; Li, B. Laser induced forward transfer isolating complex-shaped cell by beam shaping. *Biomed. Opt. Express* **2021**, *12*, 7024–7032. [CrossRef]

9. Kumar Singh, A.; Choudhary, A.K. On the electrical characterization of focused ion/electron beam fabricated platinum and tungsten nano wires. *Mater. Today Proc.* **2020**, *28*, 127–130. [CrossRef]

10. Wang, M.; Peng, Z.Y.; Huang, D.; Ning, Z.Q.; Chen, J.L.; Li, W.; Chen, J. Improving loading amount of nanodendrite array photo-electrodes on quantum dot sensitized solar cells by second electrochemical deposition. *Mater. Sci. Semicond. Process.* **2022**, *137*, 106219. [CrossRef]

11. Fang, F.Z.; Xu, Z.W.; Hu, X.T.; Wang, C.T.; Luo, X.G.; Fu, Y.Q. Nano fabrication of star structure for precision metrology developed by focused ion beam direct writing. *CIRP Ann.* **2012**, *61*, 511–514.

12. Kornbluth, Y.; Mathews, R.; Parameswaran, L.; Racz, L.M.; Velasquez-Garcia, L.F. Nano-additively manufactured gold thin films with high adhesion and near-bulk electrical resistivity via jet-assisted, nanoparticle-dominated, room-temperature microsputtering. *Addit. Manuf.* **2020**, *36*, 101679. [CrossRef]

13. Abdul-Wahed, A.M.; Roy, A.L.; Xiao, Z.M.; Takahata, K. Direct writing of metal film via sputtering of micromachined electrodes. *J. Mater. Process. Technol.* **2018**, *262*, 403–410. [CrossRef]

14. Wang, T.; Lv, L.; Shi, L.P.; Tong, B.H.; Zhang, X.Q.; Zhang, G.T.; Liu, J.Q. Microplasma direct writing of a copper thin film in the atmospheric condition with a novel copper powder electrode. *Plasma Process. Polym.* **2020**, *17*, 1–10. [CrossRef]

15. Ota, H.; Emaminejad, S.; Gao, Y. Application of 3D printing for smart objects with embedded electronic sensors and systems. *Adv. Mater. Technol.* **2016**, *1*, 1600013. [CrossRef]

16. Truby, R.L.; Lewis, J.A. Printing soft matter in three dimensions. *Nature* **2016**, *540*, 371–378. [CrossRef]

17. Bartlett, N.W.; Tolley, M.T.; Overvelde, J.T.B.; Weaver, J.C.; Mosadegh, B.; Bertoldi, K.; Whitesides, G.M.; Wood, R.J. A 3D-printed, functionally graded soft robot powered by combustion. *Science* **2015**, *349*, 161–165. [CrossRef]

18. Wehner, M.; Truby, R.L.; Fitzgerald, D.J.; Mosadegh, B.; Whitesides, G.M.; Lewis, J.A.; Wood, R.J. An integrated design and fabrication strategy for entirely soft, autonomous robots. *Nature* **2016**, *536*, 451–455. [CrossRef]

19. Muhammad, A.S.; Lee, D.G.; Lee, B.Y.; Hur, S. Classifications and Applications of Inkjet Printing Technology: A Review. *IEEE Access* **2021**, *9*, 140079–140102.

20. Dou, Y.B.; Luo, J.; Qi, L.H.; Lian, H.C.; Huang, J.G. Drop-on-demand printing of recyclable circuits by partially embedding molten metal droplets in plastic substrates. *J. Mater. Process. Tech.* **2021**, *297*, 117268. [CrossRef]

21. Zhang, D.; Qi, L.; Luo, J.; Yi, H.; Xiong, W.; Mo, Y. Parametric mapping of linear deposition morphology in uniform metal droplet deposition technique. *J. Mater. Process. Tech.* **2019**, *264*, 234–239. [CrossRef]

22. Wu, Y. Electrohydrodynamic jet 3D printing in biomedical applications. *Acta Biomater.* **2021**, *128*, 21–41. [CrossRef]

23. Park, J.-U.; Hardy, M.; Kang, S.J.; Barton, K.; Adair, K.; Mukhopadhyay, D.K.; Lee, C.Y.; Strano, M.S.; Alleyne, A.G.; Georgiadis, J.G.; et al. High-resolution electrohydrodynamic jet printing. *Nat. Mater.* **2007**, *6*, 782–789. [CrossRef]

24. Hu, Y.M.; Su, S.J.; Liang, J.S.; Xin, W.W.; Li, X.J.; Wang, D.Z. Facile and scalable fabrication of Ni cantilever nanoprobes using silicon template and micro-electroforming techniques for nano-tip focused electrohydrodynamic jet printing. *Nanotechnology* **2021**, *2*, 105301. [CrossRef]

25. Duan, Y.; Huang, Y.; Yin, Z.; Bu, N.; Dong, W. Non-wrinkled, highly stretchable piezoelectric devices by electrohydrodynamic direct-writing. *Nanoscale* **2014**, *6*, 3289–3295. [CrossRef]

26. Huang, Y.; Duan, Y.; Ding, Y.; Bu, N.; Pan, Y.; Lu, N.; Yin, Z. Versatile, kinetically controlled, high precision electrohydrodynamic writing of micro/nanofibers. *Sci. Rep.* **2014**, *4*, 5949. [CrossRef]

27. Huang, Y.; Bu, N.; Duan, Y.; Pan, Y.; Liu, H.; Yin, Z.; Xiong, Y. Electrohydrodynamic direct-writing. *Nanoscale* **2013**, *5*, 12007–12017. [CrossRef]

28. He, J.; Xia, P.; Li, D. Development of melt electrohydrodynamic 3D printing for complex microscale poly (ε-caprolactone) scaffolds. *Biofabrication* **2016**, *8*, 035008. [CrossRef]

29. Zhang, B.; He, J.; Li, X.; Xu, F.; Li, D. Micro/nanoscale electrohydrodynamic printing: From 2D to 3D. *Nanoscale* **2016**, *8*, 15376–15388. [CrossRef]

30. Qu, X.; Xia, P.; He, J.; Li, D. Microscale electrohydrodynamic printing of biomimetic PCL/nHA composite scaffolds for bone tissue engineering. *Mater. Lett.* **2016**, *185*, 554–557. [CrossRef]

31. Han, Y.; Wei, C.; Dong, J. Super-resolution electrohydrodynamic (EHD) 3D printing of micro-structures using phase-change inks. *Manuf. Lett.* **2014**, *2*, 96–99. [CrossRef]

32. Jiang, Y.; Wu, C.; Li, L.; Wang, K.; Tao, Z.; Gao, F.; Cheng, W.; Cheng, J.; Zhao, X.-Y.; Priya, S.; et al. All electrospray printed perovskite solar cells. *Nano Energy* **2018**, *53*, 440–448. [CrossRef]

33. García-Farrera, B.; Velásquez-García, L.F. Ultrathin Ceramic Piezoelectric Films via Room-Temperature Electrospray Deposition of ZnO Nanoparticles for Printed GHz Devices. *ACS Appl. Mater. Interfaces* **2019**, *11*, 29167–29176. [CrossRef] [PubMed]

34. Taylor, A.P.; Velásquez-García, L.F. Electrospray-Printed Nanostructured Graphene Oxide Gas Sensors. *Nanotechnology* **2015**, *26*, 505301. [CrossRef]

35. Onses, M.S.; Sutanto, E.; Ferreira, P.M.; Alleyne, A.G.; Rogers, J.A. Mechanisms, capabilities, and applications of high-resolution electrohydrodynamic jet printing. *Small* **2015**, *11*, 4237–4266. [CrossRef]

36. Zhang, G.; Lan, H.; Qian, L.; Zhao, J.; Wang, F. A microscale 3D printing based on the electric-field-driven jet. *3D Print. Add. Manufact.* **2020**, *7*, 37–44. [CrossRef]

37. Wang, Z.; Zhang, G.M.; Huang, H.; Qian, L.; Liu, X.L.; Lan, H.B. The self-induced electric-field-driven jet printing for fabricating ultrafine silver grid transparent electrode. *Virtual Phys. Prototyp.* **2021**, *16*, 113–123. [CrossRef]
38. Zhu, X.Y.; Xu, Q.; Li, H.K.; Liu, M.Y.; Li, Z.H.; Yang, K.; Zhao, J.W.; Qian, L.; Peng, Z.L.; Zhang, G.M.; et al. Fabrication of High-Performance Silver Mesh for Transparent Glass Heaters via Electric-Field-Driven Microscale 3D Printing and UV-Assisted Microtransfer. *Adv. Mater.* **2019**, *31*, 1902479. [CrossRef]
39. Zhu, X.Y.; Liu, M.Y.; Qi, X.M.; Li, H.K.; Zhang, Y.F.; Li, Z.H.; Peng, Z.L.; Yang, J.J.; Qian, L.; Xu, Q.; et al. Templateless, Plating-Free Fabrication of Flexible Transparent Electrodes with Embedded Silver Mesh by Electric-Field-Driven Microscale 3D Printing and Hybrid Hot Embossing. *Adv. Mater.* **2021**, *33*, 2007772. [CrossRef]
40. Liu, M.; Qi, X.; Zhu, X.; Xu, Q.; Zhang, Y.; Zhou, H.; Li, D.; Lan, H. Fabrication of embedded metal-mesh flexible transparent conductive film via electric-field-driven jet microscale 3D printing and roller-assisted thermal imprinting. *Chin. Sci. Bull.* **2020**, *65*, 1151–1162. (In Chinese) [CrossRef]
41. Morbioli, G.G.; Speller, N.C.; Cato, M.E.; Cantrell, T.P.; Stockton, A.M. Rapid and low-cost development of microfluidic devices using wax printing and microwave treatment. *Sens. Actuators B Chem.* **2019**, *284*, 650–656. [CrossRef]

Article

Numerical Analysis and Experimental Verification of Resistance Additive Manufacturing

Suli Li [1,2], Kaiyue Ma [1,2], Chao Xu [1,*], Laixia Yang [1,2] and Bingheng Lu [2]

[1] School of Mechanical Engineering, Xi'an University of Science and Technology, Xi'an 710054, China; Lsl15802949318@xust.edu.cn (S.L.); 20205224136@stu.xust.edu.cn (K.M.); yanglx@xust.edu.cn (L.Y.)
[2] State Key Laboratory for Manufacturing Systems Engineering, Xi'an Jiaotong University, Xi'an 710049, China; bhenglu@163.com
* Correspondence: Chaoxu@xust.edu.cn

Abstract: In recent years, scholars have proposed a metal wire forming method based on the Joule heat principle in order to improve the accuracy of additive manufacturing and reduce energy consumption and cost, but it is still in the theoretical stage. In this paper, a mathematical model of resistance additive manufacturing was established using finite element software, and the temperature variation of the melting process under different currents was analyzed. A suitable current range was preliminarily selected, and an experimental system was built. Through experimental study of the current and wire feeding speed, the influences of different process parameters on the forming appearance of the coating were analyzed. The results showed that the forming appearance was the best for Ti-6Al-4V titanium alloy wire with a diameter of 0.8 mm, when the current was 160 A, the voltage was 10 V, the wire feeding speed was 2.4 m/min, the workbench moving speed was 5 mm/s, and the gas flow rate was 0.7 m^3/h. Finally, the process parameters were used for continuous single-channel multilayer printing, verified the feasibility of the process at the experimental level and provided reference data for the subsequent development of this technology.

Keywords: additive manufacturing; joule thermal; mathematical model; process parameters

Citation: Li, S.; Ma, K.; Xu, C.; Yang, L.; Lu, B. Numerical Analysis and Experimental Verification of Resistance Additive Manufacturing. *Crystals* **2022**, *12*, 193. https://doi.org/10.3390/cryst12020193

Academic Editors: Hao Yi, Huajun Cao, Menglin Liu, Le Jia and Pavel Lukáč

Received: 30 December 2021
Accepted: 25 January 2022
Published: 28 January 2022

Publisher's Note: MDPI stays neutral with regard to jurisdictional claims in published maps and institutional affiliations.

1. Introduction

With the demand for high-performance, large, integrated, high-strength, and low-cost metal structural parts in the aerospace and automotive industries [1,2], metal additive manufacturing printing technology has become a hot topic for research and applications in additive manufacturing technology [3,4]. According to morphology, the raw materials of metal additive manufacturing technology can be divided into powder and wire [5]. With its advantages of high material utilization rate, high deposition rate, low cost, and low pollution, wire has been widely used in the field of metal additive manufacturing [6–8]. Current mainstream metal wire additive manufacturing technologies include Laser Additive Manufacturing (LAM), Electron Beam Additive Manufacturing (EBAM), Wire and Arc Additive Manufacturing (WAAM), and Wire and Plasma Additive Manufacturing (WPAM) [9]. LAM has high forming accuracy, high flexibility, and a short manufacturing cycle [10], but has low energy utilization; the residual stress of the formed part is large, and the cost of the laser head is high [11]. EBAM has high energy utilization and forming efficiency, but the size of the formed parts is limited by the volume of the vacuum chamber and the consequent increase in cost [12–14]. WAAM and WPAM have the advantages of low cost and high deposition efficiency [15], but the process parameters of both are variable and complex, and the dimensional accuracy of the manufactured parts is not high [16,17]. Reducing manufacturing costs, improving forming accuracy and efficiency, and ensuring forming quality are of great significance to the widespread application of metal wire additive manufacturing technology.

In recent years, scholars have proposed a new AM method [18,19]. The positive and negative terminals of a power supply are connected, respectively, with the metal wire and the metal substrate. After a current is introduced, the metal wire and the substrate contact and generate Joule heat, which melts the tip of the metal wire, forms a molten droplet, and deposits. With the movement of the workbench and wire feeding mechanism, the process can be repeated to form dense metal parts. Compared with the current mainstream metal fuse additive manufacturing technologies, Resistance Additive Manufacturing (RAM) greatly improves energy utilization through closed-loop energy transfer [20]. The smaller heat-affected zone ensures part forming accuracy [21]. The simple equipment structure reduces manufacturing costs [18]. Unlike WAAM, RAM uses a lower voltage to avoid arcing, and the contact energy transfer has a higher energy efficiency and smaller heat-affected zone. The droplet size can be adjusted by controlling the current and wire feeding speed, resulting in adjustability and higher accuracy. RAM provides a new direction for low-energy, low-cost, and high-precision additive manufacturing.

To date, some scholars have conducted exploratory research on RAM based on the principle of resistance welding. Lu et al. [22] found that the surface tension of metal droplets decreased with the increase in temperature during high-speed welding. Chen [23], from the Institute of Welding, Beijing University of Technology, studied the influence of gravity on the metal melting in the transition stage during the metal wire melting and deposition forming process of resistance heating. The results showed that, in the ground environment, the metal melt transition stage was affected by surface tension and electromagnetic contraction force, allowing it to overcome gravity to transition to the substrate.

However, all of the above studies were theoretical, and researchers have not yet printed single or multiple layers that could be better overlapped. In this paper, a RAM heat source model was established based on the Joule heat principle, and the temperature field of the melting coating process under different process parameters was simulated. The single-channel melting coating process parameters were optimized, and the feasibility of the process was verified through the experimental verification of the built RAM equipment, providing experimental information and reference data for the development of RAM.

2. Materials and Methods

The forming principle of RAM technology is shown in Figure 1. The positive and negative electrodes of the programmable power supply are connected with the wire and the metal substrate, respectively. The wire is transported by the pulsating wire feeder; when the wire is in contact with the substrate, the current flows through the metal wire, and the tip metal wire in contact with the substrate is generated by thermal resistance melting through the Joule effect. As the workbench moves, the wire continues to be fed and melts the tip metal wire, repeating this process to form dense metal parts. The forming process is protected by inert gas (Ar) to avoid oxidation. A high-speed camera system CCD and a voltage-current acquisition system are used to achieve the monitoring of the image changes and electrical signal changes during the melting of the wire. This technology uses low voltage and high current to avoid arcing during the forming process.

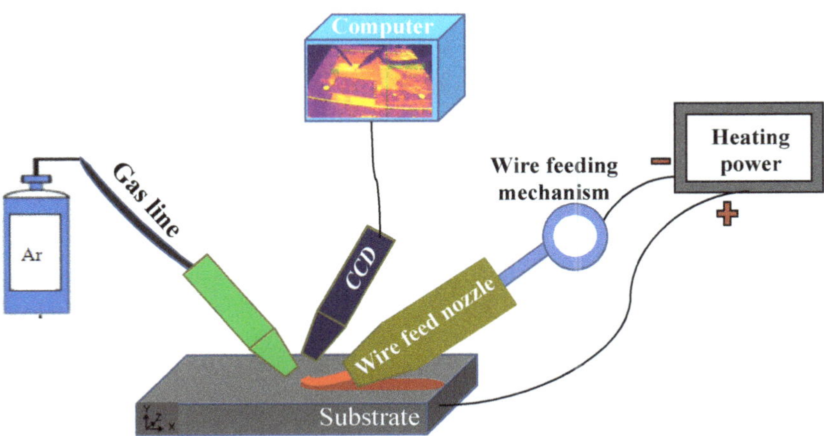

Figure 1. Forming schematic diagram of RAM.

In this paper, Ti-6Al-4V titanium alloy wire with a diameter of 0.8 mm was selected as the melting material. The substrate was a Ti-6Al-4V titanium alloy plate with a thickness of 5 mm. The physical parameters of the Ti-6Al-4V titanium alloy are shown in Table 1 [24].

Table 1. Physical parameters of Ti-6Al-4V.

Material	Density (g/cm^3)	Electrical Resistivity ($\Omega\cdot$m)	Melting Point (°C)	Thermal Conductivity (W/m·K)	Specific Heat Capacity (kJ/kg·°C)	Thermal Coefficient of Expansion (10^{-6}/K)
Ti6Al4V	4.371	5.6×10^{-7}	1678	15.2	0.52	10.8

Experimental equipment: The experimental workbench was modified from a BCH850 high-speed CNC milling machine using the programmable pulse power supply developed by ITECH company, with current range 0–300 A, voltage 0–80 V, maximum power 24,000 W, as shown in Figure 2.

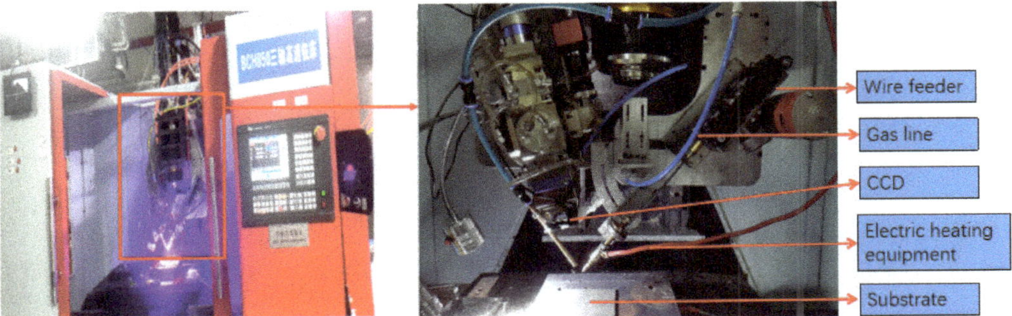

Figure 2. Experimental equipment.

3. Numerical Simulation

3.1. Deposition Principle and Establishment of Temperature Field Equation

In the process of metal melting and deposition, the metal wire flows steadily into the molten pool. The material utilization rate of the deposition process can reach almost 100%, and the unit of deposition rate is kilograms per second.

$$\dot{m} = \rho v_f \cdot \frac{1}{4}\pi D^2 = \rho v_S \cdot A \tag{1}$$

where D is the filler wire diameter, A is the cross-sectional area formed by the cladding, v_f and v_s are respectively the wire feeding speed and the laser scanning speed [25]. Under the influence of gravity, due to the limitation of the substrate, the molten metal will extend in the horizontal direction when the paste-like filament enters the molten pool. The spatial concentration distribution of mass flow is similar to the Gaussian distribution defined by the following Equation:

$$M_{wire} = \frac{\eta \dot{m}}{\pi R_m^2} exp\left(\frac{-\eta r^2}{R_m^2}\right) \tag{2}$$

where r is the radial distance from the center, and η and Rm represent the mass flow distribution factor and effective radius, respectively. The mass concentration distribution is related to the wire feed rate and viscosity. The mass increase process can be realized by adding the following external mass source term to the conservation of the mass Equation:

$$\dot{S}_{mass} = M_{wire} \cdot \delta(x - x_m) \tag{3}$$

where δ is the incremental function and x_m is the position of the source at a fixed distance from the bottom of the molten pool.

In metal additive manufacturing, the laser forms a molten pool on the surface of the substrate, which is then thermally diffused throughout the substrate. The contact between the substrate and the wire will generate convective heat transfer, which in turn affects the temperature field [26,27].

The temperature field establishment of a three-dimensional model needs to be based on the Fourier law of heat conduction and the law of conservation of energy, and then the governing equation of the heat transfer problem is established: the temperature T function in three coordinates x, y, z and time t as parameters should satisfy the following Equation:

$$\rho c \frac{\partial T}{\partial t} \pm \overline{Q} = \frac{\partial}{\partial x}\left(k\frac{\partial T}{\partial x}\right) + \frac{\partial}{\partial y}\left(k\frac{\partial T}{\partial y}\right) + \frac{\partial}{\partial z}\left(k\frac{\partial T}{\partial z}\right) \tag{4}$$

where ρ, c, and k are the density, specific heat capacity, and thermal conductivity of the wire, respectively; T (x, y, z, t) is the temperature distribution function; Q is the latent heat of fusion released during the wire forming process, and the heat release process is taken as "$-$", absorb heat to get "$+$".

3.2. Geometric Model and Meshing

The single-channel melting layer model shown in Figure 3a was established by Ansys and meshed (shown in Figure 3b). The melting layer was 30 mm in length, 0.8 mm in diameter, and 0.6 mm in height. Using hexahedral cells to mesh improved the computational efficiency. The total number of elements was 1350 and the number of nodes was 7567.

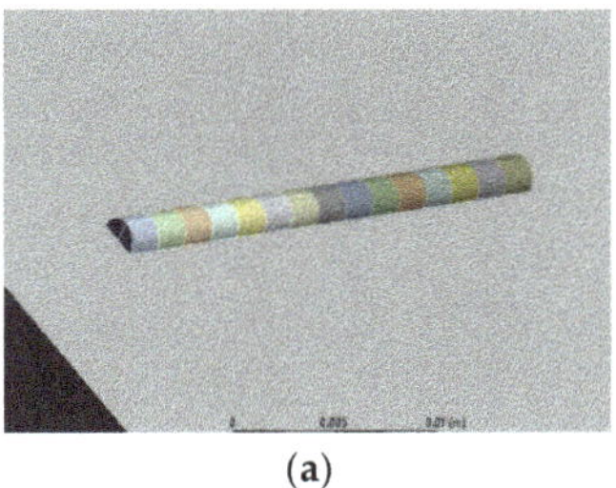

(a)

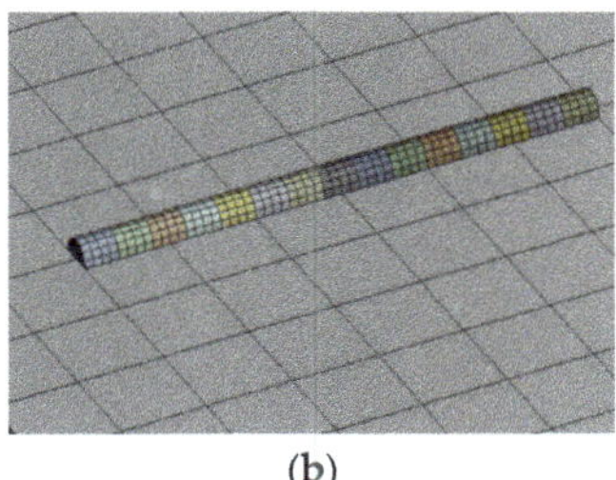

(b)

Figure 3. Geometric model and meshing. (**a**) Geometric model; (**b**) Geometric model with grids after meshing.

In the process of simulation analysis, in order to improve the efficiency of calculation and reduce the complexity of multifactor interaction in the actual melting process, the simulation process was reasonably simplified and the following assumptions were made:

(1) The heat transfer of materials in solid and molten states is continuous, and the parameters such as thermal conductivity and specific heat capacity involved in heat transfer are treated according to the continuous heat transfer.

(2) The loaded resistance heat source is the body heat source, which is loaded in the form of command flow generated by the expression of heat flux at the classical Ansys interface in the transient thermal analysis.

(3) It is assumed that the thermal absorption rate and thermal conductivity of the material do not change in the printing process.

(4) The effect of latent heat on temperature field is ignored.

3.3. Material Assignment and Boundary Setting

3.3.1. Material Assignment

The melting material and substrate were Ti-6Al-4V titanium alloy. The physical parameters are shown in Table 1. The initial condition was ambient temperature (22 °C).

3.3.2. Boundary Conditions

The boundary condition was the heat transfer condition on the surface of the temperature field.

$$Q = hf(T_s - T_b) \tag{5}$$

where h is convective heat transfer coefficient, T_s is surface temperature, and T_b is air temperature.

3.3.3. Insertion of Life and Death Unit

It was assumed that the length of a single melting coating layer was L = 30 mm, and it was divided into 15 segments. Each segment corresponded to a life and death unit, and each analysis step corresponded to a life and death unit, which was then divided into 15 steps in total. The scanning speed was set to V = 5 mm/s. The scanning time of the first layer was calculated as 6 s according to T = L/V, and then the total time t = 6 s was divided by the number of steps (15) of the analysis step according to StepTime = T/Substeps, and the resulting time substep was 0.4 s.

3.3.4. Selection of Heat Sources and Application of Loads

The characteristic of resistance heating is that the heat flux of the heat source acts not only on the surface of the wire, but also on the depth direction of the part. At this time, it should be regarded as a volume distribution heat source. Here, the double ellipsoid heat source in the proposed volume heat source was suitable to describe the molten pool shape when the heat source moved in the depth direction. In order to reduce the time of the simulation process and simplify the calculation, this paper used a similar semi-ellipsoid distribution heat source.

As shown in Figure 4, the elliptic half axis was the maximum heat flux at the center of the heat source, i.e., the origin of the coordinate system (0,0,0), and the function expression of the heat flux distribution was as follows:

$$q(x,y,z) = q_m \cdot exp(-Ax^2 - By^2 - Cz^2) \tag{6}$$

A, B, and C are the heat flux volume distribution parameters through a series of derivations:

$$q(x,y,z) = \frac{6\sqrt{3}Q}{a_h b_h c_h \pi^{1.5}} \cdot exp(\frac{-3x^2}{a_h^2} + \frac{-3y^2}{b_h^2} + \frac{-3z^2}{c_h^2}) \tag{7}$$

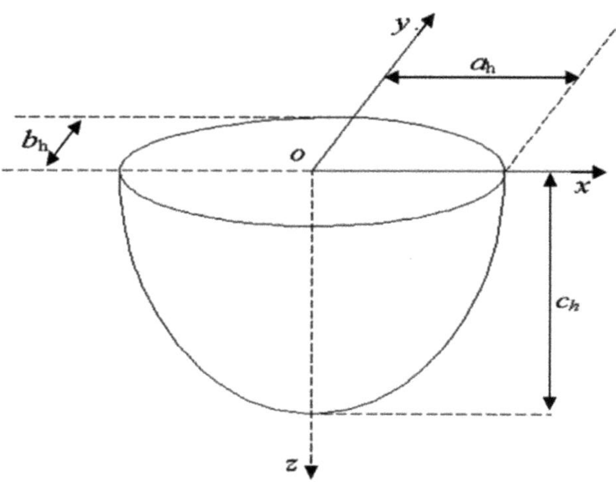

Figure 4. Schematic diagram of semi-ellipsoid domain.

3.4. Parameter Setting

In this paper, care was taken to ensure the consistency and comparability of the calculations. It was found through previous experiments that sparks will occur in the printing process when the voltage is too large (Figure 5a), that the wire cannot be effectively melted when the voltage is too small, and that good results without sparks can be achieved when the voltage is 10 V (Figure 5b). When the workbench moves too slowly, the conductive nozzle becomes blocked (Figure 5c); if it moves too fast, the wire cannot be melted (Figure 5d), but it is stable at 5 mm/s. When the gas flow rate is too low, the oxidation is more significant (Figure 5e). When the gas flow rate reaches 0.7 m^3/h, oxidation can be effectively prevented (Figure 5f). In this paper, the current was used as a single variable to analyze whether the wire would melt better under different currents. The current values were 130 A, 160 A, and 190 A, respectively. The wire feeding speed was initially set as 2.4 m/min. The parameter selection is shown in Table 2.

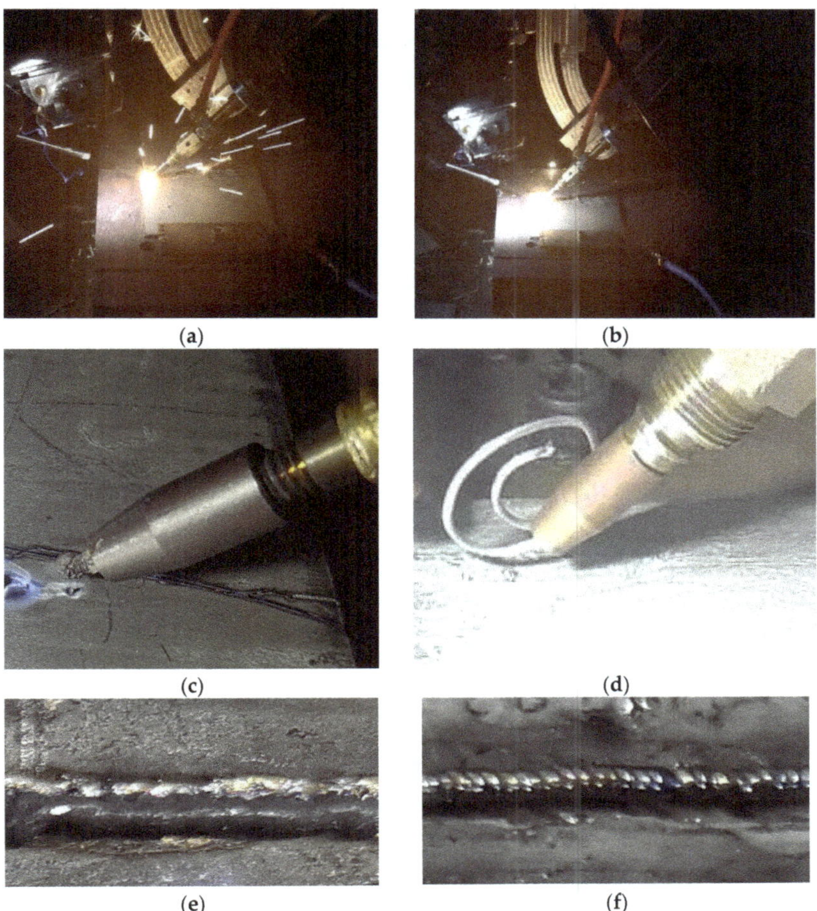

Figure 5. Comparison of parameters and effects. (**a**) The voltage is too large. (**b**) The voltage is 10 V. (**c**) The workbench moves too slowly. (**d**) The workbench moves too fast. (**e**) The gas flow rate is too low. (**f**) The gas flow rate is 0.7 m^3/h.

Table 2. Process parameters.

Number	Electric Current (A)	Gas Flow (m^3/h)	Wire Feeding Speed (m/min)	Voltage (V)	Workbench Moving Speed (mm/s)
1	130	0.7	2.4	10	5
2	160	0.7	2.4	10	5
3	190	0.7	2.4	10	5

Monitoring the real-time temperature change of the melting coating process is very beneficial to the study of the melting coating forming process and the optimization of the later process parameters. In this paper, the temperature probe monitoring points A, B, C, and D were selected to obtain the temperature–time change images of different process parameters, as shown in Figure 6.

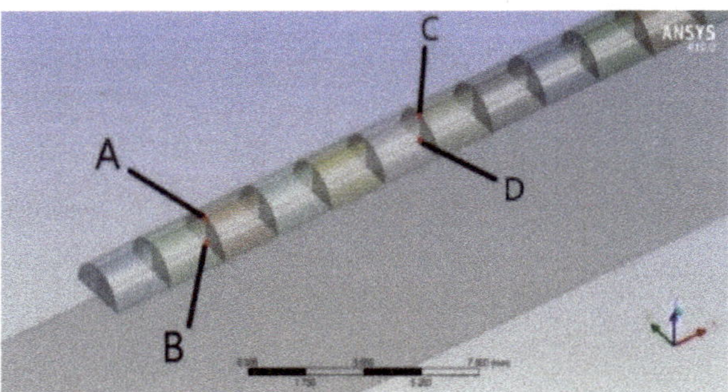

Figure 6. Monitoring points for 4 temperature probes.

4. Results

4.1. Results of Simulation

The temperature–time curves under three groups of currents obtained by simulation are shown in Figure 7, where A, B, C, and D are the temperatures measured at the four positions shown in Figure 6.

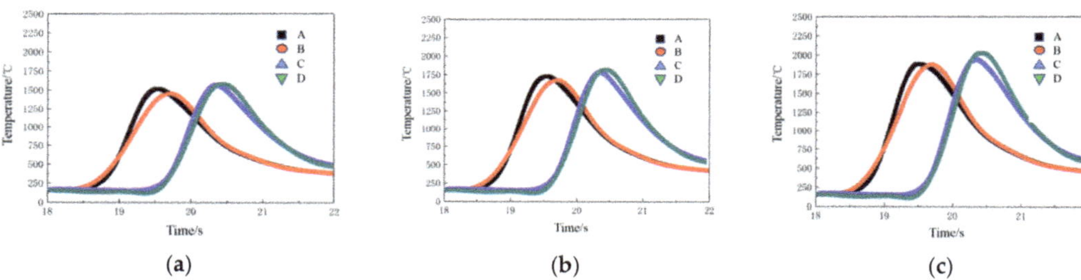

(a)   **(b)**   **(c)**

Figure 7. Temperature–time curves. (**a**) 130 A; (**b**) 160 A; (**c**) 190 A.

As shown in Figure 7a, when the current was 130 A, the temperatures of the four points were lower than or close to the melting point of the material, which meant that the coating layer could not adhere to the substrate. It can be seen from Figure 7b that when the current was 160 A, the temperature of the four points exceeded the melting point; it can be seen from Figure 7c that when the current increased to 190 A, the temperatures of the four points exceeded the melting point more. It can be seen from Figure 7b,c that under these two current values, the wire could be melted, and the coating layer could be well combined with the substrate, but excessively high temperatures may cause melting to occur too fast and affect the forming appearance.

4.2. Influence of Current on Forming Appearance

When other parameters remained unchanged, the influence of current on the forming appearance was experimentally studied, as shown in Table 3. When the current was 130 A, the generated resistance heat was insufficient to melt the wire, resulting in the coiling of the wire and its failure to adhere to the substrate, resulting in poor appearance. When the current was 160 A, the droplet size was small and uniform, the fusion layer was firmly adhered to the substrate, and the sample is good. When the current was 190 A, the resistance heat was too large, resulting in excessive melting speed and uneven droplet size, which could not form continuous deposition on the substrate. Thus, the current had a great influence on the shape of the RAM.

Table 3. Influence of current on forming appearance.

Number	Morphology	Electric Current (A)	Wire Feeding Speed (m/min)
1		130	2.4
2		160	2.4
3		190	2.4

4.3. Influence of Wire Feeding Speed on Forming Appearance

When other parameters remained unchanged, the influence of wire feeding speed on the forming appearance was studied experimentally, as shown in Table 4. When the wire feeding speed was 2.0 m/min, the wire feeding speed was too slow to form continuously, and the appearance was not good. Under the same current, when the wire feeding speed was 2.4 m/min, dense continuous forming was realized, which was firmly adhered to the substrate and had good shape. When the wire feeding speed was 2.8 m/min, the melting speed of the wire could not keep up with the wire feeding speed and the wire could not be completely melted, and the shape of the formed sample was distorted. It can be seen that the wire feeding speed had a great influence on the shape of the formed RAM.

Table 4. Influence of wire feeding speed on forming appearance.

Number	Morphology	Electric Current (A)	Wire Feeding Speed (m/min)
1		160	2.0
2		160	2.4
3		160	2.8

From the above experimental results, it can be concluded that when the current is 160 A, the wire feeding speed is 2.4 m/min, the voltage is 10 V, the gas flow is 0.7 m^3/h, and the workbench moving speed is 5 mm/s, the appearance is the best. The frequency of the current measured by oscilloscope under this process parameter was 36 Hz (as shown in Figure 8). One frequency forms a molten droplet; the higher the frequency is, the smaller the droplet that can be formed, and the higher the forming accuracy.

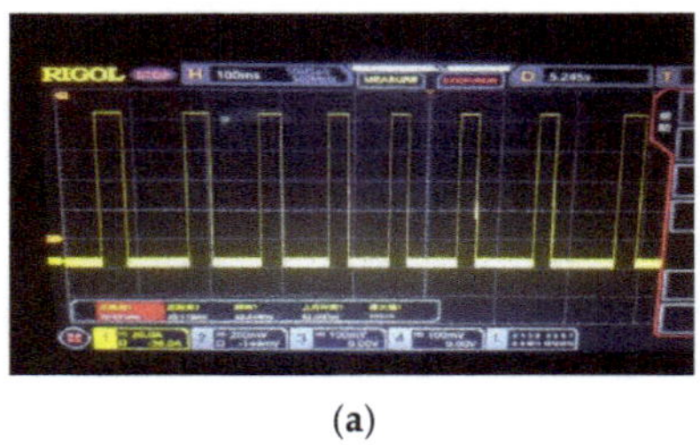
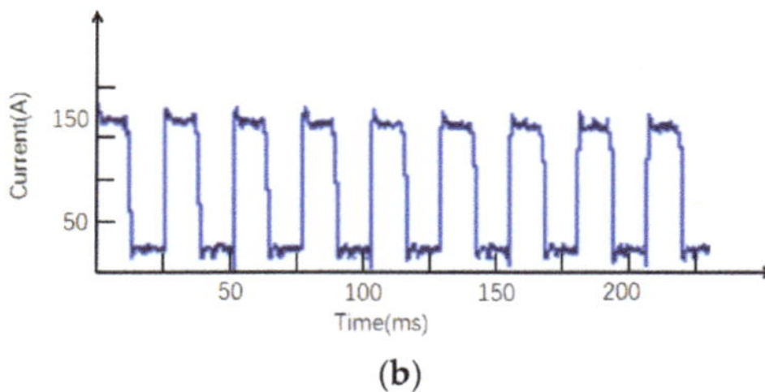

(**a**)   (**b**)

Figure 8. Current frequency. (**a**) Measurement. (**b**) Analysis.

4.4. Forming and Density Test of Single-Pass Multilayer

Using the optimized single-layer process parameters for single-pass multilayer fusion, the forming appearance shown in Figure 9 can be obtained.

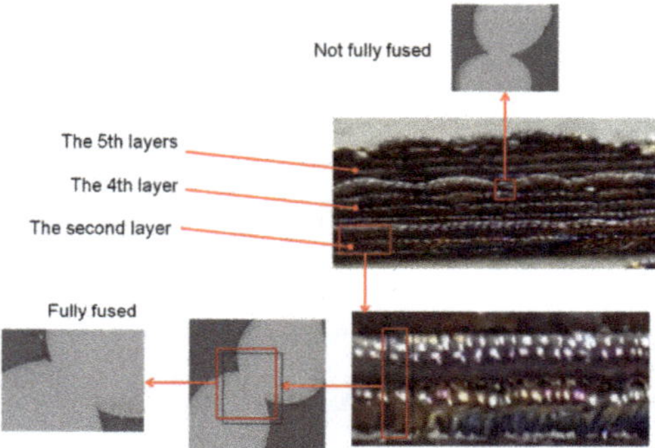

Figure 9. Single pass multi-layer forming.

Relative density is an important index of the internal quality of a metal part. In this paper, the density of a part was measured using Archimedes' principle. After degreasing and cleaning, the samples were weighed using a balance in air and then, according to Equation (5), submerged in water to obtain the density of the samples.

$$\rho = \frac{m_1}{m_1 - m_2}\rho_1 \tag{8}$$

where m_1 is the mass of the sample to be tested measured in air, m_2 is the mass of the sample to be tested fully immersed in water, ρ is the density of the tested sample, and ρ_1 is the density of water.

The balance used in the experiment was an AG204 electronic balance produced by Mettler Toledo, Zürich, Switzerland, with a measurement accuracy of 0.1 mg and a maximum range of 210 g. The relative density of the sample produced via the RAM method was determined to be 97.7% (results shown in Table 5). It can be seen that the denseness of the part was high and that under better process parameters, the layers melted sufficiently to achieve a metallurgical bond between the overlapping layers of material.

Table 5. Results of relative density.

Material	m_1 (g)	m_2 (g)	ρ (g/cm^3)	ρ_0 (g/cm^3)	Relative Density (%)
Ti-6Al-4V	9.0851	6.9577	4.2705	4.371	97.7

where ρ_0 is the theoretical density of Ti-6Al-4V.

5. Conclusions

In this paper, a RAM simulation model was established to study the temperature change of the melted layer under different currents, and the current parameter range of the melted wire was preliminarily obtained. An experimental system was built, and the correctness of the simulation results was verified. The influence of current and wire feeding speed on the forming appearance was studied, and single channel and multilayer continuous printing were carried out, verifying the feasibility of the process. The main conclusions are as follows:

(1) In the RAM process, the main influencing factors are current, wire feeding speed, voltage, workbench moving speed, and gas flow. Current and wire feeding speed mainly affect the appearance of the molten coating layer. Voltage affects whether spark splash occurs in the printing process. Platform moving speed affects whether the printing process is at the correct temperature. Gas flow affects the oxidation degree.

(2) When the material is Ti-6Al-4V, the current is 160 A, the voltage is 10 V, the wire feeding speed is 2.4 m/min, the moving speed of the workbench is 5 mm/s, and the gas flow rate is 0.7 m^3/h, the forming morphology of the single-channel coating layer is the best.

(3) By optimizing the process, a single-channel five-layer sample was printed, and its relative density was 97.7%. Good overlap between layers was achieved, which verified the feasibility of RAM at the experimental level and provided reference data for the subsequent development of this technology.

Author Contributions: Conceptualization, S.L. and B.L.; methodology, K.M.; software, C.X.; validation, S.L. and K.M.; formal analysis, L.Y.; investigation, S.L. and K.M.; resources, S.L. and B.L.; data curation, K.M. and C.X.; writing—original draft preparation, S.L. and K.M.; writing—review and editing, S.L. and K.M.; visualization, S.L.; supervision, K.M.; project administration, L.Y.; funding acquisition, C.X. All authors have read and agreed to the published version of the manuscript.

Funding: This research was funded by State Key Laboratory for Manufacturing Systems Engineering, China. Funding number: sklms2021016.

Institutional Review Board Statement: Not applicable.

Informed Consent Statement: Not applicable.

Data Availability Statement: Not applicable.

Acknowledgments: Suli Li would like to thank Xueping Yan, Haimin Zhou, Jie Xiong, Zhuang Gao, Cong Wang and Jingsheng Zhao for providing materials used for experiments and academic guidance.

Conflicts of Interest: The authors declare no conflict of interest.

References

1. Lu, B. Additive manufacturing—Current situation and future. *China Mech. Eng.* **2020**, *31*, 19.
2. Calignano, F. Additive Manufacturing (AM) of Metallic Alloys. *Crystals* **2020**, *10*, 704. [CrossRef]
3. Frazier, W.E. Metal additive manufacturing: A review. *J. Mater. Eng. Perform.* **2014**, *23*, 1917–1928. [CrossRef]
4. Sun, C.; Wang, Y.; McMurtrey, M.D.; Jerred, N.D.; Liou, F.; Li, J. Additive manufacturing for energy: A review. *Appl. Energy* **2021**, *282*, 116041. [CrossRef]
5. Jackson, M.A.; Van Asten, A.; Morrow, J.D.; Min, S.; Pfefferkorn, F.E. A comparison of energy consumption in wire-based and powder-based additive-subtractive manufacturing. *Procedia Manuf.* **2016**, *5*, 989–1005. [CrossRef]
6. Węglowski, M.S.; Błacha, S.; Jachym, R.; Dworak, J.; Rykała, J. Additive manufacturing with wire–Comparison of processes. *AIP Conf. Proc.* **2019**, *2113*, 150016.
7. Vimal, K.E.K.; Srinivas, M.N.; Rajak, S. Wire arc additive manufacturing of aluminium alloys: A review. *Mater. Today Proc.* **2021**, *41*, 1139–1145. [CrossRef]
8. Hwang, M.J.; Cho, J. Laser Additive Manufacturing Technology Review. *J. Weld. Join.* **2014**, *32*, 354–357.
9. Yilmaz, O.; Ugla, A.A. Shaped metal deposition technique in additive manufacturing: A review. *Proc. Inst. Mech. Eng. Part B J. Eng. Manuf.* **2016**, *230*, 1781–1798. [CrossRef]

10. Galati, M.; Calignano, F.; Viccica, M.; Iuliano, L. Additive manufacturing redesigning of metallic parts for high precision machines. *Crystals* **2020**, *10*, 161. [CrossRef]

11. Ding, Y.; Akbari, M.; Kovacevic, R. Process planning for laser wire-feed metal additive manufacturing system. *Int. J. Adv. Manuf. Technol.* **2018**, *95*, 355–365. [CrossRef]

12. Fuchs, J.; Schneider, C.; Enzinger, N. Wire-based additive manufacturing using an electron beam as heat source. *Weld. World* **2018**, *62*, 267–275. [CrossRef]

13. Miyata, Y.; Okugawa, M.; Koizumi, Y.; Nakano, T. Inverse columnar-equiaxed transition (CET) in 304 and 316 L stainless steels melt by electron beam for additive manufacturing (AM). *Crystals* **2021**, *11*, 856. [CrossRef]

14. Węglowski, M.S.; Błacha, S.; Pilarczyk, J.; Dutkiewicz, J.; Rogal, Ł. Electron beam additive manufacturing with wire–analysis of the process. *AIP Conf. Proc.* **2018**, *1960*, 140015.

15. Rodrigues, T.A.; Duarte, V.; Miranda, R.M.; Santos, T.G.; Oliveira, J.P. Current status and perspectives on wire and arc additive manufacturing (WAAM). *Materials* **2019**, *12*, 1121. [CrossRef]

16. Derekar, K.S. A review of wire arc additive manufacturing and advances in wire arc additive manufacturing of aluminium. *Mater. Sci. Technol.* **2018**, *34*, 895–916. [CrossRef]

17. Artaza, T.; Suárez, A.; Veiga, F.; Braceras, I.; Tabernero, I.; Larrañaga, O.; Lamikiz, A. Wire arc additive manufacturing Ti6Al4V aeronautical parts using plasma arc welding: Analysis of heat-treatment processes in different atmospheres. *J. Mater. Res. Technol.* **2020**, *9*, 15454–15466. [CrossRef]

18. Klobcara, D.; Balosb, S.; Busicc, M.; Duricd, A.; Lindica, M.; Scetineca, A. WAAM and Other Unconventional Metal Additive. *Manuf. Technol.* **2020**, *45*, 1–9.

19. Lee, C.Y.; Taylor, A.C.; Nattestad, A.; Beirne, S.; Wallace, G.G. 3D printing for electrocatalytic applications. *Joule* **2019**, *3*, 1835–1849. [CrossRef]

20. Vafadar, A.; Guzzomi, F.; Rassau, A.; Hayward, K. Advances in metal additive manufacturing: A review of common processes, industrial applications, and current challenges. *Appl. Sci.* **2021**, *11*, 1213. [CrossRef]

21. Khan, M.S.; Sanchez, F.; Zhou, H. 3-D printing of concrete: Beyond horizons. *Cem. Concr. Res.* **2020**, *133*, 106070. [CrossRef]

22. Lu, Z.; Huang, P.; Chen, S.; Li, Y. Mechanism of undercut in high speed welding based on moveless TIG welding. *Trans. Jwri* **2010**, *39*, 197–198.

23. Chen, S.; Yuan, C.; Jiang, F.; Yan, Z.; Zhang, P. Study on Heat Generation Mechanism and Melting Behavior of Droplet Transition in Resistive Heating Metal Wires. *Acta. Metall. Sin.* **2018**, *54*, 1297–1310.

24. Dai, S.; Dai, X. Interpretation of the latest edition of GB/T 3620.1-2016 Titanium and Titanium Alloy grades and chemical Composition. *Sci. Technol. Vis.* **2019**, 107–108. [CrossRef]

25. Wei, S.; Wang, G.; Nie, Z.; Huang, Z.; Rong, Y. Microstructure evolution of martensitic stainless steel in laser hot wire cladding with multiple heating passes. In *TMS 2016 145th Annual Meeting & Exhibition*; Springer International Publishing: Berlin/Heidelberg, Germany, 2016; pp. 191–198.

26. Shrestha, S.; Rauniyar, S.; Chou, K. Thermo-fluid modeling of selective laser melting: Single-track formation incorporating metallic powder. *J. Mater. Eng. Perform.* **2019**, *28*, 611–619. [CrossRef]

27. Peyre, P.; Dal, M.; Pouzet, S.; Castelnau, O. Simplified numerical model for the laser metal deposition additive manufacturing process. *J. Laser Appl.* **2017**, *29*, 022304. [CrossRef]

Article

Study on the Properties of Coated Cutters on Functionally Graded WC-Co/Ni-Zr Substrates with FCC Phase Enriched Surfaces

Shidi Li [1], Xiangyuan Xue [1], Jiaxing Chen [1], Tengxuan Lu [1], Zhe Zhao [1], Xin Deng [1,*], Zhongliang Lu [2,3], Zhongping Wang [4], Zhangxu Li [4] and Zhi Qu [5]

[1] School of Electromechanical Engineering, Guangdong University of Technology, Guangzhou 510006, China; 2112001395@mail2.gdut.edu.cn (S.L.); xuexiangyuan123@163.com (X.X.); c626443994@163.com (J.C.); 18320636910@163.com (T.L.); just4ujust4all@126.com (Z.Z.)

[2] State Key Laboratory for Manufacturing Systems Engineering, Xi'an Jiaotong University, Xi'an 710049, China; zllu@xjtu.edu.cn

[3] Guangdong Xi'an Jiaotong University Academy, Foshan 528300, China

[4] Heyuan Fuma Cemented Carbide Co., Ltd., Heyuan 517583, China; wzp@fuma-carbide.com (Z.W.); 13825375320@139.com (Z.L.)

[5] Guangdong Jin Ci San Wei Tech-Ltd., Foshan 528225, China; qz@fhzl3dp.com

* Correspondence: dengxin@gdut.edu.cn; Tel.: +86-020-39322925

Citation: Li, S.; Xue, X.; Chen, J.; Lu, T.; Zhao, Z.; Deng, X.; Lu, Z.; Wang, Z.; Li, Z.; Qu, Z. Study on the Properties of Coated Cutters on Functionally Graded WC-Co/Ni-Zr Substrates with FCC Phase Enriched Surfaces. *Crystals* **2021**, *11*, 1538. https://doi.org/10.3390/cryst11121538

Academic Editors: Hao Yi, Huajun Cao, Menglin Liu and Le Jia

Received: 20 November 2021
Accepted: 6 December 2021
Published: 9 December 2021

Publisher's Note: MDPI stays neutral with regard to jurisdictional claims in published maps and institutional affiliations.

Abstract: Currently, the research on mechanical behavior and cutting performance of functionally graded carbides is quite limited, which limits the rapid development of high-performance cemented carbide cutting tools. Based on WC-Co-Zr and WC-Ni-Zr, this study synthesized two kinds of cemented carbide cutters, i.e., the cemented carbide cutters with homogeneous microstructure and functionally graded carbide (FGC) cutters with FCC phase ZrN-enriched surfaces. Furthermore, TiAlN coating has been investigated on these carbide cutters. Mechanical behavior, friction, wear performance, and cutting behavior have been investigated for these coated carbides and their corresponding substrates. It was found that, as compared with coated cutters on WC-Co/Ni-Zr carbide substrates with homogeneous microstructures, the coated cutters on WC-Co/Ni-Zr FGC substrates with FCC phase-enriched surfaces show higher wear resistance and cutting life, and the wear mechanism during cutting is mainly adhesion wear.

Keywords: additive manufacture; cemented carbide; functionally graded structure; adhesive wear

1. Introduction

Cemented carbides, as ceramic matrix composites consisting of carbide matrix (mainly WC, TiC, and TaC, etc.) and metal binders, have been widely used as cutting tools, molds, and mining and petroleum drilling tools due to their high hardness, strength, wear resistance, excellent oxidation resistance, and thermal stability [1–7]. With the rapid development of the modern manufacturing industry, more crucial requirements, such as simultaneously high wear resistance and high toughness, are being put forward on the performance of cemented carbides [8,9]. In order to obtain better properties, current development on cemented carbides is mainly focused on functionally graded carbides (FGCs) and cemented carbide coatings [10–14]. The FGCs shows gradient microstructures from surface to internal portion, which provides both higher wear resistance (surface) and toughness (internal portion) and hence better overall performance than conventional carbide with homogeneous microstructure [15–19]. The FGCs mainly include dual-phase FGC [20,21], the FGCs with binder phase-enriched surfaces [22,23], and the FGCs with face center cubic (FCC) phase-enriched surfaces [24]. The main preparation method of the FGC with an FCC phase-enriched surface is the nitriding method [25], which converts N in the sintering atmosphere and the elements (such as Ti, Zr, etc.) in carbides into nitride or carbon–nitride compounds of face-centered cubic crystals by thermal–chemical coupling, and then obtains the functionally graded carbide with FCC phase-enriched surface [26].

At present, there are quite limited study on the cutting performance of the FGCs with FCC phase-enriched surface, especially on the cutting performance of coated cutters on FGC substrates with FCC phase-enriched surface. This paper focuses on the microstructure, mechanical properties, and cutting properties of coated cutters on FGC substrates with FCC phase-enriched surface. The high-performance carbide cutters investigated in this study has potential applications in future additive/subtractive hybrid manufacturing.

2. Materials and Methods

The composition, gradient structure, and coating types of WC-Co/Ni-Zr cemented carbide are shown in Table 1. The feedstock powders of WC, Co, Ni, Zr, and others with an average particle size of 1 μm were weighed according to the composition in Table 1 and then put into a planetary ball mill with 2 wt.% paraffin wax and an appropriate amount of n-heptane and mixed for 24 h. The wet powder mixture slurry was dried in a rotary evaporator, and then pressed at 300 MPa into 18 mm $\times$ 18 mm $\times$ 6 mm cutter billets. Afterwards, the cutter billets were dewaxed in a tube furnace at 400–550 °C in an argon atmosphere. The dewaxed cutter billets were sinter-hipped in a two-step process. Firstly, the dewaxed parts were sintered at 1450 °C for 30 min in an argon atmosphere at the pressure of 6 MPa. After furnace cooling, the pre-sintered parts were machined into a cutter shape of 12.5 mm $\times$ 12.5 mm $\times$ 5 mm and then polished. Secondly, all the machined parts were further sintered at 1450 °C in a nitrogen atmosphere at a pressure of 6 MPa for 60 min to prepare the functionally graded cemented carbides (FGCs) with FCC phase-enriched surface. The dewaxing and sintering processes are shown in Figure 1. Two-step sintering is critical to the mechanical and frictional properties of FGCs. The first step of sintering with high pressure Argon results in the full density of carbides and the second step of sintering with high pressure nitrogen leads to the formation of gradient microstructure of FGCs. Upon two-step sintering, the FGCs achieve the high wear resistance of surface and the high toughness of the inner portion.

Table 1. The composition, microstructure, and coating type of WC-Co/Ni-Zr cemented carbides.

Carbides No.	Composition (wt.%)	Sintering Process	Microstructure	Coating
FC5-1	WC-10Co-5Zr-0.5VC-2W	One-step/Ar	Homogeneous	Nothing
FC5-2	WC-10Co-5Zr-0.5VC-2W	Two step /(Ar+N$_2$)	Gradient	Nothing
FC5-3	WC-10Co-5Zr-0.5VC-2W	One-step/Ar	Homogeneous	TiAlN
FC5-4	WC-10Co-5Zr-0.5VC-2W	Two step /(Ar+N$_2$)	Gradient	TiAlN
FN1-1	WC-10Ni-5Zr-0.5VC	One-step/Ar	Homogeneous	Nothing
FN1-2	WC-10Ni-5Zr-0.5VC	Two step /(Ar+N$_2$)	Gradient	Nothing
FN1-3	WC-10Ni-5Zr-0.5VC	One-step/Ar	Homogeneous	TiAlN
FN1-4	WC-10Ni-5Zr-0.5VC	Two step /(Ar+N$_2$)	Gradient	TiAlN

In order to analyze the difference in properties between FGCs and conventional cemented carbides with homogeneous microstructure, the same WC-Co-Zr and WC-Ni-Zr parts were sintered in one-step sinter-hip process at 1450 °C for 90 min with 6 MPa argon atmosphere to prepare the conventional cemented carbide with homogeneous microstructure. Finally, TiAlN coating was deposited on the above the FGCs and the conventional carbides with homogeneous microstructure by PVD method.

Zeiss Merlin high-resolution field emission electron microscope and energy spectrometer (Zeiss, Oberkochen, Germany) were used for microstructure and composition analysis. Nanoindentation was used to measure the coating hardness with a load of 10 mN. The bonding strength between coating and carbide substrate was measured by the RST3 Large Load Scratcher [27,28]. The UMT-TriboLab friction and wear tester (Bruker, Billerica, MA, USA) was used to test the wear resistance, friction, and wear performance of carbides.

The GCr15 ball was used for the friction test of uncoated carbides with a load of 50 N, the Si$_3$N$_4$ ceramic ball was used for the friction test of coated cutters with a load of 10 N, and both friction tests had a turning radius of 4 mm and a speed of 400 r/min. The ETC3650h CNC lathe (SMTCL, Shenyang, China) was used for cutting test and an HT250, a gray cast iron with yield strength of 250 MPa, was used as the workpiece. The cutting speed was 200 m/min, the back cutting depth was 0.3 mm, and the feed rate was 0.1 mm/r.

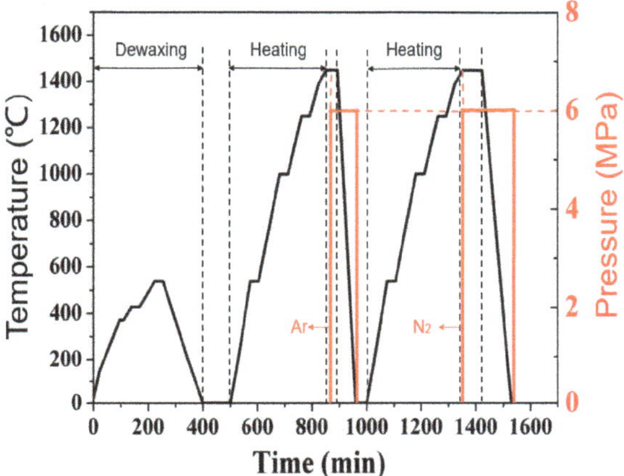

Figure 1. Dewaxing and two-step sinter-hip process for FGCs.

3. Results and Discussion

3.1. Functionally Graded Cemented Carbides

3.1.1. Microstructure of FGCs

Figures 2 and 3 show the microstructure and EDS element distribution of WC-Co-Zr and WC-Ni-Zr FGCs, respectively. The gradient layer thickness of WC-Co-Zr is significantly lower than that of WC-Ni-Zr. Both carbide surfaces contained mainly Zr and N, and the FCC phase ZrN was formed during the high-pressure nitriding process [15]. Figure 4 shows the XRD of both FC-5 and FN-1, which confirms the formation of FCC phase ZrN during high pressure nitriding process.

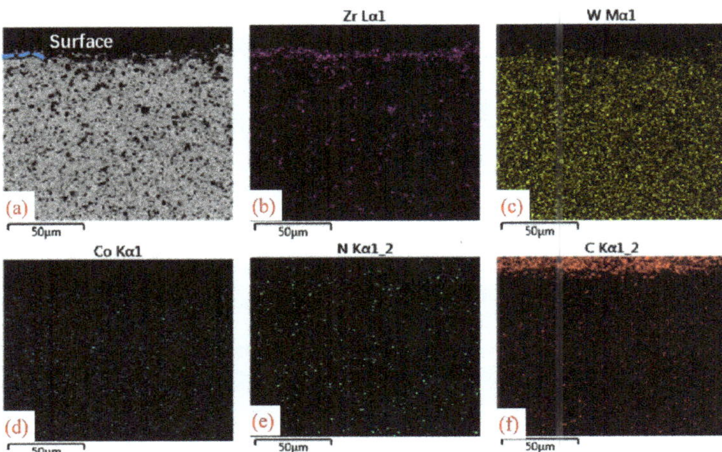

Figure 2. Microstructure and EDS element mapping of FC5-2, (**a**) microstructure, and (**b–f**) respective element mappings of Zr, W, Co, N, and C.

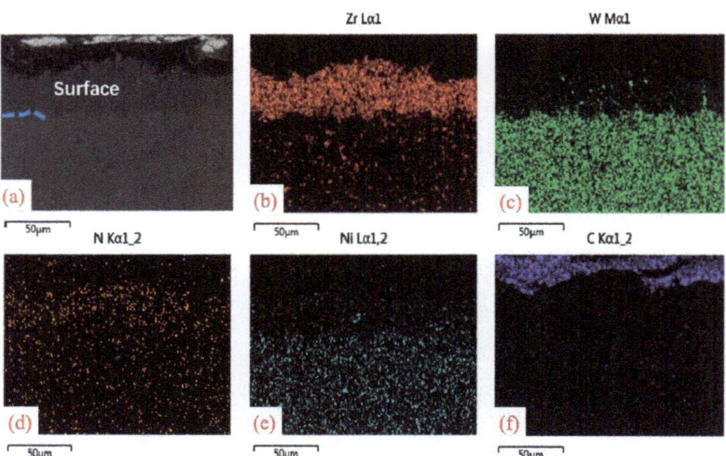

Figure 3. Microstructure and EDS element mapping of FN1-2, (**a**) microstructure, and (**b–f**) respective element mappings of Zr, W, N, Ni, and C.

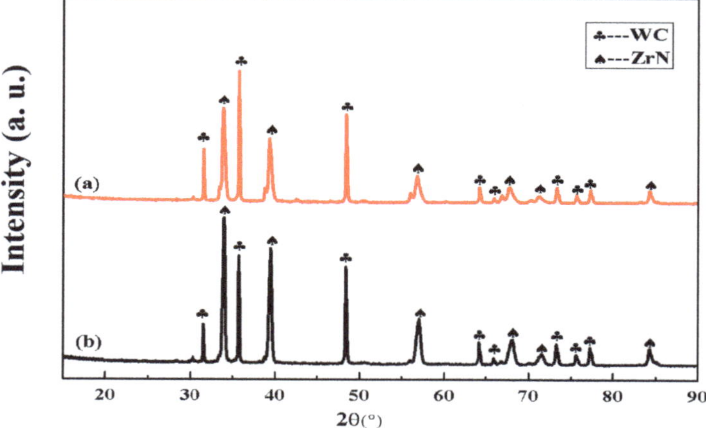

Figure 4. XRD of the surfaces of (**a**) FC5-2 and (**b**) FN1-2.

3.1.2. Friction and Wear Properties of FGCs

Figure 5 shows the friction coefficient curves of four cemented carbides. The curves can be divided into two stages: the running-in wear stage and the normal friction stage. The curves C and D in Figure 5 represent friction coefficient curves of two FGCs with FCC phase-enriched surfaces (FC5-2, FN1-2). Compared with the carbides with homogeneous microstructure, i.e., A (FC5-1) and B (FN1-1), the friction coefficient curves of the FGCs show abrupt drop in the normal friction stage, which is due to the complete wearing off of the ZrN surface layer and the exposure of the inner part of FGCs. Figures 2 and 3 show that the thickness of the surface gradient layer of WC-Ni-Zr is significantly thicker than that of WC-Co-Zr, so it takes a longer time before the mutation of friction coefficient for WC-Ni-Zr, as compared with that for WC-Co-Zr. Figure 5 also shows that quite different from the friction behavior of conventional carbides, the formation of ZrN on the FGC surface can effectively reduce the friction coefficient.

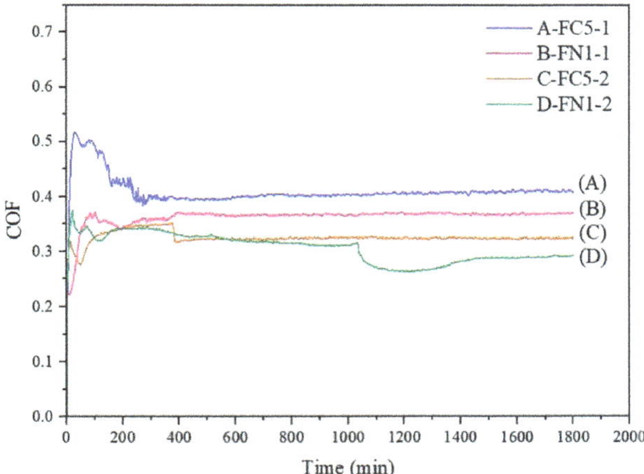

Figure 5. The friction testing curves for both FGCs and conventional carbides.

Figure 6 shows the comparison of wear rates of four cemented carbides. It can be seen that the wear rate of the FGCs is significantly lower than that of conventional carbides with homogeneous microstructure, indicating that a ZrN layer formed on the surface of the FGCs can significantly improve the wear resistance. Further, the wear rate of the WC-Co-Zr FGCs is slightly lower than that of the WC-Ni-Zr FGCs.

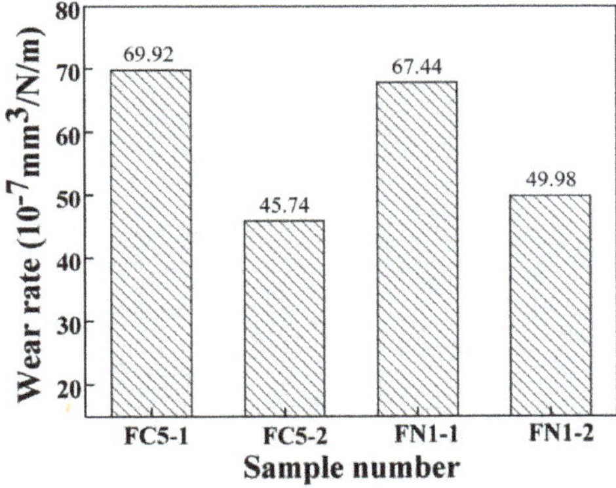

Figure 6. Wear rates of FGCs and conventional carbides.

3.2. Coated Cutters with FGC and Conventional Carbide Substrates

3.2.1. Microstructure of Coated Cutters

Figure 7 shows the cross-sectional microstructures of four coated carbides. Two types of coated cutters on conventional carbide substrates with homogeneous microstructure have no cracks around the interface between the coating and the substrate, but there are some micro-cracks for the coated cutters on the WC-Ni-Zr FGC substrate (FN1-4).

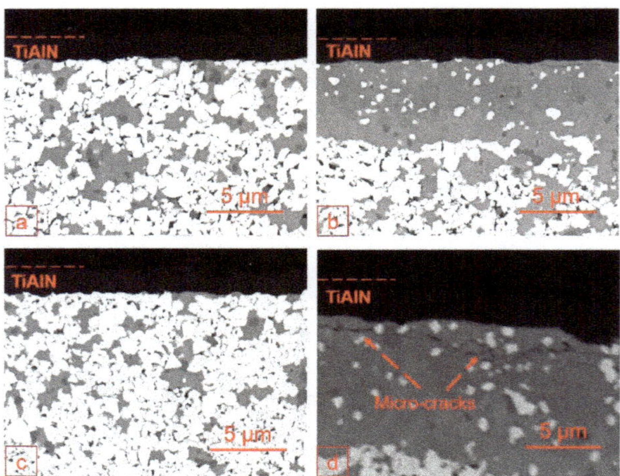

Figure 7. Cross-sectional microstructures of four coated cutters on FGC and conventional carbide substrates, (**a**) FC5-3, (**b**) FC5-4, (**c**) FN1-3, and (**d**) FN1-4.

3.2.2. Bonding Strength between Carbide Substrate and Coating

Figure 8 shows the bonding strength and scratch morphology of the coated cutters. It shows that for WC-Co-Zr, the bonding strength between coating and the FGC substrate is about 15% higher than that between coating and the conventional carbide substrate. However, for WC-Ni-Zr carbide, bonding strength shows the opposite trend, mainly due to the obvious microcracks between the FGC substrate and the coating. The average hardness of the coatings, measured by nanoindentation, is 32.17 GPa.

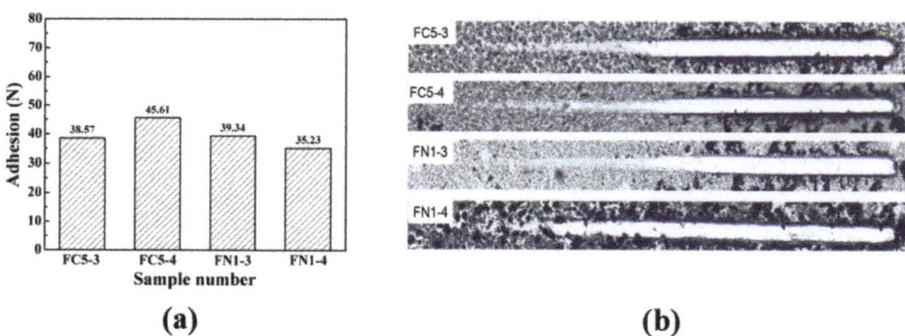

Figure 8. Bonding between carbide substrates and coatings, (**a**) bonding strength, and (**b**) scratching tests.

3.2.3. Friction and Wear Performance of Coated Carbides

Figure 9 is the friction coefficient curves of four coated carbides. It can be observed that for both coated WC-Co-Zr with either conventional carbide substrate (curve A) or FGC substrate (curve B), the friction coefficient shows substantial fluctuation, and the friction coefficient of coated WC-Co-Zr with FGC substrate is noticeably lower than that of coated WC-Co-Zr with conventional carbide substrate. For coated WC-Ni-Zr, the friction coefficient corresponding to FGC substrate (curve C) is quite similar to that corresponding to conventional carbide substrate, while the former has wider fluctuation, which is mainly caused by the apparent more micro-cracks of the former, as shown in Figure 7d.

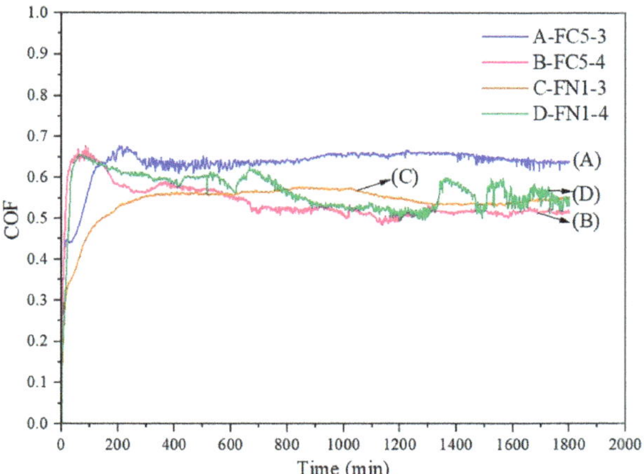

Figure 9. Friction coefficient curves of four coated carbides.

Figure 10 shows the wear rates of four coated carbides. The coated WC-Co-Zr with FGC substrate shows the lowest wear rate, while the coated WC-Ni-Zr with FGC substrate have the highest wear rate due to the existence of micro-cracks.

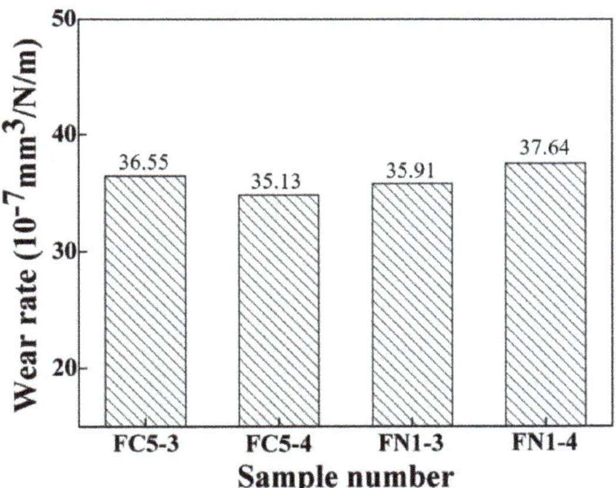

Figure 10. Wear rate of four coated carbides.

3.3. Cutting Performance of Coated Carbide Cutters

3.3.1. Carbide Cutters

The wear and cutting behavior of FGCs is quite different from those of conventional carbides [17,29]. In this study, both the FGCs and conventional carbides listed in Table 1 were used as the cutters and HT250 was used as the workpiece for cutting experiments. The cutting test parameters are shown in Section 1.

3.3.2. Cutting Life

According to ISO-3685, the criterion for cutter failure is average wear of 0.3 mm on flank face of cutters. Figures 11 and 12 show the dimension of wear vs. time on the flank face of four carbide cutters during cutting test of HT250. It can be seen that the cutting life of WC-Co-Zr is significantly higher than that of WC-Ni-Zr. For WC-Co-Zr, the cutting life of uncoated FGC cutter is about 1.23 times that of conventional carbide cutter with homogeneous microstructure, and the cutting life of the coated cutter with FGC substrate is 1.11 times that of the coated cutter with conventional carbide substrate. While for WC-Ni-Zr, the corresponding cutting life improvement of uncoated and coated FGC vs. conventional carbide is 1.54 times and 1.19 times, respectively.

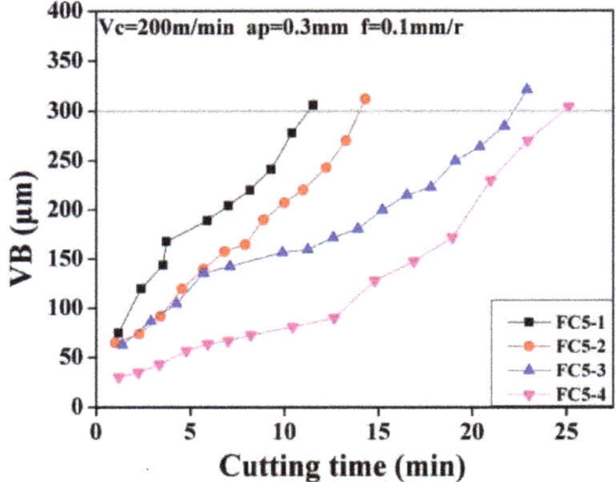

Figure 11. The flank wear VB vs. cutting time for WC-Co-Zr.

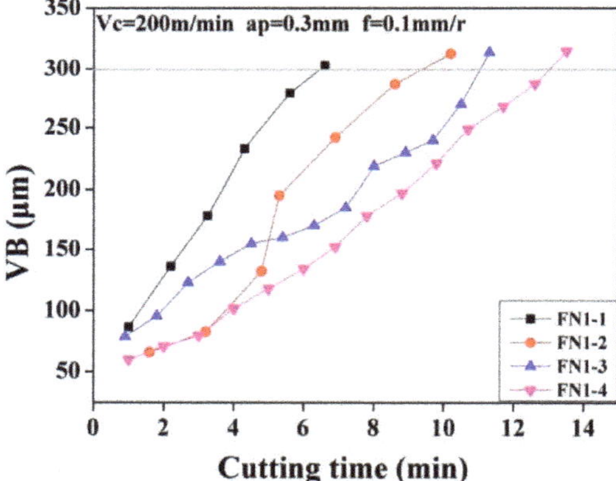

Figure 12. The flank wear VB vs. cutting time for WC-Ni-Zr.

3.3.3. Wear Mechanism Analysis

The wear of flank face is commonly used to assess the failure modes of carbide cutters and can be classified as adhesive wear, diffusion wear, abrasive wear, and oxidation wear [30–32].

Figure 13 shows the flank face of WC-Co-Zr cutters after the cutting test. The failure modes are mainly adhesive wear, a certain degree of spalling of coating, and abrasive wear of the cutter tip. Table 2 is the EDS results at Point 1 and Point 2 on the flank face in Figure 13. EDS analysis shows that the gray region (Point 1) is the adhesive material (HT250), and the bright region (Point 2) is the exposed carbide substrate after the coating is worn off. The low content of oxygen indicates limited oxidation wear at the high cutting temperature. Furthermore, it can be seen in Figure 13 that there is much less adhesion on flank face of the coated cutters than that on flank face of the uncoated cutters.

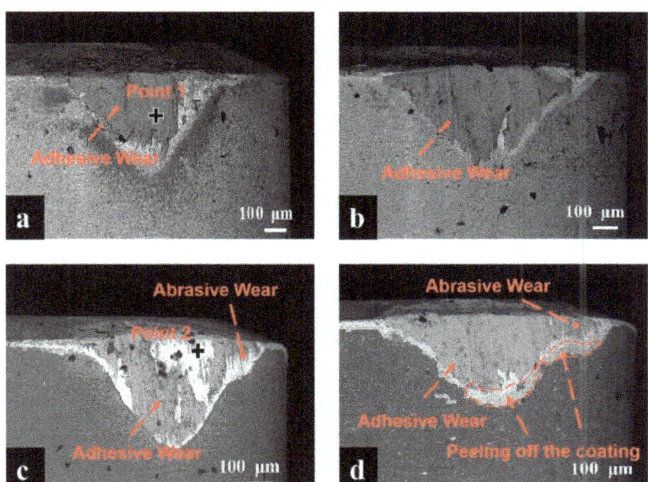

Figure 13. Morphology of flank wear for WC-Co-Zr cutters, (**a**) FC5-1, (**b**) FC5-2, (**c**) FC5-3, and (**d**) FC5-4.

Table 2. The EDS results of Points 1 and 2 in Figure 12.

Area	Element Mass Fraction/%							
	C	O	Si	Zr	W	Fe	Mn	Co
Point 1	7.06	1.77	2.27	-	-	88.91	-	-
Point 2	11.5	1.98	-	11.58	71.41	2.78	-	1.2

Figure 14 shows the morphology of flank face of WC-Ni-Zr after cutting test. The wear modes of WC-Ni-Zr cutters also include mainly adhesive wear, which is the main wear mode of cemented carbide cutters at low and medium cutting speeds [30]. The spalling of coating also occurs and is more severe than that for WC-Co-Zr cutters. Table 3 shows the EDS analysis of the points in Figure 14. Compared with the WC-Co-Zr, the oxidation wear for WC-Ni-Zr is more serious.

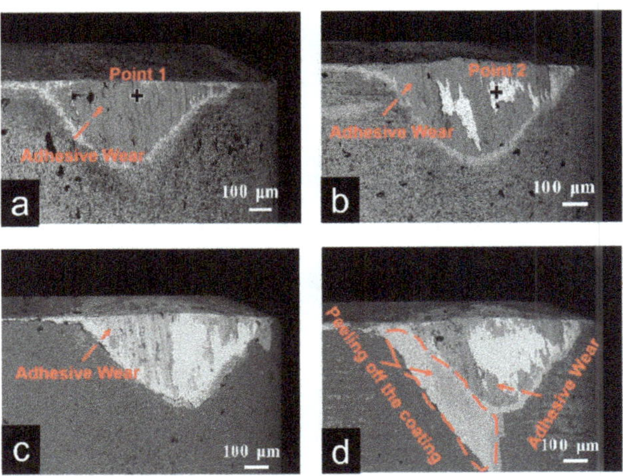

Figure 14. Morphology of flank wear for WC-Ni-Zr cutters, (**a**) FN1-1, (**b**) FN1-2, (**c**) FN1-3, and (**d**) FN1-4.

Table 3. The EDS results of Points 1 and 2 in Figure 13.

Area	Element Mass Fraction/%							
	C	O	Si	Zr	W	Fe	Mn	Co
Point 1	7.03	1.97	0.18	-	-	89.12	-	-
Point 2	9.63	2.62	-	-	76.72	9	-	2.02

4. Conclusions

In this study, WC-Co/Ni-Zr cemented carbides with homogeneous microstructure (conventional carbides) and the corresponding FGCs with ZrN (face centered cubic phase)-enriched surfaces were successfully synthesized using a high-pressure nitriding process. TiAlN coatings were prepared on these carbide substrates by a PVD process, and the mechanical and friction behaviors and cutting test were carried out on these coated and uncoated carbides. It was found that for WC-Co-Zr, friction coefficient and wear rate of coated and uncoated FGCs were, respectively, 20% and 17% lower than those of coated and uncoated conventional carbides and for WC-Ni-Zr; uncoated FGC also shows 14% lower friction coefficient and 26% lower wear rate than uncoated conventional carbide, while the coated FGC shows wider fluctuation of friction coefficient and 5% higher wear rate as compared with coated conventional carbide. For WC-Co-Zr, the bonding strength between the coating and the FGC substrate is 18% higher than that between the coating and the conventional carbide substrate, while for WC-Ni-Zr, the bonding strength between the coating and the FGC substrate, due to micro-cracks, is 10% lower than that between the coating and the conventional carbide substrate. The cutting tests showed that the cutting life of WC-Co-Zr cutters is about 85% higher than that of WC-Ni-Zr cutters, and the coated FGCs show 16% higher cutting life than the coated conventional carbides, followed by the uncoated FGCs, and the uncoated conventional carbides. The cutting test for HT250 shows the main wear mechanism is adhesive wear for all the carbide cutters.

Author Contributions: Conceptualization, X.D. and Z.L. (Zhongliang Lu); methodology, S.L.; investigation, X.X., J.C. and T.L.; formal analysis, Z.Z.; data curation, Z.L. (Zhangxu Li); resources, Z.W.; writing—original draft preparation, S.L.; writing—review and editing, X.D. and Z.L. (Zhongliang Lu); supervision, Z.Q.; project administration, X.D. and Z.Q.; funding acquisition, Z.L. (Zhongliang Lu), X.D., Z.W. and Z.Q. All authors have read and agreed to the published version of the manuscript.

Funding: This research was funded by Science and Technology Project of Guangdong Province, China (Grant No. 2017B090911006), Jihua Laboratory Project (Grant No. X190061UZ190), Foshan Science and Technology Innovation Team project (FS0AA-KJ919-4402-0023), and Heyuan Science and Technology Project (Grant No. HEKE 000781).

Institutional Review Board Statement: Not applicable.

Informed Consent Statement: Not applicable.

Data Availability Statement: Not applicable.

Conflicts of Interest: The authors declare no conflict of interest.

References

1. García, J.; Ciprés, V.C.; Blomqvist, A.; Kaplan, B. Cemented carbide microstructures: A review. *Int. J. Refract. Met. Hard Mater.* **2018**, *80*, 40–68. [CrossRef]
2. Ortner, H.M.; Ettmayer, P.; Kolaska, H. The history of the technological progress of hardmetals. *Int. J. Refract. Met. Hard Mater.* **2014**, *44*, 148–159. [CrossRef]
3. Fang, Z.Z.; Koopman, M.C.; Wang, H. Cemented Tungsten Carbide Hardmetal—An Introduction. *Compr. Hard Mater.* **2014**, *1*, 123–137.
4. Upadhyaya, G.S. Materials science of cemented carbides-an overview. *Mater. Des.* **2001**, *22*, 483–489. [CrossRef]
5. Emani, S.V.; Ramos dos Santos, A.F.C.; Shaw, L.L.; Chen, Z. Investigation of microstructure and mechanical properties at low and high temperatures of WC–6 wt.% Co. *Int. J. Refract. Met. Hard Mater.* **2016**, *58*, 172–181. [CrossRef]

6.	Lu, Z.Y.; Du, J.; Sun, Y.J.; Su, G.S.; Zhang, C.Y.; Kong, X.M. Effect of ultrafine WC contents on the microstructures, mechanical properties and wear resistances of regenerated coarse grained WC-10Co cemented carbides. *Int. J. Refract. Met. Hard Mater.* **2021**, *97*, 105516. [CrossRef]
7.	Roulon, Z.; Missiaen, J.M.; Lay, S. Shrinkage and microstructure evolution during sintering of cemented carbides with alternative binders. *Int. J. Refract. Met. Hard Mater.* **2021**, *101*, 105665. [CrossRef]
8.	Shaw, L.L.; Luo, H.; Zhong, Y. WC-18 wt.% Co with simultaneous improvements in hardness and toughness derived from nanocrystalline powder. *Mater. Sci. Eng. A* **2012**, *537*, 39–48. [CrossRef]
9.	Yang, G.J.; Gao, P.H.; Li, C.X.; Li, C.J. Simultaneous strengthening and toughening effects in WC–(nanoWC–Co). *Scr. Mater.* **2012**, *66*, 777–780. [CrossRef]
10.	Salmasi, A.; Andreas, B.; Henrik, L. Geometry effects during sintering of graded cemented carbides: Modelling of microstructural evolution and mechanical properties. *Results Mater.* **2019**, *24*, 8–11. [CrossRef]
11.	Chen, J.; Zhou, L.; Liang, J.X.; Liu, B.Y.; Liu, J.Y.; Chen, Y.; Deng, X.; Wu, S.H.; Huang, M.J. Effect of initial WC particle size on grain growth behavior and gradient structure formation of bilayer functionally graded cemented carbides. *Mater. Chem. Phys.* **2021**, *271*, 124919. [CrossRef]
12.	Liu, Y.; Wang, H.B.; Yang, J.G.; He, Y.H.; Long, Z.Y. Relationship between structure and properties of graded cemented carbide. *Mater. Sci. Eng. Powder Met.* **2005**, *6*, 356–360.
13.	Li, X.F.; Liu, Y.; Liu, B.; Zhou, J.H. Effects of submicron WC addition on structures, kinetics and mechanical properties of functionally graded cemented carbides with coarse grains. *Int. J. Refract. Met. Hard Mater.* **2016**, *56*, 132–138. [CrossRef]
14.	Zhang, L.; Zhong, Z.Q.; Qiu, L.C.; Shi, H.D.; Layyous, A.; Liu, S.P. Coated cemented carbide tool life extension accompanied by comb cracks: The milling case of 316L stainless steel. *Wear* **2019**, *418*, 133–139. [CrossRef]
15.	Frykholm, R.; Andrén, H.O. Development of the microstructure during gradient sintering of a cemented carbide. *Mater. Chem. Phys.* **2001**, *67*, 203–208. [CrossRef]
16.	Tang, S.W.; Liu, D.S.; Li, P.N.; Jiang, L.L.; Liu, W.H.; Chen, Y.Q.; Niu, Q.L. Microstructure and mechanical properties of functionally gradient cemented carbides fabricated by microwave heating nitriding sintering. *Int. J. Refract. Met. Hard Mater.* **2016**, *58*, 137–142. [CrossRef]
17.	Garcia, J.; Reinhard, P. The role of cemented carbide functionally graded outer–layers on the wear performance of coated cutting tools. *Int. J. Refract. Met. Hard Mater.* **2013**, *36*, 52–59. [CrossRef]
18.	Ruys, A.J.; Sutton, B.A. 9-Metal-ceramic functionally graded materials (FGMs). In *Elsevier Series on Advanced Ceramic Materials, Metal-Reinforced Ceramics*, 1st ed.; Ruys, A.J., Ed.; Woodhead Publishing: Cambridge, UK, 2021; pp. 327–359.
19.	Carneiro, M.B.; Machado, I.F. Sintering and Model of Thermal Residual Stress for Getting Cutting Tools from Functionally Gradient Materials. *Procedia CIRP* **2013**, *8*, 200–205. [CrossRef]
20.	Xu, X.L. Research on the Production of Gradient Dual-Phase Cemented Carbide by Single Sintering Process. *Cem. Carbide* **2008**, *25*, 12–18.
21.	Zhang, L.; Chen, S.; Xiong, X.J.; He, Y.H.; Huang, B.Y.; Zhang, C.F. Phase composition, transition and structure stability of functionally graded cemented carbide with dual phase structure. *J. Cent. South Univ. Technol.* **2007**, *14*, 149–152. [CrossRef]
22.	Zhang, J.F.; Xiong, J.; Li, Y.; Guo, L. The Structure and Performance of Gradient Cemented Carbide for Diamond Coating. *Cem. Carbide* **2003**, *20*, 200–203.
23.	Wang, H.B.; Lu, H.; Song, X.Y.; Yan, X.F.; Liu, X.M.; Nie, Z.R. Corrosion resistance enhancement of WC cermet coating by carbides alloying. *Corros. Sci.* **2019**, *147*, 372–383. [CrossRef]
24.	Chen, J.; Deng, X.; Gong, M.F.; Liu, W.; Wu, S.H. Research into preparation and properties of graded cemented carbides with face center cubic-rich surface layer. *Appl. Surf. Sci.* **2016**, *380*, 108–113. [CrossRef]
25.	Barbatti, C.; Garcia, J.; Sket, F.; Kostka, A.; Pyzallaet, A.R. Influence of nitridation on surface microstructure and properties of graded cemented carbides with Co and Ni binders. *Surf. Coat. Technol.* **2008**, *202*, 5962–5975. [CrossRef]
26.	Konigshofer, R.; Eder, A.; Lengauer, W.; Dreyer, K.; Kassel, D.; Daub, H.W.; Berg, H. Growth of the graded zone and its impact on cutting performance in high-pressure nitrogen modified functionally gradient hardmetals. *J. Alloys Compd.* **2004**, *366*, 228–232. [CrossRef]
27.	Akhter, R.; Zhou, Z.F.; Xie, Z.H.; Munroe, P. TiN versus TiSiN coatings in indentation, scratch and wear setting. *Appl. Surf. Sci.* **2021**, *563*, 150356. [CrossRef]
28.	Gross, K.A.; Lungevics, J.; Zavickis, J.; Pluduma, L. A comparison of quality control methods for scratch detection on polished metal surfaces. *Measurement* **2018**, *117*, 397–402. [CrossRef]
29.	Mohammadpour, M.; Abachi, P.; Pourazarang, K. Effect of cobalt replacement by nickel on functionally graded cemented carbonitrides. *Int. J. Refract. Met. Hard Mater.* **2012**, *30*, 42–47. [CrossRef]
30.	Chang, K.S.; Dong, Y.J.; Zheng, G.M.; Zheng, G.M.; Jiang, X.L.; Yang, X.H.; Cheng, X.; Liu, H.B.; Zhao, G.X. Friction and wear properties of TiAlN coated tools with different levels of surface integrity. *Ceram. Int.* **2021**, *10*, 105. [CrossRef]
31.	Fang, S.Q. Wear Assessment of Cemented Carbide Tools (WC-Co) with Defined Cutting Edges under Grinding-like Service Conditions. *Wear* **2021**, *476*, 203744. [CrossRef]
32.	Neves, D.; Diniz, A.E.; Lima, M.S.F. Microstructural analyses and wear behavior of the cemented carbide tools after laser surface treatment and PVD coating. *Appl. Sur. Sci.* **2013**, *282*, 680–688. [CrossRef]

Article

The Fracture Behavior of 316L Stainless Steel with Defects Fabricated by SLM Additive Manufacturing

Hui Li [1],* and Jianhao Zhang [2]

1   Science and Technology on Reactor System Design Technology Laboratory, Nuclear Power Institute of China, No. 328 Changshun Avenue, Chengdu 610200, China
2   International Research Center for Computational Mechanics, State Key Laboratory of Structural Analysis for Industrial Equipment, Faculty of Vehicle Engineering and Mechanics, Dalian University of Technology, Dalian 116024, China; 1049066751@mail.dlut.edu.cn
*   Correspondence: lhcnpi@163.com

Abstract: In this paper, the fracture behaviors of 316L stainless steel with defects fabricated by the Selective Laser Melting (SLM) additive manufacturing are studied by a peridynamic method. Firstly, the incremental formulations in the peridynamic framework are presented for the elastic-plastic problems. Then, the pairwise force of a bond for orthotropic material model is proposed according to both the local and the global coordinate systems. A simple three-step approach is developed to describe the void defects that generated in the processing of the SLM additive manufacturing in the numerical model. Next, some representative numerical examples are carried out, whose results explain the validation and accuracy of the present method, and demonstrate that the orthotropic features, micro-cracks and voids of the materials have a significant influence on the ultimate bearing capacity, crack propagation and branching of the corresponding structures. It is also revealed that the crack initiations are induced actively by the defects and the crack branching is contributed to the complex multiple-crack propagation. Finally, the achievements of this paper lay a foundation for the engineering applications of the SLM additive manufacturing materials.

Keywords: peridynamic method; multiple-crack propagation; SLM additive manufacturing; defects

Citation: Li, H.; Zhang, J. The Fracture Behavior of 316L Stainless Steel with Defects Fabricated by SLM Additive Manufacturing. *Crystals* **2021**, *11*, 1542. https://doi.org/10.3390/cryst11121542

Academic Editors: Ana Pilar Valerga Puerta, Severo Raul Fernandez-Vidal, Zhao Zhang and Umberto Prisco

Received: 14 November 2021
Accepted: 7 December 2021
Published: 9 December 2021

Publisher's Note: MDPI stays neutral with regard to jurisdictional claims in published maps and institutional affiliations.

1. Introduction

Additive manufacturing techniques are on the basis of the design data and constitute a novel manufacturing method to produce solid structures by overlaying the raw materials automatically, which is attributed to its complex construction capability with high accuracy and good surface quality. At the same time, it cuts down greatly the time needed design and developing for the new products. Therefore, additive manufacturing techniques are promising to break through the limitations of traditional casting methods and provide the opportunity to realize the transformation and upgrading of the manufacturing industries [1–3]. According to different kinds of requirements, heat sources and materials, a great deal of additive manufacturing methods have been proposed, such as the Fused Deposition Modeling (FDM) method, the Electron Beam Melting (EBM) method, the Selective Laser Melting (SLM) method, and so on. Of these, it is pleasing that the SLM additive manufacturing method has great potential in the fields of aerospace, automotive and nuclear energy because of the good mechanical properties of the metal materials fabricated by this additive manufacturing method.

It is revealed that the process parameters of the SLM additive manufacture method, such as the layer thickness, the hatch spacing, the laser power, the scanning direction and speed, directly determine the temperature distribution, the grain growth, and the deformation of the corresponding materials [4]. Meanwhile, the defects of the micro-cracks and the voids are usually caused by the reasons of the nodulizing of the deposited metal, the high temperature gradient and the partially melted zone in the process, which has also attracted much attention from economists [5,6]. Until now, a lot of the related methods have been proposed to study the effects of different kinds of defects on the mechanical behaviors of the SLM additive manufacturing materials. For example, Zhang et al. [7] summarized the recent research outcomes on defect findings and classification in SLM, analyzed formation mechanisms of the common defects, such as porosities, incomplete fusion holes, and cracks. Jiang et al. [8] and Wu et al. [9] studied the effect of process parameters on defects, melt pool shape, microstructure and tensile behavior of the materials fabricated by the SLM. Zhang et al. [10] investigated the process parameters and post processing on fracture toughness and fatigue crack growth of the selective laser melting Ti-6Al-4V. Weber et al. [11] studied the fracture behavior and mechanical properties of a support structure for additive manufacturing of Ti-6Al-4V based on the experimental method. Glodez et al. [12] investigated the high-cycle fatigue and fracture behaviors of the selective laser melting SlSi10Mg alloy with various kinds of defects. Wen et al. [13] developed a collective gray value method to automatically identify a class of defects of the additive manufacturing materials based on the different features of the defects in CT images. Lozanovski et al. [14] proposed a computational modelling for the mechanical analysis of the SLM manufactured lattice structures with hypothetical defects. In conclusion, the properties of the static strength, fatigue and fracture of the materials are significantly influenced by the defects of the micro-cracks and voids generated in the processing of the SLM additive manufacture, and extensive experimental methods have been usually applied to study these fracture behaviors. However, the experiments often have the disadvantages of long testing cycle and expensive cost, and the traditional numerical method is difficult to deal with the analysis of the complex fracture problems for the SLM fabricated materials with defects. Therefore, it is of great significance to develop an effective numerical method to study the complex fracture behaviors of materials with defects fabricated by SLM additive manufacturing.

In the most recent two decades, a non-local theory of the peridynamics proposed by Silling [15] has been continuously developed by many researchers [16–20], including the conventional bond-based model, the advanced ordinary/non-ordinary state-based model, and the dual-horizon model, etc. Meanwhile, peridynamics has been widely used in the mechanical fields [21–29]. Comparing with the continuum theory, the peridynamics is expressed by an integral equation in place of the differential equation, which is profitable to overcome the problems of the stress singularity at the tip of the cracks. For instance, Katiyar et al. [21,22] developed a peridynamic formulation for the pressure driven convective fluid transport in porous media with crack. Lai et al. [23,24] studied the geomaterial fragmentation by blast loads with the peridynamic method. Hu et al. [25] proposed a peridynamic model for dynamic fracture in unidirectional fiber-reinforced composites. Huang et al. [26–28] studied the quasi-static fracture and crack propagation problems of the brittle materials by an improved peridynamic approach. Oterkus and Madenci [29] presented a peridynamic model for the analysis of fracture problems in a fuel pellet. In a word, the peridynamics can easily handle the complex fracture problems, in which the crack path and branching are achieved to be a natural outcome of the simulation without any complicated numerical algorithms for crack path tracking. Consequently, the aim of this paper is to develop a peridynamic method for the analysis of the fracture behaviors of the SLM fabricated metal materials with defects, taking the 316L stainless steel for example.

This paper is organized as follows. Section 2 gives out the bond-based peridynamic formulations for the SLM additive manufactured orthotropic material and the fracture model. It also provides the way to present the pre-existing micro-cracks and voids. Section 3 shows several simple but illustrative numerical examples to verify the present method and to study the fracture behaviors of the 316L stainless steel with different kinds of defects. Finally, some conclusions are investigated in Section 4.

2. Peridynamic Model

2.1. Equation of Motion in Elastic-Plastic Theory

With the bond-based peridynamic theory and the meshless method [16,17], the domain of the interest is discretized into uniform material points, i.e., $\{\mathbf{x}^i, i = 1, 2, 3, \cdots N\}$, $\mathbf{x}$ is the coordinate of the material point and N is the total number of the material points. Then, the peridynamic expression of motion for any material point $\mathbf{x}^i$ can be expressed as

$$\rho \ddot{\mathbf{u}}\left(\mathbf{x}^i, t\right) = \int_{H_{\mathbf{x}^i}} \mathbf{f}(\boldsymbol{\xi}, \boldsymbol{\eta}) dV_{\mathbf{x}^j} + \mathbf{b}\left(\mathbf{x}^i, t\right) \tag{1}$$

in which, ρ is the mass density, $\ddot{\mathbf{u}}(\mathbf{x}^i, t)$ and $\mathbf{u}^i$ are the acceleration and displacement of material point $\mathbf{x}^i$ at time t, respectively. $H_{\mathbf{x}^i} = \{\mathbf{x}^j \mid \parallel \mathbf{x}^j - \mathbf{x}^i \parallel \leq \delta\}$ is the neighborhood of $\mathbf{x}^i$ with a horizon of δ, $V_{\mathbf{x}^j}$ is the volume of $\mathbf{x}^j$, $\mathbf{b}(\mathbf{x}^i, t)$ is a prescribed body force density, and $\mathbf{f}(\boldsymbol{\xi}, \boldsymbol{\eta})$ is a pairwise force (per unit volume squared) that the material point $\mathbf{x}^j$ exerts on the material point $\mathbf{x}^i$. The bond $\boldsymbol{\xi} = \mathbf{x}^j - \mathbf{x}^i$ and $\boldsymbol{\eta} = \mathbf{u}^j - \mathbf{u}^i$ present the relative position and displacement of the material points $\mathbf{x}^i$ and $\mathbf{x}^j$ in the initial configuration and current configuration, respectively.

On the basis of the mechanical considerations of the conservation of linear and angular momenta, the pairwise force $\mathbf{f}(\boldsymbol{\xi}, \boldsymbol{\eta})$ can be expressed as

$$\mathbf{f}(\boldsymbol{\xi}, \boldsymbol{\eta}) = \underline{f} \frac{\boldsymbol{\xi} + \boldsymbol{\eta}}{\parallel \boldsymbol{\xi} + \boldsymbol{\eta} \parallel} \tag{2}$$

To solve the elastic-plastic problems, the pairwise force is transferred into incremental form and the solutions are expressed as [18]

$$\dot{\underline{f}} = \begin{cases} c\overline{\omega}(\parallel \xi \parallel) \cdot \dot{s}, & \text{if } \left| \underline{f} \right| < f_y \text{ or } \underline{f} \cdot \dot{s} < 0 \\ 0, & \text{otherwise} \end{cases} \tag{3}$$

where c denotes the micro-modulus in a bond, $\overline{\omega}(\parallel \xi \parallel)$ is the weight function that presents the distribution of the force along the radius direction, the elongation of the bond s is defined as

$$s = \frac{\parallel \boldsymbol{\xi} + \boldsymbol{\eta} \parallel - \parallel \boldsymbol{\xi} \parallel}{\parallel \boldsymbol{\xi} \parallel} \tag{4}$$

In addition, the value of pairwise force corresponding to the material yield can be obtained by [18];

$$f_y = \begin{cases} \dfrac{2\sigma_y}{A\delta^2}, & 1D \\[2mm] \dfrac{12\sigma_y}{\pi h \delta^2}, & 2D \\[2mm] \dfrac{6\sigma_y}{\pi \delta^4}, & 3D \end{cases} \tag{5}$$

where A is the cross-sectional area, h is the thickness, σ_y is the yield stress of the material in continuum mechanics framework.

2.2. Orthotropic Material and Fracture Model

To consider the orthotropic characteristics of the materials fabricated by the SLM additive manufacturing, the orthotropic material model in peridynamics is improved in this section. Distinguishing from the traditional casting materials, the SLM additive manufacturing materials are built layer by layer on the basis of a planning scheme until it is complete, which leads the orthotropic features of the materials in mechanics. Meanwhile, the tensile strength along the laser scanning direction on a layer is greater than that along the orthotropic direction, that is the layer-to-layer orientation. Therefore, the pairwise force of a bond for the orthotropic material model can be expressed as

$$\dot{\mathbf{f}}(\xi, \eta) = \begin{cases} \overline{\omega}(\| \zeta \|) \cdot \dot{s}\mathbf{CHT}, & \text{if } \left|\underline{f}\right| < f_y \text{ or } \underline{f} \cdot \dot{s} < 0 \\ 0, & \text{otherwise} \end{cases} \tag{6}$$

where $\mathbf{C} = \begin{bmatrix} c^{\parallel} & 0 \\ 0 & c^{\perp} \end{bmatrix}$ is the orthotropic elasticity matrix, $c^{\parallel}$ and $c^{\perp}$ are the micro-moduli along the laser scanning direction on a layer and the layer-to-layer direction, respectively. $\mathbf{H}$ is a matrix for the coordinate transformation from the local coordinate system to the global coordinate system. $\mathbf{T}$ is a direction vector of the bond ξ in the local coordinate system. As shown in Figure 1, $O'x'y'$ is the local coordinate system that related to the laser scanning direction on a layer and Oxy is the global coordinate system. Thus, the coordinate transformation matrix $\mathbf{H}$ and the direction vector $\mathbf{T}$ can be written as

$$\mathbf{H} = \begin{bmatrix} cos\theta & -sin\theta \\ sin\theta & cos\theta \end{bmatrix} \tag{7}$$

and,

$$\mathbf{T} = \begin{bmatrix} cos\alpha \\ sin\alpha \end{bmatrix} \tag{8}$$

in which, θ denotes the angle between the laser scanning direction on a layer and the x axis in the global coordinate system, α denotes the angle between the bond ξ and the x' axis in the local coordinate system.

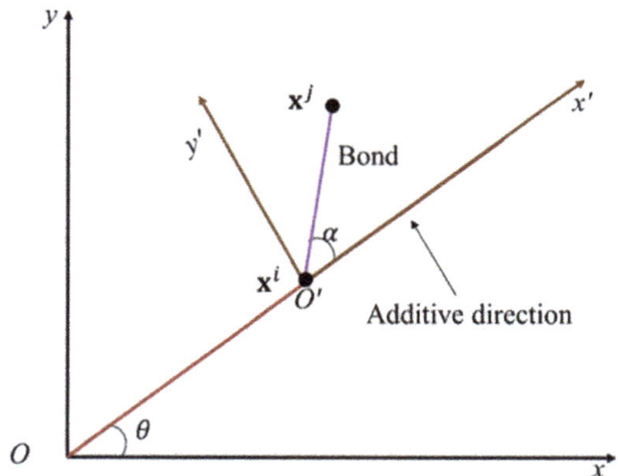

Figure 1. Schematic diagram of the local the global coordinate systems.

In addition, a scalar μ that used to define the failure state of the bond is introduced into Equation (6) to develop the fracture model in PD. Then, Equation (6) can be rewritten by

$$\dot{\mathbf{f}}(\xi, \eta) = \begin{cases} \overline{\omega}(\| \zeta \|) \cdot \dot{s}\mu\mathbf{CHT}, & \text{if } \left|\underline{f}\right| < f_y \text{ or } \underline{f} \cdot \dot{s} < 0 \\ 0, & \text{otherwise} \end{cases} \tag{9}$$

where the scalar μ for the linear and bilinear material can be expressed as [28]

$$\mu = \begin{cases} 1, & s < s_0, & \text{undamge} \\ 1 - d\frac{s_{max}-s_0}{s_{max}}, & s < s_{max} \text{ and } s_0 < s_{max} < s_c, & \text{unloading or reloading} \\ 1 - d\frac{s-s_0}{s_{max}}, & s = s_{max} \text{ and } s_0 < s_{max} < s_c, & \text{maxstreth increasing loading} \\ 0, & s \geq s_c, & \text{failed bond} \end{cases} \tag{10}$$

in which s_0 and s_c are the critical stretches of the bond at the linear elastic limit and at the failure state, respectively, and $s_{\max}$ is the maximum value of the stretch in the deformation history. s_c can be obtained by the critical energy release rate of the material and is calculated by [30]

$$s_c = \sqrt{\frac{G_c}{(\frac{6}{\pi}v + \frac{16}{9\pi^2}(\kappa - 2v))\delta}} \tag{11}$$

where G_c is the critical energy release rate, κ and v are the bulk modulus and shear modulus, respectively.

Further, a scaler field $\varphi(\mathbf{x}^i, t)$ is utilized to evaluate the failure state of the material, i.e.,

$$\varphi\left(\mathbf{x}^i, t\right) = 1 - \frac{\int_{H_{\mathbf{x}^i}} \mu dV_{\mathbf{x}^j}}{\int_{H_{\mathbf{x}^i}} dV_{\mathbf{x}^j}} \tag{12}$$

It can be seen from Equation (12) that the scaler field $\varphi(\mathbf{x}^i, t)$ is a damage factor and denotes the ratio between the broken bonds of the material point $\mathbf{x}^i$ and the whole bonds in its neighborhood at time t. Therefore, the values of $\varphi(\mathbf{x}^i, t)$ are $0 \le \varphi(\mathbf{x}^i, t) \le 1$, with $\varphi(\mathbf{x}^i, t) = 0$ presenting an undamaged state while $\varphi(\mathbf{x}^i, t) = 1$ indicating a complete separate state.

2.3. Characterization of Micro-Cracks and Voids

In the processing of the SLM additive manufacturing, the elaborate mechanism comprising of the materials may result in the defects of the micro-cracks and voids due to the melting, solidification and phase transition, which has a great influence on the behaviors of the strength, fatigue and fracture of the materials. In this paper, on the basis of the shapes and sizes of the voids obtained by the measurement methods [13], the defects of the voids in the geometric model will be generated with a three-step approach:

Step 1: Discretize the whole region into material points with the meshless method.

Step 2: Generate the center coordinates of the voids with the uniform and random distribution methods according to the number and volume fraction of the voids in the whole region.

Step 3: Define the volume fraction of the voids Φ and the critical threshold r. Here, r presents the distances between the centers of the voids and the material points. In the case of spherical voids, the material points will be removed when the distances between the material points and the center of sphere is less than the radius of the sphere.

In this paper, the fictitious defects are presented. Figures 2 and 3 show the void defects with different sizes and volume fractions using uniform and random distribution methods, respectively, in which the blue spots denote the void defects.

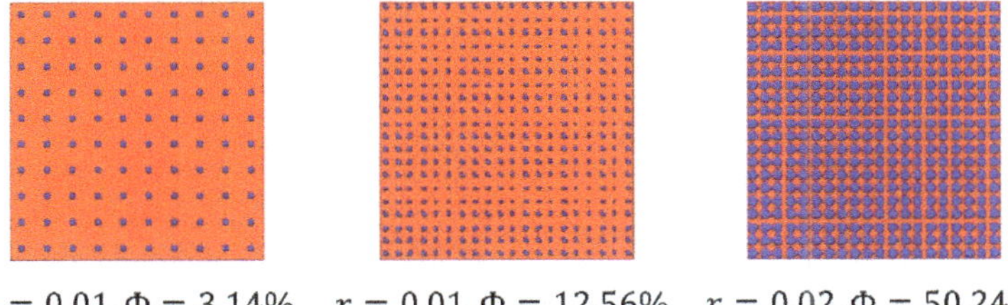

$r = 0.01, \Phi = 3.14\%$ $r = 0.01, \Phi = 12.56\%$ $r = 0.02, \Phi = 50.24\%$

Figure 2. Uniform distribution void defects with different volume fractions.

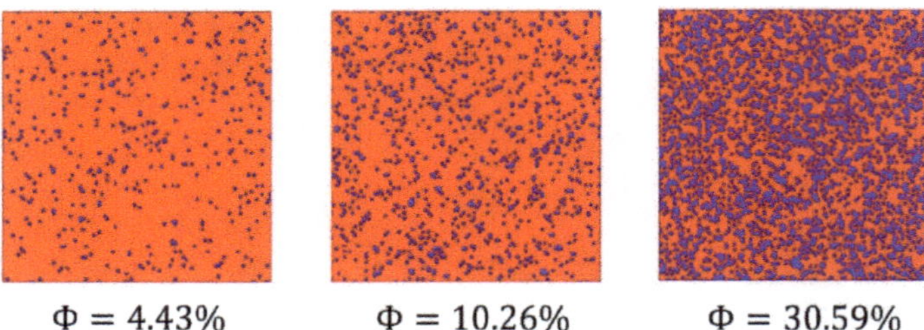

Figure 3. Random distribution void defects with different volume fractions.

3. Numerical Examples

A classical benchmark analyzed by Zaccaritto et al. [31] and Carpinteri and Colombo [32] is considered to verify the validation and accuracy of the present method for the analysis of the crack propagation problems. As shown in Figure 4, the curves of the reaction force versus the imposed displacement obtained by the present method fit well with the reference solutions and the overall slopes of the curves are well captured, which demonstrates the validation and accuracy of the present peridynamic method. Next, the proposed method will be used to analyze the effect of the pre-existing micro-crack, the orthotropic feature, and the void defects on the fracture behaviors of the SLM additive manufacturing 316L stainless steel.

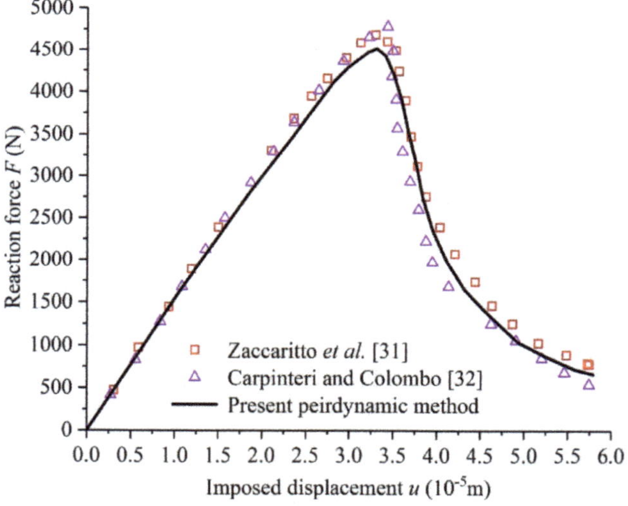

Figure 4. Curves of the reaction force versus imposed displacement obtained by the present method and reference solution.

3.1. A Three-Point Bending Beam with a Pre-Existing Micro-Crack

Firstly, a three-point bending beam with a pre-existing micro-crack is taken into account to study the effect of the location and size of the pre-existing micro-crack on the fracture behaviors of the SLM additive manufacturing 316L stainless steel. Figure 5 gives out the geometric dimensioning and the boundary conditions of the numerical model, in which the size of the beam is $1.0 \times 0.4 \times 0.0033$ m^3 with a vertical pre-existing notch at the bottom. The plane stress is assumed. For the spatial discretization, the beam is discretized into the uniform material points with the same sizes of $\Delta x = \Delta y = 0.0033$ m and the horizon of the neighborhood of the material points is $\delta = 3\Delta x$. For the material parameters, the Young's modulus is $E = 2.12 \times 10^{11}$ Pa, the Poisson's ratio is $\nu = 0.3$, and the critical stretch of the bond is $s_0 = 0.01$. The velocity of the imposed loading is 10 m/s and the time step of the integration is 1.0×10^{-8} s. Four kinds of pre-existing micro-cracks are considered: (1) $L = 0$, $a = 0.05$ m, (2) $L = 0$, $a = 0.1$ m, (3) $L = 0.1$ m, $a = 0.05$ m and (4) $L = 0.1$ m, $a = 0.1$ m. Figure 6 plots the curves of the reaction force versus imposed displacement for different kinds of pre-existing micro-cracks, which explains that the location and size of the pre-existing micro-crack arising in the process of the SLM additive manufacturing will reduce the load-carrying limit of the corresponding structures. Figure 7 illustrates the crack patterns obtained by the present method at time $t = 3.0 \times 10^{-4}$ s and $t = 6.0 \times 10^{-4}$ s for these four loading cases, which shows that the pre-existing micro-crack also has a great influence on the path and velocity of the crack propagation.

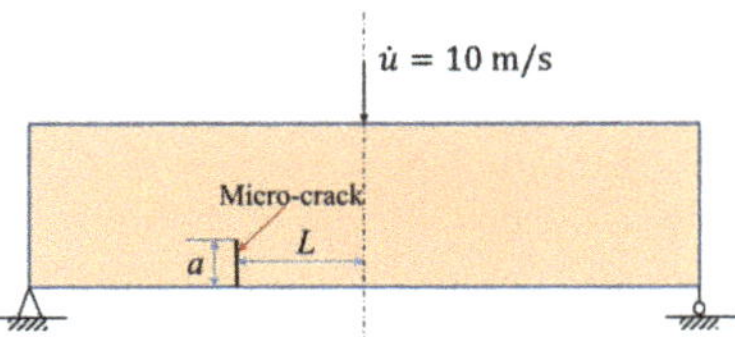

Figure 5. Schematic diagram of a three-point bending beam with a pre-existing micro-crack.

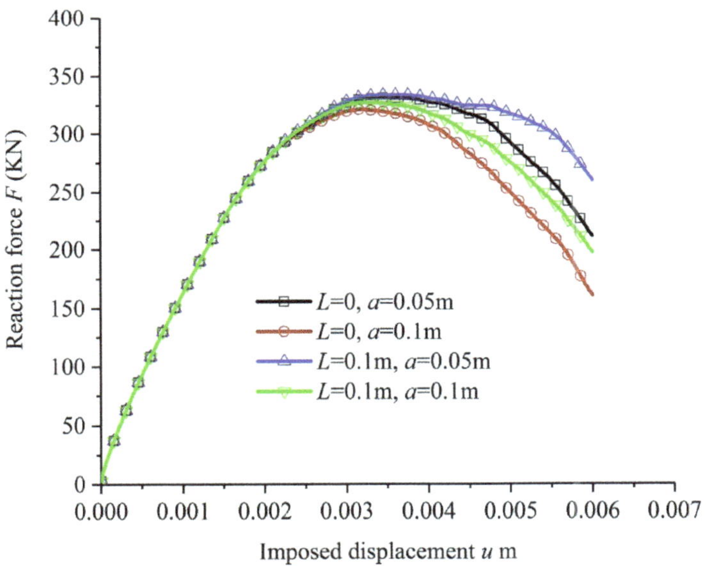

Figure 6. Curves of the reaction force versus imposed displacement for different kinds of pre-existing micro-cracks.

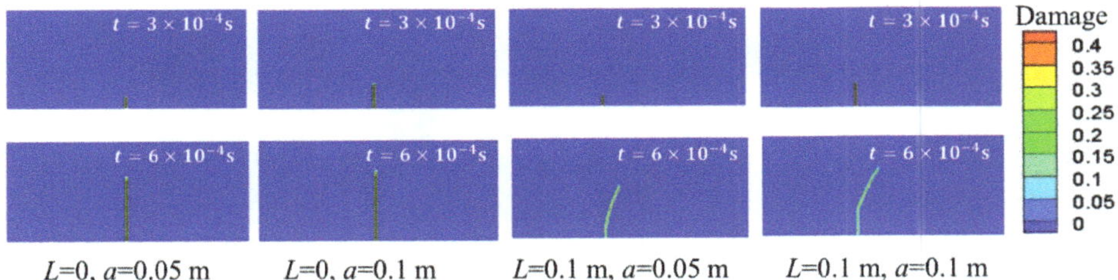

Figure 7. Crack patterns for these four loading cases at time $t = 3.0 \times 10^{-4}$ s and $t = 6.0 \times 10^{-4}$ s.

3.2. A Three-Point Bending Beam with Orthotropic Moduli

In this section, a three-point bending beam with orthotropic moduli, whose geometric and material parameters and whose boundary conditions are similar to those in Section 3.1, is considered to investigate the effect of the orthotropic features of the SLM additive manufacturing materials on their fracture behaviors. The initial micro-crack is $L = 0.1$ m, $a = 0.05$ m. Different kinds of ratios of the orthotropic Young's moduli, i.e., $\frac{E_\perp}{E_\parallel} = \lambda \leq 1.0$, are constructed, in which $E_\parallel = 2.12 \times 10^{11}$ Pa, the subscripts $\parallel$ and $\perp$ denote the laser scanning direction on a layer and the layer-to-layer direction, respectively. Figure 8 shows the curves of the reaction force versus the applied displacement for different ratios of the orthotropic Young's moduli. The results reveal that the smaller the Young's moduli ratios of the material (i.e., the Young's moduli along the layer direction is smaller), the smaller the load-carrying limit of the corresponding structure, which means that the performance of the additive manufactured material is related to the laser scanning strategy and the layer-to-layer bond strength. Figure 9 exhibits the contour plots of the crack propagation of the three-point bending beam at time $t = 6.0 \times 10^{-4}$ s, from which it is announced that the orthotropic features of the SLM additive manufacturing materials will also contribute to the path and velocity of the crack propagation.

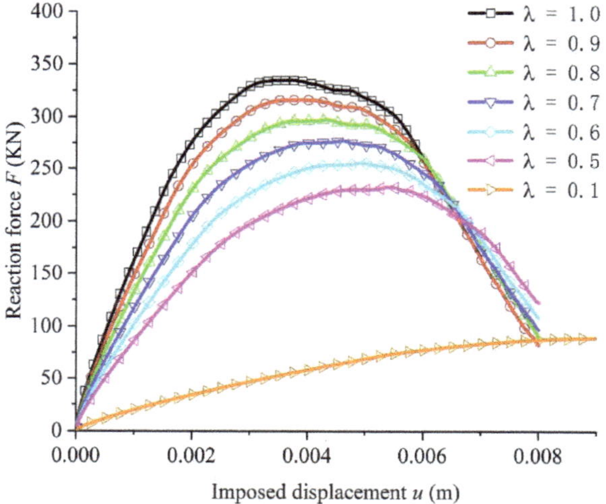

Figure 8. Curves of the reaction force versus imposed displacement for different ratios of the orthotropic Young's moduli.

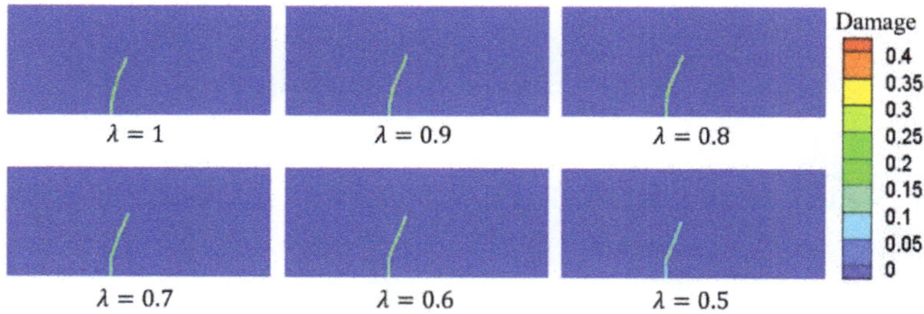

Figure 9. Contour plots of the crack paths for the additive manufactured orthotropic materials at time $t = 6.0 \times 10^{-4}$ s.

3.3. A Tensile Test of Plate Specimen with the Defects of Voids

In order to analyze the impact of the defects of the micro-voids on the behaviors of the multiple crack growth for the SLM additive manufacturing 316L stainless steel, a tensile test of a plate specimen with random distribution defects of micro-voids is considered. The volume fraction of the micro-voids in the numerical model is 19.88%. As shown in Figure 10, the size of the plate specimen is $0.4 \times 0.4 \times 0.002$ m^3 and the radius of two holes is 0.03 m. The Young's modulus is $E = 2.12 \times 10^{11}$ Pa. The critical stretch s_0 of the bond is 0.01. The interesting domain is discretized uniformly into 200×200 material points and the horizon δ of the neighborhood of the material point is 0.006 m. The velocity of the imposed loading is 10 m/s and the time step of the integration is 1.0×10^{-8} s. Figure 11 depicts the curves of reaction force versus imposed displacement obtained by the present method, from which it can be seen that the peak value of the reaction force for the case with random distribution defects is smaller that of the case with no defects, that is, the strength of the plate reduces because of the defects, and multiple-crack propagation will appear near the micro-voids. Furthermore, Figure 12 shows the paths of the multiple crack propagation of the plate specimen under tensile loading at time $t = 0$ s, $t = 0.8 \times 10^{-4}$ s, $t = 1.6 \times 10^{-4}$ s and $t = 2.4 \times 10^{-4}$ s. The results demonstrate that the micro-voids will actively induce the complex crack initiation and branching, which is contributed to the complex multiple crack propagation.

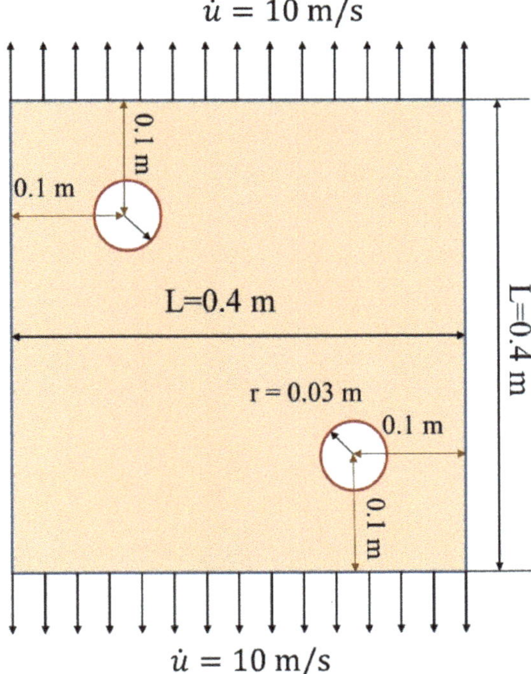

Figure 10. Schematic diagram of a tensile test of a plate specimen.

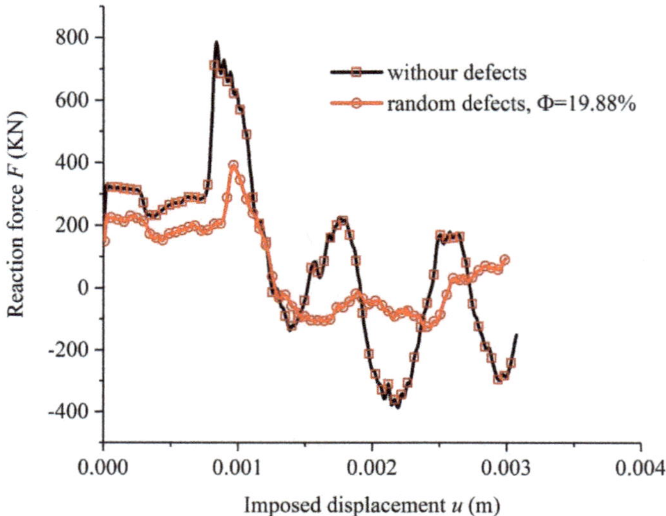

Figure 11. Curves of the reaction force versus imposed displacement.

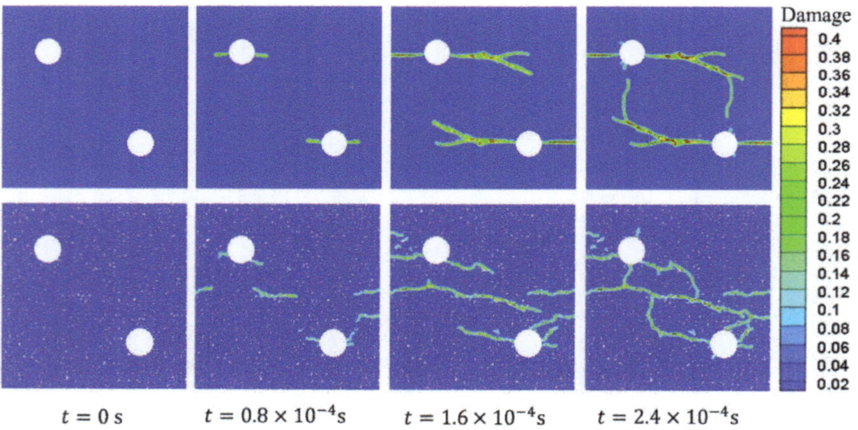

Figure 12. Paths of the multiple-crack propagation of the plate specimen under tensile loading at different time.

4. Conclusions

This paper presents a peridynamic method for the analysis of the fracture behaviors of 316L stainless steel with micro-cracks and voids fabricated by the SLM additive manufacturing. In this framework, the peridynamic formulation for the elastic-plastic theory was presented in incremental form. The pairwise force of a bond for the SLM additive manufacturing orthotropic material model was proposed according to the local and global coordinate systems. Then, a three-step approach was applied to describe the void defects in the geometric model, which was generated in the processing of the additive manufacturing. Next, three representative numerical examples were carried out and the results demonstrated that the present method can deal well with the complex multiple-crack problems of the additive manufacturing materials. Meanwhile, the results also indicated that the mechanical orthotropy, micro-cracks and voids had a significant influence on the ultimate bearing capacity, crack path and branching of the additive manufacturing materials. Finally, the achievements of this paper lay thew foundation for forming the evaluation methodology and criteria for the engineering applications of the SLM additive manufacturing materials.

Author Contributions: Conceptualization, H.L. and J.Z.; Formal analysis, H.L.; Methodology, H.L. and J.Z.; Validation, J.Z.; Writing—original draft, H.L.; Writing—review & editing, J.Z. All authors have read and agreed to the published version of the manuscript.

Funding: This work was supported by the National Natural Science Foundation of China [No. 12102416].

Institutional Review Board Statement: Not applicable.

Informed Consent Statement: Not applicable.

Data Availability Statement: Not applicable.

Conflicts of Interest: The authors declare no conflict of interest.

References

1. Lu, B.H. Additive manufacturing-current situation and future. *China Mech. Eng.* **2020**, *31*, 19–23.
2. Zhang, X.J.; Tang, S.Y.; Zhao, H.Y.; Guo, S.Q.; Li, N.; Sun, B.B.; Chen, B.Q. Research status and key technologies of 3D printing. *J. Mater. Eng.* **2016**, *44*, 122–128.
3. Shamsaei, N.; Yadollahi, A.; Bian, L.; Thompson, S.M. An overview of direct laser deposition for additive manufacturing; Part II: Mechanical behavior, process parameter optimization and control. *Addit. Manuf.* **2015**, *8*, 12–35. [CrossRef]
4. Attar, H.; Bönisch, M.; Calin, M.; Zhang, L.C.; Scudino, S.; Eckert, J. Selective laser melting of in situ titanium-titanium boride composites: Processing, microstructure and mechanical properties. *Acta Mater.* **2014**, *76*, 13–22. [CrossRef]
5. Heigel, J.C.L.; Michaleris, P.; Reutzel, E.W. Thermo-mechanical model development and validation of directed energy deposition additive manufactureing of Ti-6Al-4V. *Addit. Manuf.* **2015**, *5*, 9–19.
6. Liu, X.B.; Xiao, H.; Xiao, W.J.; Song, L.J. Microstructure and crystallographic texture of laser additive manufactured nickel-based superalloys with different scanning strategies. *Crystals* **2021**, *11*, 591. [CrossRef]
7. Zhang, B.; Li, Y.; Bai, Q. Defect formation mechanisms in selective laser melting: A review. *Chin. J. Mech. Eng.* **2017**, *30*, 515–527. [CrossRef]
8. Jiang, H.Z.; Li, Z.Y.; Feng, T.; Wu, P.Y.; Chen, Q.S.; Feng, Y.L.; Chen, L.F.; Hou, J.Y.; Xu, H.J. Effect of process parameters on defects, melt pool shape, microstructure, and tensile behavior of 316L stainless steel produced by selective laser melting. *Acta Metall. Sini.* **2021**, *34*, 495–510. [CrossRef]
9. Wu, H.Y.; Zhang, D.; Yang, B.B.; Chen, C.; Li, Y.P.; Zhou, K.C.; Jiang, L.; Liu, R.P. Microstructural evolution and defect formation in a powder metallurgy nickel-based superalloy processed by selective laser melting. *J. Mater. Sci. Technol.* **2020**, *36*, 7–17. [CrossRef]
10. Zhang, H.Y.; Dong, D.K.; Su, S.P.; Chen, A. Experimental study of effect of post processing on fracture toughness and fatigue crack growth performance of selective laser melting Ti-6Al-4V. *Chin. J. Aeronaut.* **2019**, *32*, 2383–2393. [CrossRef]
11. Weber, S.; Montero, J.; Petroll, C.; Schäfer, T.; Blechmann, M.; Paetzold, K. The fracture behavior and mechanical properties of a support structure for additive manufacturing of Ti-6Al-4V. *Crystals* **2020**, *10*, 343. [CrossRef]
12. Glodez, S.; Klemenc, J.; Zupanic, F.; Vesenjak, M. High-cycle fatigue and fracture behaviours of SLM AlSi10Mg alloy. *Trans. Nonferrous Met. Soc. China* **2020**, *30*, 2577–2589. [CrossRef]
13. Wen, E.T.; Gao, T.T.; Zhang, Y.Y. 3D visualization method for complex lattice structure defects in 3D printing. *Acta Metrol. Sin.* **2020**, *41*, 1077–1081.
14. Lozanovski, B.; Leary, M.; Tran, P.; Shidid, D.; Ma, Q.; Choong, P.; Brandt, M. Computational modelling of strut defects in SLM manufactured lattice structures. *Mater. Des.* **2019**, *171*, 107671. [CrossRef]
15. Silling, S.A. Reformulation of elasticity theory for discontinuities and long-range forces. *J. Mech. Phys. Solids* **2000**, *48*, 175–209. [CrossRef]
16. Silling, S.A.; Lehoucq, R.B. Peridynamic theory of solid mechanics. *Adv. Appl. Mech.* **2010**, *44*, 73–168.
17. Bessa, M.; Foster, J.T.; Belytschko, T.; Liu, W.K. A meshfree unification: Reproducing kernel peridynamics. *Comp. Mech.* **2014**, *53*, 1251–1264. [CrossRef]
18. Bobaru, F.; Foster, J.T.; Geubelle, P.H.; Silling, S.A. *Peridynamic Modeling*; Springer: New York, NY, USA, 2017.
19. Han, F.; Lubineau, G.; Azdoud, Y.; Askari, A. A morphing approach to couple state-based peridynamics with classical continuum mechanics. *Comput. Methods Appl. Mech. Engrg.* **2016**, *301*, 336–358. [CrossRef]
20. Ren, H.L.; Zhuang, X.Y.; Rabczuk, T. Dual-horizon peridynamics: A stable solution to varying horizons. *Comput. Methods Appl. Mech. Eng.* **2017**, *318*, 762. [CrossRef]
21. Katiyar, A.; Foster, J.T.; Ouchi, H.; Sharma, M.M. A peridynamic formulation of pressure driven convective fluid transport in porous media. *J. Comput. Phys.* **2014**, *261*, 209–229. [CrossRef]
22. Ouchi, H.; Katiyar, A.; York, J.; Foster, J.T.; Sharma, M.M. A fully coupled porous flow and geomechanics model for fluid driven cracks: A peridynamic approach. *Comput. Mech.* **2015**, *55*, 561–576. [CrossRef]

23. Lai, X.; Ren, B.; Fan, H.F.; Li, S.F.; Wu, C.T.; Regueiro, R.A.; Liu, L.S. Peridynamics simulations of geomaterial fragmentation by impulse loads. *Int. J. Numer. Anal. Methods Geomech.* **2015**, *39*, 1304–1330. [CrossRef]
24. Fan, H.F.; Bergel, G.L.; Li, S.F. A hybrid peridynamics-SPH simulation of soil fragmentation by blast loads of buried explosive. *Int. J. Impact Eng.* **2016**, *87*, 14–27. [CrossRef]
25. Hu, W.; Ha, Y.D.; Boraru, F. Peridyanmic model for dynamic fracture in unidirectional fiber-reinforced composites. *Comput. Methods Appl. Mech. Eng.* **2012**, *217*, 247–261. [CrossRef]
26. Huang, D.; Lu, G.D.; Qiao, P.Z. An improved peridynamic approach for quasi-static elastic deformation and brittle fracture analysis. *Int. J. Mech. Sci.* **2015**, *94*, 111–122. [CrossRef]
27. Li, H.; Zhang, H.W.; Zheng, Y.G.; Ye, H.F.; Lu, M.K. An implicit coupling finite element and peridynamic method for the dynamic problems with crack propagation in solids. *Int. J. Appl. Mech.* **2018**, *10*, 1850037. [CrossRef]
28. Zhang, H.W.; Li, H.; Ye, H.F.; Zheng, Y.G. A coupling extended multiscale finite element and pridynamic method for modelling of crack propagation in solids. *Acta Mech.* **2019**, *230*, 3667–3692. [CrossRef]
29. Oterkus, S.; Madenci, E. Peridynamic modeling of fuel pellet cracking. *Eng. Fract. Mech.* **2017**, *176*, 23–37. [CrossRef]
30. Madeci, E.; Oterkus, E. *Peridynamic Theory and Its Applications*; Springer: New York, NY, USA, 2014.
31. Zaccariotto, M.; Luongo, F.; Sarego, G.; Galvanetto, U. Examples of applications of the peridynamic theory to the solution of static equilibrium problems. *Aeronaut. J.* **2015**, *119*, 677–700. [CrossRef]
32. Carpinteri, A.; Colombo, G. Numerical analysis of catastrophic softening behavior (snap-back instability). *Comput. Struct.* **1989**, *31*, 607–636. [CrossRef]

MDPI

St. Alban-Anlage 66

4052 Basel

Switzerland

Tel. +41 61 683 77 34

Fax +41 61 302 89 18

www.mdpi.com

Crystals Editorial Office

E-mail: crystals@mdpi.com

www.mdpi.com/journal/crystals